Geology of North-West Borneo
Sarawak, Brunei and Sabah

Geology of North-West Borneo
Sarawak, Brunei and Sabah

By

Charles S Hutchison

Professor emeritus
Department of Geology, University of Malaya
50603 Kuala Lumpur, Malaysia

ELSEVIER

Amsterdam • Boston • Heidelberg • London • New York • Oxford
Paris • San Diego • San Francisco • Singapore • Sydney • Tokyo

ELSEVIER B.V.
Radarweg 29
P.O. Box 211, 1000 AE Amsterdam
The Netherlands

ELSEVIER Inc.
525 B Street, Suite 1900
San Diego, CA 92101-4495
USA

ELSEVIER Ltd
The Boulevard, Langford Lane
Kidlington, Oxford OX5 1GB
UK

ELSEVIER Ltd
84 Theobalds Road
London WC1X 8RR
UK

First edition 2005

Library of Congress Cataloging in Publication Data
A catalog record is available from the Library of Congress.

British Library Cataloguing in Publication Data
A catalogue record is available from the British Library.

ISBN-13: 978 0 444 51998 6
ISBN-10: 0 444 51998 X

♾ The paper used in this publication meets the requirements of ANSI/NISO Z39.48-1992 (Permanence of Paper).
Printed in The Netherlands.

The Geological Survey, Borneo Region, Malaysia, began its existence in 1949 as the Geological Survey Department, British Territories in Borneo. This book is dedicated to all the geologists who, through their pioneering work in difficult nearly inaccessible terrain, made it one of the greatest in the world.

Acknowledgements

I am extremely grateful to Robert Hall for considerable assistance during the preparation of this book. I have also benefited from discussion with Christopher Morley and Robert Tate.

I am also extremely grateful to Timothy for frequently upgrading my computer and for helping me when there were problems with the programmes.

I am grateful to the following organizations for permission to modify several of their published illustrations for incorporation in this book: Brunei Shell Petroleum Company, Geological Society of London, Geological Society of Malaysia, Jabatan Mineral dan Geosains Malaysia, Oxford University Press and Petroliam Nasional Berhad (Petronas).

Contents

Introduction

HISTORY OF GEOLOGICAL INVESTIGATION

Early exploration

The earlier, mostly non-geological, exploration reports of Borneo have been listed by Worth (1940). The first reliable geological observations on Sabah were by Hatton (1885), a young mineral explorer employed by the Chartered Company of North Borneo. He met a tragic death while hunting along the Segama River. The most significant publication was that of Posewitz (1892), who summarized the earlier writings of geologists who had journeyed through Borneo. Although, he never set foot in Sarawak or Sabah, but had journeyed through Kalimantan. He collected the literature and compiled a geological map of the known parts of Borneo. The work was unsystematic and is now obsolete. Rutter (1922) described the journeys of some mineral prospectors in his account of Sabah, but he did not mention the real geological explorers.

Netherlands East Indies geological and mining department

The geological survey of the Netherlands East Indies was established 99 years before that of the British Territories in Borneo. With geological investigations directed from Bandung and mineral exploration from Jakarta (then Batavia), the systematic exploration of Kalimantan was already well advanced before that of Sarawak and Sabah (then North Borneo). The international border between Kalimantan and Sarawak is not of geological significance, and van Bemmelen (1939) studied and mapped the geology of western Borneo, and in broad outline the geology of much of Sarawak was shown to be an extension of that of Kalimantan. Details were published usually in *Jaarboek van het Mijnwezen*, published in Batavia. The remote eastern border between Kalimantan and Sabah was not traversed and the geology there remained unknown until much later.

The extensive outcrops of phyllite and slate were already well known to Molengraaf (1902), who referred to them as the ‘Old Slate Formation’. Bruggen (1935) later summarized their geology in Kalimantan, Sarawak, and Sabah under the name ‘Eocene Phyllite Formation’. His generalizations were criticized by Zeijlmans van Emmichoven and Ubaghs (1936), who applied the name ‘Embaloeh Complex’ to the supposed pre-Tertiary part. The major compilation by Zeijlmans van Emmichoven (1939) also continued the mapping of the Ketungau Basin of Kalimantan into Sarawak, and he also introduced the term ‘Engkilili Beds’.

The oil company era

In 1948, the Shell Oil Company appointed Max Reinhard and Eduard Wenk of Basel University to compile a comprehensive report on the geology of North Borneo. The publication (Reinhard and Wenk, 1951) appeared as Bulletin 1 of the newly formed Geological Survey Department. It represents an important milestone in the regional geological literature.

In 1956, an agreement was made between the Royal Dutch Shell Group and the Geological Survey Department to make a joint compilation of the geology of Sarawak, Brunei and part of North Borneo, under the senior authorship of P. Liechti, who had been in charge of the field investigations of Brunei Shell Petroleum Company since 1952. This agreement resulted in the second major milestone in the regional geological literature (Liechti et al., 1960). On the title page, it is stated that the publication was 'compiled from work of the Royal Dutch Shell Group of companies in the British Territories in Borneo and from various published accounts'.

The unpublished reports are those of Sarawak Shell Oilfields Limited (SSOL), Brunei Shell Petroleum Company Limited (BSPC) and the Shell Company of North Borneo Limited (SCNB). Their work in Borneo dates back to 1910, when oil was first discovered in Miri. From the very beginnings of the Geological Survey, there was a very close working relationship with SSOL, BSPC, and SCNB. An earlier unpublished compilation by Waite (1940) proved extremely comprehensive and valuable in the writing of Liechti et al. (1960).

The Geological Survey (European era 1949–1968)

The Geological Survey Department, British Territories in Borneo, was established on 16 March 1949 with money provided from British Colonial Development and Welfare funds. The staff were stationed either in the Kuching or Jesselton (now Kota Kinabalu) offices depending on whether their field areas were in Sarawak or North Borneo (now Sabah), respectively. An office was opened in Brunei Town (now Bandar Sri Begawan) in 1957. From the beginning, the director of the survey stationed himself in Kuching, and the deputy director, once appointed, was stationed in Jesselton. This was a productive 19 years during which the whole country was systematically mapped on a reconnaissance scale, and characterized by regular publication of bulletins, memoirs, maps, and annual reports, upon which the present-day knowledge of the country is predominantly based.

The first director, F. W. Roe, was transferred from the Federation of Malaya in March 1949. F. H. Fitch joined him in December 1949, also on transfer from the Federation. The following new appointees arrived during 1949—N. S. Haile and G. E. Wilford in October and P. Collenette in November. One month later, he was transferred to take charge of the Jesselton office. In early 1950, F. H. Fitch was transferred to the Jesselton office and promoted to the post of Deputy Director in

April 1952. In January 1954, H. J. C. Kirk began his new appointment. E. A. Stephens also arrived the same month, but left permanently in February 1957. E. B. Wolfenden took up appointment in December 1956. R. A. M. Wilson was transferred from Cyprus and arrived in the Jesselton office in May 1958. F. H. Fitch was promoted as Director in May 1960 to succeed F. W. Roe and he moved from Jesselton to Kuching. R. A. M. Wilson was appointed as Deputy Director in 1961 and was stationed in Jesselton. A. C. Pimm arrived in Sarawak in October 1961.

C. H. Kho was the first Malaysian appointee, to the post of assistant geologist. He assumed duty in Kuching in August 1961. N. Y. P. Wong was the second Malaysian to be appointed, also to the post of assistant geologist. He assumed duty in Jesselton in October 1961.

The Geological Survey Department, British Territories in Borneo became an independent branch of the Geological Survey of Malaysia on the 16 September 1963, now to be known as the Geological Survey, Borneo Region, Malaysia, with headquarters in Kuching and branch office in Jesselton, soon to be renamed Kota Kinabalu. An active policy of Malaysianization was now implemented to phase out expatriate staff.

R. A. M. Wilson was promoted to the post of director in December 1963 following the retirement of F. H. Fitch. N. S. Haile retired from the geological survey in October 1964 to take up the chair of geology at the University of Malaya. J. Newton-Smith arrived in Jesselton in October 1964 to begin his appointment. C. H. Kho and N. Y. P. Wong were both promoted to the positions of senior assistant geologist in 1964.

In 1965, a scheme was implemented whereby University of Malaya staff could conduct post-graduate research in Sabah and be financially and materially supported by the Geological Survey. The scheme supported the work of Dhonau, Hutchison, Stauffer, and Koopmans in 1965 and Stauffer again in 1967. E. B. Wolfenden retired from the geological survey in October 1965 and R. A. M. Wilson retired as director in December 1966. P. Collenette was promoted to the post of director to succeed him and transferred to the Kuching office. Both C. H. Kho and N. P. Y. Wong were promoted in April 1966 to the post of geologist.

G. E. Wilford was promoted to the post of deputy director in October 1966. A. C. Pimm left Malaysia in September and J. Newton-Smith in September 1967. D. T. C. Lee was appointed as geologist in January and K. M. Leong in July 1967.

H. J. C. Kirk retired from the geological survey in January 1968. P. Collenette retired as director in November and G. E. Wilford retired as deputy director in December 1968, thus bringing to an end the European era. From 1968, the administration came directly under control of the headquarters in Ipoh. C. H. Kho was in charge of the Kuching office and N. P. Y. Wong of the Kota Kinabalu office. In 1968 and 1969, G. Jacobson was attached to the Kota Kinabalu office under the Australian Volunteers Abroad Programme.

The Geological Survey (Malaysian era 1969–onwards)

Sarawak and Sabah became separate divisions within the Malaysian Geological Survey. In 1985, Charlie Kho Chin Heng and David Lee Thien Choi were confirmed as directors of the Sarawak office in Kuching and the Sabah office in Kota Kinabalu, respectively. In 1986, Chen Shick Pei became director in the Kuching office on retirement of C. H. Kho. In 1998, Alexander Unya anak Ambun became director in Kuching when Chen Shick Pei was promoted to the post of Director General of the Geological Survey and transferred to the Kuala Lumpur headquarters.

Regional Tectonic Setting

Our knowledge of the South China Sea marginal basin begins with the compilation of Hamilton (1979), significant for insightful understanding of a heretofore little known region. His analysis resulted from unprecedented access to unpublished oil company data. Taylor and Hayes (1983) made this sea their major research interest and their main conclusions have stood the test of time. They identified and documented a sequence of magnetic anomalies 11 through 5d in the zone of sea-floor spreading. A later reassessment by Briais et al. (1993), supported the earlier interpretation that the pattern extends from 11 to 5c. (32 to 16 Ma) (Lower Oligocene to early Middle Miocene). However, the identification of these anomalies has yet to be confirmed by direct drilling (Hutchison, 1996b). Their identified pattern is illustrated in Figure 1.

The South China Sea marginal basin formed by rifting of the continental lithosphere of Sundaland. This peninsular continental protrusion from Eurasia was characterized by an extensive Palaeocene landmass that extended into western Sarawak and northeastwards as far as the West Baram Line (Figure 1). The contrast across the West Baram Line is dramatically demonstrated by average geothermal gradients >41°C km^{-1} to the west (upon continental crust) as compared with gradients <28°C km^{-1} eastwards of the West Baram Line (Hutchison, 1989), as measured in offshore wells. Sabah and Brunei Darussalam are not underlain by the Sundaland continental crust, and the evidence from Sabah is that the basement is of Mesozoic oceanic lithosphere with local occurrences of continental 'microcontinents'.

The Northwest Borneo Trough extends northeastwards from the West Baram Line, parallel to the Sabah coast. It has traditionally been interpreted as a fossil trench (Hamilton, 1979), but in this book it is regarded as a foredeep with the actual trench located beneath the thick turbiditic Crocker Formation.

The major onland feature that forms the mountainous backbone of Borneo is formed by the Upper Cretaceous through Oligocene to Lower Miocene Rajang Group flysch (Figure 1). In Sarawak, it was uplifted during the Sarawak Orogeny to form part of the landmass by the end of the Eocene. However, its extension into Sabah is complicated for turbidite sedimentation continued predominantly throughout the Oligocene into the Lower Miocene and spectacular uplift in the Middle Miocene created the mountainous Western Cordillera of Sabah.

Southwards from the Rajang turbidite basin, the Sundaland Palaeocene landmass continues through Kalimantan as far as middle Java. The Rajang turbidite basin therefore was bounded both on the north and the south by continental Sundaland. Evidence for its eastern extension remains obscure.

The regional analysis of Hall (1996), slightly modified in Hall (2002), is remarkable in that so many elements of Southeast Asia have been successfully integrated into the synthesis. A controversial item may be his acceptance of the anti-clockwise

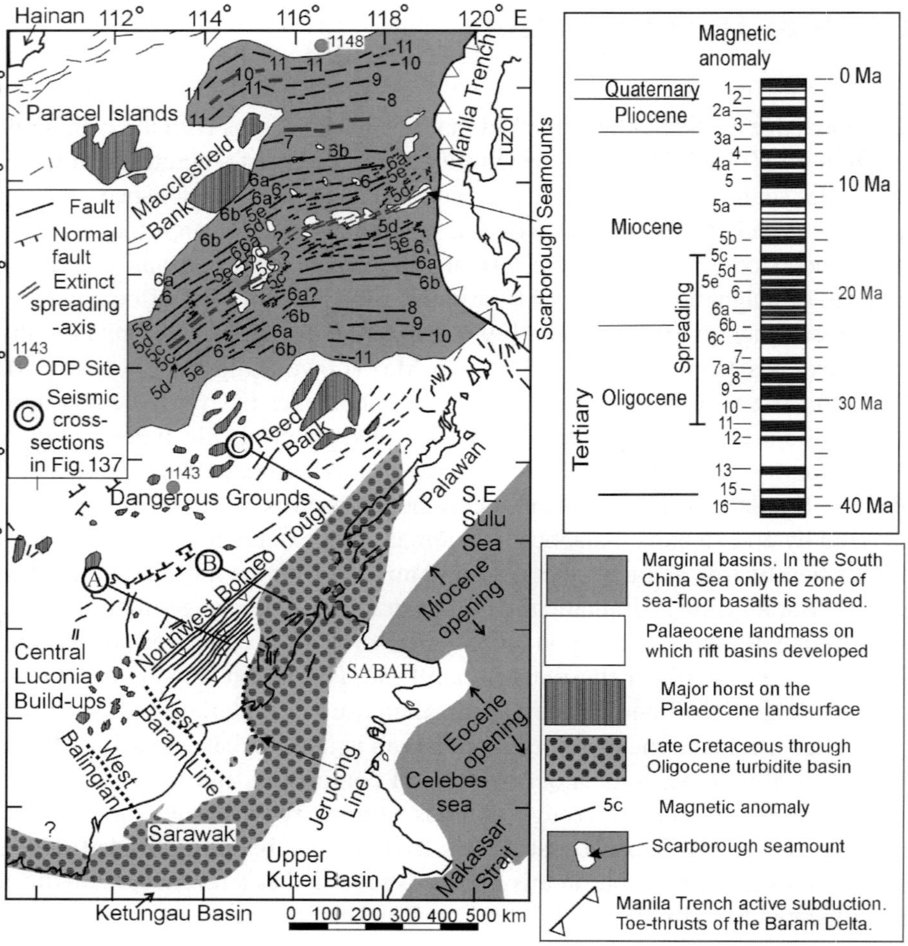

Figure 1. The major geological features of the southern South China Sea. Magnetic anomalies are from Briais et al. (1993). The rifted terrain enveloping the zone of sea-floor spreading on the west and south (unshaded area) developed initially on the Palaeocene Sundaland landmass (Hutchison 1992b). A, B & C are seismic sections of Figure 137.

rotation of Borneo beginning ~24 Ma (Late Oligocene) and ending with the present-day orientation ~11 Ma (Middle Miocene). The rotation has been achieved without a comparable movement of relatively stable Sundaland resulting in an inadequately addressed geometrical problem. The present book infers that western Borneo, as far eastwards as the West Baram Line, is an integral part of Sundaland. Hall (2002) does not show the West Baram Line on any of his computer drawings. The anti-clockwise rotation, as suggested by palaeomagnetic data, will continue to be controversial and difficult to integrate successfully into a regional evolutionary synthesis. There are significant differences between the timing of events as portrayed by Hall (1996,

2002) and those of Hutchison (2004), as summarized in Figure 55 of this book, but there is no conflict in the overall scenario.

An alternative reconstruction of Southeast Asia, incorporating a non-rotating Borneo, has been developed by Murphy (1998). He concludes that "the evolution of Southeast Asia has by no means been satisfactorily handled".

Part A

Sarawak

A large part of this section is similar to Hutchison et al. (in press), which deals only with the geology and tectonics of Sarawak. The latter needs to be included in this book so that the whole Northwest Borneo region, stretching from Sarawak to Sabah, could be incorporated in this regional synthesis.

Chapter I

Regional Geology Concepts

Before the advent of plate tectonics, Haile (1969) tried to understand the tectonic make-up of Sarawak. He fitted the region into the well-established geosynclinal theory. The term 'Northwest Borneo Geosyncline' was then well accepted. The 'eugeosynclinal furrow' was to become the Sibu Zone, and the 'miogeosynclinal furrow' was the Miri Zone. Although, Haile (1969) is now entirely of historic value, it is significant to point out that the organizational pattern, according to geosynclinal theory, required that there is a 'foreland' or continent offshore to the north and west of Sarawak. But instead, there is the South China Sea. With the advent of plate tectonics, encouraging very active geological exploration of the marine part of the region, it soon became established that this part of the South China Sea is indeed underlain by foundered continental crust, attenuated by Early Tertiary extension (Hutchison et al., 2000). This is indeed the Sundaland 'foreland'; buried beneath the South China Sea. The four-fold divisions of Haile (1974) continue to be valid, for they are based on a careful recognition of the geological elements and their tectonic significance. The most northerly division is the Miri Zone, dominated by what Haile (1974) called miogeosynclinal (shelf) and molasse (late- or post-orogenic) strata, deposited upon older continental crust (Figure 2).

The second subdivision is the Sibu Zone, dominated by eugeosynclinal flysch, or a thick monotonous sequence of shale/sandstone turbidite, most likely to have been deposited upon oceanic crust. The turbidite flysch was uplifted in Late Eocene then unconformably overlain by outliers of molasse (Figure 2).

The third division is the Kuching Zone, which Haile (1974) described as composed of basement complex, Jurassic–Cretaceous shelf deposits, and molasse and related deposits. Hundreds of small 'post-basement intrusions' cut through the rocks of this zone. They are dated Late Oligocene to Miocene and known as the 'Sintang Intrusives' from their greatest concentration in the Ketungau and Melawi basins of Kalimantan (Williams and Harahap, 1987).

The fourth and most southerly division was named by Haile (1974) as the 'West Borneo Basement', a rather misleading term, for the region, although it does contain outcrops of Carboniferous–Permian basement rocks, is dominated by Cretaceous volcanic and plutonic rocks that constitute the Schwaner Mountains (Williams et al., 1988). Since Haile (1974) used the names of the largest town for the other zones, it is more logical to name this the 'Pontianak Zone' (Figure 2).

The boundary between zones 2 and 3 has been named the Lupar Line by Tan (1979) and that between zones 1 and 2 the Tatau–Mersing Line (Hutchison, 1989). Both of these lines, or narrow zones, contain fragmented ophiolite and mélange, and may represent former plate boundaries. However, the Kuching Zone contains

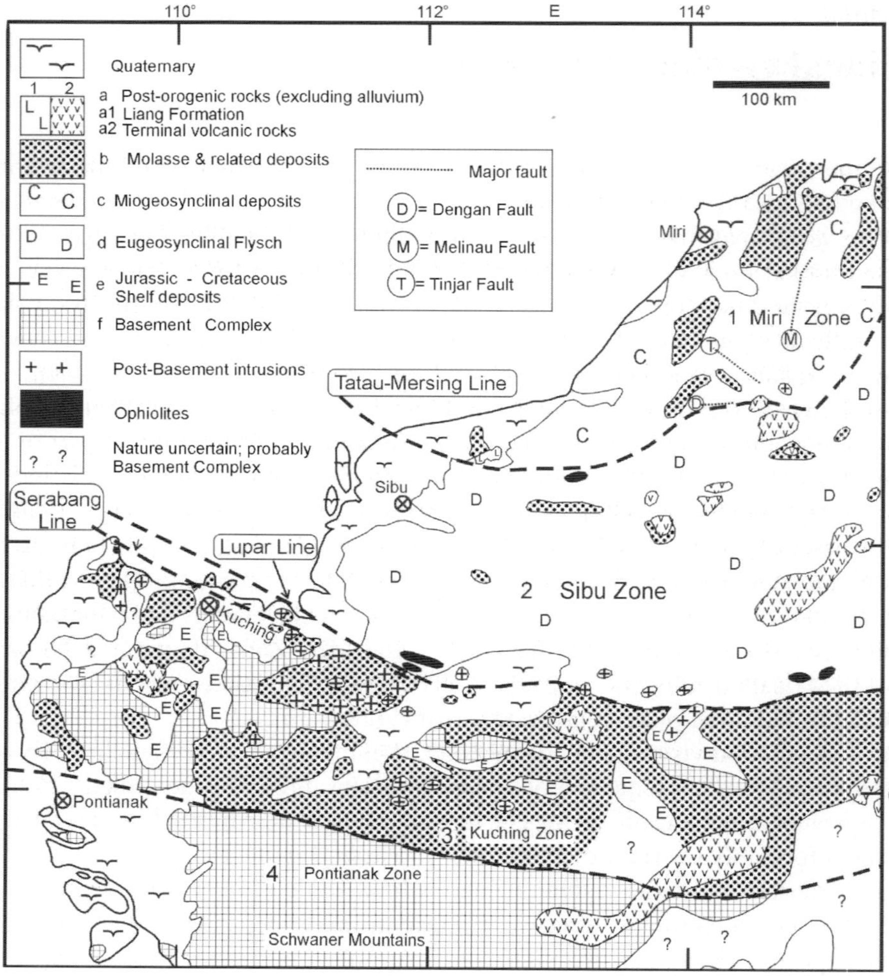

Figure 2. Structural zones of northwest Borneo, (modified after Haile, 1974). Note that Publications from within Indonesia use 'Kalimantan' to mean the whole of Borneo. Other publications, including this one, use 'Kalimantan' to mean Indonesian Borneo. (by permission of the Geological Society of London)

another such zone, the Boyan Mélange that lies parallel to, but south of the Lupar Line (Williams et al., 1988).

Chapter II
Palaeomagnetism of Sarawak

A considerable amount of palaeomagnetic research has been carried out on rocks collected from Borneo (Fuller et al., 1999), and in particular from Sarawak. The Sarawak results were first comprehensively reported by Schmidtke et al. (1990).

The most complete and coherent work comes from northwest Borneo, predominantly western Sarawak, but with specimens collected from contiguous Kalimantan. This study has become the standard for the whole of Borneo. The results are relatively straightforward. The Miocene high-level intrusions are increasingly anti-clockwise rotated with age, reaching a maximum of 51° in an intrusion dated by K:Ar to be 25.8 ± 1.9 Ma. The Late Eocene Silantek Formation gives 41° of anti-clockwise rotation. The Mesozoic samples from the Bau Limestone, Kedadom and Pedawan formations in Sarawak, and from the Jurassic–Cretaceous and Triassic sites across the border in Kalimantan, all indicate a strong rotation of about 90° that took place sometime after 93 Ma, the age of the Cretaceous dyke. These results indicate that anti-clockwise rotation, at least of northwest Borneo, and perhaps of a greater part of Borneo, has been taking place and was completed by about 10 Ma ago, when Borneo achieved its present orientation. The mechanism for anti-clockwise rotation remains enigmatic and many geologists continue to be sceptical because it is difficult to satisfactorily incorporate into regional tectonics.

Chapter III

Geomorphology

III.1. MESA TOPOGRAPHY

Plateaus built of Late Cainozoic volcanic and pyroclastic rocks are characteristic of a remote interior region of the Sibu Zone, between longitudes 113.5° and 115° E; latitudes 1.5° to 3.2° N. Usun Apau is the best example, with an area of about 900 km^2. It consists of an Eastern Tableland of about 1000 m elevation, and a smaller Western Tableland about 760 m high (Figure 3). Three mountains, Bukit Batu Mabun, Bukit Selidang, and Bukit Kenawang, stand out more than 300 m above the tablelands, and are the relicts of former volcanoes. The Linau-Balui Plateau has an area of about 290 km^2, consisting of a Northern Tableland of 1100 m elevation and a larger Southern Tableland of about 820 m elevation. There are other smaller plateaus.

The mesas have strongly dissected margins with precipitous cliffs and deep marginal embayments. The rivers plunge over the plateau edges in high cataracts and waterfalls. The most spectacular are the falls of the head of the Julan River Valley on the northern side of the Usun Apau Plateau.

Deeply dissected mesas occur at the Hose and Nieuwenhuis Mountains. The thick pyroclastic deposits of the Hose Mountains have been deeply dissected to form a cluster of closely grouped peaks, with many summits over 1200 m high, culminating in Bukit Batu, 2006 m. The Nieuwenhuis Mountains, which straddle the border with Kalimantan, resemble the Hose Mountains, but are less rugged and of lesser elevation, about 1500 m.

III.2. KARST TOPOGRAPHY

Limestone with well-developed karst topography and cave systems is found scattered from west to east Sarawak, but concentrated in certain regions. The best known occurrences have been described by Wilford (1964). In the west, the best karst hills are in the Bau district (Figure 9), with extension towards Serian. Southwards, near the Kalimantan border, is Gunung Selabor.

Gunung Subis, near Batu Niah, is the most readily accessible famous karstic hill. The Niah Great Cave is a well-known locality for the harvesting of birds' nests, and human remains have been excavated and described from 1954 onwards, dating back at least 50,000 years, by Tom Harrison, former curator of the Sarawak Museum in Kuching. A separate, and higher cave is famous for its pre-historic cave paintings depicting hunting scenes.

The type locality of the Melinau Limestone, which unconformably overlies the Mulu Formation, presents spectacular karst topography on a grand scale. The best

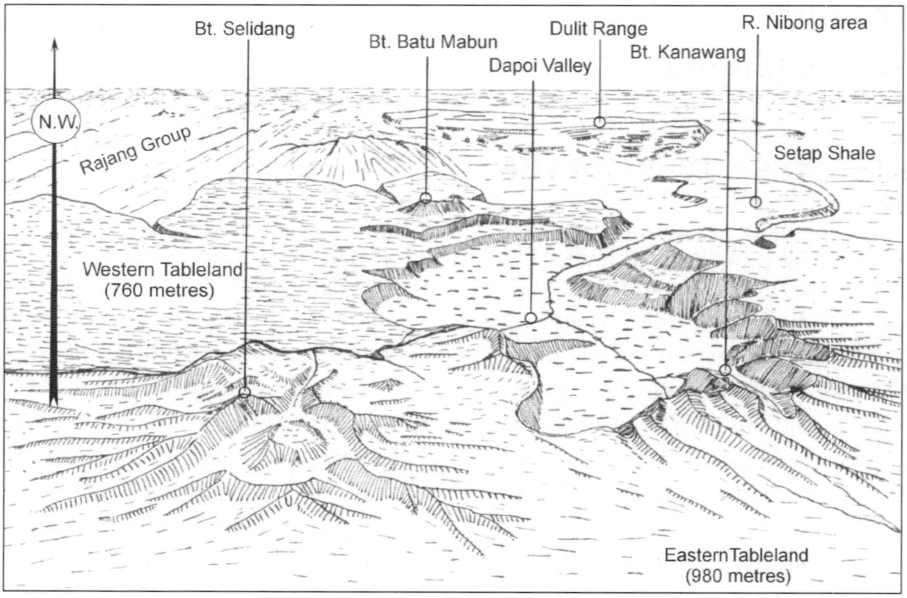

Figure 3. Sketch of the geomorphology of the Usun Apau Tablelands, as seen from the air looking north-westwards. Drawing by C.J. Campbell in 1956 (Kirk, 1957). With permission from Minerals and Geoscience Department, Malaysia.

known of its caves, and considered to be the largest in the world, is Deer Cave (also known as Gua Payau or Lobang Rusa). It is situated in a very steep limestone hill at the SW end of the main outcrop, about 5 km from the Tutoh River. The cave entrance is from the Melinau Paku River (Wilford, 1964).

III.3. RAJANG GROUP INLIERS IN MIRI ZONE

The inliers make isolated mountain ranges. The Mulu Range is a remarkable assemblage of chaotic steep ridges on either side of the Tutoh Gorge. The highest peak is Gunung Mulu (2374 m). The local inhabitants call it Gunung Ubong. The steep ridges do not follow the strike of the Mulu Formation (Haile, 1962).

III.4. SYNCLINES OF SANDY FORMATIONS

Thick sandstone beds of the Nyalau and Meligan formations form impressive and rugged mountain ranges bordered by steep scarps. The most interesting mountain ridge is the Dulit Range, made of a syncline of Nyalau Formation. It has been bent into a distinctive poorly understood right-angled knee.

III.5. MUD VOLCANOES

Cold mud volcanoes occur where the country rock is rich in mudstone. They are not numerous but occur about 2 km south of the main road between Batu Niah and Miri. Large bubbles of presumably methane gas slowly break the surface and a mud cone is slowly built up. Wild animals frequent these mud volcanoes, so there must be an enrichment in salts, though they are not actually saline. Mud volcanoes are very dependent on rainfall, and may cease activity during a very dry season. There is no connection with petroleum, but the actual cause, whether it may be a relationship to an underlying fault, remains unknown.

Chapter IV
The Kuching Zone

The stratigraphy of the Kuching Zone is illustrated in Figure 4. The zone may be conveniently divided into two sectors—the main inland sector and a northern coastal sector, which contains poorly understood formations, named the Serabang, Sejingkat, and Sebangan shown diagrammatically as fault-bounded.

IV.1. BASEMENT SCHISTS

The name Kerait Schist was introduced by Pimm (1965) for deeply weathered road cuts SE of Serian along the Sri Aman road, where it crosses the Sungai Kerait (Figure 5). The outcrops appear as windows within the Upper Triassic Sadong Formation, and no contacts have been observed. The relationship to the Sadong Formation is obscure. Lithologies include quartz mica schist quartz schist, and quartz tremolite schist. Quartz veining is common, but all outcrops are weathered. The age is unknown but inferred to be pre-Upper Triassic and possibly as old as Carboniferous.

The Tuang Formation (Hon, in manuscript; Tate and Hon, 1991) forms small outcrops of greenschist facies schists as windows within the Pedawan Formation. When weathered, they are easily mistaken for the Pedawan Formation. The type locality is in the Sungai Tuan, 17 km SE of Kuching. The schist and phyllite are composed of quartz, muscovite, carbonaceous material, and chlorite, with or without albite and calcite. There are also quartz-epidote and quartz-actinolite schists and meta-sandstone lithologies. The less schistose basic schists occur as a major mass exposed for ~200 m across the general NE trend, and as thinner beds and lenses within the pelitic schist (Hon, in manuscript). At the Kim Hin ceramics factory, 7 km from Kuching, the meta-sandstone beds are strongly boudinaged. Banded quartzitic schists are associated with quartz-veined chlorite-epidote schist. No dating has been carried out and there is no direct evidence of their age, except for a 'dubious fossil tentaculid', found by Tan (in manuscript), suggesting a pre-Carboniferous age. Disharmonic and superimposed folds have been observed in the schists and the foliation planes and minor-fold axes indicate more than one episode of folding (Hon, in manuscript).

Schists are exposed sporadically from Sungai Pasir to Biawak in the Lundu area, and across the border into Kalimantan (Allen, 1952). Near Tanjung Duta, biotite–cordierite hornfels occur near the granite contact. The metamorphic rocks are weathered and converted into hornfels in proximity to the Gunung Pueh granite. The schist types are: muscovite–biotite, hornblende–quartz–plagioclase, staurolite–biotite–muscovite–quartz, chlorite–quartz, and graphite–quartz–muscovite. There is also muscovite–quartz gneiss. There is no direct indication of age, but Zeijlmans van Emmichoven (1939) considered them to be pre-Carboniferous.

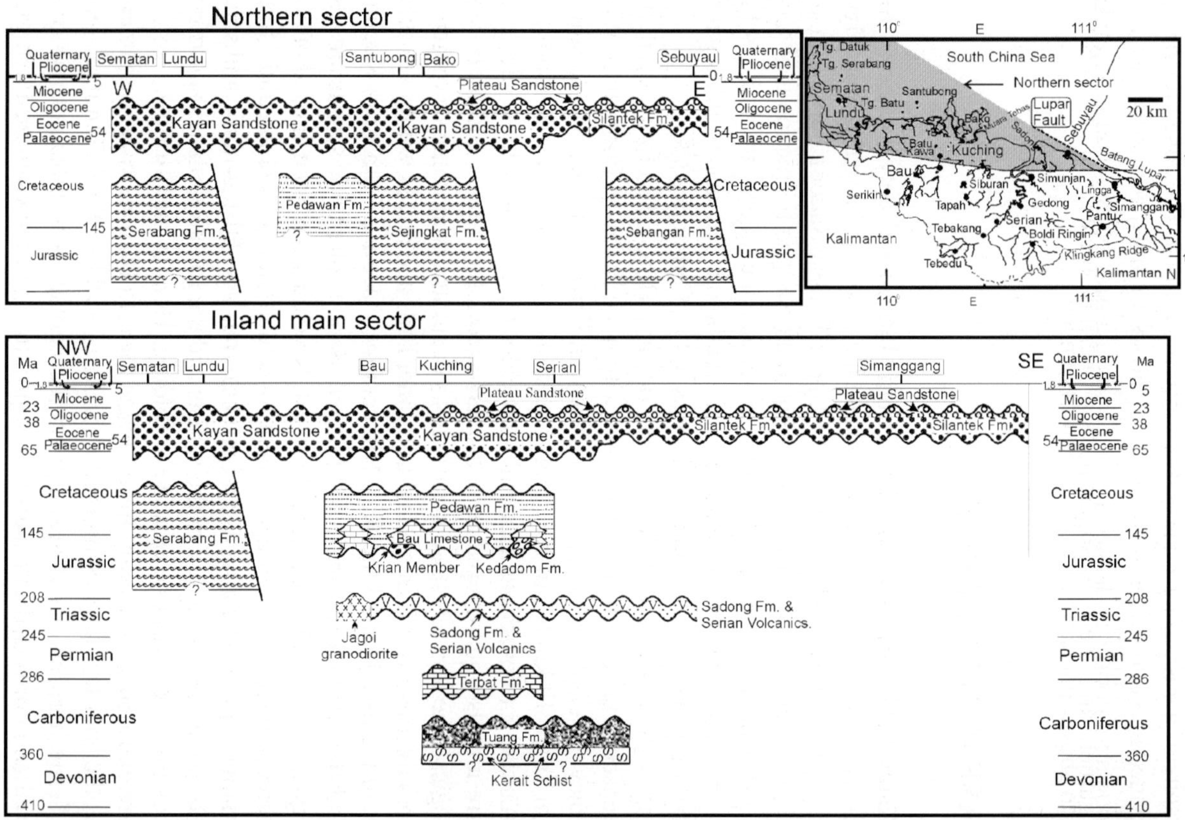

Figure 4. Chrono-stratigraphy of the Kuching Zone. The nature of the boundary between the two sectors is unresolved. After Hutchison et al. (in press).

IV.1.1. Correlatives

The oldest rocks of Kalimantan are the Pinoh Metamorphics, which outcrop in a 50-km-wide latitudinal belt extending discontinuously from near Pontianak eastwards to the margin of the West Kutei Basin. The type locality is in the Nangapinoh quadrangle (Amiruddin and Trail, 1989). Tate (1991) considers them to be the correlatives of the Tuang Formation and Kerait Schist of Sarawak.

Their grade of metamorphism is higher in Kalimantan—greenschist facies, but containing garnet, and there is superimposed thermal metamorphism from the Cretaceous granitoids of the Schwaner Mountains. The age of the Pinoh Metamorphics is uncertain (Tate, 1991), but Lower Carboniferous to pre-Carboniferous is generally held (Pieters and Supriatna, 1990).

IV.2. TERBAT FORMATION

The formation crops out at Gunung Selabor and Gunung Storib, an isolated precipitous mountain with twin karstic peaks rising to 400 m (Wilford and Kho, 1965). The type section (Pimm, 1965) lies in the Sungai Kedup, near Terbat Bazaar, about 5 km south-east of Gunung Selabor and only 5 km from the Indonesian border (Figure 5). Access is by road southwards from Serian. The Terbat Formation outcrop is of very restricted extent. It appears to occupy the partly faulted core of a tight NW–SE trending anticline in the Upper Triassic Sadong and Serian Volcanic formations (Pimm, 1965).

IV.2.1. Thickness and relationships

Its base is nowhere seen and it represents the oldest formation in this neighbourhood. Its contact with the overlying Sadong Formation is not exposed, but conglomerates near the Sadong Formation base near Gunung Selabor contain clasts of fossiliferous Terbat limestone, thus inferring an unconformity. Shale sequences near Terbat have not yielded fossils and, where they occur, the mapped boundary between the Sadong and Terbat formations is uncertain. The age gap between the youngest known Terbat Formation (Lower Permian) and the oldest known Sadong Formation (Upper Triassic) point to an important unconformity. The actual contact is presumed to be faulted. A N- to NNE-trending strike is common in Gunung Selabor and most dips are between 70 and 90°. Assuming a monoclinal section, the width of the outcrop would infer a formation thickness of around 600 m (Wilford and Kho, 1965). However, near Terbat Bazaar, the strike is predominantly NW, parallel to the overlying Sadong Formation (Pimm, 1965), and a thickness of at least 920 m is inferred.

The Terbat formation outcrops have been uplifted and actively eroded to form a provenance for conglomerates of the Late Triassic Sadong Formation, so that the original stratigraphical extent into the Permian is unknown.

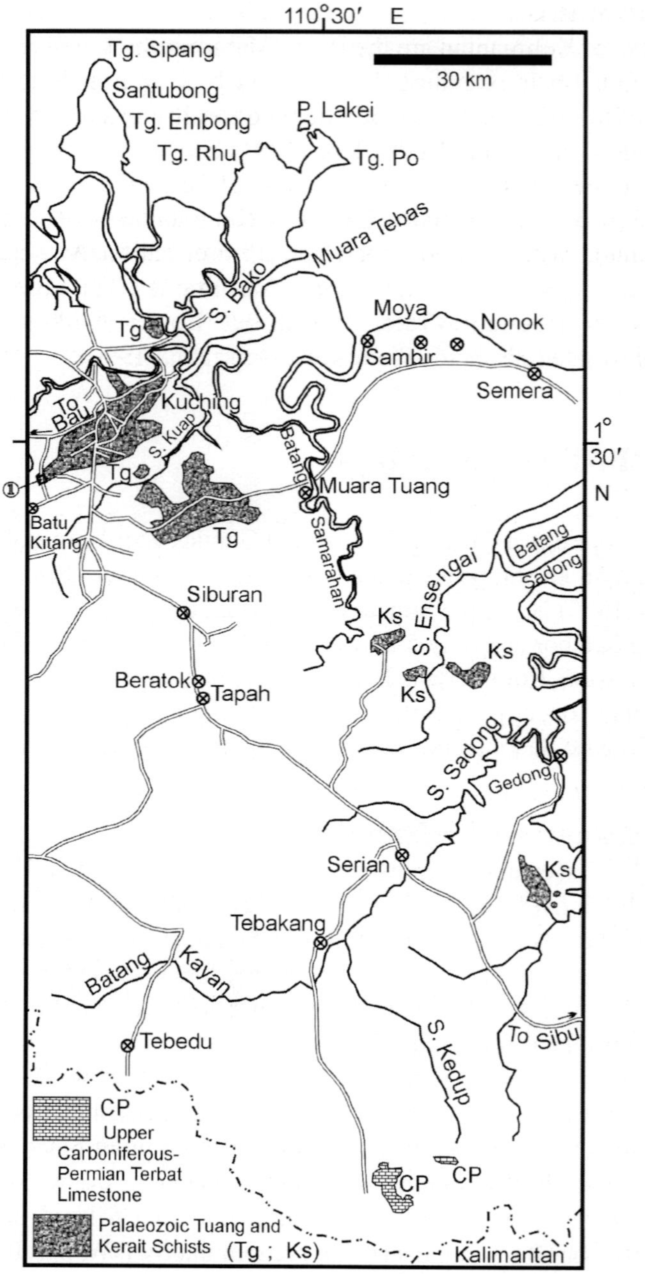

Figure 5. Map of Kerait, Tuang Schists and Terbat Limestone. After Hutchison et al. (in press).

IV.2.2. Lithology

The three main lithologies are limestone, chert and shale. The limestone is gray and massive and recrystallized to different degrees. Calcilutite, calcarenite and argillaceous limestone have been identified (Pimm, 1965). The Terbat limestone is commonly a wackestone or packstone, rich in bryozoa. Fragments of crinoids are frequent but corals appear to be absent (Fontaine, 1990). Dark grey chert beds, a metre or so thick, and nodules a few centimetre in diameter, define the steeply dipping bedding in Gunung Selabor (Wilford and Kho, 1965).

Dark grey shale is of similar appearance to shale of the Sadong Formation, but has been assigned to the Terbat Formation either because it is calcareous or is closely associated with limestone or chert (Pimm, 1965). A shale sequence near Terbat, which may exceed 300 m thickness, has been assigned to the Terbat Formation (Pimm, 1965). The Fusulinids indicate that the water of deposition was warm in shallow shelf seas. The accumulation of calcareous mud was built up as a result of the binding role of the bryozoa.

IV.2.3. Palaeontology and age

The limestone micro-fauna is sparse and frequently in a poor state of preservation. It has been studied by Cummings (1962), who described *inter alia* the Fusulinids: *Pseudoschwagerina (Zellia) heritschi heritschi, Pseudoschwagerina uber*, *Paraschwagerina* cf. *gigantea* and *Schwagerina* sp. proving the Wolfcampian (= Asselian) epoch of the Lower Permian.

Sanderson (1966) published a preliminary note on his studies of the Fusulinids. He promised a later detailed analysis of the fauna, but to-date it has not appeared. He concluded that the western part of Gunung Selabor contains Upper Carboniferous Fusulinids of Moscovian and possibly Late Bashkirian age. The eastern part of the hill contains only Permian Fusulinids.

Metcalfe (1985) described a sparse conodont assemblage from Gunung Selabor and near Terbat Bazaar. Indeed these are the first record of conodonts from Borneo. The identified species *Streptognathodus* indicate an age range from Upper-most Carboniferous to early Lower Permian. A more comprehensive study of the Foraminifera was carried out by Vachard (1990). He identified the following biozones:

Lower Permian:	
Langella ex gr. perforata	Artinskian
(no specimens found)	Sakmarian
Ozawainella angulata, Boultonia willsi Lee, *Schwagerina subnathorsti* and *Occidentoschwagerina fusulinoides*	Asselian
Upper Carboniferous:	
Pseudofusilina ex gr. kljasmica Sjomina, Nodosaria bella Lipina and *Dutkevichites aff. ramovsi*	Gzhelian
Beedeina lanceolata, Komia elegans Kordé; *Pseudostaffella greenlandica* and *Goksuella lunaensis*	Moscovian

Apparent discontinuities and gaps between the biozones are probably due to insufficient field sampling.

IV.2.4. Correlatives

Tate (1991) believes that the equivalent of the Terbat Formation in the Sanggau quadrangle of Kalimantan (Supriatna et al., 1989) is the Balaisebut Group that comprises carbonaceous slate and phyllite, schist and quartzite, with minor limestone, marble and chert. The rocks are cleaved and the grade of metamorphism is higher than in Sarawak. Zeijlmans van Emmichoven (1939) identified Permo-Carboniferous Fusulinids as well as Late Triassic faunas, demonstrating the difficulty of distinguishing the equivalents of the Terbat and Sadong formations.

During the Late Carboniferous, there were similarities between the Terbat Formation and East Peninsular Malaysia, Eastern Thailand and Vietnam, though an exactly identical limestone has yet to be found (Fontaine, 1990). In all these Cathaysian regions sedimentation was shallow marine, in warm tropical water and calcareous sedimentation was widespread (Fontaine, 1990). By strong contrast, the Terbat Formation has no similarity with the Sinoburmalaya terrains of Perlis, Kedah and Peninsular Thailand.

IV.3. UPPER TRIASSIC FORMATIONS

The basement rocks (Kerait Schist) and Terbat Formation are unconformably and extensively overlain by an Upper Triassic sedimentary sequence, known as the Sadong Formation, that is closely associated with and intercalated by a well developed basalt–andesite–rhyolite sequence, known as the Serian Volcanic Formation (Figure 6). The Sadong Formation represents an apron of sediments derived from a landmass that was dominated by the active Serian Volcanic arc, and rapidly deposited in adjacent neritic seas.

IV.3.1. Serian Volcanic Formation

From the type locality at Gunung Semuja, near Serian, the volcanic rocks form ridges trending NW almost to the Bau area and SE to the Gunung Selabor area, continuing 10 km across the border into Kalimantan as the Sekadau Volcanics. The greatest thickness of volcanic rocks occurs in the Semuja and Ampungan hills, possibly attaining a maximum thickness of 1500 m, but there are no control sections where the numbers can be quantified.

IV.3.1.1. Chemistry

When plotted on a Peccerillo and Taylor (1976) diagram (Figure 7), the Serian Volcanic Formation (Table 1) spans the range of basalt to rhyolite, mainly of both the calc-alkaline and the high-K calc-alkaline series using the analyses of Kirk (1968) and Hon (1976). Basalts and basaltic-andesites are the most common rock types, but there are also dacitic and rhyolitic differentiates. The trends displayed

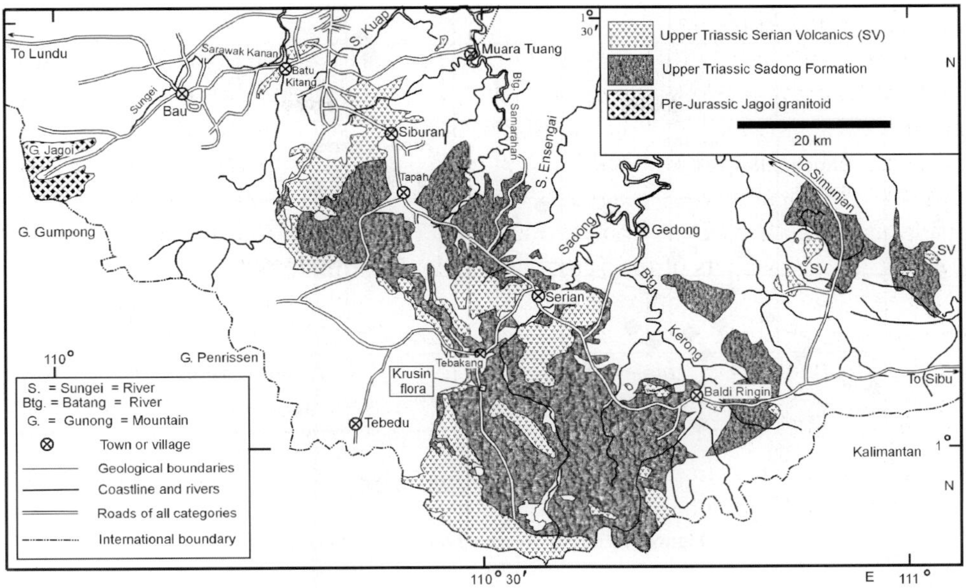

Figure 6. Map of the Upper Triassic Sadong Formation and Serian Volcanics. Showing also the Jagoi granitoid (after Hutchison et al., in press).

show some similarity to those of the Mediterranean region (Ewart, 1982), typical of subduction-related volcano-plutonic arcs, but display more closely the complexity of the active Indonesian (Sunda) volcanic arc (Hutchison, 1982). Using other petrographic criteria, Hon (1976) held that the volcanic rocks of the Kuap area are best interpreted as tholeiitic—the Serian Volcanics are characteristically both quartz and hypersthene normative, but this does not prove a tholeiitic affinity as shown by the more commonly used K_2O vs. SiO_2 plot of Figure 7.

IV.3.1.2. Lithology

The most common rock type is basalt to basaltic-andesite, only about 10% of which is porphyritic (Pimm, 1965). Between 5 and 10% of the basalt-andesite contain amygdales usually of 0.5–5 mm diameter (Pimm, 1965). They are filled with chalcedony, opal, chlorite, epidote, prehnite and calcite. The basalt–andesite shows moderate to strong hydrothermal alteration. The plagioclase is saussuritized and the ferromagnesian minerals altered to chlorite, which has also resulted from devitrification of the matrix glass, giving the whole rock a green colour. Volcanic breccia and lapilli tuff are locally significant. The acid differentiates, dacite and rhyolite, form fewer outcrops but are important in the Semabang Member.[1]

[1]This was incorrectly named the Semabang Trachyte Member by Wilford and Kho (1965) and Pimm (1965) because of its common trachytic or flow texture, but compositionally is not a trachyte.

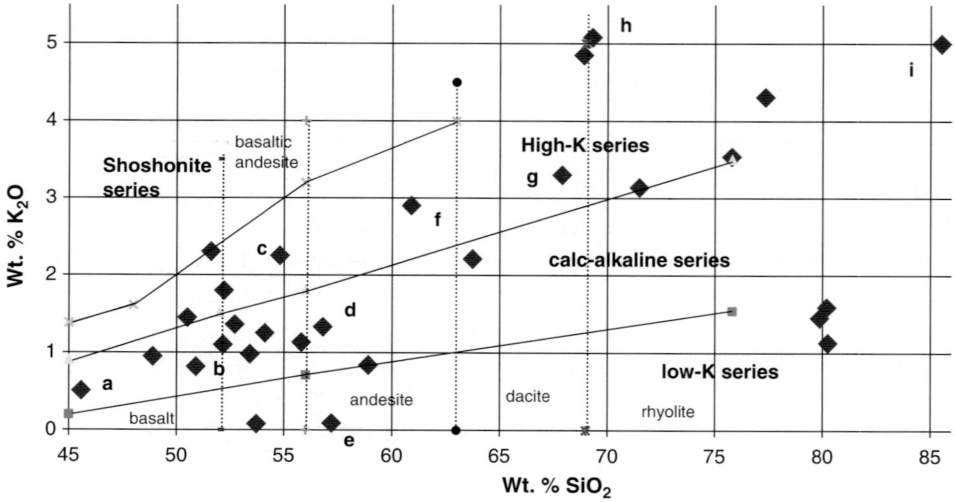

Figure 7. Serian Volcanic Formation K_2O vs. SiO_2.

Table 1. Chemical analyses of volcanic rocks of the Serian Volcanic Formation

wt%	a	b	c	d	e	f	g	h	i
SiO_2	45.58	50.90	54.80	56.80	57.20	60.90	67.90	69.30	85.50
TiO_2	2.37	0.95	1.15	0.88	1.31	0.82	0.75	0.80	0.37
Al_2O_3	14.13	16.60	14.90	17.54	8.85	16.00	11.50	12.48	6.35
Fe_2O_3	6.07	2.20	1.05	2.39	1.57	1.85	1.05	2.32	0.11
FeO	8.62	6.30	7.25	4.17	6.60	5.40	5.30	3.37	1.27
MnO	0.30	0.18	0.15	0.12	0.16	0.09	—	—	—
MgO	5.06	6.40	4.55	3.53	9.30	1.02	1.03	1.84	0.10
CaO	10.06	9.80	6.50	7.12	7.40	3.05	2.35	0.94	0.17
Na_2O	2.46	2.75	3.60	3.97	2.95	5.15	1.90	2.05	0.09
K_2O	0.51	0.82	2.25	1.33	0.09	2.90	3.30	5.08	5.00
H_2O+	2.60	2.40	2.45	1.27	2.70	2.10	2.40	1.59	0.62
H_2O-	1.77	0.18	0.78	0.16	1.14	0.43	0.29	0.34	0.07
CO_2	0.07	0.21	0.14	0.72	0.55	0.15	1.95	0.04	0.26
P_2O_5	0.67	0.18	0.15	0.34	0.13	0.19	0.20	0.24	0.06
Total	100.27	99.87	99.72	100.34	99.95	100.05	99.92	100.39	99.97

a = calc-alkaline basalt (V469), near Kampong Sta'ang, Kuap area (Hon, 1976).
b = calc-alkaline basalt (S11576), Sungai Idi, Kedup Valley (Wilford, 1965).
c = high-K calc-alkaline basalt (S13041), Gunung Semuja, Serian area (Pimm, 1965).
d = calc-alkaline andesite (V73), Kuching-Serian road, Kuap area (Hon, 1976).
e = low-K calc alkaline andesite (S13092), Gunung Semuja, Serian area (Pimm, 1965).
f = flow-textured high-K andesite (S13017), Gunung Semuja, Serian area (Pimm, 1965).
g = flow-textured high-K dacite (S11306), north of Gunung Selabor (Wilford, 1965).
h = high-K calc-alkaline rhyolite (V465), Kampong Stra'ang, Kuap area (Hon, 1976).
i = high-K rhyolite (devitrified obsidian) (NB11535), Sungai Paku, south of Gunung Selabor (Kirk, 1965).

Non-porphyritic basalt-basaltic andesite rocks are dark greenish grey, mainly composed of andesine–labradorite, augite, chlorite and iron oxides. The randomly oriented feldspar laths are usually 0.1–0.2 mm long and the pyroxene crystals

0.2–0.3 mm across (Pimm, 1965). The groundmass is normally rich in chlorite or may be glassy or cryptocrystalline. The texture is glassy to ophitic.

Porphyritic basalt-basaltic andesite rocks are dark greenish to black. The phenocrysts are predominantly of andesine–labradorite, with lesser augite, chlorite and iron oxides. The plagioclase phenocrysts are usually 1–2-mm long and the anhedral augites up to 2 mm across (Pimm, 1965). Some rocks contain altered olivine, while some contain only labradorite. The more potassic varieties contain phenocrysts of basaltic hornblende. The groundmass is microcrystalline to glassy and usually rich in chlorite and feldspar.

Breccia and tuff The greenish-grey breccias consist of clasts of glassy to holocrystalline porphyritic lava in a quartzo-feldspathic or glassy matrix. The tuffs are fine-grained and vitric (Kirk, 1968). The fragments in the breccia and lapilli tuffs include porphyritic and non-porphyritic basalt-andesite and amygdaloidal basaltic-andesite, basaltic tuff and volcanic glass. The clasts in the breccia are angular and commonly 5–15 cm across. The lapilli tuff contains smaller rounded more glassy fragments. The tuffs contain angular broken crystals of quartz, plagioclase, pyroxene and shards of volcanic glass up to 0.3 mm across (Pimm, 1965).

Acid differentiates The greatest occurrence of acid differentiates has been assigned by Wilford and Kho (1965) to the Semabang Member, which is thought to occur near the top of the volcanic succession, but such rocks also occur in minor amounts throughout the Serian Volcanic Formation. The rhyolite is a light grey microcrystalline aggregate of quartz and feldspar. Amygdales filled with chalcedony, chlorite and quartz are common, reaching a diameter of 5 mm.

The *pyroclastic rocks* are of grey lapilli tuffs composed of small fragments of porphyritic quartz dacite with grains of feldspar, quartz and chlorite in a dusty matrix. Fine-grained tuffs are largely composed of devitrified volcanic dust. Breccias are mauve-green rocks composed of angular fragments of porphyritic dacite in a fine-grained quartz–feldspar matrix.

Intrusive rocks Stocks, dykes and sills of diorite and micro-diorite cut the Sadong and Serian Volcanic formations (Wilford and Kho, 1965; Pimm, 1965).

IV.3.2. Jagoi Granodiorite

About 10 km SW of Bau, the oldest outcrops are represented by I-type granitoids of the Jagoi and Kisam hills (Ting, 1992), which extend across the border into Kalimantan (Rusmana and Pieters, 1989). The granodiorite has yielded a K:Ar date of 195±2 Ma (Bladon et al., 1989). Bignell (1972) had failed to obtain fresh biotite from the Gunung Jagoi granitoid. He extracted a hornblende-bearing xenolith from it, which yielded an unacceptably low K–Ar age of 112 Ma (Table 5). The outcrops range from quartz-diorite to granite, but granodiorite is dominant. It is medium grained and non-porphyritic. Hornblende and biotite are the characteristic mafic minerals and some specimens contain relicts of fayalitic olivine (Ting, 1992). The northern margin of the Jagoi Granodiorite is sharply downthrown to the north along an E–W fault and shear zones occur to the south in Sarawak and Kalimantan. The 195 Ma K:Ar date is therefore thought to have resulted from argon loss (Tate, 1991). Nevertheless even an emplacement age of 195 Ma is acceptable to place the Jagoi Granodiorite within the Serian volcano-plutonic suite, following Hamilton (1979) and Tate (1991).

IV.3.3. Sadong Formation

Strata of the Sadong Formation have a strong geographical relationship with the volcanic edifices, suggesting a genetic relationship with the contemporaneous Serian Volcanic Formation. The main outcrop of the Sadong formation (Figure 6) is transected by the Sri Aman Road southeast of Serian, described by Pimm (1965) and by the Pedawan and Tebedu roads to the west of Serian, described by Wilford (1965).

IV.3.3.1. Thickness and relationships

The base of the Sadong Formation is not exposed, but coarse conglomerate is common in the Sadong Formation where it borders the Terbat Formation. At the western foot of Gunung Selabor, the conglomerate clasts are predominantly of calcareous chert and silicified limestone containing Carboniferous (Terbat Formation) crinoids (Wilford and Kho, 1965). Although the contact may be locally faulted, there can be little doubt of a strong unconformity between the Sadong and the Terbat formations.

The Sadong Formation is unconformably overlain by the Upper Jurassic Kedadom Formation and the Upper Jurassic part of the Bau Limestone Formation north of Batang Kayan, and by the Lower Cretaceous part of the Pedawan Formation south of Batang Kayan (Wilford and Kho, 1965).

Because of insufficiently resolved structure, the Sadong Formation thickness is unknown and it must vary locally because it is bounded by unconformities. However, the maximum estimated along the line of the Pedawan Road, where southerly dips predominate, assuming an undisturbed southwards younging, is 2285 m. Minimum thickness for the Sadong Formation is suggested to be 1525 m and for the Serin Arkose Member 765 m (Wilford and Kho, 1965). Pimm (1965) estimates the Serin Arkose Member to be 915 m thick, similar to that of the conglomerate and coarse sandstone section of the Sadong Formation. He questioned the estimate of Wilford and Kho (1965) and pointed out that in some river sections over 3050 m of thickness are exposed (Pimm, 1965). In the absence of good structural control, this figure may be excessive.

IV.3.3.2. Lithologies

Feldspathic sandstone is the dominant lithology, occurring as beds averaging 3–6 m, occasionally 30 m thick. The most typical outcrops are of alternations of feldspathic sandstone, sandy shale and shale (Wilford and Kho, 1965). The shale and sandy shale beds are usually sheared, suggesting rapid de-watering. They occur as beds of a metre or so thickness, but may be up to 15 m thick (Wilford and Kho, 1965). Other subordinate lithologies are conglomerate, limestone, chert and intermediate to acid tuff. A distinct sequence of feldspathic sandstone with subordinate shales has been mapped separately near the base of the formation as the Serin Arkose Member.

Sandstones are typically thick bedded to massive, commonly cross-bedded, medium to coarse grained. Quartz forms 50–85% of the coarse-grained sandstones. It is invariably sub-angular and interpreted to be of metamorphic or volcanic origin (Wilford and Kho, 1965). Abundant shard-like quartz grains and mica give fissility to the finer-grained sandstones.

All sandstones are feldspathic, and some are arkosic. Both orthoclase and plagioclase occur in variable proportions, remarkably fresh and only little sericitized. Slightly chloritized biotite and muscovite are common. The sandstones commonly contain rock fragments such as mica schist and chert, but remarkably clasts of acid volcanic rock and plutonic rocks are relatively rare. Carbonaceous matter is common, as fine disseminations or concentrated in laminae and lenses.

Conglomerates occur as lenses within the sandstone beds. They contain well-rounded granules, pebbles and a few cobbles, typically about 50% of the clasts are of gneissose mica granite and the remainder includes mica schist, chlorite schist, quartzite, carbonaceous phyllite, porphyritic lava, and chert. The grains are of orthoclase, microcline, and quartz. One sample from near Mount Selabor contains silicified limestone containing a Terbat Limestone fauna.

Shales are grey in colour and commonly slickensided and veined by quartz (Wilford and Kho, 1965). Most samples are silty and contain splinters of quartz, sub-angular feldspar and abundant muscovite, hence may be interpreted as crystal tuffs. Carbonaceous matter is finely disseminated

Dark grey *limestone* lenses, as thick as 2 m, grade into and are interbedded with calcareous carbonaceous shale, and occur locally in the Sadong Formation. They also contain flakes of muscovite and splinters of quartz (Wilford and Kho, 1965).

Pale grey or green *chert* occurs locally as boulders, but never yet seen as actual outcrops.

Pyroclastic rocks Thin beds of agglomerate, tuff, tuffaceous sandstone and shale are interbedded with feldspathic sandstone and shale, but they form only a minor part of the Sadong Formation. The fragments in the tuff range from 0.1 to 0.2 mm and include feldspar, quartz, basalt, glass and rarely ferromagnesian minerals (Pimm, 1965).

Thermally metamorphosed rocks Sandstone and shales of the Sadong Formation have been thermally metamorphosed to hornfels adjacent to small intrusions of diorite and tonalite in the Serian area (Pimm, 1965). Diopside is a characteristic mineral in the hornfelsed rocks.

IV.3.3.3. Serin Arkose Member

Wilford and Kho (1965) introduced the term for outcrops predominantly of feldspathic sandstone and arkose occurring along the Sungai Serin, transected by the Pedawan Road, and Pimm (1965) mapped an extension into a low-lying area north and northeast of Serian and the Sri Aman Road.

The typical feldspathic sandstone, which contains a few thin beds of hard grey shale, is a hard grey massive well-jointed medium to coarse-grained poorly sorted rock, composed mainly of angular to sub-angular quartz, sericitized sodic plagioclase and lesser orthoclase (Wilford and Kho, 1965). Both muscovite and partly chloritized biotite occur. Epidote is a minor constituent and small contents of hornblende and pyroxene have been identified (Pimm, 1965). The rock is cut by irregular quartz veins. The interbedded shale beds are impersistent and rarely exceed 3 m thickness.

Rock clasts occur very commonly. They include microgranite, mica schist and phyllite, chlorite sericite quartz rock and dacite-rhyolite (Wilford and Kho, 1965).

IV.3.3.4. Palaeontology and age

Plants Road cuts at Krusin village, 3.6 km southwards along the Terbat road from its junction with the Tebedu road, expose a thin argillaceous sandstone, from which Kon'no (1972) systematically described a small flora of 17 species, interpreted to be Upper Carnian rather than Norian. He named the assemblage the 'Krusin florula'. The plant material was not rooted in the strata, but rapidly washed in, as suggested by various fern fronds embedded obliquely in the rock matrix. The following were identified from the dark coloured muddy sandstone: Sphenopsida; *Annulariopsis Hashimotoi* KON'NO sp. nov.; *Neocalamites carrerei* (ZEILLER) HALLE; Neocalamostachys takehashii (KON'NO) BOUREAU; Equisetum spp. (2); Pteropsida; Clathropteris meniscoides BRONGNIART; *Dictyophyllum* cf. *nilssoni* (BRONGNIART) var. *genuinum* NATH; *Cladophlebis haiburnensis* (LINDLEY and HUTTON) BRONGNIART; *Cladophlebis* cf. *haiburnensis* (LINDLEY and HUTTON) BRONGNIART; Cladophlebis Ishiiana KON'NO sp. nov.; *Sphenopteris* (*Todites ?*) sp.; *Todites Katoi* KON'NO sp. nov.; *Todites sarawakensis* KON'NO sp. nov.; *Todites Tamurae* KON'NO sp. nov.; *Cycadopsida*; *Dictyozamites krusinensis* KON'NO sp. nov.; *Otozamites* sp.; and *Problematicum*. The Krusin flora has nothing in common with the flora described from Bintan Island by Jongmans (1951), and which is probably much younger than the Krusin flora (Kon'no, 1972).

On the other hand, the Krusin flora is closely comparable to the Tonkin flora of the Hongay coal series of northern Vietnam, as described by Zeiller (1902). Silicified *Araucarioxylon* wood and plant remains within chert occur commonly where the Semabang Volcanic Member is interbedded with the Sadong Formation (Pimm, 1965).

Bivalves, especially *Halobia* sp., have been found in many localities (Wilford and Kho, 1965). The following bivalves have been identified at the British Museum, reported by Wilford and Kho (1965) and all are consistent with a Carnian or Norian Upper Triassic age: *Asoella* aff. *confertoradiata* (TOKUYAMA); *Asoella* sp.; *Chlamys* sp.; *Entolium* sp.; *Entolium* aff. *hallensis* (WOHRMANN); *Entomonotis* sp. nov.; *Grammatodon* or *Parallelondon* sp.; *Gryphaea* aff. *keilhaui* BOHM; *Halobia* cf. *molukkana* WANNER; *Halobia ? talauana* WANNER; *Halobia* sp.; *Indopecten* sp.; *Monotis inaequivalvis* BRONN; *Monotis salinaria* BRONN; *Nucula* sp.; *Oxytoma* sp.; *Plicatula* aff. *hekiensis* NAKAZAWA; *Pseudolimea* spp.; *Pseudomonotis* sp.; and *Unionites* sp.

Of the above list, *Halobia*, which is Carnian or Lower Norian, occurs most commonly. The other bivalves confirm a Norian and possibly a partly Carnian age.

Radiolaria Long-ranging *Cenosphaera* and *Dictyomitra* radiolaria have been identified in some of the Sadong Formation tuffs (Pimm, 1965). Radiolaria extracted from a dacitic radiolarian tuff near Piching (Basir et al., 1996) indicate a Lower Jurassic age (Pliensbachian to Toarcian). Originally ascribed to the Kedadom Formation, the dacitic tuff is now considered to belong to the upper part of the Sadong Formation–Serian Volcanics, where it is unconformably overlain by the Upper Jurassic Kedadom Formation. The identified species are: *Canoptum anulatum* Pessagno and Poisson; *Canoptum rugosum* Pessagno and Poisson; *Canutus indomitus* Pessagno and Whalen; *Canutus izeensis* Pessagno and Whalen; *Pantallium sanrafaelense* Pessagno and Blome; *Parahsuum simplum* Yeh; *Parahsuum takarazawaensis* Sashida; *Praeconocaryomma decora* Yeh; *Praecococaryomma media* Pessagno and Poisson.

These are the youngest fossils determined for the upper part of the Sadong Formation, now shown to extend from the Upper Triassic to the Lower Jurassic (Basir Jasin, 2000).

Other fossils A crushed ammonite was identified as probably a *Pinaceratid* of Middle to Upper Triassic age. A probable new species of *conchostraca* was provisionally assigned to *Isaura (Euestheria)*, which is well represented in the Upper Triassic (Wilford and Kho, 1965). Long-ranging Foraminifera have been discovered, but they have no value in assigning a formation age. Crinoid stem fragments occur in calcareous sandy shale. *Zonotrichites sp.* algae have been identified in chert and given a Triassic age.

IV.3.3.5. Provenance of the Sadong Formation

Kirk (1968, p. 9) concluded: "The general scarcity of volcanic detritus among the Sadong Formation sedimentary rocks indicates that penecontemporaneous erosion of the volcanic rocks was slight, and subsidence appears to have kept pace with deposition, preventing the formation of large volcanic islands". Nevertheless the Sadong Formation, including the Serin Arkose Member, may be interpreted as having been eroded predominantly from nearby active volcanic highlands of the contemporaneous Serian Volcanic Formation, which also contained outcrops of basement country rocks of Permo-Carboniferous and older age. The sandstones and conglomerates are typically volcanic arkose, characterized by variable proportions of plagioclase and orthoclase. Quartz grains are angular, fractured, and described in some cases as 'shard-like'. The common contents of detrital biotite, epidote, hornblende and pyroxene and the clasts of microgranite, rhyolite and porphyritic lavas demonstrate rapid erosion from volcanoes of an arc in its dacite–rhyolite eruptive phase. Bulk chemical analyses of arkosic sandstones (Table 2) plot directly in the high-K dacite and rhyolite fields of the adjacent Serian Volcanics, as would be expected from rapid erosion and deposition with minimal weathering of the products. An arid climate was therefore not needed for preservation. Dacitic–rhyolitic phases of volcanic arcs are rapidly eroded and transferred to be re-deposited nearby as volcanic arkosic sedimentary

Table 2. Chemical analyses of arkosic sandstones

wt%	a	b	c
SiO_2	78.3	67.9	64.1
Al_2O_3	10.75	14.5	14.1
Fe_2O_3	0.84	1.63	0.55
FeO	1.55	2.65	3.60
MgO	0.41	1.59	1.74
CaO	0.31	1.38	5.00
Na_2O	2.48	3.05	3.55
K_2O	3.52	2.65	2.45
H_2O-	0.17	0.78	0.20
H_2O+	0.91	2.65	2.15
CO_2	0.22	0.37	1.67
TiO_2	0.25	0.55	0.52
P_2O_5	0.06	0.09	0.13
MnO	0.02	0.07	0.06
Total	99.79	99.90	99.80

a = S8548, loc. Sungai Bukar (Sadong Formation).
b = S13201, loc. Sungai Tarat (Serin Arkose Member).
c = S13131, loc. Sungai Sebengkong (Serin Arkose Member).

aprons adjacent to the arc, which eventually exceed the volume of dacite and rhyolite remaining in the volcanic pile. Thus, most of the sandstones and conglomerates were deposited while the arc was in its acid phase of eruption. By contrast, while the arc was in its basaltic–andesitic phase, its contribution to the sedimentary apron should be predominantly muddy and mudstones and shales were shown by Pimm (1965) to form the major part of the Sadong Formation, outcrops of which are poor in comparison with the arenaceous facies.

The volcanic edifices were well vegetated, so that plant debris are commonly incorporated in the sandstones, as a result of rapid washing into detrital alluvial fans that accumulated subaerially on the volcanic slopes or into the adjacent neritic seas.

Other clasts such as phyllite and mica schist and calcareous chert, containing Terbat Formation fauna near Mount Selabor, indicate that the volcanic highlands were built on a basement of Kerait Schist and Terbat Formation. The detrital quartz of the Sadong Formation arenaceous rocks has been consistently interpreted by Wilford and Kho (1965) and Pimm (1965) to be predominantly of metamorphic or vein origin, indicating erosion from a metamorphic terrain.

The shales of the Sadong Formation are described as silty because they contain angular feldspar and quartz grains, commonly described as ‘splinters’. The shales are therefore crystal tuffs, deposited in neritic conditions and were receiving pyroclastic infall from the not too distant volcanoes.

The succession also contains many horizons of intermediate to acid tuffs, which contain pyroxene crystals, lava fragments and glass shards (Wilford and Kho, 1965).

The cross sections of Wilford and Kho (1965) and Pimm (1965) suggest that the Serian Volcanic Formation was not yet in existence at the beginning of the Sadong Formation and that the Serin Arkose Member was pre-volcanic. All sandstones and conglomerates, whether of the Serin Member or higher in the succession are similarly arkosic, so that the environment of proximity to an active volcano-plutonic arc persisted throughout the Sadong Formation.

IV.3.3.6. Correlatives

The equivalent to the Sadong Formation is the Bengkayang Group, which outcrops extensively in NW Kalimantan

IV.3.4. Regional palaeogeography

The search for a suitable Late Triassic ensialic volcano-plutonic arc, with which to attach the Serian Volcanic arc and its Sadong Formation apron, was earlier conducted by Pimm (1967a), who demonstrated the dissimilarity in age and geochemistry between the Serian Volcanic Formation and the Pahang Volcanics of Peninsular Malaysia. A more successful attempt was conducted by Gatinsky and Hutchison (1986) and Gatinsky et al. (1984), summarized by Hutchison (1989, pp. 119, 128, 130). They concluded that the most appropriate positioning of the Serian Volcanic arc in Late Triassic time was adjacent to the Precambrian Kontum Massif of the eastern coastline of Vietnam, southwards from Da Nang, through Qui Nhon, and across the Hon Khoi peninsula to the district of Dalat in the south.

This central and southeastern district of Vietnam is characterized by a Precambrian metamorphic massif on which Mesozoic formations are widely distributed within fault-bounded grabens, especially around the south, western and northern margins. Triassic volcanic rocks form an important part of the stratigraphy of these grabens. They are commonly rhyolite, dacite and andesite (Fontaine and Workman, 1978), and may be correlated with the Serian Volcanic Formation.

The Upper Triassic Nongson Formation occupies a large E–W graben near Da Nang on the northern Kontum Massif. It is composed of thick sandstones with shale intercalations, conglomerate and coal beds, and both in age and lithology may be equated with the Sadong Formation of Sarawak. A good Norian–Rhaetic fossil flora has been described by Vozenin-Serra (1977). The flora contains many species common to the Tonkin flora of North Vietnam and the Krusin flora of Sarawak—*Neocalamites*, *Equisetum*, *Cladophlebis*, and *Dictyophyllum*.

The Upper Permian to Middle Triassic Mangziang Formation, of rhyolite, tuff, sandstone, siltstone and mudstone, fills several N–S depressions along the southern margin of the Kontum Massif. The Mangziang Formation volcanic rocks are associated with sub-volcanic granites of Late Permian to Triassic age. The Phu Son igneous complex is of gabbro, gabbro-norite and granitoids, dated 250–190 Ma (Hutchison, 1989).

The continental shelf of eastern Vietnam is anomalously narrow and strongly faulted. Between 10° and 11.5° N, the normal Cenozoic faults trend NE–SW, whereas from 11.5° to 16° N the trend of the normal faults is N–S (Wirasantosa et al., 1992). The change in fault direction occurs at an inflexion point offshore Phan Rang, mimicking the change in trend of the coastline. The N–S and NE–SW directions suggest two arms of a triple junction and that continental lithosphere has been lost by rifting from the Vietnam coast. There is a belt of Late Cretaceous to Palaeogene leucocratic sub-alkaline granites through Phan Rang, which are related to the Dongzuong Volcanic Formation. They probably indicate the time of widespread rifting of the continental shelf of Vietnam (Hutchison, 1989).

This discussion does not prove that the west Sarawak terrain was attached to southeastern Vietnam. Indeed the Luconia microcontinent is more likely to have been attached here, and the Serian Volcanic Formation terrain would have occupied a position to the south of it. There is a basement of Upper Triassic limestone in the southern part of the Malay Basin, sampled at Sotong B-1 well, offshore Trengganu (Fontaine et al., 1990). It is clear that the Serian Volcanic Formation–Sadong Formation terrain is not exotic to this region—characterized by numerous linear grabens filled by strata containing Carnian and Norian fauna and flora of a regionally characteristic type, closely associated with volcanic rocks. All of these grabens, occurring now in regions as far apart as western Sarawak, the central basin of Peninsular Malaysia, and eastern and northern Vietnam, have similar type igneous and sedimentary sequences, and are all products of the Triassic Indosinian Orogeny, whose effects were widespread throughout the region (Hutchison, 1989).

IV.4. UPPER JURASSIC AND CRETACEOUS FORMATIONS

IV.4.1. Kedadom Formation

The Kedadom Formation is of restricted occurrence, lying west of and overlying the upper part of the Serian Volcanic Formation west and northwest of Pichin along the Tebedu road (Figure 8). It occurs in a similar geological setting to the Semabang Member of the Serian Volcanic Formation (Wilford and Kho, 1965). The latter is taken as Upper Triassic to Lower Jurassic. The Kedadom Formation unconformably overlies the Serian Volcanic Formation and the Sadong Formation on the east, but the time represented by the unconformity may be short. On the west it is overlain conformably by, or passes laterally into the Upper Jurassic Pedawan and Bau Limestone formations. The formation thins rapidly northwards and southwards, and its maximum estimated thickness is about 760 m (Wilford and Kho, 1965).

The formation is predominantly of massive to thick-bedded sandstone with thin layers of dark-coloured carbonaceous sandy shale, commonly sheared and slickensided. Conglomerate is common towards the base. Dacitic radiolarian tuff also occurs near the base. Limestone lenses, as much as 60 m thick, occur near the base and towards the top. The limestone is a dark grey fine-grained rock with bedding emphasized by laminae of shale and carbonaceous material. Microfossils are sparse, but the rocks contain gastropods and bivalves.

The basal part of the Kedadom Formation consists of a basal conglomerate, carbonaceous sandstone and shale with thin beds of dark grey fine-grained limestone, dacitic radiolarian tuff and conglomerate (Wilford and Kho, 1965). More than 50 m

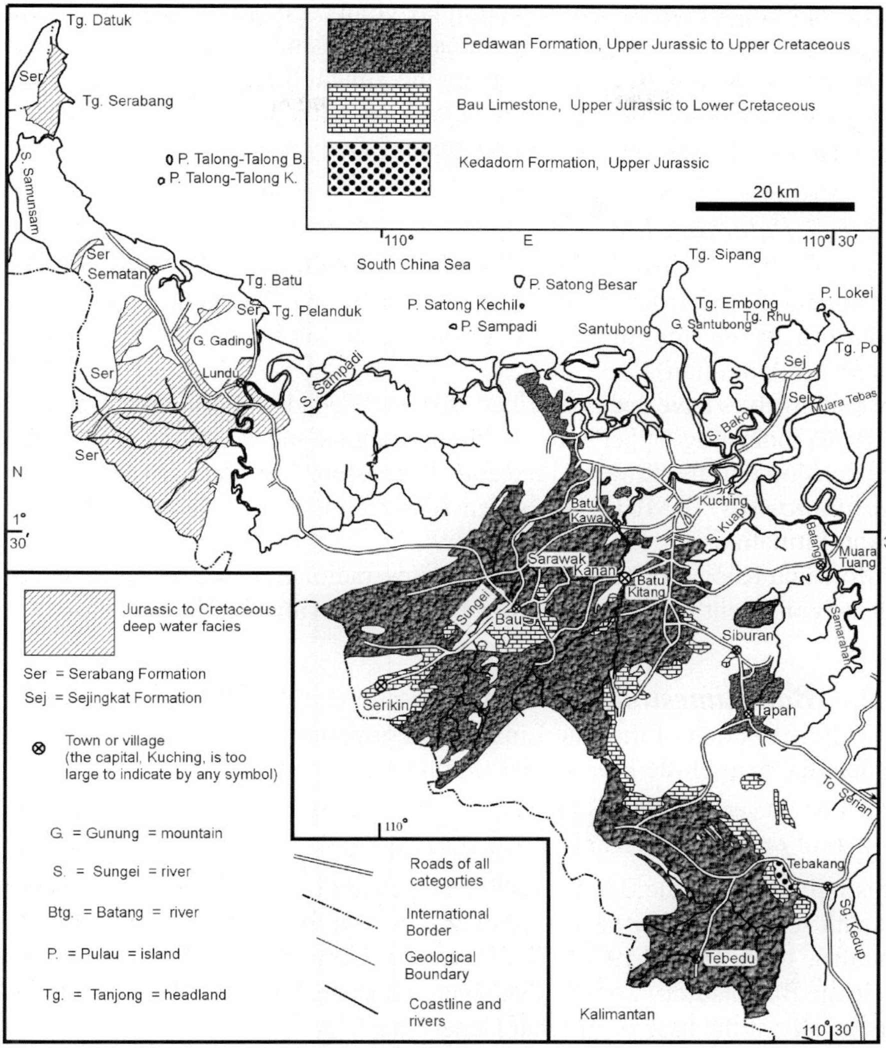

Figure 8. Map of the Jurassic and Cretaceous formations of western Sarawak (based on Hutchison et al., in press).

of dacitic tuff is exposed 2.5 km west of Piching along the Tebedu road from Serian, which has yielded Lower Jurassic radiolaria (Basir et al., 1996). It is a vitric tuff composed glass shards, feldspar crystals and radiolaria. The sandstones near the base are composed predominantly of volcanic rock fragments derived from the underlying Serian Volcanic Formation. The sandstones, which occur commonly higher up the section (westwards), are composed of angular to sub-angular quartz, clasts of sandstone, acid lava and mica schist, with a few grains of completely sericitized feldspar, chert and shale clasts.

The conglomerates in the basal section are composed predominantly of pebbles of intermediate to acid lavas, angular grains of quartz and feldspar. A clast of Terbat Limestone was found in one outcrop along the Sungai Rembus. The conglomerates higher in the succession are composed of well-rounded pebbles and boulders up to 0.6 m diameter of sandstone, conglomerate and sandy shale.

IV.4.1.1. Palaeontology and age

Ishibashi (1982) described the ammonites *Berriasella* sp., *Neolissoceras* sp. and *Proniceras* sp. from the Kedadom Formation. The ammonites span the boundary between the uppermost Jurassic to lowermost Cretaceous (Late Tithonian–Valanginian).

Several bivalves have been described from the Kedadom Formation (Tamura and Hon, 1977) indicating Upper Jurassic Kimmeridgian and Tithonian ages with possible extension into the Lower Cretaceous Berriasian. *Nuculana* (*Praesaccella*) sp. cf. *yatsushiroensis* TAMURA has been found both in the Kedadom and in the Pedawan formations.

Wilford and Kho (1965) list other ammonites, radiolaria, bivalves and gastropods, which are consistent with a Kimmeridgian to Lower Tithonian Upper Jurassic age.

IV.4.2. Bau Limestone Formation

The areal distribution of the Bau Limestone Formation is shown in Figure 8 and a more detailed map of the Bau district (Figure 9).

IV.4.2.1. Lithology

The main lithology of the Bau Limestone Formation is massive pale grey pure limestone with some dark grey bedded argillaceous limestone (Wilford and Kho, 1965; Wolfenden, 1965; Pimm, 1967). The limestone forms impressive karstic hills with caves in the Bau district. A standstone-shale sequence, known as the Krian Member, occurs locally at the base of the Bau Limestone.

IV.4.2.2. Palaeontology

The Foraminifera of the Bau Limestone Formation have been identified by Bayliss (1966) from extensive sampling from the Bau and Penrissen areas. The fauna is of restricted nature and marked uniformity over a wide area. It comprises surprisingly few species. The fauna indicates a general Upper Jurassic age, probably Kimmeridgian. The most frequently occurring species are: *Miliolidae*, *Textulariidae*, *Valvulinae*, Pseudocyclammina lituus (YOKOYAMA), Nautiloculina oolithica MÖHLER, *Ammomarginulina* spp., *Trocholina* spp., and *Rotaliform* species (*Protopeneroplis ?).*

The following Terebratulid brachiopods have been described by Yanagida and Lau (1978) from the Bau Limestone Formation: *Neumayrithyris torinosuensis*

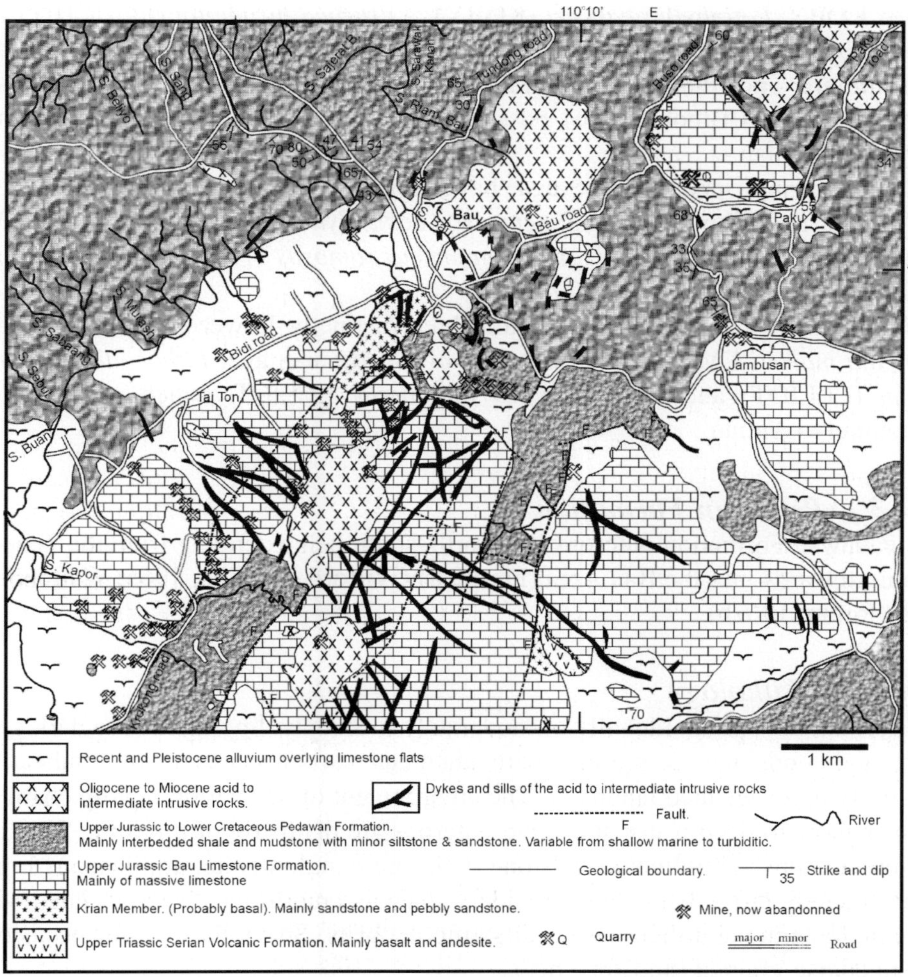

Figure 9. Geological map of the Bau district, of western Sarawak (after Hutchison et al., in press).

TOKUYAMA, which is Upper Jurassic, and found at Paku, 4 km E of Bau, and *Ornatothyris bauensis* YANAGIDA and LAU, sp. nov., which is determined to be Lower Cretaceous from the accompanying assemblage, found at Gunung Stulang, 13 km SW of Bau.

Although corals are not prolific and never attained the status of reef builders, a well preserved fauna has been obtained from the Bau and Penrissen area and identified by Beauvais and Fontaine (1990), as follows: *Cuneiphyllia somaensis* (EGUCHI), *Amphiastraea cf. gracilis* KOBY, *Donacosmillia cara* (ELIASOVA), *Microsolena sp.*, *Latomeandra ramosa* (KOBY), *Epistreptophyllum cylindratum* MILASCHEWITCH, *Microphyllia* cf. *undans* (ETALLON), *Thamnoseris frotei* THURMANN & ETALLON, *Adelocenia bacciformis* (MICHELIN), *Cladophyllia rumea* KOBY, *Latiphyllia cartieri* (KOBY), *Astraraea huzimotoi* (EGUCHI), and *Litharaeopsis fontainei* BEAUVAIS. Together with the accompanying algae, Foraminifera and rudists, the corals are consistent with an Upper Jurassic (Kimmeridgian to Tithonian age), possibly extending into the Lower Cretaceous up to the Valanginian.

The following Upper Jurassic algae have been identified by G. F. Elliot and listed by Wilford and Kho (1965): *Clypeina* sp. nov., *Cylindropella arabica* ELLIOT, *Lithocodium japonicum* ENDO, *Nipponophycus ramosus* YABE & TOYAMA, and *Salpingoporella annulata* CAROZZI.

Wilford and Kho (1965) also listed three distinctly Lower Cretaceous algae: *Clypeina marteli* EMBERGER, which is of Valanginian age, *Lithocodium aggregatum* (ELLIOT) and *Permocalculus inopinatus* ELLIOT, which ranges from Barremian to Aptian.

IV.4.3. Pedawan Formation

The Padawan Formation outcrops in a north–south belt extending from the Tebedu area northwards into the Bau Area (Figure 8) (Wilford and Kho, 1965; Wolfenden, 1965; Pimm, 1967b).

IV.4.3.1. Lithology

The formation is predominantly of moderately to steeply dipping marine dark grey shale and mudstone, commonly with abundant carbonaceous matter indicating proximity to a vegetated landmass. The environment of deposition rapidly changed from shallow to deep water and some outcrops are distinctly of turbidite. The formation contains subordinate sandstone with rare conglomerate, argillaceous limestone and radiolarite. Pebbly and bouldery shale and mudstones are associated with the conglomerate and there are local slump deposits. Shale clasts occur occasionally in coarse-grained sandstone beds.

Some outcrops in the Bau region indicate that the conglomerate, sandstone, pebbly mudstones and slump deposits form large-scale channels that transported the coarser grained sediments into the main basin.

The Tambang Tuff Member is predominantly dacitic, and there are associated lavas ranging to andesitic (Wilford and Kho, 1965).

IV.4.3.2. Palaeontology and age

The Lower Division of the Pedawan Formation was set by Wilford and Kho (1965) to be stratigraphically below the first occurrence of *Orbitolina*. The Lower Division is poor in fossils. The shales contain long-ranging arenaceous Foraminifera such as *Bathysiphon* sp., *Haplophragmoides* spp. and *Glomospira* sp. Radiolaria are poorly preserved in the shales but better preserved in radiolarite, which occurs near the base of the Lower Division, probably indicating a Lower Cretaceous age. *Pseudocyclammina* sp. has been identified in limestones indicating a similar age to the

Bau Limestone Formation, from which it may have been reworked. Algae have been identified but have little age discriminating value.

Ishibashi (1982) described the following ammonites from the Pedawan Formation: *Neocomites* sp., *Limaites* sp., *Phylloceras* sp., *Thurmanniceras* sp., *Micracanthoceras* sp., *Phanerostephanus* sp., *Virgatosphinctes* sp. and *Paraboliceras jubar* (BLANFORD). The latter especially points to an Upper Jurassic Tithonian age. Sarkar (1973) identified fragmentary ammonites from Pedawan shales: *Berriasella* sp., *Microcanthoceras* sp. and *Thurmanniceras* sp. Based on these he concluded an Upper Tithonian–Lower Valanginian age.

The Middle Division is known to range from Barremian or Aptian to Cenomanian because of the presence of the age-diagnostic foram *Orbitolina lenticularis* (BLUMENBACH), which occurs in the lower part of the Middle Division of the Pedawan Formation (Wilford and Kho, 1965), some distance above the contact with the Bau Limestone Formation. Its presence was also determined by Hashimoto and Matsumaru (1977) from a slumped horizon in the Bau district and two distinctly different forms have been dated Lower Cretaceous (uppermost Barremian and Upper Aptian). *Hedbergella* sp. also suggests an Aptian or Albian age. This division also contains possible Albian to Lower Cenomanian pollen.

The Upper Division was formerly considered by Wilford and Kho (1965) to range up to the Maastrichtian because of the foram assemblage, but Nuraiteng and Kushairi (1987) cast doubt on the previous identification (Wilford and Kho, 1965) of *Globotruncana tricarinata* and near Tepoi they identified abundant *Marginotruncana coronata, Marginotruncana angusticarenata* and *Dicarinella carinata*, suggesting that the known upper range does not go beyond the Upper Santonian. This upper range is in agreement with the palynological determinations of Muller (1968).

Plants Muller (1968) made a comprehensive study of the palynology of the Upper Division of the Pedawan Formation and divided it into three floral zones:

A. *Caytonipollenites* zone (provisional). This zone is characterized by a low frequency of *Caytonipollenites pallidus*, which does not occur in younger zones. *Classopollis* sp. cf. *Classopollis classoides* forms the dominant element of the microflora. This zone was found only in the Lundu-Kayan area. The age of the zone is pre-Turonian, based on the presence of *Caytonipollenites pallidus*.

B. *Cicatricosisporites* zone. A high frequency of *Cicatricosisporites* sp. cf. *Cicatricosporites dorogensis* and *Retitricolpites vulgaris* typifies this zone. Of special interest is the abundance of *Exesipollenites tumulus*. The age of this zone is calibrated by the co-existence of planktonic Foraminifera in the Penrissen area, taken to be Albian–Cenomanian.

C. *Araucariacites* zone. This zone is characterized by the sudden appearance of a high frequency of *Triorites minutipori* and a marked increase in the abundance of *Psilatricolporites acuticostatus*. The zone is characterized by an abundance of *Araucariacites australis* and *Ephedripites spp*. The top of the zone is characterized by planktonic Foraminifera of Turonian to Upper Santonian age. The age range is interpreted to be Cenomanian to Senonian (Muller, 1968).

Radiolaria Fifty-three taxa of radiolaria were identified from 10 chert samples collected from the Tubeh and Pang Bau areas (Basir, 2000). They are very different from those identified by G. F. Elliot (Wilford and Kho, 1965). The important taxa for age allocation are: *Loopus primitivus* (Matsuoka and Yao), *Angulobracchia (?) rugosa* Jud, *Cinguloturris cylindrica* Kemkin and Rudenko, *Artocapsa (?) amphorella* Jud, *Hsuum raricostatum* Jud, *Obesacapsula rusconensis* umbriensis Jud, *Syringocapsa longitubus* Jud and *Parapodocapsa furcata*.

The age of the cherts therefore straddles the Jurassic–Cretaceous boundary and ranges from Late Tithonian (uppermost Jurassic) to Berriasian (lowermost Cretaceous).

IV.5. CRETACEOUS ACCRETIONARY COMPLEXES

Three complexes outcrop along the northern coastal sector of western Sarawak. Only the first presents good coastal outcrops, inland outcrops are poor and the rocks commonly deeply weathered and poorly understood. The fourth is the Lupar Formation–Pakong Mafic Complex–Lubok Antu Mélange that forms the border between the Kuching and Sibu zones, constituting the Lupar Line, named and identified as a suture by Hutchison (1975).

IV.5.1. Serabang Formation

The best outcrops of this formation form cliffs and coastal platforms northwards from Kuala Samunsam along the northwestern peninsula of Sarawak (Wolfenden and Haile, 1963). The Serabang Formation is characterized by steep dips and a regional NW–SE strike. The thickness is about 32 km perpendicular to the strike. Two coastal sections have been selected from Wolfenden and Haile (1963) as the type localities (Figures 10 and 11). This formation also occupies low-lying areas in the Lundu area, where it is usually deeply weathered.

IV.5.1.1. Metasedimentary and mélange rocks

The formation is metamorphosed in the greenschist facies and commonly cut by thin quartz veins. The most important lithology is slate and cleaved mudstone, in places mylonitized and brecciated. Pelitic hornfels is the commonest rock type near to the granite intrusions. The hornfels contains biotite, cordierite and muscovite. Andalusite and garnet occur less frequently (Wolfenden and Haile, 1963). There are beds of greywacke ranging from a few metres to about 150 m in thickness. They are lithic and contain feldspar and are hornfelsed by the granites. Biotite and amphibole are common, cordierite and andalusite less common.

There are rare conglomerates containing clasts up to 10 cm diameter. The pebbles are of chert, siltstone, slate, vein quartz and altered igneous rocks. Most of the clasts are thermally metamorphosed.

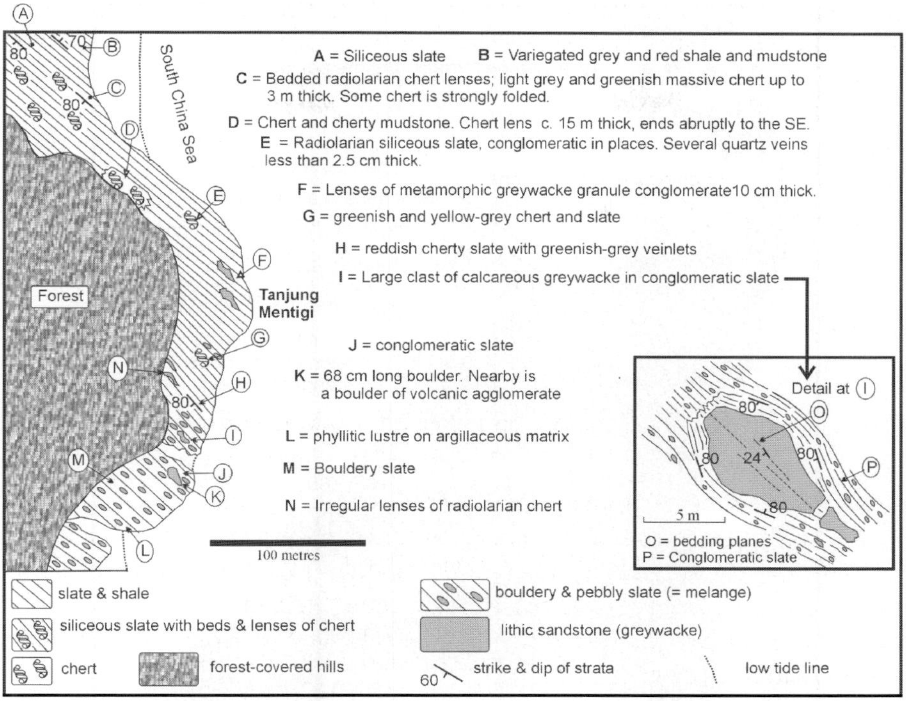

Figure 10. Type locality at Tanjung Mentigi of the Serabang Formation (after Wolfenden and Haile, 1963). Based on Hutchison et al. (in press). With permission from Minerals and Geoscience Department, Malaysia.

Lenses of chert are common. The chert and contained radiolaria are recrystallized. Thin beds and small lenses of calcareous rocks occur within the pelitic hornfels. The marble or calc-silicate rocks contain diopside, garnet, tremolite, clinozoisite, wollastonite and biotite.

Bouldery and pebbly slate is common, interpreted as metamorphosed mélange. At Tanjung Mentigi (Figure 10), vertically dipping bouldery slate has an exposed thickness of 98 m. It is composed of rounded to sub-angular boulders, pebbles of greywacke and chert in a matrix of pebbly sandy slate. The largest boulders are 8 m long, contained in a matrix of slate containing small clasts. The foliation of the matrix exhibits 'flow-banding' around the clasts (Wilford, 1955), a feature typical of mélanges. Conglomeratic slate contains clasts up to 4 mm across in a slate matrix. The common grains are of radiolarian chert and plagioclase crystals. Near the granites, the slates are hornfelsed.

IV.5.1.2. Palaeontology and age

The shale of the mélange matrix has yielded radiolaria, and 11 taxa have been identified (Basir and Aziman, 1996; Basir Jasin, 2000). The taxa identified by G. F. Elliott (Wolfenden and Haile, 1963) have not been reconfirmed, so that the extension downwards into the Jurassic is unproven. Radiolaria in the chert near Lundu

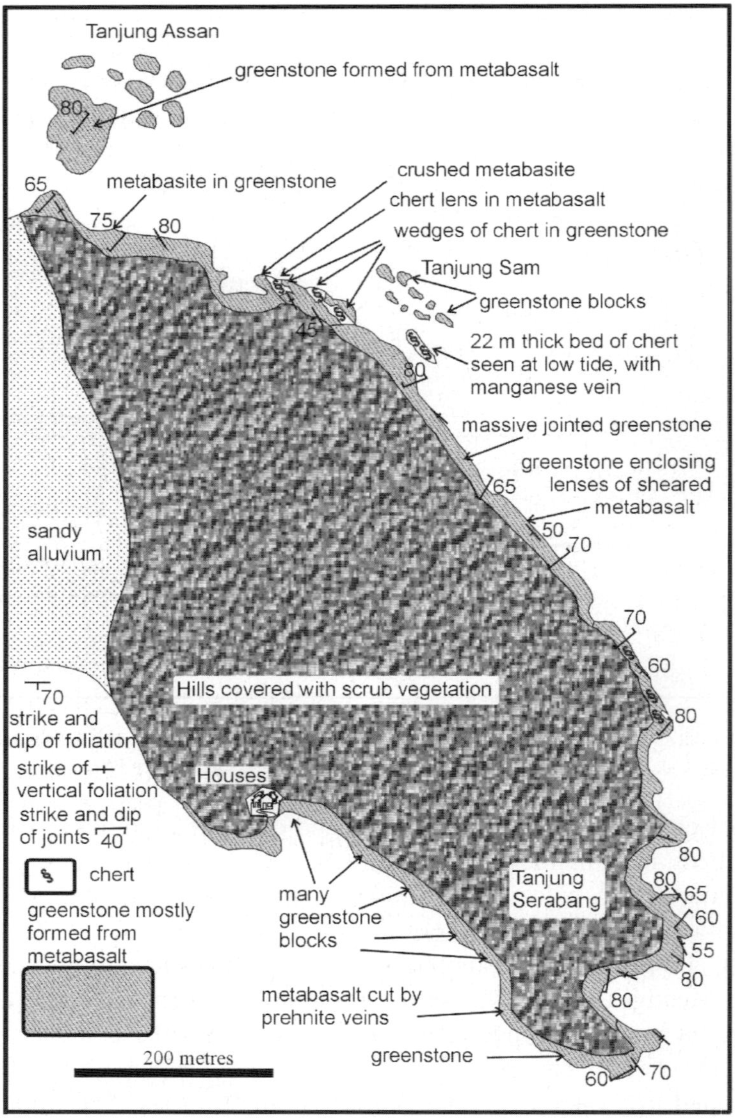

Figure 11. Detailed geology of coastal outcrops of the Serabang Formation around Tanjung Serabang (after Wolfenden and Haile, 1963). With permission from Minerals and Geoscience Department, Malaysia.

are completely recrystallized. Only one deeply weathered and oxidized slate yielded moderately well-preserved radiolaria (Basir and Aziman, 1996): *Hemicryptocapsa* sp.; *Hemicryptocapsa* cf. *prepolyhedra* Dumitrica; *Archaeodictyomitra lacrumula* (Foreman); *Archaeodictyomitra vulgaris* Pessagno; *Archaeodictyomitra* sp. A; *Archaeodictyomitra* sp. B; *Pseudodictyomitra puga* (Shaaf); *Pseudodictyomitra cf. puga* (Shaaf); *Thanarla conica* (Aliev); *Thanarla pulchra* (Squinabol); *Pawicingula* sp.; and *Xitus* sp.

The occurrence of *Archaeodictyomitra lacrimula* indicates a Lower Cretaceous (Valanginian to Aptian) age. Tumanda et al. (1993) reported that only Lower Cretaceous radiolaria occurred in the Serabang Formation, but details of the assemblage were not given. The argillite matrix therefore has an identical age to some of the chert clasts in the Lubok Antu Mélange. The Lubok Antu–Lupar Formation complex is therefore genetically related to the Serabang Formation.

IV.5.1.3. Ophiolitic suite

Metabasic rocks form an important part of the Serabang Formation. In addition to occurring as boulders and smaller clasts within the slate, they also form large bodies and whole headlands, Tanjung Serabang being the best example (Figure 11). All are presumed to have been an imbricated part of the oceanic basement to the chert–argillite of the Serabang Formation, an integral part of the Proto-South China Sea, which became disrupted at the active plate margin. These ophiolitic rocks are therefore the metamorphosed equivalent of the Pakong Mafic Complex of the Lupar Line.

Banded amphibolite is the most common of the ophiolite suite. It is a fine-grained metabasite formed of sodic plagioclase and actinolite with a pronounced cataclastic texture. In places they have been described as mylonite. The abundant shear planes trend NW–SE parallel to the metasediments of the Serabang Formation. The term greenstone has been used by Wolfenden and Haile (1963). Within the amphibolite, outcrops are small lenses of less-deformed meta-gabbro and metabasalt, named epidiorite by Wolfenden and Haile (1963). The original igneous fabric has been destroyed by ubiquitous shearing, but the igneous andesine and pyroxene persist as relicts.

Silicified serpentinite forms a small hill SW of Lundu in the Buloh tributary of Sungai Melaban. Hornblende pyroxenite and mica peridotite occur on Gunung Putting in the Samunsam valley.

Whole rock chemical analyses (from Wolfenden and Haile, 1963) are given in Table 3, and a Pecerrillo and Taylor (1976) type plot of the K_2O vs. SiO_2 given in Figure 12. The suite has distinct ophiolitic chemistry.

IV.5.2. Sejingkat Formation

This is a poorly defined formation, the more resistant lithologies of which make hills north of Kuching along the Sarawak River (Liechti et al., 1960). The main lithology is a black phyllite, commonly carbonaceous that is locally silicified. Other lithologies are chert and feldspathic sandstone. The chert contains radiolaria that are poorly preserved and cannot be properly assigned an age (Basir Jasin, 2000). Liechti et al. (1960) have stated that most of the rocks mapped as Sejingkat Formation resemble the Sadong Formation.

IV.5.3. Sebangan Formation

Chert, hornfels and amphibolite, thermally metamorphosed by granite near Sebuyau, have been assigned to the Sebangan Formation. No fossils have been found in these rocks.

Table 3. Ophiolitic igneous and meta-igneous rocks of the Serabang Formation

wt%	a	b	c	d	e	f	g	h	i
SiO_2	43.66	47.01	49.58	49.80	50.59	51.31	51.68	53.66	61.60
TiO_2	0.02	2.63	2.12	1.70	1.36	0.30	0.81	0.52	0.64
Al_2O_3	—	13.43	15.52	13.35	15.19	2.17	15.64	18.64	16.00
Fe_2O_3	4.14	3.44	1.70	2.87	1.02	3.15	1.23	1.57	1.02
FeO	4.90	11.29	11.31	10.54	9.25	3.36	7.64	6.36	6.15
MnO	0.26	0.28	0.22	0.18	0.16	0.16	0.17	0.33	0.15
MgO	41.43	7.92	4.84	6.53	6.76	18.28	7.50	3.70	2.40
CaO	0.00	8.62	8.20	11.35	10.03	18.04	10.00	6.29	4.30
Na_2O	0.49	2.74	4.11	2.16	3.78	0.48	3.13	3.56	4.85
K_2O	0.14	0.41	0.27	0.09	0.28	0.10	0.30	0.87	0.53
H_2O+	3.03	1.85	1.96	1.44	1.52	0.86	2.10	2.87	1.93
H_2O-	0.30	0.11	0.09	0.20	0.06	0.32	0.07	0.19	0.14
CO_2	1.11	n.d.	0.04	n.d.	n.d.	0.48	n.d.	1.36	n.d.
P_2O_5	0.38	0.28	0.17	0.18	0.10	0.02	0.06	0.11	0.12
S	—	0.53	0.38	—	0.44	0.42	0.10	0.05	n.d.
Total	100.36	100.34	100.32	100.39	100.32	99.61	100.38	100.06	99.83

a = peridotite, Gunung Putting (S9547), Cr_2O_3 = 0.31%, NiO = 0.19%.
b = amphibolite, Gunung Putting (S7917), Cr_2O_3 = trace.
c = altered gabbro, Tanjung Serabang (S7749).
d = altered basalt, Tanjung Pelandok (S7417).
e = greenstone, Tanjung Serabang (S7985).
f = pyroxenite, Gunung Putting (S7929), Cr_2O_3 = 0.37%.
g = meta-tuff, Tanjung Assan (S7757).
h = basalt, small islands SE of Pulau Badar (S7871).
i = sheared tonalite, Sungai China (S9615). n.d. = not determined. — = zero.

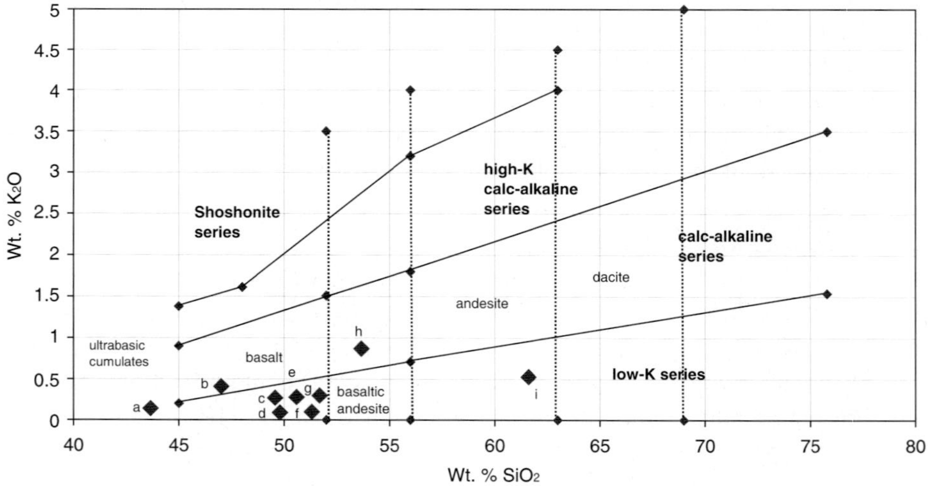

Figure 12. Serabang Formation metamorphosed ophiolite rocks K_2O vs. SiO_2 diagram.

IV.5.4. Lupar Line complex

This accretionary complex was originally known as the Danau Formation (Molengraaf, 1902). He used it for the complex of ophiolitic spilitic rocks and associated sedimentary rocks. The rocks of the complex (Figure 13) have been interpreted

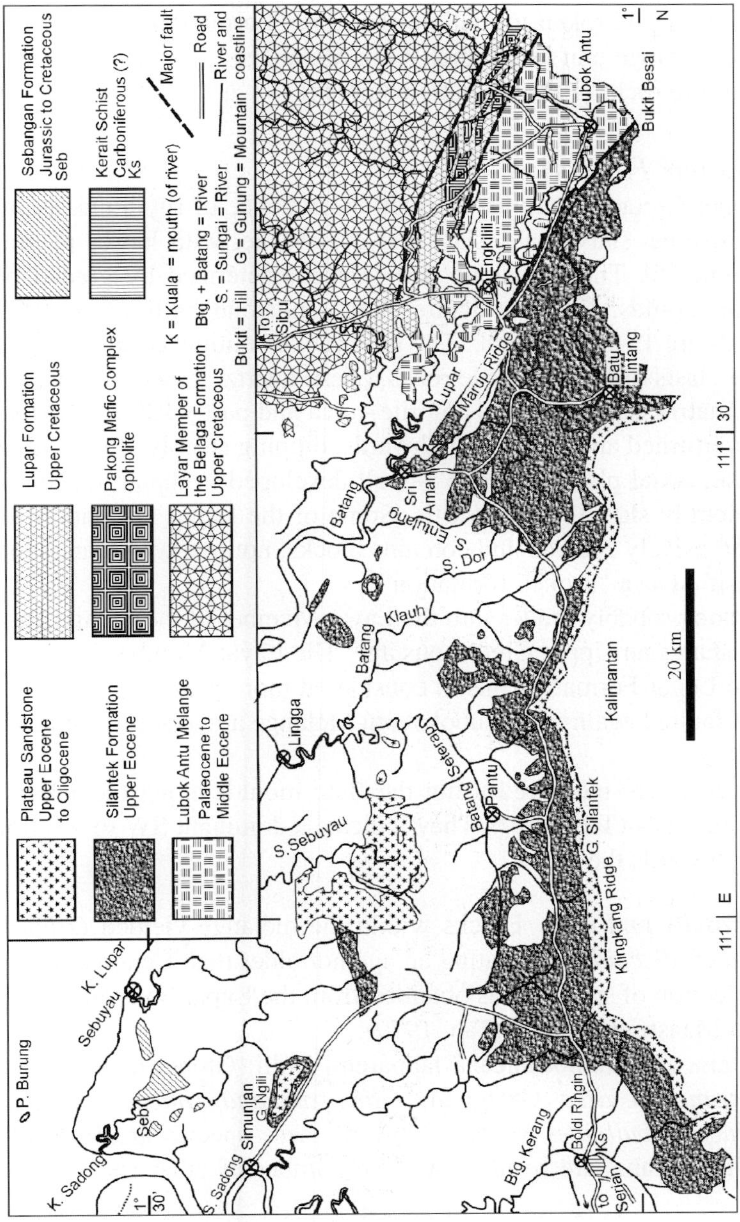

Figure 13. Distribution of Silantek Formation and Plateau Sandstone and geology around the Lupar Line.

by Honza et al. (2000) as an accretionary prism facing and younging northeastwards, and the structures attributed to imbricate thrusting. The individual components were named individually by Tan (1979) as:

A turbidite flysch broken formation [the Lupar Formation]
A chaotic mélange unit [the Lubok Antu Mélange]
The ophiolitic rocks [Pakong mafic complex]

IV.5.4.1. Lupar Formation

This is an Upper Cretaceous Lupar Formation sequence of rhythmically interbedded shale, mudstone, slaty shale, slate and greywacke, with lenses of granule to pebble conglomerate. The sandstone beds are characterized by graded bedding, load and flute moulds, flame structures and sandstone balls (Tan, 1979). The greywacke beds are 15–30 cm thick. The greywacke is lithic, composed of angular to subrounded clasts of metamorphic rocks, chert, quartz, lesser volcanics and plagioclase. The matrix is fine-grained chlorite–mica and partly siliceous. Most of the sequence is overturned and youngs southwards, dipping steeply NNE with a dominant SSE strike. Axial plane cleavage is well developed in the argillaceous rocks, which are generally slaty to subphyllitic. Some of the Lupar Formation is boudinaged. It is pervasively sheared and contains blocks, none of which are exotic. It is therefore classified as a “broken formation”.

The formation probably grades into the Layar Member of the Belaga Formation, with which it shares an Upper Cretaceous age. The Layar Member is more argillaceous than the Lupar Formation that is considered more proximal to the source. It is sheared and faulted against the Lubok Antu Mélange and occurs as blocks within it near Engkilili.

Palaeocurrents were measured using the flute moulds along the Jakar-Saratok and Lubok Antu roads (Tan, 1979). They indicate a dominant SW to NE flow from a source lying towards the SW.

IV.5.4.1.1. Fossils and age Blocks within the mélange yielded *Orbitolina* sp. and *Orbitolina* cf. *discoidea* indicating an age not older than Cenomanian.

A large collection of species was obtained from the Lupar Formation, indicating a Santonian to Maastrichtian age (Tan, 1979):

Globotruncana sp.; Globotruncana lapparenti Bolli; Globotruncana falsostuarti Sigal; *Globotruncana arca* Cushman; *Globotruncana stuarti* (de Lapparent) and *Globotruncana bulloides* Volger. Some of these species are from blocks in mélange. All the forams are benthic. A large number of other species have been identified.

Radiolarian chert pebbles from pebbly sandstone of the Lupar Formation yielded nine species (Basir, 2000): *Acaeniotyle umbilicata*; *Thanarla conica*; *Archaeodictyomitra vulgaris*; *Archaeodictyomitra lacrimula*; *Archaeodictyomitra* sp.; *Eucyrtis micropora*; *Eucyrtis* sp.; *Sethocapsa* sp. and Xitus *spicularius*.

This assemblage indicates an age of Hauterivian to Barremian, belonging to assemblage II of that found in chert clasts within the Lubok Antu Mélange (see below). The cherts are therefore of the same origin.

IV.5.4.2. Lubok Antu Mélange

The mélange belt is on an average 10.5 km wide (Figure 13). The rock fragments and blocks range from a few centimetre to a few kilometer in size. They are of a variety of lithologies: mudstone, sandstone, shale, hornfels, chert, conglomerate; basalt, gabbro (and their metamorphic equivalents); limestone and serpentinite. The clasts are randomly contained in a highly cleaved chloritized pervasively sheared pelitic matrix. The clasts are mostly angular, some are subrounded.

IV.5.4.2.1. Age of formation The mélange argillite matrix has yielded Lower Eocene fossils, but also with reworked Upper Cretaceous coccoliths (Tan, 1979). The following forams indicate a Lower Eocene age: *Ammodiscus* sp. *Bolivina* sp.; *Praeglobolulimina pupoides* (d'Orbigny);

Globigerina gravelli Bronnimann and *Globigerina linaperta* Finlay. The Lower Eocene age is also confirmed by a complete nannofossil assemblage of the *Discoaster lodoensis* zone. The fossils yielded by the Lubok Antu Mélange indicate a marine inner neritic environment of deposition.

IV.5.4.2.2. Age of the enclosed chert blocks E. A. Pessagno (in Tan, 1978) extracted good radiolaria assemblages from 5 chert blocks collected along the Lubok Antu road, Batang Lupar and Batang Ai. The commonly identified species are: *Thanarla conica* (Aliev), *Parvacingula* sp. and *Archaeodictyomitra* sp. He ascribed a Lower Cretaceous age (Valangian to Aptian) to the radiolarian assemblages, distinctly older than the age ascribed by Basir (in Basir and Haile, 1993). However, it is a common feature of suture zones to find cherts of a range of ages, representing the deep marine deposits of the ocean before it was extinguished.

Chert from a road cut 5 km N of Lubok Antu had the following common radiolaria extracted by Basir (Basir and Haile, 1993): *Holocryptocanium tuberculatum*; *Holocryptocanium barbui*; Crypthamporella conara; *Thanarla praeveneta*; *Thanarla elegantissima* and *Xitus spicularius*. These forms suggest the chert has an age range of Albian to Cenomanium.

The most recent work by Basir (1996, 2000) indicates that blocks of chert in the mélange belong to three distinct radiolarian assemblages:

Assemblage I is of 17 taxa, including *Homoeoparonaella gigantea*, *Ristola altissima* and *Parvicingula excelsa*, indicating a Kimmeridgian to Tithonian latest Jurassic age.

Assemblage II consists of 21 species, including: *Cerops septemporata,* and *Archaeodictyomitra lacrimula*, indicating a middle Valanginian to Barremian early Cretaceous age.

Assemblage III contains 18 species, including: Obecapsula somphedia; Holocryptocanium barbui; Squinabollum fissilis; Pseudodictyomitra pseudomacrocephala; Novixitus weyli; Novixitus mclaughlini; Rhopalosyringium majuroensis; Stichomitra communis; Holocryptocanium tuberculatum and Thanarla praeveneta, indicating a late Albian–Cenomanian age.

The conclusion from the foregoing papers is that cherts of three distinct ages: Upper Jurassic, Lower and Upper Cretaceous, occur as blocks in the Lubok Antu Mélange. This is a feature in common with other suture zones, and indicates that the proto-South China Sea was extant during that time, and received chert deposition.

IV.5.4.3. Engkilili Formation

Although Tan (1979) abandoned the use of the Engkilili Formation and incorporated it within the Lubok Antu Mélange, Haile (1996) re-emphasized the need to maintain the term Engkilili Formation, as defined by Liechti et al. (1960). The formation forms a belt of restricted occurrence, only 15×3 km extending upstream from Engkilili, and lying along the southern margin of the Lubok Antu Mélange belt. The reasons for maintaining the formation as separate from the Lubok Antu Mélange are:

Limestone blocks, up to 3 m diameter, appear to be confined to the Engkilili Formation. The calcareous shale of the Engkilili Formation is unlike the pervasively sheared matrix of the Lubok Antu Mélange. Instead, it is unaltered, unsheared grey shale containing some concretions and at one place sandy burrows, flaser bedding and fine ripple marks. There are no blocks of radiolarian chert in the Engkilili Formation, which characterize the Lubok Antu Mélange.

Both the limestone blocks and the mudstone matrix have yielded Foraminifera of the same age: Mid-Palaeocene to Middle Eocene. Unlike the Lubok Antu Mélange, the Engkilili Formation does not contain exotic clasts, and appears to be a single stratal sequence, which has been broken or disrupted.

IV.5.4.3.1. Palaeontology and age The limestone blocks have yielded a good Foraminifera fauna indicating a Late Palaeocene to Middle Eocene age (Tan, 1979). The matrix has yielded Lower Eocene aged nannofossils belonging to the Lower Eocene *Discoaster lodoensis* zone, but with reworked Upper Cretaceous coccoliths (Tan, 1979): *Ericsonia ovalis*; *Ericsonia Formosa;* Ericsonia cava; *Discoaster lodoensis*; and *Discoaster kuepperi.*

The following forams also confirm a Lower Eocene age (Tan, 1979): *Ammodiscus glabratus* Cushman & Jarvis; *Bolivina* sp.; *Praeglobobulimina pupoides* (d'Orbigny); *Psammosiphonella carapitana* (Hedberg); *Osangularia culter* (Parker & Jones); *Eggerella bradyi* (Cushman); *Globigerina gravelli* Bronnimann and *Globigerina linaperta* Finlay.

The fossils indicate that in the Lower Eocene, shallow marine conditions prevailed. The following early Middle Palaeocene planktonic Foraminifera have been identified from the mudstone matrix of mélanged Engkilili Formation (Basir

Jasin and Taj Madira, 1995): *Subbotina triloculinoides* (PLUMMER); *Subbotina velascoensis* (CUSHMAN); *Globorotalia quadrilocula* BLOW; *Globorotalia pseudobulloides* (PLUMMER); *Morozorella uncinata* (BOLLI); *Morozorella trinidadensis* (BOLLI) and *Morozorella praecursoria* (MOROZOVA).

The following Middle Eocene species were illustrated: *Morozorella aragonensis* (NUTTALL), *Morozorella naussi* (MARTIN), *Acarinina bulbrooki* (BOLLI), *Subbotina frontosa boweri* (BOLLI), and *Globanomalina indiscriminata* (MALLORY).

The authors suggested that the mudstone matrix contains a block of basal Silantek Formation, but this is an unlikely interpretation because of the younger (Upper Eocene) age of the Silantek Formation.

The Engkilili Formation contains the same palynomorphs as the upper zones E and F of the Kayan Sandstone, and Muller (1968) assumed a Palaeocene to Middle Eocene age for them.

IV.5.4.4. Ophiolite

The ophiolite suite occurs within the Lupar Formation and is known as the Pakong mafic complex (Tan, 1979). Blocks ranging from a few centimetre to 2 km of spilite, basalt, gabbro, and their metamorphic equivalents, occur also within the Lubok Antu Mélange. The Pakong mafic complex is an incomplete dismembered ophiolite, named from rapids at Wong Pakong, where the spilite and basalt outcrops are pillowed. The rounded pillows are 15 cm to 2 m diameter. Coarse gabbro is exposed at Wong Irup on the Batang Ai (Figure 14)

The analyses (Table 4) are plotted on a Peccerillo and Taylor (1976) diagram (Figure 15). The rock types are typical of the ophiolite suite, but the two samples of altered basalt (e and f) have been strongly modified from their original composition: e has been enriched in potassium and f in silica.

IV.6. UPPER CRETACEOUS — TERTIARY FORMATIONS

IV.6.1. Kayan Sandstone Formation

Formerly the term Plateau Sandstone Formation was applied to all feature-forming dominantly arenaceous strata, but detailed palynological work by Muller (1968) on the Penrissen and Lundu areas necessitated a fundamental revision. The Plateau Sandstone Formation, which forms a spectacular scarp along the Klingkang Range along the border between Sarawak and Kalimantan, is known to conformably overlie the Eocene Silantek Formation (Tan, 1979). The predominantly sandy formation, which unconformably overlies the Pedawan Formation in the Penrissen and Lundu areas (Figure 16), probably ranges from Upper Cretaceous (Senonian) to Eocene or even younger. In part, therefore, it correlates with the Silantek Formation, but is older than the Plateau Sandstone Formation of the Klingkang Range. This necessitated a renaming of the western outcrops and the name Kayan Sandstone Formation was adopted (Figure 4).

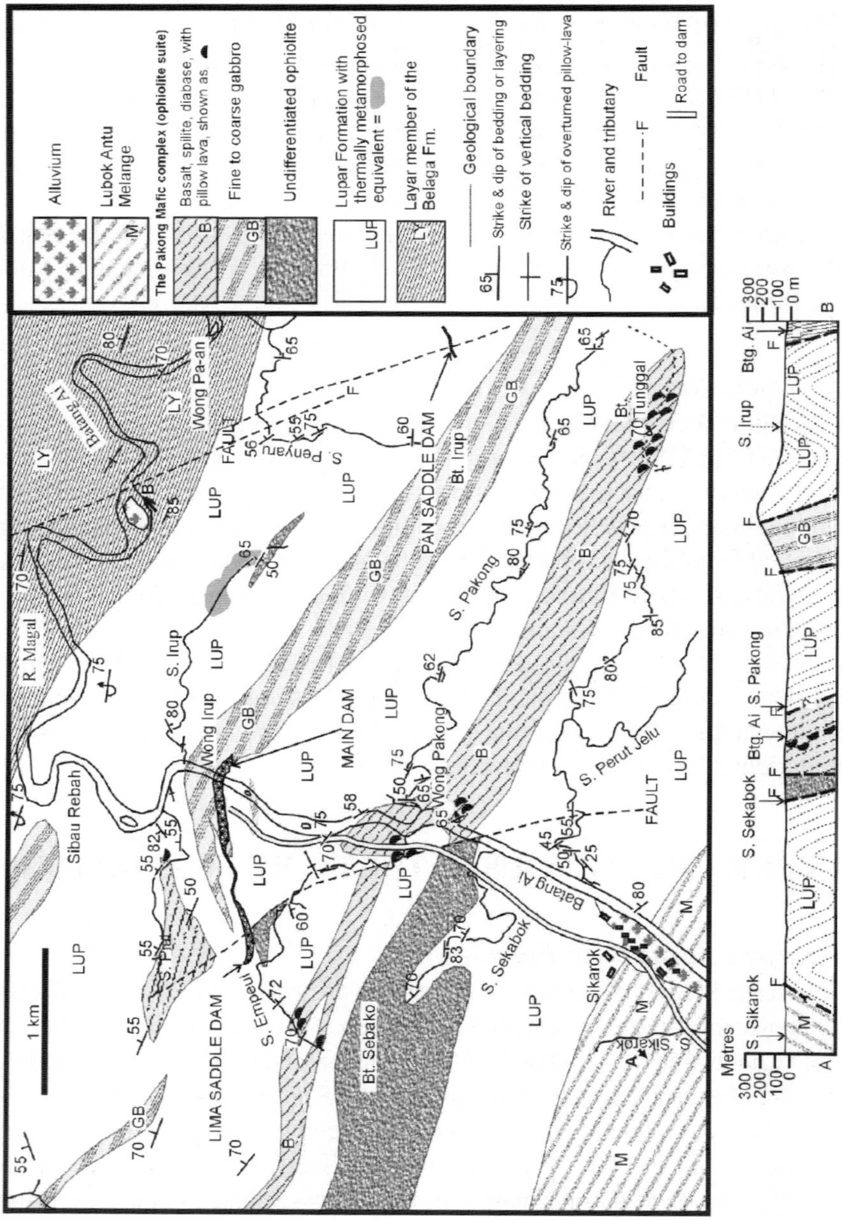

Figure 14. Outcrop sketch map of the Batang Ai dam site (Redrawn after Tan and Hon, 1984). With permission from Minerals and Geoscience Department, Malaysia.

Table 4. Chemical compositions of the ophiolite suite from the Pakong Mafic Complex

wt%	a	b	c	d	e	f	g	h	i	j	k	l
SiO_2	47.04	49.58	51.44	53.90	54.71	60.98	46.61	46.93	46.15	46.87	51.08	51.28
TiO_2	0.55	1.00	1.34	0.66	2.20	0.54	0.89	1.52	0.62	1.10	1.02	1.81
Al_2O_3	16.49	15.58	14.34	15.55	11.90	13.64	18.35	15.70	17.61	14.64	14.42	12.62
Fe_2O_3	2.63	2.12	2.63	2.95	5.94	2.68	0.96	2.08	2.45	1.68	3.02	3.31
FeO	6.53	7.14	7.68	5.16	6.81	3.90	6.67	9.21	6.59	10.05	7.19	10.96
MnO	0.14	0.09	0.14	0.18	n.d.	0.14	0.13	0.15	n.d.	0.21	0.16	0.22
MgO	10.04	7.62	6.90	6.16	4.20	4.48	8.01	7.51	8.84	7.73	8.16	4.39
CaO	10.70	11.24	9.44	8.88	5.41	5.86	12.99	11.57	11.87	11.83	10.78	7.80
P_2O_5	0.06	0.14	0.16	0.22	0.52	0.08	0.21	0.23	0.23	0.19	0.12	0.26
Na_2O	1.87	2.26	3.05	1.99	4.25	2.35	2.31	2.00	2.04	2.67	1.96	3.81
K_2O	0.24	0.15	0.34	0.66	1.78	0.85	0.49	0.47	0.31	0.69	0.13	0.11
H_2O+	3.10	2.14	2.06	2.71	2.20	3.76	2.37	2.68	3.19	2.53	1.20	3.26
H_2O-	0.77	0.34	0.39	0.78	0.40	0.07	0.21	0.20	0.58	0.22	0.30	0.24
CO_2	0.02	0.07	0.10	0.10	n.d.	0.07	n.d.	n.d.	n.d.	n.d.	0.07	0.07
S	0.26	0.17	0.14	0.08	n.d.	0.06	n.d.	n.d.	n.d.	n.d.	0.04	0.29
Total	100.44	99.64	100.15	99.98	100.32	100.09	100.20	100.25	100.48	100.41	99.65	100.43

a = spilite (K6703), Bukit Batu.
b = spilite (K6072), ms.10 Lubok Antu road.
c = spilite (K6635), Wong Pakong, Batang Ai.
d = olivine basalt (S2763), Bukit Besai.
e = albitized basalt (S2846), Bukit Tras.
f = altered basalt (K6631), Bukit Suai.
g = ophitic diabase (S2931), Wong Pakong, Batang Ai.
h = diabase (S2929), Wong Pakong, Batang Ai.
i = olivine gabbro (S2849), Batang Skrang.
j = gabbro (S2928), Wong Irup, Batang Ai.
k = gabbro (K6709), Bukit Batu.
l = gabbro (K6761), Bukit Batu.
All K specimens are from Tan (1979) and S specimens from Haile (1957). n.d. = not determined.

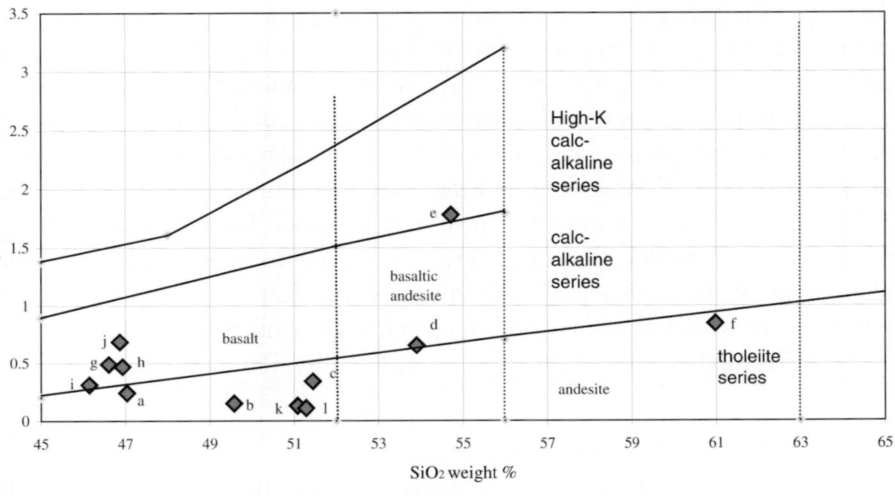

Figure 15. Pakong Mafic Complex ophiolite K_2O vs. SiO_2 diagram.

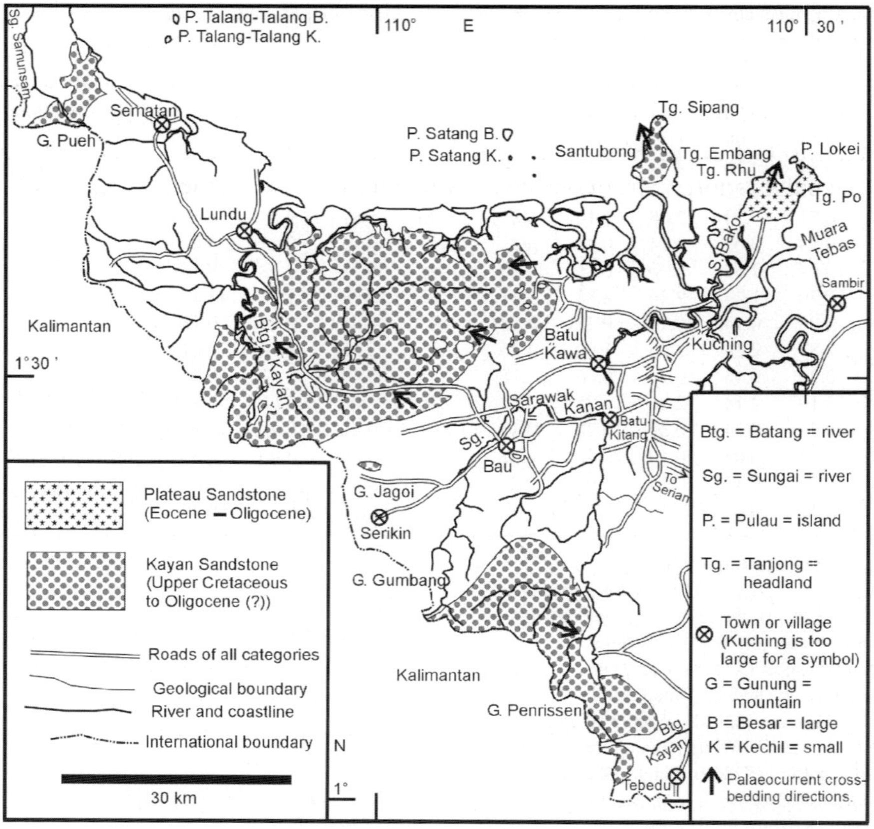

Figure 16. Distribution of the Kayan Sandstone and Plateau Sandstone in western Sarawak.

IV.6.1.1. Lithology

The Kayan Sandstone is predominantly of massive and cross-bedded sandstone and conglomerate with thin beds of black, red, and grey mudstone (Wilford, 1955). The sandstones are buff to white and often contain fossil wood. Layers of well-rounded pebbles are common in the sandstones. They are of quartzite, vein quartz, black and red chert, and agate. Coarse conglomerate contains similar pebbles and in addition some of acid and basic volcanics and metamorphic rocks. The mudstone beds are silty and are less than 3 m thick. Carbonaceous shales and thin coal seams occur at Gunung Undan. None of the coals exceeds 22 cm thickness. Rare lenticular limestone beds 0.3 m thick occur on Pulau Sampadi in a black shale sequence (Wilford, 1955).

A fluvial to deltaic environment was suggested by Tan (1984).

IV.6.1.2. Palaeontology and age

Muller (1968) erected three palynological zones for the Kayan Sandstone Formation. From the base upwards they are:

D. *Rugubivesiculites* Zone. There is a striking floral change across the boundary with the underlying Pedawan Formation, suggesting a hiatus. The base is defined by the increase of *Alisporites similis*, *Rugubivesiculites reductus*, *Triorites minutipori* and *Verrutriporites lunduensis* (Muller, 1968). *Apiculatisporis ferox* is restricted to this zone. Although there are problems in correlating the age significance of palynomorphs from one country to another without the calibration of other kinds of fossils, Muller (1968) made the reasonable conclusion of a Senonian–Palaeocene age for this zone. In the Penrissen area, there is a rather striking floral change from Zone C to D, suggesting a hiatus.

During deposition of the *Rugubivesiculites* Zone, the proximity of a mountain range is indicated by a high proportion of montane pollen types that are virtually absent from the Pedawan Formation and from the younger parts of the Kayan Sandstone Formation and the Engkilili Formation (Muller, 1968). The mountain chain may have come into existence during the stratigraphic hiatus at the base of the Kayan Sandstone Formation, as indicated by increased sand influx into the sedimentary environment. A collisional orogenic event may be inferred to have formed the uplift.

E. *Proxapertites* Zone. The base of this zone is marked by the sudden increase in *Proxapertites operculatus* and *Proxapertites cursus*, which Morley (1991) pointed out became extinct in the Late Eocene. Certainly it is absent from the overlying zone and from all younger formations of Borneo. Muller (1968) also identified palynomorphs from the Engkilili Formation from samples that have yielded Palaeocene planktonic Foraminifera. Therefore, he concluded that this zone ranges from Upper Senonian to Upper Palaeocene.

F. *Retitriporites variabilis* Zone (provisional). This zone is marked by the sudden increase in *Retitriporites variabilis* and extinction of *Proxaperites operculatus* and *Proxaperites cursus*. The suggested Eocene age is therefore most appropriate (Muller, 1968). The regular presence of *Discoidites borneensis* indicates a lacustrine facies.

It is important to note that *Florchuetzia trilobata* is very important in the Southeast Asian region from Mid-Eocene through to Mid-Miocene (Morley, 1991), but it is completely absent from the Kayan Sandstone Formation zones of Muller (1968), indicating that his Zone F cannot be younger than Eocene.

Wilford (1955) listed the identified tree species found as fossil wood in the sandstones: *Palmoxylon* cf. *lacunosum* (Unger), *Palmoxylon* sp., *?Ficoxylon* sp. and *Phyllocladoxylon* sp.

IV.6.1.3. Palaeocurrents and provenance

A detailed programme of palaeocurrent measurement by Tan (1984) gave the following summary (Figure 16): Santubong Peninsula—small-and medium-scale cross-bedding indicates a unimodal current direction towards 338°. The prevalent currents were therefore from the SSE. Other measured areas indicate: Snibong, towards 303°; Stinggang, towards 303°; Gunong Moi, towards 290°; Gunong Serapi, towards 265° and Sungai

Tobia, towards 110°. There are local currents directed towards the NW, N, NNE and ESE. The overall analysis is that the Kayan Sandstone was provenanced from erosion of the Bau–Kuching–Serian area. The problem is that this area could not have provided the predominance of quartz necessary. However, highlands of unknown composition have been eroded to give the provenance for the Kayan and Plateau Sandstone.

IV.6.2. Silantek Formation

This thick formation of Upper Eocene to Oligocene age outcrops between Serian and the Lupar Line (Figure 13). It has to lie unconformably upon the Sadong Formation and Serian Volcanics, although no contact has ever been seen. It grades conformably into the overlying Plateau Sandstone. The environment of deposition is near shore and estuarine at the base, becoming fluviatile, lacustrine and terrestrial upwards. It is the equivalent of the Melawi and Kantu Beds of Kalimantan. The Silantek Formation represents a continuation of the Ketungau Basin of Kalimantan into Sarawak.

IV.6.2.1. Divisions

There is a three-fold division of the Silantek Formation (Tan, 1979):

Upper Redbed Member, which merges conformably into the overlying Plateau Sandstone. It is predominantly of red micaceous mudstone and shale (occasionally mottled grey), red siltstone and sandstone. The mudstone beds contain layers of hard ferruginous nodules.

Temudok Member is of bedded 3–150-cm-thick lenticular sandstone, yellowish-white to grey, variable from fine- to coarse-grained. The lenticular sandstones occur at various stratigraphic levels. The sandstone is interbedded with thin grey siltstone, silty shale and mudstone. This member has no conglomerates and is more thinly bedded than the Basal member.

Basal Sandstone Member forms the Marup Ridge, which is bounded along its NE side by the Lupar Line. The strata of the Marup Ridge are steeply dipping to vertical (Haile, 1957). The Member is of yellow or white polymict to quartzose sandstone beds up to 15 m thick, interbedded with thin silty shale and mudstone, and containing lenses of polymict conglomerate and pebbly sandstone.

Miocene stocks and dykes of acid to intermediate composition intrude the Silantek Formation and are surrounded by narrow thermal aureoles. These high-level hypabyssal intrusions are characteristic of the whole Ketungau Basin, where they are known as the Sintang Intrusives.

IV.6.2.2. Rock types

The sandstones are predominantly lithic arenites with 25–50% rock fragments and up to 5% feldspar grains. Quartz is predominant. The Basal Sandstone Member conglomerate lenses, up to 1 m thick, contain rounded clasts up to 1.5 cm across. The clasts are of quartz, chert, hornfels, metagreywacke, argillite and mafic volcanics

(Tan, 1979; Haile, 1957). In the Strap and Sadong valleys, Haile (1954) described basal conglomerates overlying the Serian Volcanics (no actual contact exposed), containing clasts up to 15 cm of sandstone, quartzite, vein quartz, volcanic rocks and hornfels. The coarse conglomerates pass up into finer conglomerates and sandstones containing pebbles of vein quartz, chert and phyllite.

Rare coal seams, <1 m thick, are exposed along the Batu Lintang road and in Sungai Besai. Only the last has economic significance and, in 1974, the Utah Pacific Company prospected the area south of Bukit Selanjan, where there are two seams about 1 m thick (Tan, 1979). The coal at Silantek and Abok in the Klingkang Range scarp and at Batu Besai has been intruded by Miocene stocks (Haile, 1954). The coals are low volatile bituminous, black and shiny. However, they have been cindered by the Miocene intrusions. Cindered coal, found adjacent to the intrusions, does not burn (Haile, 1954). This is strange, for in Germany intrusions generally increase the coal grade from bituminous to anthracite.

IV.6.2.3. Palaeocurrents and provenance

Silantek Formation sandstones are characterized by ripple marks and cross-bedding and Tan (1979) made the following deductions from their measurement. The predominant current direction from cross-bedding was from NE and NNE towards the SW and SSW. Asymmetric ripple marks confirm this, with a predominant direction towards 228°. The provenance of the Silantek Formation, taking into account the sandstone and conglomerate petrology, can therefore be deduced to have been the Upper Cretaceous Belaga Formation, Lupar Formation and Pakong Mafic Complex ophiolites (Rajang Group) that lies to the NE and NNE across the Lupar Fault. The Rajang Group was uplifted to form an eroding landmass during the Late Eocene Sarawak Orogeny (Hutchison, 1996a).

IV.6.2.4. Palaeontology and age

Haile (1954) described *Cyrena (Batissa) subtrigonalis* Krause and *Thiara (Melania)* sp. nov. from the Middle Silantek Beds. Kanno (1978) described fossils from a roadcut of the Lower Sandstone Member of the Silantek Formation, 10 km south of Sri Aman, 100 m north of the bridge over the Sungai Entulang: *Geloina hashimotoi* Kanno n. sp., *Corbula (Tenuicorbula) dajacensis* Krause, and "*Paludomus*" *gracilis* (Krause). They are all brackish water forms such as live in a tropical mangrove swamp, and indicate an Upper Eocene age, equivalent to the Melawi fauna of Kalimantan.

Haile (1957) described the following Foraminifera from calcareous mudstone of the basal sandstone near Lubok Antu, indicating an Upper Eocene age.

Actinocyclina sp., *Nummulites spp., Heterostygina* sp. and *Rotalia* sp. The Basal Sandstone Member is Upper Eocene. The rest of the Silantek is unfossiliferous or has yielded undiagnostic fossils. The great thickness may suggest that the formation extends into the Oligocene.

The coal beds of the Klingkang Range have been palynologically studied at the laboratory of Sarawak Shell (Haile, 1954) and have yielded a good assemblage of pollen and spores. Regretfully, Haile (1954) gives no details but quotes only letter classifications for the floral groups. Palynology is important to determine the age of the upper continental part of the Silantek Formation.

IV.6.3. Plateau Sandstone

The type locality of the Plateau Sandstone is the Klingkang Range that marks the border between Sarawak and Kalimantan (Figure 13). The plateau scarp is an impressive geomorphological feature visible from far and rising to 854 m at Bukit Mansul. The geology has been described by Haile (1954) and Tan (1979). There are several outliers to the north overlying the Silantek Formation, such as Gunung Ngili (Figure 13), where 153 m of flat-lying Plateau Sandstone conformably overlies coal-bearing Silantek Formation.

Zeijlmans van Emmichoven (1939) has shown that the Plateau Sandstone dips southwards to form the lower unit of the Ketungau Basin, where its thickness is estimated to be 1800 m, and subsequently named the Tutoop Sandstone. The fossils indicate a shallow marine to brackish environment with freshwater influence, but they are not age indicative. The Plateau Sandstone has also been mapped within the Melawi Basin and named the Dangkan Sandstone.

IV.6.3.1. Lithology

The main lithology is thick-bedded massive and cross-bedded sandstone with minor intercalations of grey to red mudstone (Tan, 1979). Conglomerate is absent on the eastern Klingkang Range (Tan, 1979), but increases in importance westwards (Haile, 1954). Grey to red mudstone forms minor intercalations. Palaeocurrent analysis was impractible because the sandstones form steep cliffs and waterfalls.

The sandstones are light grey becoming brownish-yellow and friable on weathering. They are fine- to medium-grained, moderately well sorted and contain mud clasts of red mudstone. Clast compositions are quartz (24–63%), rock fragments (21–27%), and feldspar (4–13%). The siliceous cement forms 1–10%. Sericite forms 2–8% and opaque minerals 1–4% by volume (Tan, 1979). The rock fragments are shale, phyllite, hornfels, and occasionally porphyritic volcanic rocks and chert.

The conglomerates are composed of well-rounded clasts of white vein quartz, some chert and hornfels (Haile, 1954). The red mudstones indicate a terrestrial environment and the formation may range through fluvial to estuarine.

IV.6.3.2. Palaeontology and age

Haile (1954) recorded the following fossils from about 46 m above the Main coal seam at Gunung Ngili of uncertain age significance: *Cyrena subrotundata* Krause, *Clementia papyracea* Gray, and *Veneridae* sp.

The evidence from the Kalimantan Ketungau basin is that the Plateau Sandstone is Upper Eocene, perhaps extending into the Lower Oligocene (Doutch, 1992). The suggestion of Tan (1979) of a Miocene age is not likely.

IV.6.3.3. Bako National Park

Horizontally bedded sandstones and conglomerates form the coastal headland north of Kuching with a maximum thickness >290 m. The formation is ascribed to the Plateau Sandstone, but the Santubong headland to the west has been ascribed to the Kayan Sandstone (Figure 16). The two formations are identical and the formation naming is a problem of nomenclature. The Bako headland has been studied by Johansson (1999). The Bako Formation sits unconformably upon the Sejingkat Formation and the basal beds dip N to NE at 5–15°. Palaeocurrent directions measured from cross-bedded sandstones are predominantly directed NNE, with a mean value of 20° (Tan, 1984).

IV.6.3.3.1. Age *Angiosperm*-derived *Tricolporate/tricolpate* pollen was found and suggest an age not older than Upper Cretaceous. Contained kerogen shows an apparently high maturity so that a Miocene age is not possible and an Eocene age is most likely (Johansson, 1999). However, the Plateau Sandstone is intruded in places by the Lower Miocene Sintang intrusives, indicating that a Miocene age for the Plateau Sandstone is unlikely.

IV.6.3.3.2. Lithology and depositional environment The main lithology is thick-bedded to cross-bedded polymict and quartzose sandstones with lenses of conglomerate and intercalations of grey to red mudstones. The conglomerates make lenticular channels commonly 2 m high. They show basal scouring into the underlying rocks. The following facies have been described by Johansson (1999): Pebble conglomerates are structureless and poorly sorted. The clasts are sub- to well- rounded and they are matrix-supported. The clasts are of quartz, quartzite, chert, together with lesser argillite, hornfels and schist. Pebbly sandstone is coarse- to medium-grained containing isolated scattered pebbles. Some are cross-bedded. The sandstones are thick-bedded and range from 15 to 260 cm in thickness. There are also laminated and cross-bedded sandstones showing trough and tabular morphology. There are rare interbedded fine sandstones and siltstones often a few centimetre thick. Mudstones are rare. Shale clasts are uncommon and occur at the channel bases.

The sandstones were rapidly deposited and water escape features are common. The depositional environment is interpreted as an alluvial sandy braided channel system with a high bed load (Johansson, 1999).

IV.7. CRETACEOUS PLUTONISM

Western Sarawak Cretaceous plutonic rocks form an expanded calc-alkaline series, which includes gabbro, but the suite is predominantly acidic (Table 6, Figure 17). There are extensive zones where the granites have invaded and hybridized the gabbros. K:Ar Cretaceous ages have been obtained from granitoids of Tanjung Datu (the most northerly headland of western Sarawak), the Lundu district, and Tinteng Bedil (SW of Lingga on the Sungai Strap) (Table 5). The tectonic significance of this Mid- to Late-Cretaceous calc-alkaline series is unknown, but granitoids of this age are widespread around the South China Sea region.

In comparison with the Miocene intrusives, the Cretaceous granitoid bodies are large. The largest is the Pueh adamellite, of outcropping area ~160 km^2. It straddles the western border with Kalimantan. Second in size is the Tinteng Bedil – Bukit Tabong, east of Serian.

In the Sematan area, reaction between adamellite and gabbroic enclaves has yielded a wide range of hybrid rocks. It is interpreted that the gabbros are cumulate rocks, resulting from differentiation, which have been incorporated in the granitic magma and carried upwards as enclaves, often with diffuse contacts. Gabbro, dolerite and hybrid rocks occur in the Sematan area, at Gunung Gebong, along the coast north of Lundu at Tanjung Pelanduk, and on the Talang Talang islands (Kirk, 1968). The Gebong gabbro lies on the NW flank of the large Gading granodiorite. It is cut by adamellite on its NE side and locally hybridized by granite veins from the Gading intrusion. The gabbro is of labradorite, augite, hypersthene and variable amounts of olivine. On Talang Talang islands, dolerite is cut and hybridized by granite veins. The hybrid rocks typically show partial alteration of pyroxene to hornblende.

In the Sematan area, hybrid rocks have been produced by extensive reaction between gabbro and the Cretaceous intrusions. Most of the hybridization occurs on the N and NE sides of the Gading granodiorite. Irregular granite veining and heterogeneity indicate that the basic and acid rocks are genetically related and are products of differentiation.

Seven separated outcrops of granodiorite form low hills S and W of Sebuyau, at Triso Darat headland, and Pulau Triso in the Lupar estuary. They are petrographically similar and deduced to be outcrops of the same pluton in this rather swampy area.

The Tinteng Bedil adamellite body is about 19 km long in a NE elongation in the Strap Valley. There are two main parts, separated by alluvium: Tinteng Bedil and Bukit Tabong. The Pueh intrusion forms a mountain range of biotite adamellite along the western border of Sarawak. Tourmaline pegmatite veins are common. There is a porphyritic facies along the SE margin. The Gunung Gading stock forms

Table 5. K:Ar radiometric dating of West Sarawak rocks (Kirk, 1968; Bignell, 1972; Prouteau et al., 2001)

Number	Rock	Locality	Mineral	K %	Age (Ma)	Period
Bau97-10	Adakite	Bukit Plandok	$K_2O^{\dagger} = 1.39$		6.4 ± 0.2	Upper Miocene
Bau97-9	Adakite	Gunung Murong	$K_2O = 1.57$		9.3 ± 0.2	Upper Miocene
Bau97-2	Adakite	Gunung Serambu	$K_2O = 1.38$		11.3 ± 0.3	Middle Miocene
Bau97-8	Adakite	Gunung Orad	$K_2O = 1.53$		11.6 ± 0.3	Middle Miocene
Kuc97-6	Adakite	Kuching	$K_2O = 1.22$		12.2 ± 0.3	Middle Miocene
Kuc97-1	Adakite	Stabar	$K_2O = 1.29$		12.9 ± 0.3	Middle Miocene
Kuc97-2	Adakite	Penkuari	$K_2O = 1.77$		14.6 ± 0.4	Middle Miocene
S11658	Tonalite	Gunung Rawan	Biotite	7.94	16 ± 4	Lower Miocene
S5812	Diorite	Pulau Satang	biotite	5.94	19 ± 3	Lower Miocene
Kuc97-14	Diorite	Gunung Tiram	K_2O wt.% = 1.92		21.9 ± 0.5	Lower Miocene
Kuc97-10	Diorite	Gunung Buah	K_2O wt.% = 2.02		22.3 ± 0.5	Lower Miocene
Kuc97-7	Diorite	Pulau Salak	K_2O wt.% = 2.53		22.9 ± 0.6	Lower Miocene
Kuc97-7	Diorite	Pulau Salak	K_2O wt.% = 2.53		23.7 ± 0.6	Lower Miocene
S25	Porphyritic dacite	Bau-Lundu road	Hornblende	0.333	27 ± 2	Upper Oligocene
S7984	Adamellite	Tanjung Datu	Biotite	6.82	75.6 ± 4	Cret. (Senonian)
S15	Biotite granite	Kg. Abok	Biotite	4.20	76 ± 2	Cret. (Senonian)
S2	Hbl.-bio. granite	Gunung Buri	Biotite	4.75	77.3 ± 2.1	Cret. (Senonian)
S9677	Adamellite	Lundu	Biotite	7.49	77 ± 4	Cret. (Senonian)
S7285	Granodiorite	Lundu	Biotite	7.56	78 ± 5	Cret. (Senonian)
S6299	Adamellite	Tinteng Bedil	Biotite	6.12	79 ± 5	Cret. (Senonian)
S1900	Granodiorite	Sebuyau	Biotite	5.62	101 ± 5	Cretaceous (Gallic)
S24	Amphib. xenolith	Sg. Duyoh	Hornblende	0.343	112 ± 4	Cretaceous (Gallic)
The following samples are from **Indonesian** islands offshore Sarawak (from Bignell, 1972)						
JM9.19	Biotite granite	Tambelan island	Biotite	4.01	84 ± 2	Cret. (Senonian)
JM15.4	Biotite granite	Natuna island	Biotite	6.46	73 ± 2	Cret. (Senonian)
The following Rb:Sr isochron is from Priem et al. (1975)						
Karimata island granite 74 Ma with an initial isotope ratio of 0.7101. Associated amphibolite K:Ar age of 78 Ma.						
Rb–Sr isochron (Bignell, 1972)						
S8	Biot.-hblde. granite	Sungai Tapang (Bt.Tabong)	These samples from the Bukit Tabong – Tinteng Bedil granite define a Rb–Sr isochron of 82 ± 17 Ma (Senonian), with an initial Sr isotope ratio of $R_i = 0.7043$, This confirms the K:-Ar ages of 76–79.			
S9	Biot.-hblde. granite	Sungai Ular (Bt. Tabong)				
S14c	Biotite granite	Tinteng Bedil				
S14d	Biot.-hblde. granite	Tinteng Bedil				
S15	Biotite granite	Kampong Abok (Abok)				

† Specimens for which K_2O contents are given were K:Ar dated on the whole rock. K_2O is in wt%.

Table 6. Chemical analyses (wt%) of Cretaceous plutonic rocks from West Sarawak (Kirk, 1968)

Oxide	a	b	c	d	e
SiO_2	49.58	68.80	70.21	71.07	72.11
TiO_2	0.30	0.58	0.32	0.55	0.24
Al_2O_3	18.88	14.31	14.55	13.61	12.80
Fe_2O_3	0.86	0.56	0.60	0.62	1.51
FeO	4.99	4.03	2.63	3.29	1.29
MnO	0.12	0.08	0.07	0.08	0.04
MgO	8.19	1.41	1.09	1.06	0.71
CaO	14.20	2.32	3.31	2.32	1.77
Na_2O	2.08	1.96	3.60	2.64	4.46
K_2O	0.13	3.48	2.76	3.34	4.14
H_2O+	0.61	1.58	0.59	1.05	0.72
H_2O-	0.03	0.23	0.28	0.06	0.34
CO_2	—	nil	—	0.06	—
P_2O_5	—	0.13	0.24	0.13	0.02
Total	99.97	99.97	100.25	99.88	100.15

a = olivine gabbro (S 6820) from Gunung Gebong, near Sematan.
b = adamellite (S 28) from Tanjung Datu.
c = granodiorite (S 546) from Batang Mambai, near Sebuyau.
d = granodiorite (S 7285) from Gunung Gading, Lundu.
e = adamellite (S 1822) from Tinteng Bedil, near Lingga.

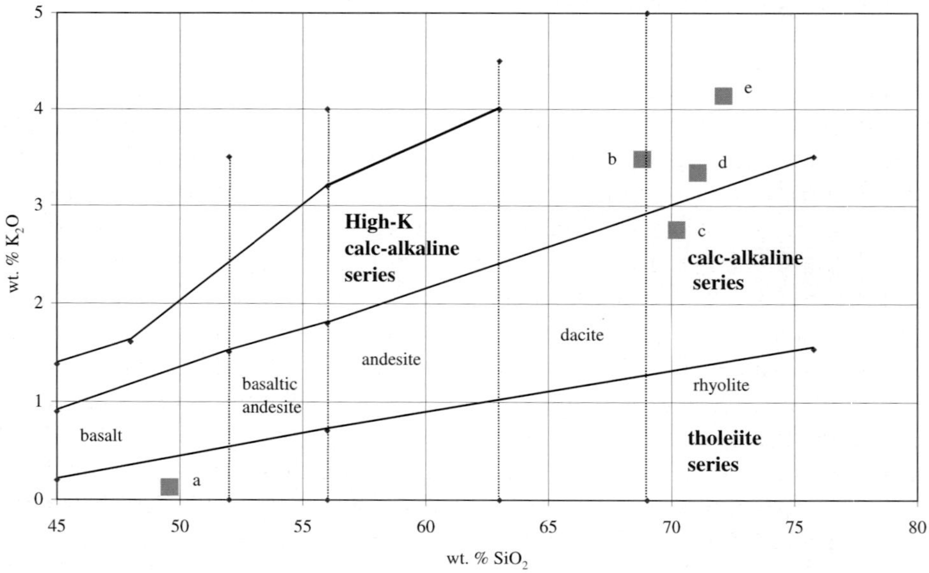

Figure 17. Late Cretaceous intrusives of West Sarawak K_2O vs. SiO_2.

a mountain 880 m high, south of Lundu. Along its N and NE margin, it hybridizes gabbro, and there are excellent outcrops along the coast near Lundu. The Tanjung Datu stock is only 10 km^2 in area.

IV.8. TERTIARY HIGH-LEVEL INTRUSIVES

Geographically, the Sintang intrusive suite occurs as hundreds of stocks, sills and dykes in the region of NW Borneo extending from the westernmost part of Sarawak 109.5° to 112.5° E and the zone extends southwards from 2° N to almost as far as 0.5° S. The main occurrences in Sarawak are shown in Figures 18 and 19. Most of the intrusions cut through the complete sedimentary sequence of the Ketungau and Melawi Tertiary basins. They also intrude the older formations that lie adjacent to the Tertiary basins. They frequently form high inselbergs, which in Indonesia may have a height of 1000 m, rising from the surrounding flat alluvial plain. This attests to the high erosion rates that have occurred since the early Miocene. No lava flows have been found, but they could well have fed lava formations that have been eroded.

In western Sarawak, two kinds of intrusions can be distinguished (Prouteau et al., 2001). High-K calc-alkaline to medium-K calc-alkaline subduction-related diorites and micro-diorites occur in the northern part, in Salak Island and Santubong Peninsula. Microtonalites and dacites occur near Kuching and in the Kuap and Bau areas to the south. They have a restricted chemistry and are known as adakites, which are related to trondhjemite and interpreted to have resulted from melting of a subducted slab within the upper mantle. For a definition and discussion, see Drummond et al. (1996) and Defant and Drummond (1990).

IV.8.1. The Bau district

The intrusives are very well known from the Bau district, where they are genetically related to the mineral deposits and have been exposed in the large number of mines, predominantly now abandoned (Figure 9). The stocks are of microgranodiorite porphyry (dacite) and the dykes and sills are of porphyritic dacite (Pimm, 1967b). Basaltic dykes are rare. The dacite phenocrysts are of plagioclase, pyroxene and hornblende in a glassy or microcrystalline matrix. The microgranodiorite porphyry contains phenocrysts of quartz, plagioclase, hornblende in a fine-grained matrix of quartz and feldspar. The rocks are mostly hydrothermally altered, calcitized and sericitized. Some are altered to hydrothermal clay and disseminated pyrite is common (Wolfenden, 1965; Pimm, 1967a).

Thin dykes extend more than 1.5 km along fractures in the limestone, but have a short extent in other host rocks. The dykes range from 10 cm to 30 m thick, but commonly are about 1.5 m thick. They are irregular—pinch and swell, branch and change direction and enclose limestone blocks. Because of the hydrothermal alteration, the dykes erode more readily than the country rocks, resulting in steep-sided gorges in limestone. Sills are 1.5–4.5 m thick, but do not persist in length.

The high-level intrusives have no thermal aureoles. Weak contact metamorphism has converted limestone at the contact zones to wollastonite, grossularite, rare diopside and vesuvianite (Wolfenden, 1965).

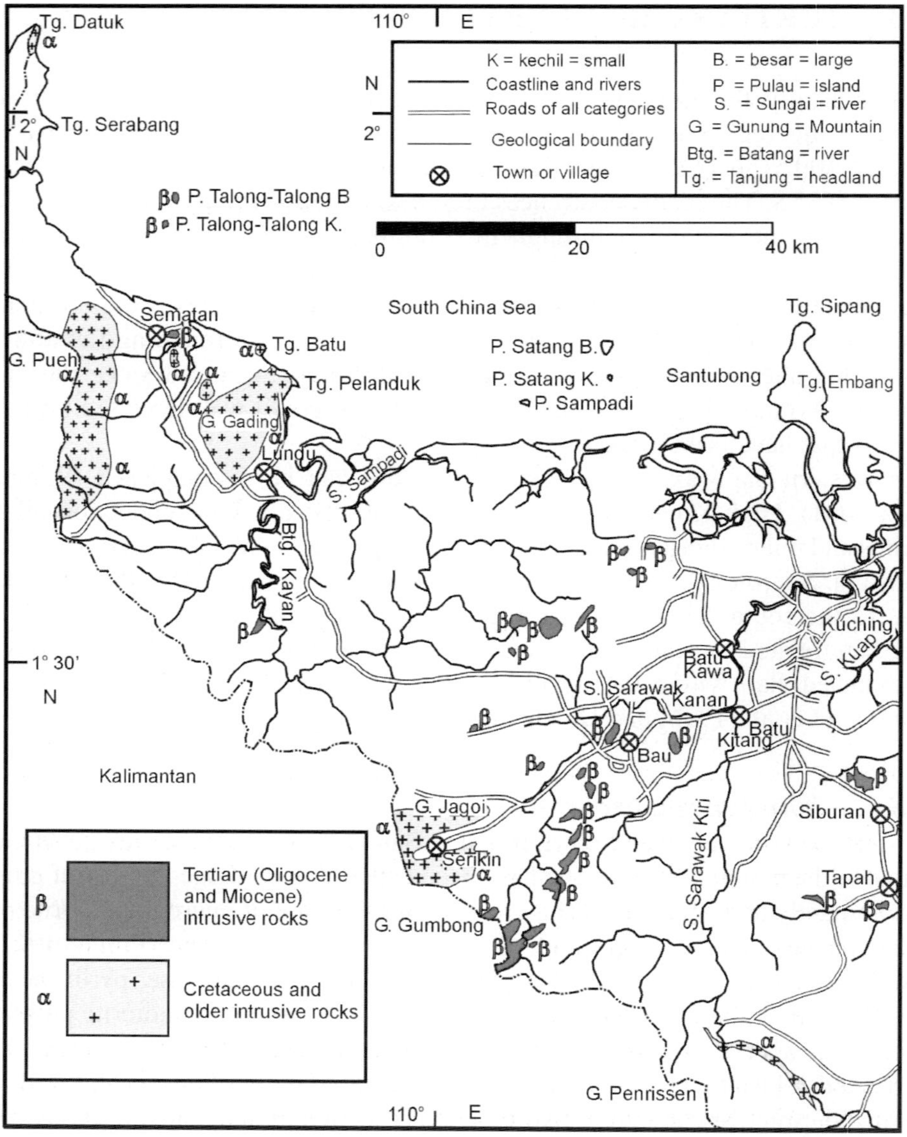

Figure 18. Plutonic igneous rocks of Sarawak west of Kuching.

IV.8.2. Chemistry

The rock suite of Sarawak, including the Sintang Intrusives of contiguous northwest Kalimantan, exhibit a calc-alkaline trend and have distinct I-type characteristics (Williams and Harahap, 1987). Selected chemical analyses of the high-level intrusives are listed in Tables 7 and 8 and the complete set of analyses (from Kirk, 1968; Prouteau et al., 2001) are plotted on a Peccerillo and Taylor (1976) diagram (Figure 20). The suite of rocks, which form the high-level stocks and dykes, extends over

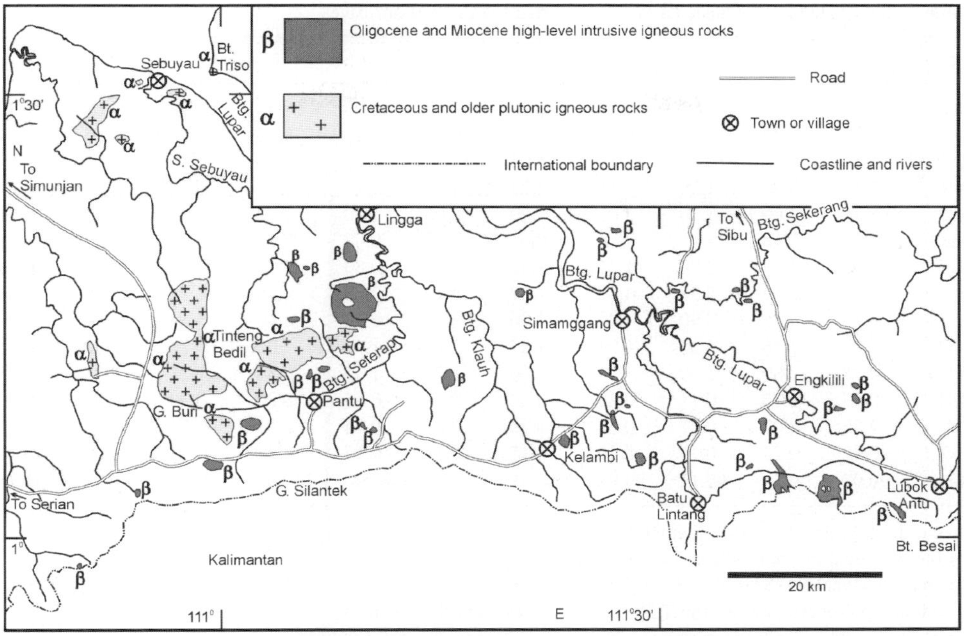

Figure 19. Plutonic igneous rocks east of Serian, western Sarawak.

the complete range from basaltic to rhyolitic. The suite is calc-alkaline and the large majority of the rocks lies in the dioritic or andesitic field and has been commonly called microgranodiorite porphyry.

Major and trace element data (Prouteau et al., 2001) show that the Lower Miocene diorites show all the usual characteristics of subduction-related magmas (Table 8). The Middle to Upper Miocene microtonalites and dacites share some of these characteristics, but in addition they display the following typical adakitic features: they are SiO_2-rich (65.5–70%), sodic ($Na_2O/K_2O>2$), Al_2O_3 contents exceed 14.5%. In trace elements, they have Th (3.5–6.8 ppm) and Nb (4–7 ppm). They are always low in Y (10.7–5.9 ppm) and in HREE (Yb = 0.39–0.85 ppm). These characteristics are typical of adakites (Defant and Drummond, 1990; Drummond et al., 1996). Figure 21 shows a plot of Sr/Y vs.Y (ppm) on which the field of adakites is taken from Drummond et al. (1996). The Sarawak and Kalimantan adakites fit well into the defined field. Of course, the data from Prouteau et al. (2001), summarized in Table 8, must follow a curve that is asymptotic to both the X- and Y-axis (because of the mathematical nature of plotting Y vs. $1/Y$). The main value of Figure 21 is to show the ranges of Y, which do not overlap, and become depleted in more silica-rich magmas.

Pyroxene is rare or lacking: it is usually replaced by early-crystallized amphibole. The rocks have very low Y and HREE contents, suggesting a garnet presence in their source. This leads to their characteristically high La/Yb and Sr/Y ratios. Their titanomagnetite–hemoilmenite associations reflect equilibrium features indicating moderate temperatures (< 900°C) and highly oxidizing crystallizing conditions.

Table 7. Selected chemical analyses (wt%) of Tertiary high-level intrusives

Oxide	a	b	c	d	e	f	g	h	i	j	k
SiO_2	50.59	53.50	62.70	64.20	66.78	67.60	68.0	68.7	70.60	73.60	77.30
TiO_2	0.49	0.79	0.54	0.48	0.40	0.37	0.25	0.35	0.14	0.05	0.14
Al_2O_3	13.19	18.50	16.40	17.10	16.03	15.30	16.70	15.30	15.30	14.70	13.50
Fe_2O_3	2.15	3.80	1.20	1.10	1.25	1.08	1.75	0.22	0.58	0.16	0.29
FeO	5.18	4.00	3.30	3.05	1.99	2.06	1.10	1.21	2.02	1.62	0.72
MnO	0.16	0.18	—	—	0.05	0.04	0.08	0.03	0.06	0.23	—
MgO	8.66	2.65	2.70	1.27	1.47	1.23	0.86	1.12	—	0.33	0.18
CaO	7.78	5.55	5.30	4.42	4.52	4.10	2.75	4.45	1.77	0.39	0.44
Na_2O	2.19	4.75	3.90	4.45	4.00	3.72	4.90	4.40	5.69	2.90	0.10
K_2O	1.17	1.16	1.50	1.94	1.27	2.06	1.46	2.05	1.77	2.50	3.55
H_2O+	2.38	2.80	1.00	1.24	2.00	1.17	1.37	0.85	1.24	2.05	2.85
H_2O-	0.77	1.75	0.14	0.26	0.20	0.83	0.55	0.30	0.30	0.25	0.29
CO_2	4.55	0.07	0.27	0.01	0.20	0.03	0.04	0.91	0.29	1.15	0.29
P_2O_5	0.71	0.24	0.16	0.26	0.15	0.14	0.15	0.12	0.13	0.08	0.08
Total	99.97	99.74	99.11	99.78	100.31	99.73	99.96	100.01	99.89	100.01	99.73

a = basalt (dolerite) (S11989). Silantek.
b = basaltic andesite (microdiorite) (S13583), Sungai Tada, Penrissen area.
c = andesite (tonalite) (S11658), Gunung Rawan, Penrissen area.
d = dacite (microgranodiorite) (S6261), Klambi quarry, near Sri Aman.
e = dacite (microgranodiorite porphyry) (S1), Mile 7, Penrissen road, Kuching.
f = dacite (microgranodiorite) (S6260), Lanchau quarry near Silantek.
g = dacite (microgranodiorite porphyry) (S8060), Near Lundu.
h = dacite (microgranodiorite porphyry) (K322), Near Gunung Lidau, Bau area.
i = rhyolite (microgranodiorite porphyry) (S6359), Abok quarry, near Silantek.
j = rhyolite (alkali microgranite) (S14528), Sungai Retoh, near Tebakang.
k = rhyolite (alkali microgranite porphyry) (S11620), Gunung Rawang near Penrissen.

IV.8.3. Age

Around Sintang, the K:Ar ages of the high-level intrusives range from 30 to 16 Ma (Late Oligocene to Early Miocene), and comparable ages have been obtained in Sarawak (Williams and Harahap, 1987). Tonalite at Gunung Rawan in the Penrissen area gave a K:Ar biotite age of 16±4 Ma, and diorite from Pulau Satang gave a K:Ar age of 19±3 Ma (Kirk, 1968). Details are given in Table 5.

The new detailed study of western Sarawak by Prouteau et al. (2001) has shown, by whole-rock K:Ar dating, that the high-K calc-alkaline diorites were emplaced during the Lower Miocene (22.3–23.7 Ma), whereas the microtonalites and dacites were emplaced in the Middle to Upper Miocene (14.6–6.4 Ma). The separation between these two episodes was at least 8 Ma.

IV.8.4. Origin

The Lower Miocene diorites are typically subduction-related from a geochemical point of view. They were likely derived and evolved from island-arc basaltic magmas (Prouteau et al., 2001). The Middle–Upper Miocene adakitic microtonalites and dacites are of a different origin. They were likely derived from the partial melting of previously subducted basalts from a fragment of oceanic lithosphere residing

Table 8. Selected whole rock analyses of the Miocene high-level intrusives of the Kuching–Bau district

Oxide	a	b	c	d	e	f	g	h	i
SiO_2	57.8	58.3	66.3	65.5	66.8	68.0	67.5	69.1	70.0
TiO_2	0.76	1.18	0.54	0.47	0.41	0.43	0.36	0.28	0.31
Al_2O_3	16.75	16.50	15.35	15.68	16.35	14.85	15.05	14.75	15.15
Fe_2O_3	7.10	7.75	4.22	3.64	3.67	3.38	3.30	2.98	2.75
MnO	0.13	0.13	0.07	0.07	0.06	0.06	0.06	0.05	0.05
MgO	4.04	3.20	1.65	2.00	1.57	1.50	1.48	1.42	1.12
CaO	6.96	5.58	3.35	4.00	4.40	3.88	4.27	3.78	4.06
Na_2O	3.42	3.85	3.81	3.70	3.66	3.65	3.58	3.38	3.63
K_2O	1.91	2.37	3.35	1.61	1.10	1.69	1.31	1.45	1.40
P_2O_5	0.23	0.38	0.15	0.15	0.12	0.11	0.12	0.10	0.09
Loss	0.53	0.34	1.28	3.11	1.92	2.23	2.48	2.61	0.99
Total	99.63	99.58	100.07	99.93	100.06	99.70	99.51	99.90	99.55
The concentrations of the following elements are in ppm									
Sc	17.4	17.2	8.2	6.6	6.2	6.8	6.0	5.3	5.0
V	162	148	74	56	48	59	44	40	36
Ba	605	535	555	405	372	402	303	505	460
Th	11.5	8.0	16.1	5.15	3.55	6.80	4.70	5.10	3.80
Ce	70	78	67	36	26	35	32	28.5	22
Nd	31.0	38.0	29.0	14.8	12.0	15.0	13.5	12.0	9.0
Eu	1.57	1.79	1.08	0.80	0.67	0.78	0.74	0.62	0.54
Yb	1.88	3.03	1.95	0.84	0.66	0.92	0.85	0.76	0.56

Loss = Loss on ignition. Fe_2O_3 = total iron expressed as Fe_2O_3.
a = calc-alkaline diorite, KUC97-10 (Gunung Buah).
b = calc-alkaline diorite, KUC97-2 (Pulau Salak).
c = calc-alkaline diorirte, KUC97-8 (Kuching).
d = adakite, KUC97-19 (Gunung Sibanyia).
e = adakite, KU97-6 (Kuching).
f = adakite, KUC97-2 (Penkuari).
g = adakite, BAU97-10 (Gunung Plandok).
h = adakite, BAU97-4 (Gunung Truan).
i = adakite, BAU97-2 (Gunung Serambu).
(All from Prouteau et al., 2001).

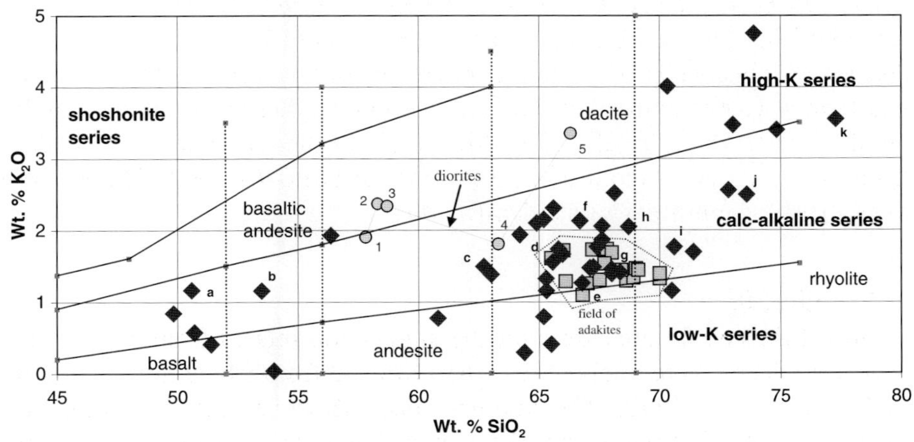

Figure 20. Tertiary high-level intrusives K_2O vs. SiO_2 diagram.

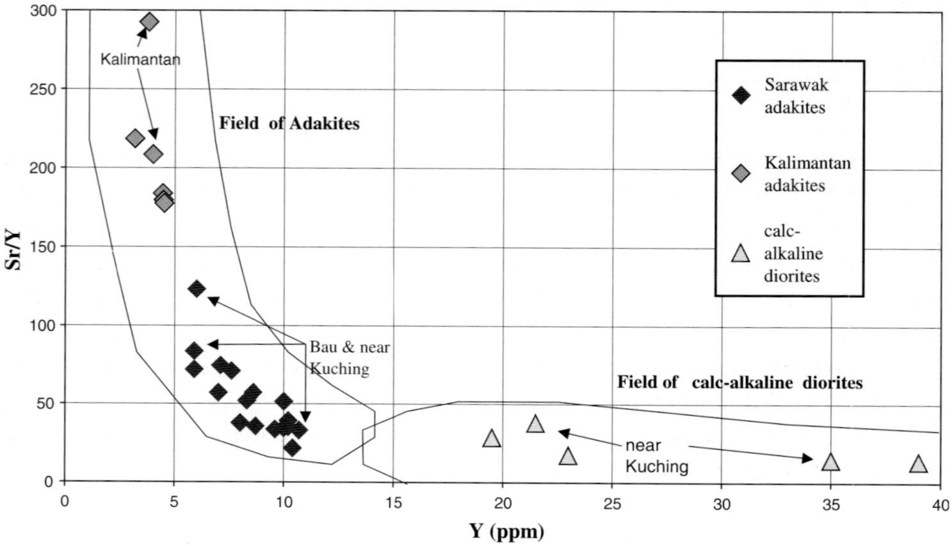

Figure 21. Y and Sr relationship in adakites and diorites of west Sarawak.

within the upper mantle beneath western Sarawak resulting from post-subduction collision. It is the adakitic rocks that are associated with the gold mineralization of the Bau district.

Chapter V

Sibu Zone

The 200 km wide Sibu Zone is predominantly of highly deformed steeply dipping low-grade metamorphic flysch, known as the Belaga Formation. It forms the greater part of what is known as the Rajang Group (Figure 22). During the Upper Cretaceous to Upper Eocene, the Belaga Formation was deposited in a deep marine basin then intensely folded, subjected to low-grade metamorphism (slate and phyllite) as a result of compression and uplifted to form an integral part of Sundaland, on which molasse formations were unconformably deposited within the Miri Zone, but also as small outliers within the Sibu Zone. The dramatic Late Eocene change from flysch to molasse sedimentation, resulting from compressive deformation and uplift, has been named the Sarawak Orogeny (Hutchison, 1996a).

V.1. BELAGA FORMATION

The Belaga Formation ranges from Upper Cretaceous to Upper Eocene and consists of a great indeterminate thickness, possibly 4.5–7.5 km, of inter-bedded argillite and greywacke sandstone. The strata show characteristics of deep-water bathyal distal turbidites. The whole formation is strongly folded, in many places isoclinally, and locally tectonically disrupted and mélanged. Dips of 90±10° are usual. The Formation has been mapped by Wolfenden (1960) and Kirk (1957) and subdivided on the basis of its palaeontological age, progressively younging northwards away from the Lupar Line (Liechti et al., 1960):

Stage I Layar Member, Upper Cretaceous (Cenomanian to Turonian)
Stage II Kapit Member, Palaeocene to Lower Eocene
Stage III Pelagus Member, Middle to Upper Eocene
Stage IV Metah Member, Upper Eocene
Stage V Bawang Member, Upper Eocene

V.1.1. The Layar Member (Stage I, Upper Cretaceous)

This member is composed predominantly of a turbidite sequence of slate and phyllite with rhythmically inter-bedded meta-greywacke laminae and thin beds, though beds as thick as 3 m occur sporadically (Tan, 1979). The strata have been intensely and tightly folded, generally overturned with dips generally ranging between 60° and 85° (Figures 13 and 14). Slaty cleavage is very well developed in the phyllite and slate. The sandstone beds show typical turbidite features such as rhythmic bedding, graded bedding and small-scale cross-laminations. Shearing along the bedding planes has

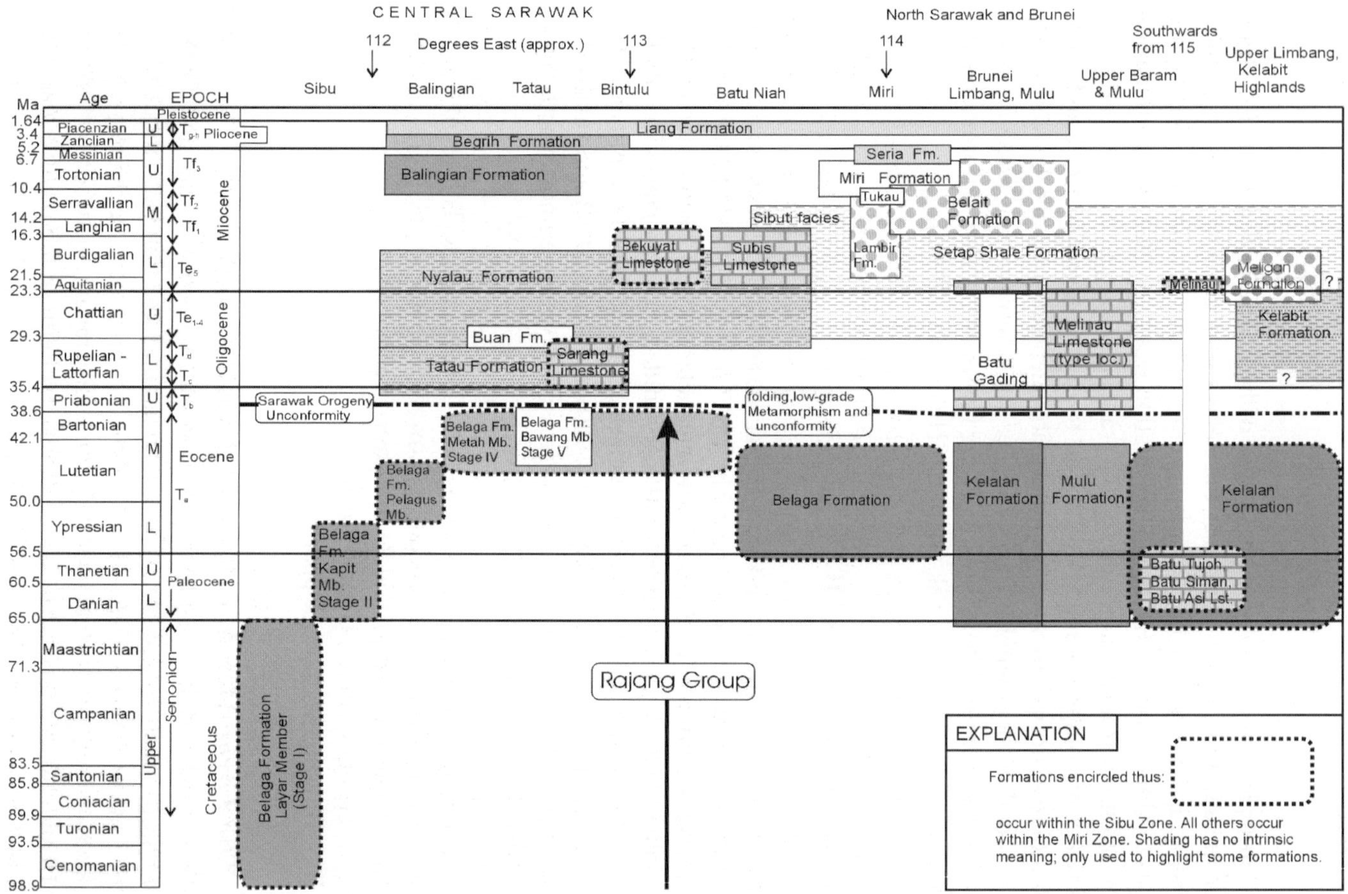

Figure 22. Chrono-stratigraphic table of the major formations of the Sibu and Miri zones. The Upper Cretaceous to Upper Eocene Rajang Group (flysch) forms the Sibu Zone and underlies the Miri Zone, outcropping locally as inliers. The Sarawak Orogeny caused low-grade metamorphism and strong folding characterised by steep dips of the Rajang Group. The overlying formations (molasses) are much less deformed, and unmetamorphosed, except in localized shear zones.

usually destroyed the flute moulds. The lower boundary of the Layer Member with the Lupar Formation is a fault. The top of the member grades into the Kapit Member.

V.1.1.1. Palaeontology and age

The following Foraminifera have been identified in Stage I (Wolfenden, 1960; Kirk, 1957): *Globotruncana* spp., *Orbitolina* sp., *Globigerina* sp., *Bathysiphon* spp. (also found in Stages II, III and IV); *Haplophragmoides* spp. (also found in Stages II, III and IV); *Textularia* spp. (also found in Stage IV); *Trochamminoides* spp. (also found in Stages II, III and IV).

The assemblage, and especially the first two, suggests a Cenomanian to Turonian age, but Liechti et al. (1960) interpret the sparse fauna to indicate an age range over the whole Upper Cretaceous (Figure 22), from Turonian to Maastrichtian, based on the association of *Globotruncana* and *Haplophragmoides.*

V.1.2. Kapit Member (Stage II, Palaeocene to Lower Eocene)

The boundary with Stage I is structurally conformable, and drawn based on palaeontology, for there is little lithological difference between the rocks. Kapit Member rocks are predominantly steeply dipping and are of low-grade argillite, slate and some phyllite. The argillaceous rocks are rhythmically inter-bedded with thin greywacke sandstone beds and rare conglomerate. Thin quartz veins are common and some contain traces of stibnite and gold in the west (Wolfenden, 1960).

The lower part of Stage II rocks may be distinguished from Stage I by containing the following: More common greywacke, which occurs as thick beds, some of which are as much as 90 m thick. Beds of greywacke conglomerate, some of which contain larger Foraminifera. Nodules of black siliceous mudstone with pyrite-rich cores. Locally developed greyish-red and green-grey argillite and slate.

At higher horizons in Stage II, the greywacke conglomerate does not occur and the greywacke beds become thinner and less common. The upper part of Stage II is a monotonous laminate sequence of argillite, slate, rare phyllite and occasional laterally persistent beds of fine-grained greywacke. Such a sequence is usually called distal turbidite.

V.1.2.1. Palaeontology and age

Stage II is dated Palaeocene to Lower Eocene (Wolfenden, 1960, Kirk, 1957) by the Foraminifera: *Alveolina* sp.; *Assilina* sp. (also occurs in Stage III); *Asterigerina* sp.; *Cassidulina* sp. (also occurs in Stage III); *Discocyclina* sp. (also occurs in Stage I); *Globoratalia wilcoxensis* Cushman & Ponton; *Nummulites* cf. *javanus* Verbeek, *Lockartia* sp., *Camerina* (*Nummulites*) *Nutalli*, *Anomalina* sp., *Bathysiphon* sp., *Cibicides* sp., *Cyclammina* sp., *Globigerina* spp., *Globorotalia* sp., *Glomospira* sp., *Trochammina* sp. and *Trochamminoides* sp.

East of the Usun Apau Plateau, about 114° 45′ to 115° E the Belaga Formation is known locally as the Kelalan or Mulu Formation (Haile, 1962). The sandstone–shale sequence contains thick layers of porcellaceous limestone (alodapic) at Batu Tujoh, Batu Asi and east of Batu Siman that have yielded a Palaeocene to Lower Eocene age (Haile, 1962; Adams, 1965). The rich Foraminifera and radiolaria fauna include the larger Foraminifera: *Miscellania miscella* (D'Archiac & Haime), *Nummulites nuttalli* Davies and *Opertorbitolites cf. Douvillei* Nuttall. Together with the alga *Distichoplax*, these indicate a Palaeocene to Lower Eocene age.

V.1.3. Pelagus Member (Stage III, Lower to Middle Eocene)

This member occurs in a belt about 40 km wide extending ESE from Sibu. It is predominantly of argillite, with common greywacke beds up to 15–90 m thick and some greywacke conglomerate. It is strongly folded and locally tectonically disrupted.

V.1.3.1. Palaeontology

Kirk (1957) has tabulated the following Foramininifera, indicating a Lower to Middle Eocene age for the Pelagus Member: *Cyclammina* spp., *Ammodiscus* sp., *Anomalina* sp., *Bathysiphon* sp., *Camerina (Nummulites)* sp., *Cibicides* sp., *Discocyclina* sp., *Globogerina* sp., *Globorotalia* sp., *Haplophragmoides* sp., *Operculina* sp., *Pellatispira* cf. *madaraszi*, *Pellatispira* cf. *glabura*, *Trochammina* sp., *Trochamminoides* sp. and *Verneuilina* sp.

V.1.4. Metah Member (Stage IV, Middle to Upper Eocene)

The Metah Member occurs as a belt to the north of and conformable with the Pelagus Member, to which it is lithologically similar. However, it is less metamorphic and the sandstone beds are sub-greywacke, rather than greywacke. Sandstone beds are commonly 15–150 m thick but are less common than in the Pelagus Member. Along the main Sibu–Tatau road, the sandstones are commonly 1 m thick or less.

A 25-m-thick redbed within the Metah Member occurs within thinly bedded turbidites. The reddening is associated with porosity enhancement and was caused by connate fluids percolating down from the unconformably overlying compacting Balingian and Begrih-Liang formations (de Silva, 1988).

V.1.4.1. Palaeontology

Kirk (1957) has tabulated the following Foraminifera, from which a Middle to Upper Eocene age is deduced for the Metah Member: *Ammodiscus* sp., *Anomalina* sp., *Bathysiphon* sp., *Camerina (Nummulites)* sp., *Cibicides* sp., *Cyclammina* sp., *Discocyclina* sp., *Globigerina* sp., *Globobulimina* sp., *Globorotalia* sp., *Gyroidina* sp., *Hantkenina* sp., *Haplophragmoides* sp., *Operculina* sp., *Pellatispira* cf. *madraszi*, *Sigmoilina* sp., *Trochammina* sp., *Trochamminoides* sp., *Uvigerina* sp., and *Vulvulina* sp.

V.2. KELALAN FORMATION

In the Upper Baram River, the Rajang Group is known as the Kelalan Formation (Haile, 1962). It is of inter-bedded sandstone and hard grey shale, in many places slaty, and rare lenses of limestone. It looks like the Setap Shale Formation, but the sandstone beds of the Kelalan Formation are thicker and the strata are more metamorphic. Along the Upper Baram River, the shales may be reddish, tuffaceous or calcareous.

There are persistent limestone beds at Batu Asi, Batu Tujoh and east of Bukit Siman. Most are fine-grained calcilutites. The lower beds of the porcellaneous limestones in the Upper Baram district have yielded *Alveolina* sp., *Nummulites* sp., *Miscellanea miscella* (d'Archaic & Haime), *Nummulites nuttalli* Davies, *Opertorbilolites* cf. Douvillei Nuttall and *Discocylina* sp., indicating a general Palaeocene to Lower Eocene age (Adams, 1965). Together with the alga *Distichoplax*, this faunal assemblage is distinctly older than anything in the Melinau Limestone.

Shales from the Kelalan Formation have yielded a fauna of small benthonic Foraminifera, but they are not age-diagnostic. However they resemble in some species that of Stage III of the Belaga Formation of the Rajang Valley (Haile, 1962).

V.3. POST-RAJANG GROUP MOLASSE FORMATIONS

There are several outliers of post-Upper Eocene unconformity shallow marine to non-marine strata that unconformably overlie the Rajang Group of the Sibu Zone. There is a tendency for carbonate sedimentation to persist in any particular locality, but the limestones at such localities contain one or more important hiatuses.

The younger carbonates of the Sibu Zone correlate with similar age carbonates within the Miri Zone, showing that both the Sibu and Miri zones were then unified into a single terrain.

V.3.1. The Tujoh-Siman Limestone

The Upper beds of the limestones are in part coralliferous conglomeratic, indicating an unconformity between them and the Lower Palaeocene–Lower Eocene beds (Adams, 1965). The upper beds contain such species as *Neoalveolina philippinensis* (Hanzawa); *Austrotrillina howchini* (Schlumberger); *Lepidocyclina (Nephrolepidina)* spp.; *Halkyardia* sp.; *Spiroclypeus* and *Heterostegina*. This assemblage suggests a Lower Miocene Aquitanian age. The unresolved problem with limestones, which overlie a major disconformity, is the degree of reworking of older forms into the active sedimentary environment. The palaeontological evidence nevertheless suggests that the upper part of the Bukit Tujoh Limestone is equivalent to the upper part of the Melinau Limestone of the Miri Zone (Figure 22).

V.3.2. *Sungai Akah (near Long Siniai) Limestones*

Fossiliferous calcilutite, calcarenite and calcirudite are inter-bedded with grey shale, calcareous sandstone and red calcareous shale in the Akah Valley. Limestone forms several beds and the thickest near Long Siniai is about 30 m. Upper Eocene (Priabonian) Foraminifera have been described from the Akah Valley near Long Siniai by Adams (1959): *Chiloguembelina* spp.; *Discocyclina* sp.; *Eorupertia* sp.; *Globigerina* spp.; *Gypsina globulus* (Reuss); *Halkyardia bikiniensis* Cole; *Nummulites* sp.; and *Operculina* sp.

V.3.3. *Bekuyat Limestones*

The limestone is fairly thick and entirely of Lower Miocene age (Liechti et al., 1960, p. 120).

V.3.4. *Bukit Sarang Limestone*

Approximately 190 km SW of the Melinau Limestone type locality is an outcrop known as Bukit Sarang. It occurs as an isolated karst hill surrounded by alluvium (Liechti et al., 1960).

V.3.5. *Nyalau Formation outliers*

V.3.5.1. Kakus Member

This member occupies the Kakus and Kemena river valleys, occurs also in the Batu Bora and Plieran rivers and underlies the Hose Mountains volcanic massif (Figure 23). It is of friable sandstone, laminated clays and coal seams, which are as much as 2.7 m thick. The sandstone layers are 4.5–9 m thick and alternate with soft carbonaceous clays (Liechti et al., 1960). At the Hose Mountains, it was formerly called the Hose Mountains Coal Formation. It unconformably overlies highly deformed Belaga Formation. It contains a poorly developed brackish-water fauna including: *Glomospira* sp.; *Trochammina* sp., both of which are common, and a less abundant *Ammobaculites* sp. They are not precisely age-diagnostic, but are Miocene generally.

V.4. LATE TERTIARY VOLCANISM

The upper Rajang area of the Sibu Zone contains large post-orogenic outcrops of volcanic rocks, totalling more than 1800 km^2. They form the most outstanding geomorphological features of the region, and Bukit Batu of the Hose Mountains, at 2006 m, is the highest in Sarawak. The age is generally taken as Pliocene–Quaternary, but in the absence of radiometric dating, the deep dissection of the Hose Mountains may suggest they may extend back to the Late Miocene. Hot springs are completely lacking, and the lesson from the Semporna Peninsula of Sabah, indicates that a volcanic arc, long held to be Pliocene–Quaternary, is now dated Miocene. This may also be the case for these volcanic edifices of interior Borneo, and a programme of radiometric dating is badly needed.

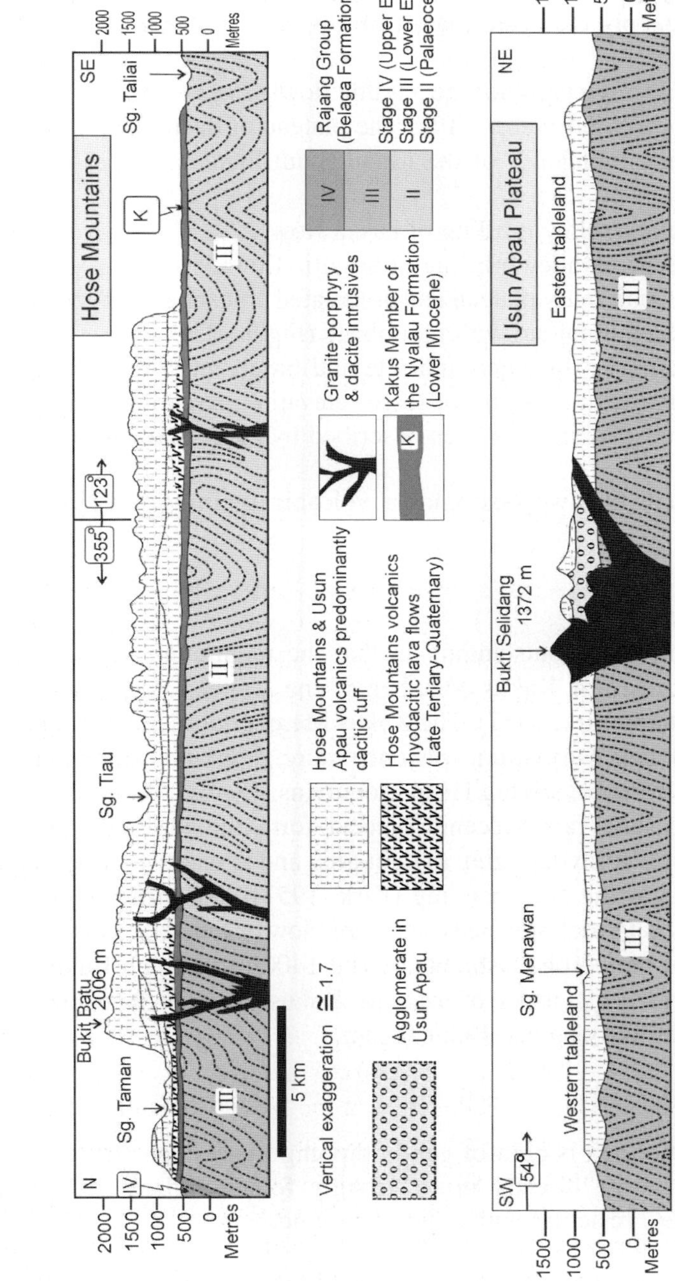

Figure 23. Cross-sections through two best-known volcanic mesas of the Upper Rajang area. Upper: Hose Mountains. Lower: Usun Apau. Age is assumed to be Upper Miocene to Quaternary. Both are calc-alkaline to high-K calc-alkaline dacitic, predominantly ignimbritic (redrawn after Kirk, 1957).

There is a distinct bimodal distribution, in common with volcanic arcs, and the dacites build the mountainous massifs, while the basalts usually fill valleys. The main localities (Kirk, 1957) are:

Hose Mountains: a deeply dissected massif of Upper Miocene–Pliocene rhyodacite and dacite.
Bukit Kajang: a plateau of Upper Pliocene basaltic flows (Figure 23).
Bukit Batu Laga (Linau Balui area): a Pliocene plateau of dacite lava and tuff.
Northern and Southern Tablelands of the Linau-Balui Plateau: Upper Pliocene basalt lava and breccias.
Linau Valley: Quaternary valley infilling of basalt flows.
Plieran Valley: Quaternary valley infilling of basaltic flows.
Usun Apau Plateau: Quaternary plateau and associated volcanoes of hypersthene dacite flows, tuff and agglomerate with ignimbrite (Figure 23).
Nieuwenhuis Mountains: A dissected andesite and basalt mesa resembling the Hose Mountains, but less rugged and of lesser elevation, straddling the border with Kalimantan. Their geology has been described by Haile and Kirk (1956).

Cross-sections through the two best-known volcanic ignimbritic edifices are shown in Figure 23.

V.4.1. Hose Mountains

The Hose Mountains massif is outstanding in that the mountainous edifice was formed upon a land surface of Kakus Member of the Upper Oligocene–Lower Member Nyalau Formation (Kirk, 1957). The Kakus Member outlier, which unconformably overlies the Belaga Formation, was once more geographically extensive, protected from erosion by the overlying Hose Mountains (Figure 23).

Rhyodacite lava, tuff and rare volcanic breccia form the base of the Hose Mountains. The lava varies from dark grey to light grey and is massive, rarely vesicular only occasionally showing flow banding (Kirk, 1957). The tuffs are strongly jointed hard white rocks. Pyroclastic beds and rare flows of dacitic composition build the higher parts of the edifice, estimated to be 1400 m thick on Bukit Batu. Hypersthene dacitic pyroclastic rocks overlie the Kakus Member, and build the mountains between Bukit Mabong and Bukit Jugam.

V.4.2. Batu Laga

The high plateau of Batu Laga is built of gently dipping hypersthene dacite lavas; agglomerate and tuff, resting directly upon strongly folded Rajang group. The dacite lavas are dark grey, vesicular with a glassy groundmass.

V.4.3. Linau-Balui Plateau

The edifice is built of basaltic rocks. Most of the rocks are non-vesicular dark grey to black, containing abundant hypersthene, lesser clinopyroxene and olivine.

V.4.4. *Bukit Kajang plateau*

This edifice is also built of basalt, mainly massive and only slightly vesicular and not porphyritic.

V.4.5. *Usun Apau Plateau*

The geology of the edifice was described by Campbell (1956). Unlike the other edifices, this one is bi-modal. The high tablelands of the central area are made of hypersthene dacite tuff and agglomerate.

Basalt lava occupies a large part of the southern mountains. It forms vertical cliffs as high as 70 m high offering excellent exposures (Figure 23). The basalt is dark grey and strongly vesicular. Columnar jointing is well developed. Olivine occurs in some specimens.

V.4.6. *Nieuwenhuis Mountains*

A thick formation of andesite and basalt overlies the Rajang group. Most of the Mountains occur across the border in Kalimantan. Zeijlmans van Emmichoven (1938) described the rocks as alkaline, predominantly basaltic. He described nepheline basalt, the only known occurrence in Borneo.

V.4.7. *Related intrusive rocks*

Intrusive rocks are of common occurrence, but small in areal extent. Granite porphyry occurs as a thick dyke near Batu Laga in the Linau-Balui area. Tonalite porphyry forms the intrusive stock at Bukit Kalulong and there is a small intrusion at Bukit Maloi in the Tinjar Valley. Dykes of hornblende andesite cut the volcanic breccias in NW Nieuwenhuis Mountains. Olivine basaltic dykes occur in the upper Balui River.

Lamprophyre intrusions occur in the Nieuwenhuis Mountains. Kersantite lamprophyre forms small plugs in the valley of the River Busang, a small tributary of the Upper Balui (Kirk, 1957). The chemistry of the intrusive rocks conforms to the volcanic rocks and offers no surprises (Figure 24), although the sampling of these remote regions is limited.

V.4.8. *Chemistry*

Whole rock chemical analyses of the Pliocene–Quaternary volcanic rocks, which overlie the Rajang Group of the Sibu Zone, are given in Table 9. They are taken from Kirk (1957). c to g are repeated in Kirk (1968). The analyses are plotted on a Peccerillo and Taylor type diagram (Figure 24) that clearly shows that the rock suite is closely characterized straddling the fields of calc-alkaline to high-K calc-alkaline. This strongly suggests that they are a subduction-related arc, but the related trench does not lie within Sarawak, and perhaps must lie in eastern Sabah.

Chemical analyses of intrusive igneous rocks associated with the volcanic massifs from the Upper Rajang area of the Sibu Zone are also given in Table 9 (from Kirk, 1957).

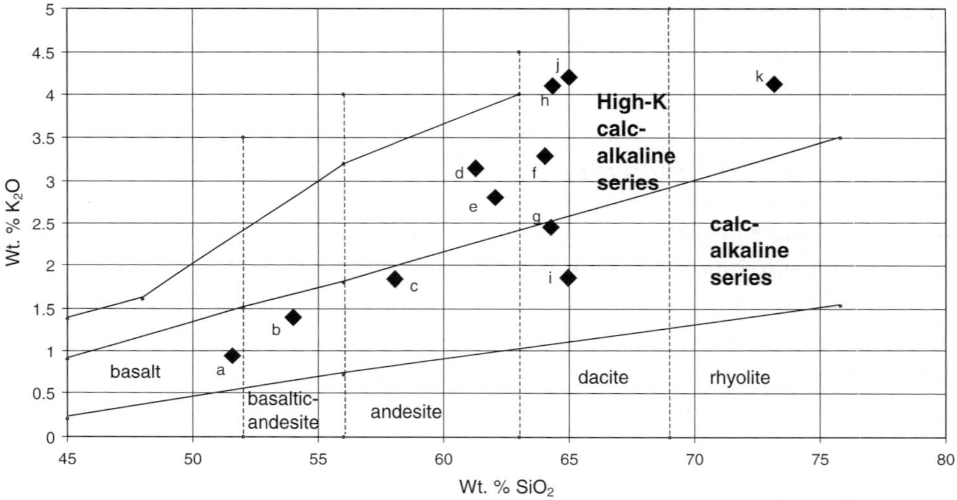

Figure 24. Late Tertiary–Quaternary volcanic and associated intrusive rocks, K_2O vs. SiO_2.

Table 9. Chemical analyses of Late Tertiary–Quaternary Sibu Zone volcanic and related intrusive rocks

Oxide	a	b	c	d	e	f	g	h	i	j	k
SiO_2	51.57	54.00	58.05	61.27	62.05	64.04	64.27	64.35	64.96	65.00	73.17
TiO_2	1.21	1.38	0.23	0.13	0.85	0.35	0.67	0.42	0.49	0.20	0.21
Al_2O_3	15.71	16.12	16.02	16.64	16.10	14.21	15.54	13.39	16.85	13.16	13.90
Fe_2O_3	4.31	2.20	4.82	3.57	0.81	0.35	2.12	0.57	2.56	0.39	0.95
FeO	7.25	6.77	3.00	2.85	4.83	2.93	2.57	3.25	0.98	3.04	0.65
MnO	—	0.12	0.10	0.10	—	0.09	0.08	0.09	0.08	0.09	—
MgO	6.72	5.34	0.10	9.12	3.19	1.10	1.93	1.16	1.64	0.95	0.54
CaO	7.79	6.92	0.80	5.10	4.33	4.90	3.98	5.40	3.18	5.10	2.37
Na_2O	2.60	3.55	5.30	3.85	2.94	5.01	4.27	5.54	4.61	6.10	3.61
K_2O	0.94	1.40	1.84	3.15	2.81	3.29	2.45	4.10	1.86	4.20	4.12
P_2O_5	0.39	0.29	1.39	0.03	0.24	0.12	0.21	1.10	0.13	1.85	0.42
H_2O+	1.19	1.14	4.36	1.62	1.66	1.62	0.48	0.65	1.10	0.39	0.36
H_2O-	0.22	0.70	0.05	1.56	0.32	1.85	1.16	0.16	0.51	0.09	0.11
CO_2	—	0.04	0.35	0.36	—	0.50	0.34	0.30	0.07	0.05	Nil
S	—	0.04	—	—	—	0.03	0.06	0.05	1.54	0.05	—
Total	99.99	99.99	100.41	100.35	100.13	100.38	100.11	100.50	99.98	100.63	100.39

a = basalt (S. 3450) from Nawei river, Southern Tableland, Linau-Balui Plateau.
b = Basaltic andesite (S. 3344) from Plieran river.
c = andesite (S. 4233) Bukit Tibang, Nieuwenhuis Mountains.
d = hypersthene dacite (S. 3849), Batu Laga, Linau-Balui area.
e = hypersthene dacite (S. 3446), Batu Laga, Linau-Balui area.
f = hypersthene dacite tuff (S. 3936), Mujan river, Usun Apau plateau.
g = hypersthene dacite (S. 3289), Bukit Batu, Hose Mountains.
h = hypersthene dacite porphyry (S. 3934), Bukit Mabun, Usun Apau Plateau.
i = rhyodacite (S. 3309), Taman river, Hose Mountains.
j = tonalite porphyry (S 3935), Bukit Kalulong.
k = granite porphyry (S. 3444), Bunut river, Linau-Balui area (from Kirk, 1957).

Chapter VI

Miri Zone

The Late Eocene and younger stratigraphy (post-Sarawak Orogeny) of the Miri Zone is wholly of molasse. The strata were deposited in non-marine to inner neritic marine conditions and local unconformities are common as a result of long-ranging thin-skinned compressional tectonics. The basement of the Miri Zone, at least in considerable part, is of Rajang Group flysch, which has been thrust up in compressional steeply dipping and complexly folded anti-formal structures to form inliers, bearing local names such as Bawang Member of the Belaga Formation, Kelalan Formation and Mulu Formation (Figure 22). They are all remarkably similar and of sandstone–shale laminite turbidite.

Everywhere these inliers are separated from the overlying molasse by the regional Late Eocene unconformity. Adams (1965) suggests that there is conformity between the Melinau Limestone and the Mulu Formation, but this is unlikely because of the strong contrast in metamorphic grade and the remarkably simpler structural style above as compared with that beneath the unconformity. Furthermore no descriptions exist of the actual contact, which may not be exposed.

VI.1. INLIERS OF RAJANG GROUP (PRE-LATE EOCENE UNCONFORMITY)

Throughout the Miri Zone there are several inliers composed of low-grade metamorphic rocks that may be ascribed to the Rajang Group on a lithological and structural basis. In many cases palaeontological control is poor, but from regional stratigraphic considerations, they are generally Palaeocene–Eocene, with possible Upper Cretaceous extension. The Rajang Group continues northwards from the Bukit Mersing Line, which is the northern limit of continuous outcrop of the Belaga Formation, to re-appear in places as inliers of flysch pushed up as anticlinal structures from beneath the blanketing molasse strata.

VI.1.1. Bawang Member of the Belaga Formation (Stage V)

The core of the Arip–Pelungau anticline, which plunges towards the ESE, is composed of the Bawang Member of the Belaga Formation. Similar Bawang Member rocks outcrop as the Tatau Horst, to the north and west of Tatau (Figure 25). From this evidence, it must be concluded that the Belaga Formation continues underneath much of the Miri Zone.

This member of the Belaga Formation consists largely of low-grade meta-pelites, including slate and rare phyllite (Wolfenden, 1960; Kirk, 1957), inter-bedded with

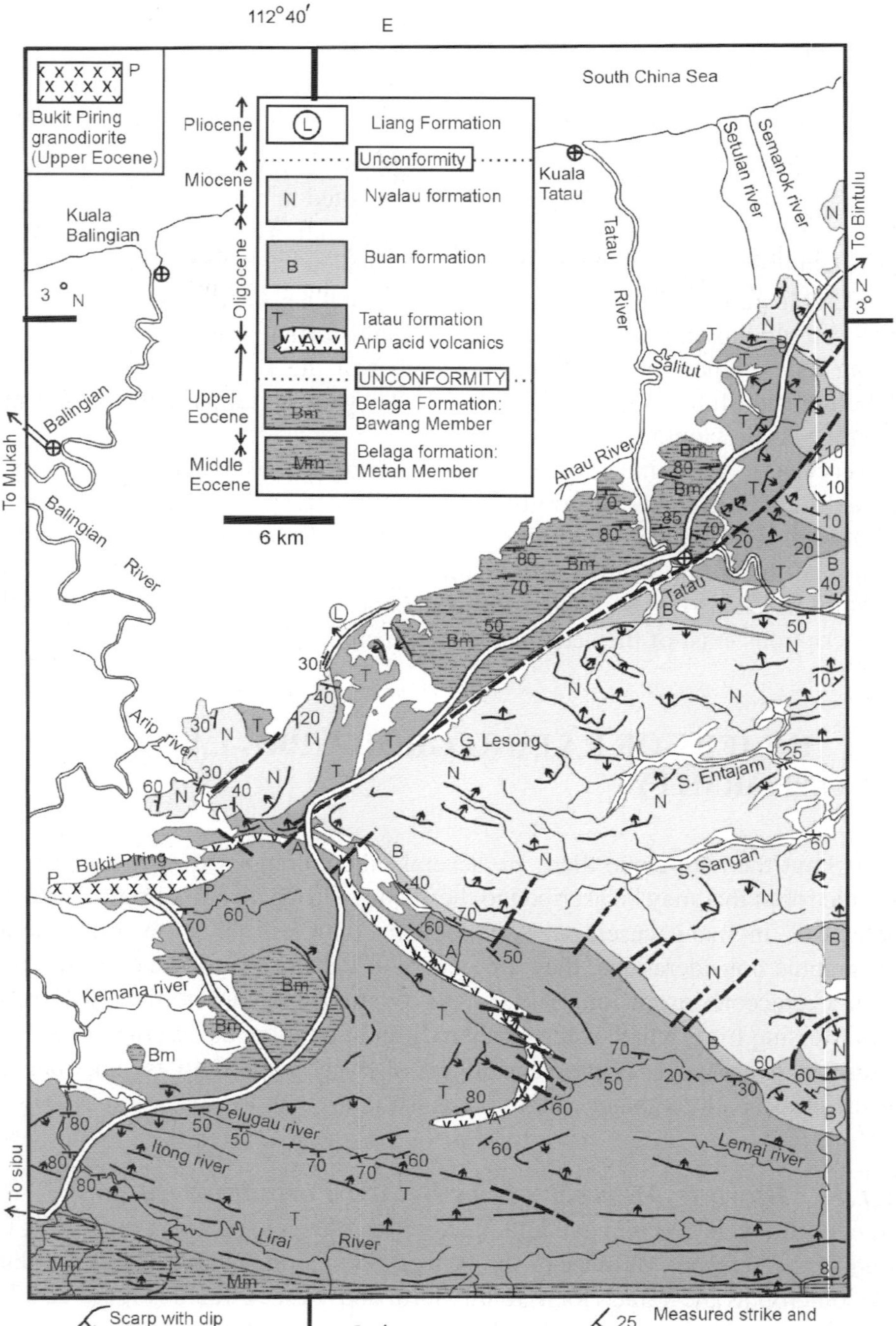

Figure 25. Geology of the Arip–Pelungau anticline and Tatau compressional horst (after Wolfenden, 1960). With permission from Minerals and Geoscience Department, Malaysia.

thinly bedded greywacke; a sequence now interpreted as distal turbidite. Sampling has not produced any definitive microfossils. The Bawang Member is characterized by thick (2–6 m) amalgamated turbidite sandstone beds separated by thinner laminites (alternations of dark grey argillite and centimetre thin turbidite sandstones, but with some thicker sequences of laminites without thick sandstones). The Bawang Member resembles the Metah Member, but the thicker sandstone beds of the Bawang are uncommon in the Metah. The dips are mostly very steep, but especially along the upthrust Tatau Horst, there is chaotic folding and disruption to give broken beds. The Bawang Member, which strikes E–W, is unconformably overlain by the Upper Eocene to Oligocene Tatau Formation that forms topographic scarps dipping away from the Belaga Formation horst and core of the plunging Arip–Pelungau anticline. Outcrops of Bawang Member are strongly folded and the sandstone beds generally boudinaged (Figure 26).

VI.1.2. Kelalan Formation

The inlier occupies the Temala anticlinorium and is exposed at Batu Gading, up river from Marudi along the Baram River (Haile, 1962). It is a turbidite sequence mainly of shale and sandstone, with subordinate limestone, tuffite and tuffaceous limestone. In many places the shale is slaty. The formation is intensely folded. The age of the limestone ranges from Upper Cretaceous to Upper Eocene. *Globotruncana* sp. has been recorded only at one locality beneath the Melinau Limestone at Bukit Besungai (Northeast) indicating an Upper Cretaceous age.

VI.1.3. Mulu Formation

This formation of sub-metamorphosed shales and slates, inter-bedded with hard sandstones, occupies the large inlier of the Mulu Anticlinorium and the highest mountain, Gunung Mulu (2376 m) (Haile, 1962). It is a monotonous succession of intensely folded turbidite composed of slaty and phyllitic shales and quartz sandstones. Bedding of the thin sandstones is commonly obliterated by well-developed cleavage. The best outcrops are in the Tutoh Gorge (Figure 31), where the sandstone beds may be as much as 150-m thick, inter-bedded with thin slates. Quartz veining is important locally. In the Tutoh Gorge, the strike is NE–SW and nearly vertical dips towards the SE predominate (Haile, 1962). Liechti et al. (1960) confidently correlate the Mulu with the Belaga and Kelalan formations.

VI.1.3.1. Palaeontology and age

The formation is poorly fossiliferous, but *Discocyclina* sp.; *Globorotalia wilcoxensis* Cushman & Ponton have been recovered. Liechti et al. (1960) have concluded that the Mulu Formation is Palaeocene to Lower Eocene in age.

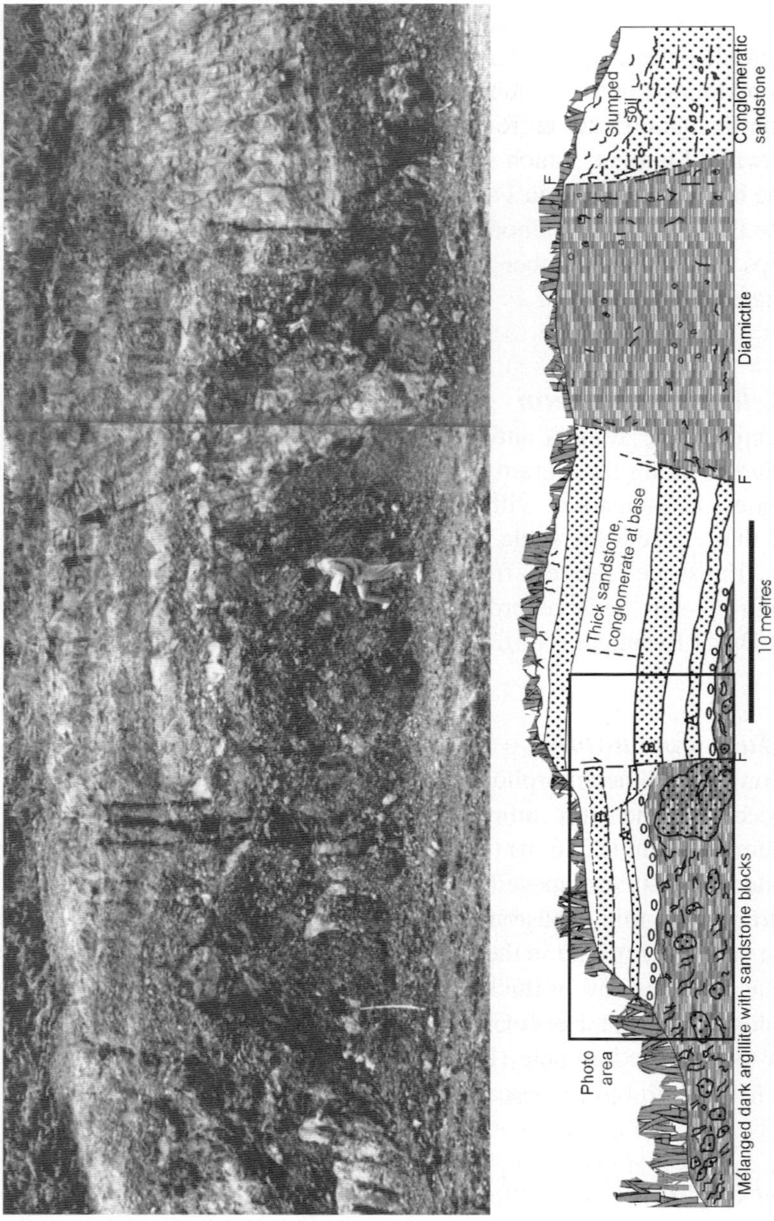

Figure 26. Undated sandstone–conglomerate sequence unconformably overlying mélange and diamictite. (after Halife and Ho, (1991).

VI.2. UNCONFORMITIES

The three most spectacular unconformities in the Tatau district are:

VI.2.1. Late Eocene

This unconformity separates the underlying strongly deformed phyllitic Bawang Member of the Belaga Formation (Rajang Group) from the unmetamorphosed Upper Eocene–Oligocene Tatau Formation at the Tatau Horst and the underlying strongly deformed Kelalan Formation flysch from the overlying Melinau Limestone at Batu Gading (Figure 30). This is a regionally powerful unconformity, resulting from the Sarawak Orogeny, synchronous with the collision of India with Eurasia and marking the general beginnings of the Tertiary basins of Sundaland. This orogenic event caused extinction of the Indian Ocean spreading axis in the Wharton Basin west of Sumatra, and the jump of active spreading to a new system between Australia and Antarctica.

VI.2.2. Oligocene

A disconformity separates the underlying Upper Eocene Melinau Limestone from the overlying Lower Miocene limestone at Batu Gading (Figure 30) and there is a total Oligocene hiatus. However, although this unconformity is widespread, there is a continuous limestone sequence throughout the Oligocene at the Melinau Limestone type locality (Figure 31).

VI.2.3. Mid-Miocene

This regionally important unconformity separates the Upper Oligocene–Lower Miocene Nyalau Formation from the Upper Miocene Balingian Formation (Figure 33), in the middle to upper Baram region. The unconformity occurs throughout the Dangerous Grounds.

The Tunggal–Ransi Conglomerate forms a discontinuous hogback along the NW border of the Tatau Horst (Liechti et al., 1960). The conglomerate occurrences extend over a distance of 25 km in the Tatau Horst area. A thickness of around 200 m has been measured. The conglomerate unconformably overlies the Tatau Formation (Liechti et al., 1960), but also overlies the Bawang Member of the Belaga Formation (Figure 26). The unconformably overlying conglomerate is tentatively correlated with the Begrih Conglomerate, which is said to form the base of the Upper Miocene–Pliocene Begrih Formation, although this is not universally accepted (Liechti et al., 1960). The conglomerate on the Tatau Horst occurs at Bukit Rangsi (Figure 27) and Bukit Tunggal, west and east, respectively, of the Tatau River. It is Barren of fossils and onlaps onto the Bawang Member of the Belaga Formation, the Tatau and Nyalau formations. In the Kaluan Valley, an outlier of the Nyalau Formation rests with angular unconformity upon steeply dipping Belaga Formation (Liechti et al., 1960). The unconformity shown in Figure 26 is of undated conglomeratic sandstone overlying dark argillite matrix

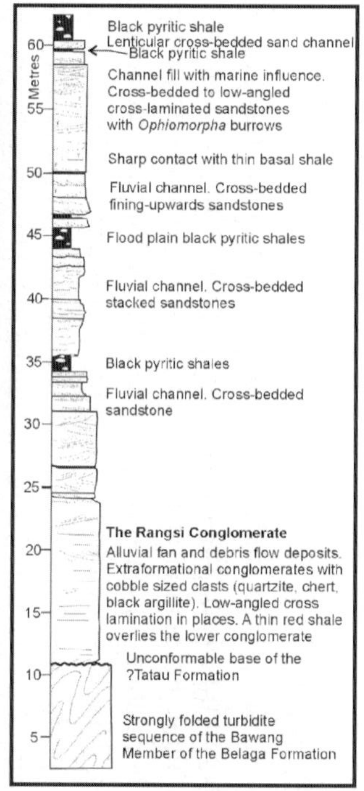

Figure 27. Angular unconformity at eastern Tatau Horst between strongly folded Bawang Member turbidites and overlying undated Rangsi Conglomerate (Figure 22). Measured section and interpretation by Boniface Bait (from Haile and Ho, 1991).

mélange and diamictite. The mélanged formation is taken by Haile and Ho (1991) as possibly Bawang Member, but it could equally be of highly tectonized Tatau Formation. The overlying conglomeratic sandstone is of unknown age, and likely to be as young as Upper Miocene.

Regional seismic mapping along the coast and offshore NW Sarawak has led to the identification of seven regional unconformities (Figure 28). The sequences are named 'Tertiary-one sequence' (T1s) onwards. The sequences are relatively easy to identify on seismic sections (Figure 29). Base Tertiary-1 and -2 boundaries appear to be irregular erosional surfaces. All the Oligocene to Late Miocene sequence boundaries are tectonically induced, showing no clear relationship to global eustatic sea-level falls (Haq et al., 1987). The Pliocene and Pleistocene unconformities, by contrast, may well be a result of eustasy (Figure 22). The cycle scheme of Ho (1978), and the modified scheme of Hageman (1987) are therefore criticized by Ismail Che Mat Zin and Tucker (1999), who proposed a sequence stratigraphy, in which boundaries can be identified on seismic sections, irrespective of sedimentary facies. The cycles of Ho (1978) and Hageman (1987) are accurate only for marine and marine-influenced environments.

VI.2.3.1. Regional interpretation of the unconformity

Prior to the Mid-Miocene Unconformity (MMU), the sedimentary formations were deposited on the coastal zone of Sarawak with a coastline directed NNW from Bintulu and water deepening towards the ENE towards bathyal conditions beyond what has become known as the West Baram Line (Figure 36). Land lay to the west in the Penian High. The sediment provenance was Sundaland on the west with fluvial systems directed ENE. The sedimentary sequence offshore, interpreted as lower coastal plain to holomarine inner neritic, is well exposed on land and known as the Upper Oligocene–Lower Miocene Nyalau Formation (Wolfenden, 1960). Farther eastwards along the coastal plain, deeper water is represented by a line of carbonates known as the Subis Limestone, and eastwards by the muddy Setap Shale (Haile, 1962). There was no Baram Delta at this time, in a region characterized by deep water. We can interpret the pattern of Figure 36 as the margin of continental Sundaland: land giving way in an ENE direction via a rifted continental margin. The West Baram Line would have been in the position of the continental slope.

Within a geologically short time of ~3 my, the palaeo-coastline changed from being directed NNW (Figure 36) to the present-day coastline. The rapid change in orientation might suggest a dramatic anti-clockwise rotation of Borneo and the contiguous shelf. The change from Figure 36H through Figure 36F is attributable to the end of rifting of the South China Sea region and uplift of the Borneo landmass during the MMU (Hutchison, 2004). The uplift of Sarawak at this time means that the Middle Miocene and younger formations outcrop only along the Sarawak coastal plain, and they dramatically thicken seawards. The post-unconformity formations, of Upper Miocene through Pliocene age, are the Balingian, Begrih and Liang

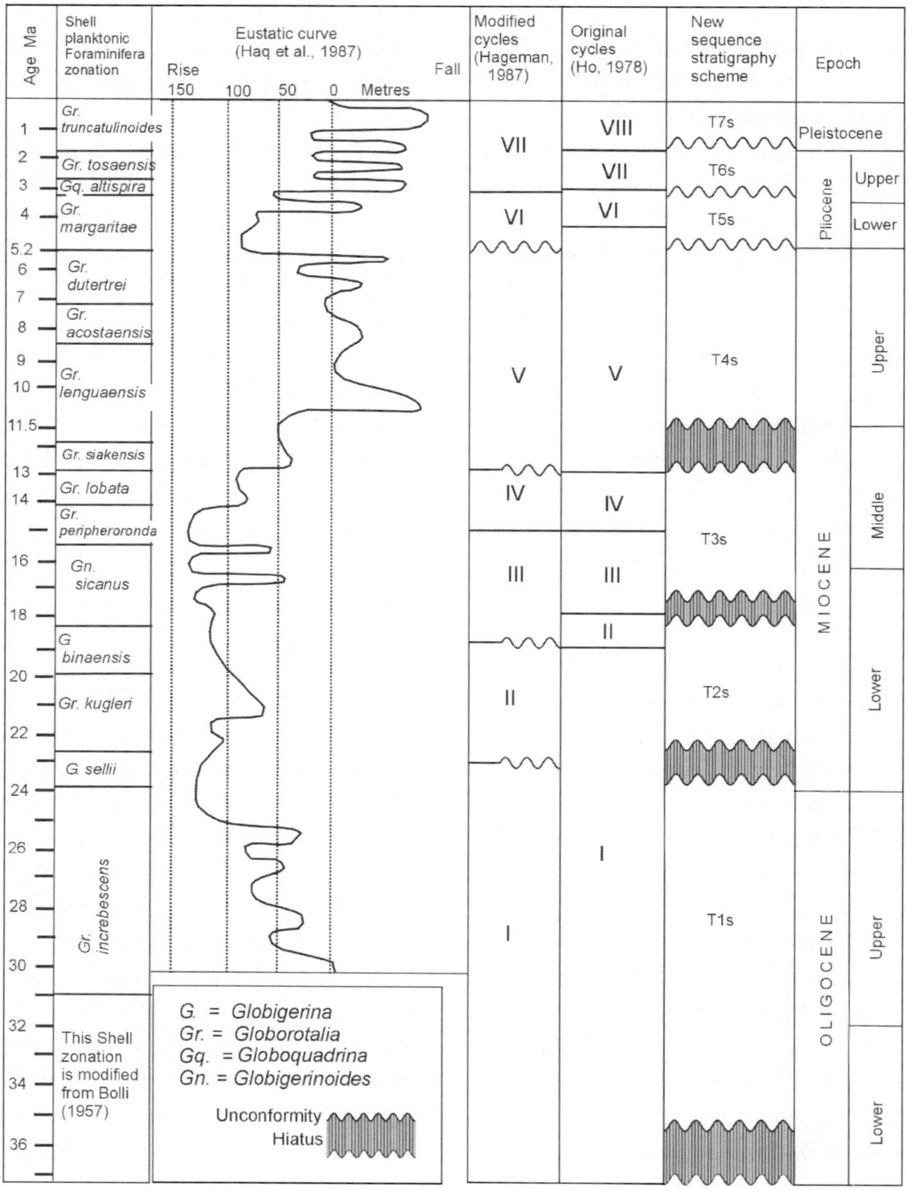

Figure 28. Relationship between the Foraminiferal zones, cycles and stratigraphic sequences of offshore and onshore northwest Sarawak (after Ismail Che Mat Zin and Tucker, 1999).

formations, and are of very restricted on land coastal zone outcrop (Wolfenden, 1960). A published seismic section extending NNW from the coast at Mukah illustrates the spectacular unconformity, albeit with excessively young age designations (Ismail Che Mat Zin and Tucker, 1999). Their paper also illustrates a half graben offshore Balingian that has been tilted seawards by the uplift of Sarawak (Figure 29).

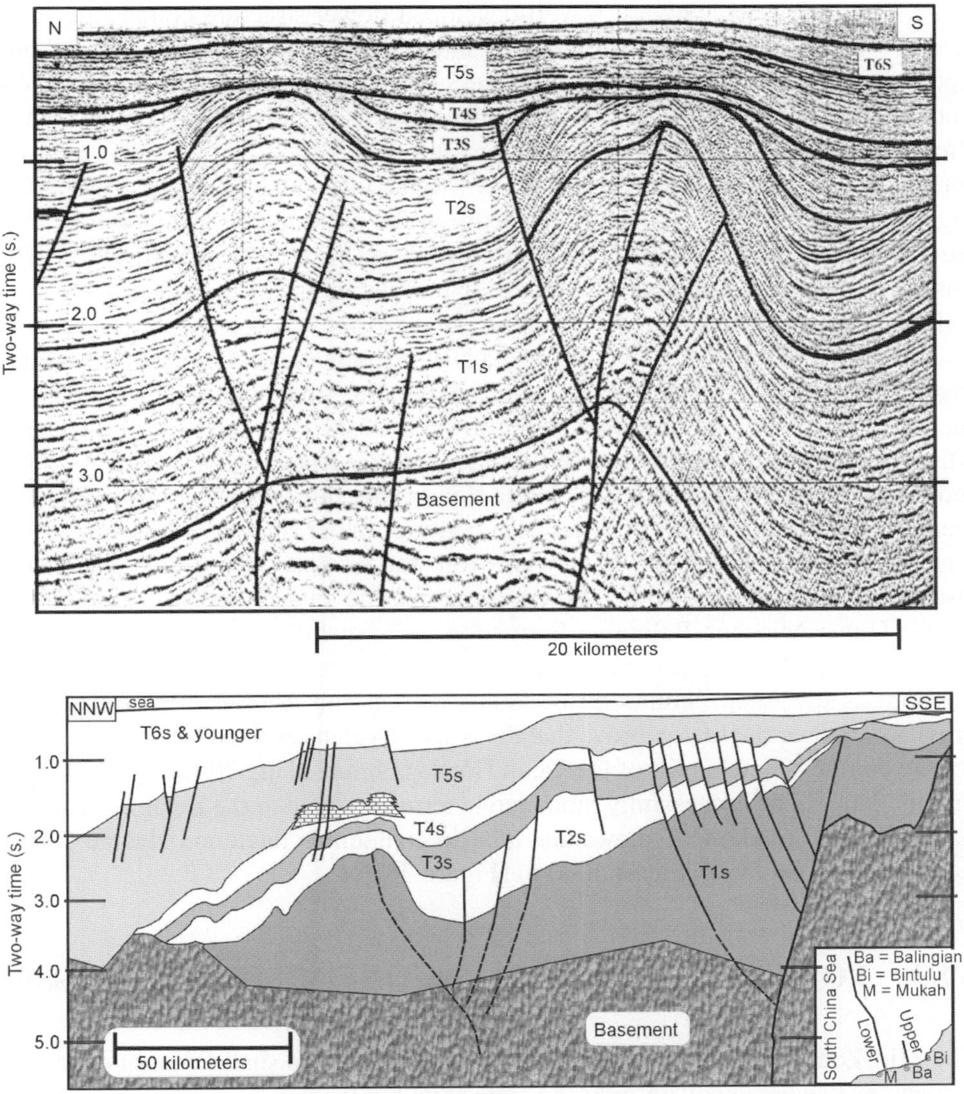

Figure 29. Stratigraphic sequences for the Miri Zone and offshore after Ismail Che Mat Zin and Tucker, 1999). Top: Unconformable sequence boundaries for base T3s, T4s and T5s. The sequence bases are characterized by erosional unconformities. Lower: Sequences T1s to T4s were deposited in restricted grabens, tilted seawards only after T4s by orogenic uplift of onland Sarawak in Pliocene (T5s) time.

The unconformity is also spectacularly exposed on the eastern Tatau 'Horst', where the Upper Miocene or Pliocene Rangsi Conglomerate sits with strong angular unconformity upon steeply dipping and folded turbidites of the Eocene Bawang Member of the Belaga Formation (Rajang Group). A seismic section across the Tatau 'Horst' into the offshore region (Ismail Che Mat Zin, 2000) clearly demonstrates that the hiatus is confined to the interval late Lower Miocene (18 Ma) and late Middle Miocene (11 Ma). A well-developed flower structure bounds the Tatau 'Horst', which more properly should be interpreted as a tightly compressed and upthrust anticline resulting from transpression. Other similar upthrust basement anticlines occur offshore in the Balingian Province and onland in the Tinjar Province, characterized by tightly compressed and faulted anticlines (Figure 47) interspersed with broad simple synclines (Mohd Idrus and Redzuan, 1999). The regional unconformity is accordingly between the Setap Shale Formation (Lower Miocene, Figure 43) and the Balingian Formation (Middle–Upper Miocene). The basement Belaga Formation is exposed at the unconformity by upthrusting of a tightly compressed anticline, followed by rapid erosion.

Ismail Che Mat Zin (2000) published a NW–SE seismic section across the Tatau district to show that the undated Rangsi Conglomerate that outcrops on the edge of the Tatau Horst, may be extrapolated northwestwards to form the base of the Upper Miocene Balingian Formation. The seismic section of Ismail Che Mat Zin (2000) indicates that the Tatau Horst is a classic flower structure, resulting from compressional tectonics as a result of strike-slip faulting, rather than as a result of extensional tectonics (Figure 33). It appears from the seismic section that the flower structure faulting ceased at the unconformity and the Tatau Horst was draped over by the Upper Miocene Balingian Formation, subsequently eroded from the Tatau Horst outcrop. The compressional Tatau Horst structure continues northeastwards offshore as the Anau–Nyalau thrust fault (Mazlan and Abolins, 1999). The aero gravity data over the region prove that the Anau–Nyalau reverse fault is responsible for the compressional uplift of the Tatau Horst (Othman et al., 2001). The observed gravity data can be successfully modelled in terms of depth to the Belaga Formation basement. The model also shows that a basin containing Tatau and Nyalau formations attains a depth of at least 5000 m (Figure 33).

VI.3. POST-UPPER EOCENE UNCONFORMITY MOLASSE

The Miri Zone is predominantly overlain by non-marine to shallow marine molasse formations that are unmetamorphosed (except within localized shear zones) and considerably less deformed than the underlying Rajang Group inliers. Regions, which showed some shoaling of the seas in pre-unconformity times and developed carbonates, characteristically continued, intermittently, with later carbonate deposition.

VI.3.1. The Melinau Limestone Formation

This is a marine biohermal limestone, which attains a maximum thickness of 1500 m. Its type locality is at Gunung Melinau on the NW flank of Gunung Mulu, where the outcrop length is 37 km and width 8 km (Figure 31). The formation details vary regionally and detailed studies have been made of the highly fossiliferous limestone.

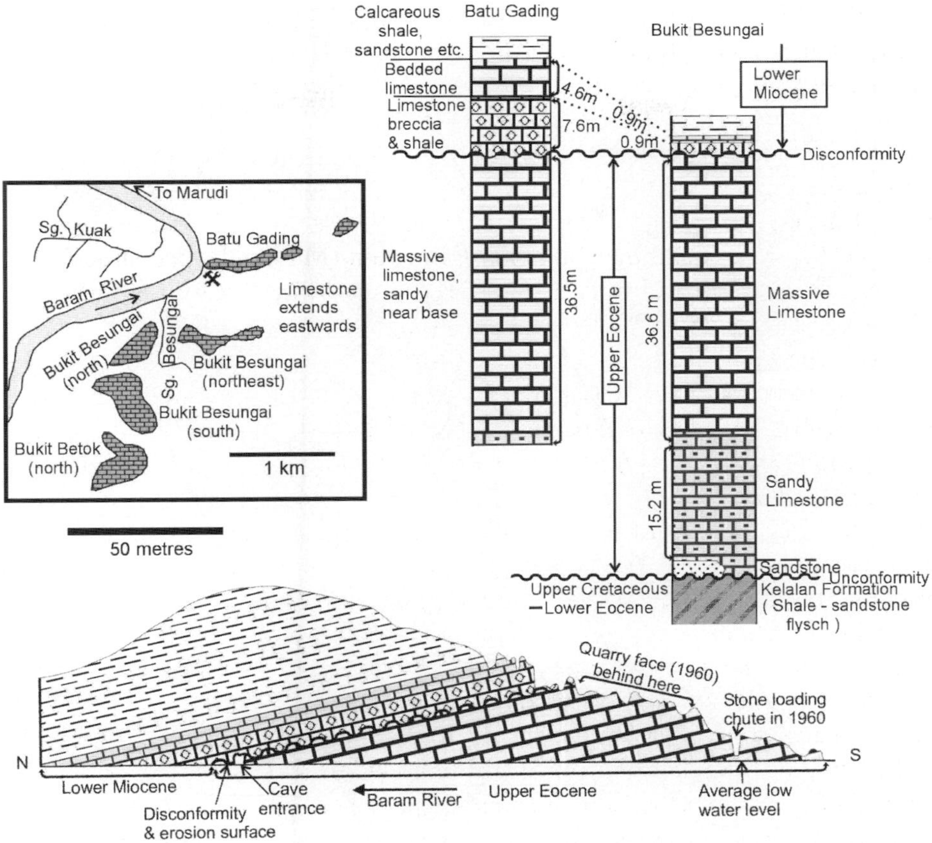

Figure 30. Stratigraphic sequence of the Batu Gading area adjacent to the Baram River (after Adams and Haak, 1962).

VI.3.1.1. Batu Gading and Bukit Besungai

The sequence, which overlies the strongly folded Kelalan Formation with angular unconformity, dips 12–15° towards the north (Figure 30). The Upper Eocene limestones are massive, mainly unbedded, of high purity, dark grey and highly fossiliferous. The principal organisms are larger Foraminifera, algae, fragmented corals, together with echinoid and bryozoan debris (Adams and Haak, 1962). The following fauna have been identified, of definite Upper Eocene age: *Pellatispira* spp. including *Pellatispira crassicolumnata* Umbgrove; *Discocyclina* sp.; *Aktinocyclina* spp.; *Nummulites* spp. including *Nummulites cf. semigloblus* (Doornink) and *Nummulites javanus* Verbeek and *Operculina* spp.

Calcareous algae occur profusely throughout the limestone, and commonly make up the bulk of the rock. They are mainly *Melobesieae* and the genus *Archaeolithothamnium* occurs in great profusion.

The overlying sequence is conformable, but there is a major hiatus; the Oligocene is totally absent! The disconformity is not always readily seen, except where the

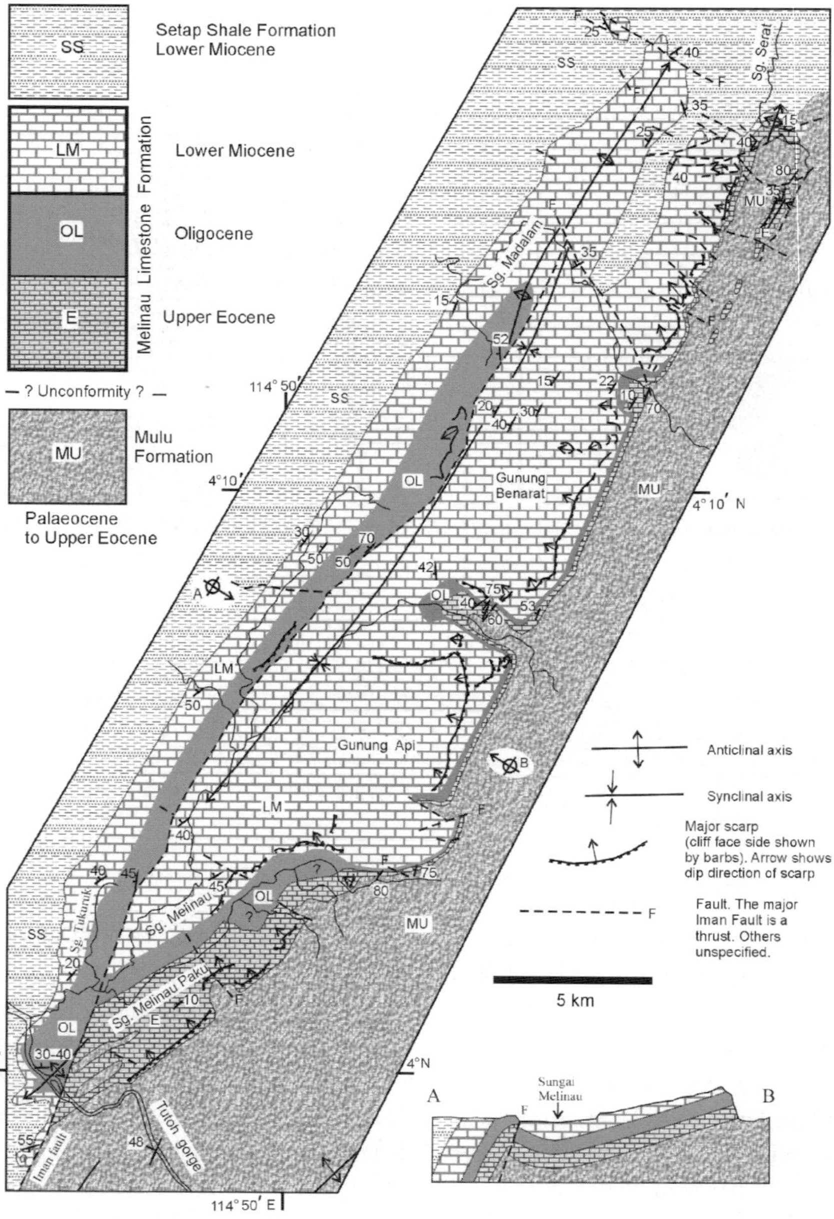

Figure 31. Geological map of the type locality of the Upper Eocene to Lower Miocene Melinau Limestone Formation. The Oligocene section is very thin on an outcrop map towards the northeast because it outcrops on a near vertical cliff face. The cross-section is diagrammatic (modified after Adams, 1965).

Eocene limestone is overlain by calcareous silty shales (Figure 30), immediately north of the cave entrance. The shales, which contain irregular clasts of Eocene limestone, have been deposited on an eroded surface of Eocene limestone. The calcareous shales contain a rich Lower Miocene fauna: *Globigerina dissimilis* Cushman & Bermudez var.; *Globigerina binaiensis* Koch; *Globigerinoides* spp.; *Globorotalia mayeri* Cushman & Ellisor and *Globoquadrina venezuelana* (Hedberg) var.

Shales are absent from the south side of the cave entrance. The lower part of the Miocene succession is of limestone breccia, which contains irregular blocks of darker Eocene limestone in a lighter matrix that contains abundant compound corals. The limestone breccia contains a definite fauna of larger Foraminifera: *Heterostegina* spp. such as *Heterostegina borneensis* Van der Vlerk; *Spiroclypeus* sp.; *Lepidocyclina* spp. of *Nephrolepidine* type; *Lepidocyclina (Eulepidina)* sp. and *Neoalveolina pygmaea* (Hanzawa) of rare occurrence only. Reworked Eocene specimens are found throughout the limestone breccia, and the fauna of the included blocks, of course, is entirely of Eocene age.

The upper sequence of the limestone is well stratified, comprising alternating fine, medium- and coarse-grained bioclastic algal-Foraminifera calcarenites. The fauna is mainly of algal, Foraminiferal and echinoidal debris. The following are typical, representing a Lower Miocene age: *Amphistegina* sp.; *Textularia* sp.; *Heterostegina* sp.; *Lepidocyclina* spp. of *Nephrolepidine* type; *Miogypsinoides* sp.; *Operculina* sp. and *Neoalveolina pygmaea* (Hanzawa).

The limestones are overlain by alternating calcareous sandstones (up to 17 cm thick) and sandy limestones. These beds are transitional into overlying shales. The following Lower Miocene fauna were identified: *Globigerina binaiensis* Koch; *Globigerina dissimilis* Cushman & Bermudez var.; *Globigerina* cf. *Ciperoensis* Bolli; *Globigerina* spp.; *Globigerinoides* spp.; *Globoquadrina venezuelana* (Hedberg) var. and *Globorotalia mayeri* Cushman & Ellisor.

VI.3.1.2. Melinau Limestone type locality

The limestone rests with apparent structural conformity upon Mulu Formation along the northwestern flank of the Mulu anticlinal inlier. It extends 37 km in a NE–SW direction with a maximum width of about 8 km (Figure 31). The formation is named from the Melinau River, a tributary of the Sungai Tutoh. The limestone sequence is about 2133-m thick and houses the world class Mulu Caves.

The limestone dips fairly steeply NW, averaging 50–80° near its base and 20–50° near its top. A syncline is present in the centre of the outcrop. A high-angled reverse fault, called variously the Melinau or Iman Fault, extends the whole length of the outcrop, with a maximum throw of 900 m. There are also a number of minor normal faults. The Melinau Limestone Formation is overlain by the Setap Shale Formation. The Melinau Limestone Formation has been extensively sampled along the river system and studied and described by Adams (1965). The limestone is massive throughout and dip and strike difficult to ascertain in the field. Its colour ranges from grey to blue-grey.

The combined Foraminifera, algal and sedimentological evidence supports the conclusion (Liechti et al., 1960; Adams, 1965) that the Melinau Limestone was laid down in shallow water some distance from a shore. It was not a reef but more like a carbonate platform. No true reef structures have been observed.

VI.3.1.2.1. ***Age and palaeontology*** Macrofossils are uncommon, though gastropods and bivalves have been locally recorded. Large algal growths (<1 m across) occur along the Tutoh River. The limestone is rich in larger Foraminifera, and an extensive sampling programme along the rivers and within the gorges has led to the detailed stratigraphy. The important genera and species identified are detailed in Figure 32.

It may be argued that the sudden disappearance of Eocene faunas indicates a hiatus or disconformity. However, if this were the case, one would expect Oligocene fauna to come in immediately. This is not the case. They come in gradually and there is no field evidence for a break in stratigraphy. The sharp faunal change within the Lower Miocene Aquitanian suggests that a disconformity exists, allowing the subdivision into a lower and upper Aquitanian (Figure 32).

VI.3.2. The Keramit and Selidong Limestones

There are two important limestone localities in the upper Limbang area; Keramit (22.5 km N of the Melinau Limestone-type locality, 40 km south of Limbang) and the Selidong Limestone (17.7 km NE of Melinau and 51.5 km south of Limbang). They are exposed along the banks of the Limbang River, and described by Adams and Wilford (1972). As a guide to interpreting the older literature on Sarawak stratigraphy, the Asian Tertiary letter classification (Adams and Wilford, 1972) is now accepted as (see Figure 22):

Tg-h	=	Upper Pliocene and younger
Tf_3	=	Upper Miocene-Lower Pliocene (Tortonian, Messinian, Zanclian)
Tf_2	=	Middle Miocene (Serravallian)
Tf_1	=	Lower-Middle Miocene (Upper Burdigaloian + Langhian)
Te_5	=	Lower Miocene (Aquitanian and Burdigalian)
$Te_{1\text{–}4}$	=	Upper Oligocene (Chattian)
Td	=	Lower Oligocene (Rupelian)
Tc	=	Lower Oligocene (Lattorfian)
Tb	=	Upper Eocene (Priabonian)
Ta	=	Palaeocene + Lower and Middle Eocene

The Keramit and Selidong limestones are not biohermal and did not grow on topographic highs. They may have been deposited in fairly deep water by turbidity currents as calcarenite beds (alodapic limestone). The Te beds are flysch, deposited in troughs on the sea-floor, derived from erosion of the main Melinau Limestone massif. This interpretation is supported by the high percentage of pelagic Foraminifera and reworked microfossils, and the presence of disconformities within the sequence (Adams and Wilford, 1972).

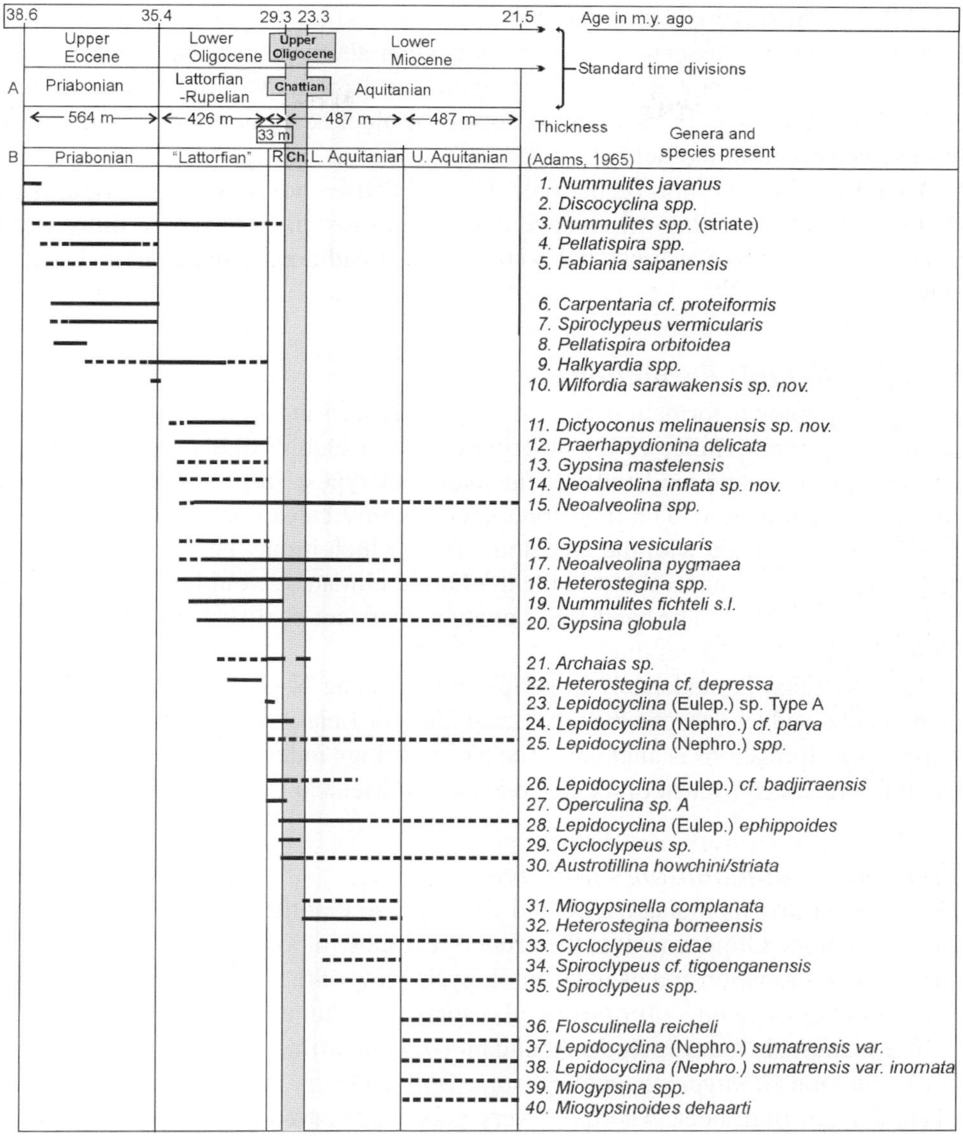

Figure 32. Range of important Foraminifera in the Melinau Limestone Formation (from Adams, 1965). R = Rupelian, Ch = Chattian. A = standard time divisions. B = time divisions from Adams (1965).

VI.3.2.1. Keramit Limestone

Lower Te (Upper Oligocene) brown and grey shales and calcarenites have a basal conglomerate that overlies, with angular unconformity, Tb and/or Tcd (Lower Oligocene) thinly bedded calcarenites and calcilutites. A disconformity separates them from underlying Tab (Palaeocene to Eocene) marls and shales.

VI.3.2.2. Selidong Limestone

Lower Te (Upper Oligocene) calcarenites contain abundant reworked older microfossils. They are separated by a stratigraphic gap from underlying Tab thinly bedded calcilutites, which in turn overlie Eocene or older Mulu Formation. The contact has not been seen in the field.

This limestone is not biohermal. The northern end of the Melinau Limestone type locality is known to have undergone erosion in lower Te time, and it may have served as the erosional source for the Selidong Limestone calcarenite turbidite.

VI.3.3. Kelabit Formation

This poorly known formation comprises mudstone, sandstone and thin lenses of impure limestone (Haile, 1962). It occurs in the Kelabit Highlands in the Upper Baram. Mudstone and sandstone are the main rock types. There are also lignite beds and rare conglomerate. The mudstone is commonly calcareous. A number of salt springs are associated with the mudstone, from which it may be suspected that salt beds occur beneath surface, but in the hot humid climate it would be impossible for salt outcrops to exist. The sandstones contain abundant coalified plant remains and lignite lenses up to 10 cm thick.

The formation is moderately to steeply folded along N–S axes, but the folding is more regular and less chaotic than the Setap Shale or Kelalan formations. The relationship to other formations is unknown. The palaeontology indicates an Oligocene age, so that it is a facies equivalent of the Setap Shale and Melinau Limestone formations.

VI.3.3.1. Palaeontology and age

The collected fossils indicate a Tc and Te_{1-4} age, so that the formation extends over the whole of the Oligocene. It is therefore of extreme importance that more study be made of this formation and its palaeontology, in view of the fact that there commonly exists an Oligocene unconformity, well displayed within the Melinau Limestone.

The planktonic fauna include: Abundant *Globoquadrina venezuelana* (Hedberg); fairly common *Globigerina ampliapertura* Bolli and very rare *Globortalia centralis* Cushman and Bermudez. There is a very poorly preserved benthonic fauna, which gives no clues as to the formation age.

An abundant fauna was identified by Sarawak Shell Oilfields Ltd.: *Ammobaculites, Ammodisus, Arenobulimina, Bathysyphon, Bolivina, Bulimina, Cassidulina, Chilostomelia, Cibicides, Clavulina, Cristellaria, Cyclammina, Ellipsogiandulina, Epinoides, Gaudryina, Globigerina, Globoquadrina, Glomospira, Gyroidina, Haplophrogmoides, Harmosina, Lagenammina, Nodosaria, Nonion, Nummilites, Operculina, Pleurostomella, Pullenia, Reophax, Rotalia, Textularia, Trochommina, Trochomminoides, Uvigeria, Verneulina* and *Vulvulina.*

The ages determined by the above fauna ranges from Tc-Td to Te_{1-4}, extending over the whole of the Oligocene.

VI.4. BUKIT MERSING

The contact between the Metah Member of the Belaga Formation (Sibu Zone) and the Tatau Formation of the Miri Zone has been named the Bukit Mersing Line by Hutchison (1975), presumed to be a narrow zone characterized by ophiolite and related rocks. However, doubt must now be cast on this interpretation. At Bukit Mersing (2° 30′ N; 113° 05′ E), the Belaga Formation contains basalt, tuff and agglomerate. The pyroclastic rocks are commonly silicified and epidotized (Kirk, 1957). The hill feature (Tau, Bukit Mersing and Bukit Dada) extends 22 km E–W and 6 km N–S. The volcanic rocks are infolded with Rajang Group strata. Spilite and pillow basalts are common. The volcanic rocks are overlain on the north by Tatau and Buan formations (Kirk, 1957). The volcanic suite is poorly known and only one chemical analysis has been published (Kirk, 1957, 1968). At 46.57 wt% SiO_2, the K_2O wt% of 2.82 is far too high for this rock to represent ophiolite; rather it plots within the shoshonite suite field (Figure 32). Such an analysis is at conflict with the thin section (S 3474) description of Kirk (1968), who describes it as: "microgranular and non-porphyritic dark grey andesite consisting of laths of albite-oligoclase and grains of augite in a fine grained matrix of augite, albite-oligoclase, abundant magnetite granules and much bright green chlorite".

Such a description is at conflict with the analysis and suggests that a potassic phase may have been overlooked, or the published analysis is wrong. A further and powerful reason for reassessment of the Bukit Mersing Line is that Richard Mani Banda (pers. comm.) has discovered that the so-called cherts mapped along this line to the east are not cherts but rhyolite, perhaps related to the Bukit–Arip suite.

VI.5. TATAU FORMATION (UPPER EOCENE–LOWER OLIGOCENE) [I]*

This formation is considered to rest unconformably on the Bawang Member of the Belaga Formation, based on the geomorphological expression and seismic section of Ismail Che Mat (2000), but an actual outcrop of the unconformity has not been seen. The most complete stratal sequence is exposed in the tributaries of the Bawang and Arip rivers on the northern flank of the eastwards-plunging Arip–Pelungau anticline (Figure 25).

The thickness is less than 3000 m, including about 450 m of volcanic rocks, which divide the formation into two different parts.

* Roman numeral in brackets [ ] indicate equivalent Shell cycles of the Baram Delta oil province (Ho, 1978).

VI.5.1. Lower sequence

It is mainly carbonaceous shale and siltstone, including beds of fine to medium grained quartz sandstone commonly about 15 m thick, rare conglomerate beds, fossiliferous limestone lenses up to 9 m thick and associated marl. The conglomerates consist of pebbles and cobbles of greywacke, chert, vein quartz and shale. The shales are intensely folded and slaty cleavage well developed. Adjacent to the Belaga Formation outcrops, many of the Tatau Formation shales cannot be distinguished from those of the Belaga Formation. Farther from the Belaga Formation outcrop, the folding is less severe and slaty cleavage of Tatau Formation shales becomes absent (Wolfenden, 1960).

VI.5.2. Volcanic sequence

The volcanic sequence occurs on the flank of the east-plunging Arip anticline (Figure 25), reaching a maximum thickness of 450 m, but thinning to the west and SE. The rocks are mainly rhyolite, including welded tuff and lava, with two flows of andesite locally at the base. A characteristic feature of the rhyolites is the intense hydrothermal alteration (Wolfenden, 1960). Chalcedony and agate commonly occur with quartz as veinlets. Chemical analyses of typical volcanic rocks are given in Table 10.

The volcanic rocks plot on a Peccerillo and Taylor-type diagram (Figure 34) as belonging predominantly to the High-K calc-alkaline series, as also does the Bukit Piring granodiorite, which is probably from the same magma source. The predominantly silicic pyroclastics have been equated with other Eocene volcanics in Kalimantan north of the Schwaner Mountains; interpreted as post-subduction and related to basin extensional processes (Hutchison, 1996a). The Arip and Bukit Piring igneous rocks belong to this category. They are an integral part of the Tatau Formation and accordingly do not represent a subduction-related volcanic arc. The tectonic cartoon of Tjia et al. (1987), in which they relate the Arip explosive volcanism to a forearc basin in which Tatau and "Bawang Beds" were deposited, and a back arc basin in which the "Belaga beds" were deposited, cannot be entertained within presently understood plate tectonic concepts. It is remarkable that their cartoon has been reproduced in Tjia (1999).

VI.5.3. Bukit Piring granite–granodiorite

This stock forms an east-trending ridge (Figure 25). It intrudes steeply dipping Upper Eocene sandstone and shale of the Tatau Formation and caused contact metamorphism to biotite hornfels (Wolfenden, 1960). The eastern part is of granophyre, the western part of granite and granodiorite. The chemistry (Table 10; Figure 34) suggests that the intrusive rocks are related to the Sungai Arip volcanics and the granophyritic texture indicates a sub-volcanic emplacement.

VI.5.4. Post-volcanic sequence

Resting upon the volcanic rocks in the Arip Valley are about 240 m of conglomerate and coarse sandstone. The conglomerate is predominantly of clasts of chert and vein quartz; but volcanic debris is rare. Above the conglomerate and sandstone is a thick

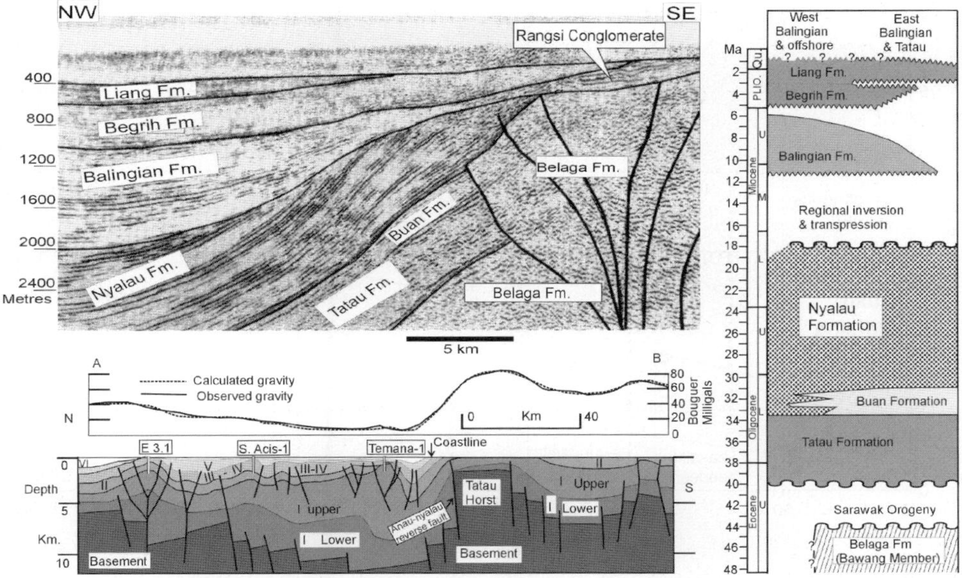

Figure 33. Seismic section from Balingian to Tatau (after Ismail Che Mat Zin, 2000). Gravity measurements and model from Othman et al. (2001). Gravity section as in Figures 25 and 27 but extended southwards beyond the compressional Tatau Horst.

Table 10. Chemical analyses of rocks associated with the Tatau Formation

wt %	1	2	3	4	5
SiO_2	46.57	54.68	74.30	77.24	60.22
TiO_2	3.03	1.37	0.24	0.21	1.17
Al_2O_3	13.08	15.48	12.66	11.19	14.40
Fe_2O_3	7.90	5.92	0.31	1.52	1.07
FeO	8.37	2.09	3.37	Tr.	8.11
MnO	0.03	0.17	Tr.	0.00	0.15
MgO	5.07	5.02	0.39	0.18	1.74
CaO	8.33	7.74	Tr.	0.26	4.80
Na_2O	2.68	3.60	4.11	2.33	3.56
K_2O	2.82	1.67	3.14	5.35	2.64
P_2O_5	0.61	0.50	0.24	0.09	0.53
H_2O+	1.54	1.51	0.29	1.00	0.94
H_2O-	0.29	0.64	0.84	0.54	0.20
CO_2	0.00	0.00	0.00	Tr.	0.48
Total	100.32	100.39	99.89	99.91	100.01

1 = Upper Eocene pillow basalt (S3474), Bukit Mersing (Kirk, 1957).
2 = andesite (S5682), Tabau Hill, Arip Valley (Wolfenden, 1960).
3 = alkali rhyolite (S5664), Muput Valley (Wolfenden, 1960).
4 = alkali rhyolite (S3461), Arip Valley (Wolfenden, 1960).
5 = hornblende biotite granodiorite (S5690), Piring Hill (Wolfenden, 1960).
Tr. = Trace

sequence of fossiliferous limestone, shale, sandstone and marl. The calcareous facies of the Tatau Formation is closely similar to that of the conformably overlying Buan

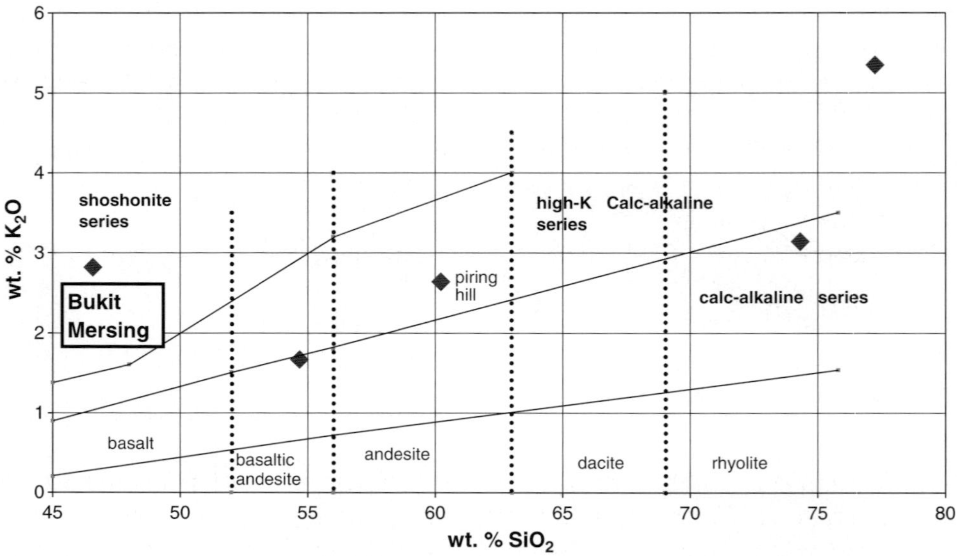

Figure 34. Tatau Formation igneous rocks, K_2O vs SiO_2.

Formation. Most of the formation near Tatau is of calcareous shale and marl. The Tatau Formation passes by transition into the largely argillaceous Buan Formation.

VI.5.5. Palaeontology and age

The limestones of the Tatau Formation contain a rich Foraminifera fauna. Most are of Upper Eocene age, but east of Tatau, the fossils indicate that the upper part of the sequence is Lower Oligocene. The following is a selection of the larger Foraminifera of *Upper Eocene* age found in the Tatau Formation limestones (Wolfenden, 1960): *Aktinocyclina alticostata* (Nuttall); *Amphistegina* sp.; *Assilina* sp.; *Biplanispira* sp.; *Discocyclina* sp.; *Eorupertia* sp.; *Fabiania* cf. *Saipanensis* Cole; *Gypsina globulus* (Reuss); *Halkyardia* cf. *minima* (Liebus); *Heterostegina* sp.; *Nummulites* cf. *javanus* Verbeek; *Nummulites kelatensis* [Carter(Douvillé)]; *Nummulites subglobula* (Doornink); *Operculina* sp.; *Pellatispira crassicolumnata* (Umbgrove); *Pellatispira* cf. *glabra* Umbgrove; *Pellatispira madaraszi* von Hantken var. *provalei* (Yabe) and *Spiroclypeus* cf. *vermicularis* Tan Sin Hock.

Oligocene larger Foraminifera found in the Tatau Formation limestones are: *Heterostegina* sp.; *Neoalveolina* sp.; *Nummulites absurda* (Doornink) and *Operculina* sp. Pelagic Foraminifera of *Upper Eocene* in Tatau Formation marls are: *Globigerina* cf. *dissimilis* Cushman and Bermudez; *Globigerinella micra* (Cole); *Globorotalia centralis* Cushman and Bermudez; *Globorotalia cerro-azulensis* Cole and *Hantkenina* cf. *alabamensis* Cushman.

Pelagic Foraminifera of Oligocene age in Tatau Formation marls are: *Globigerina* cf. *dissimilis* Cushman and Bermudez and *Globigerina* cf. *increbescens* Brady. Marls in the Penipah area contain *Globigerina* cf. *dissimilis* Cushman and Bermudez and *Globigerina* cf. *increbescens* Brady: an association of *Oligocene* age (Wolfenden, 1960). Kirk (1957) tabulated the following Foraminifera from which he deduced that the Tatau Formationm straddles the Eocene–Oligocene boundary: *Ammobaculites* sp., *Ammodiscus* sp., *Bathysiphon* sp., *Cyclammina* sp., *Gaudryina* sp., *Glomospira* sp., *Haplophragmoides* sp., *Sigmoilina* sp., *Trochammina* sp., *Trochamminoides* sp. and *Verneuilina* sp.

The Tatau Formation fossils indicate an Upper Eocene to Lower Oligocene age.

VI.6. BUAN FORMATION (UPPER OLIGOCENE TO LOWER MIOCENE) [I/II]

This predominantly shaly formation contains thin beds of siltstone and fine to medium grained sandstone. It is conformable on the underlying Tatau Formation and with the overlying Nyalau Formation and, although its Foraminifera fauna are not age diagnostic, it is confidently concluded to be of Oligocene age (Wolfenden, 1960). Its lower boundary marks a transition from the more calcareous Tatau Formation to the predominantly argillaceous Buan Formation. Its upper boundary marks a rapid transition to the more arenaceous Nyalau Formation. The shales are calcareous and contain sideritic nodules. In the Mount Buan region it attains a thickness of around 600 m.

VI.7. SHELL PALAEOFACIES MAPS, MUKAH–MIRI

In November 1986, Dr. Hageman of Shell, Lutong, completed a very important set of palaeogeographic maps of the region of northwest Sarawak, from Mukah to Miri and continuing offshore to include data from all wells drilled . A preliminary summary was earlier published (Hageman, 1987). The maps were based mainly on interpretation of the inferred life environment of the identified Foraminifera, together with rock sample analysis. The interpreted environments are:

Non-marine environments
Coastal (COL) and lower coastal plain (LCP)
Fluvial and deltaic (mixed river & marine)
Coastal fluviomarine (COF); fluviomarine inner neritic (FIN)
Fluviomarine middle neritic (FMN)
Holomarine
Holomarine inner neritic (HIN); Holomarine middle neritic (HMN)
Holomarine outer neritic (HON); Bathyal (BAT)

The first three palaeofacies maps are shown in Figure 35. Since the maps cover the area both offshore and onshore, there is strong relevance to the outcropping on land formations, which helps in their interpretation. The selected time frame for the eleven maps is given at the top of Figure 38.

VI.7.1. Upper Oligocene, 27 Ma (Chattian) (Cycle I)

The palaeogeographical depositional zones are at right angles to the present-day coastline (Figure 35K). A non-marine environment (land) extends north-eastwards as far as midway between Bintulu and Batu Niah. North-eastwards from this, the sea progressively deepened towards Miri and Brunei.

The region designated from well data to be 'lower coastal plain' and 'coastal' is represented by extensive outcrops of Nyalau Formation. To the west of Bintulu, the formation contains significant coal seams. Eastwards from Bintulu, a coastal environment is indicated by well-developed ripple marks, indicating that the tidal currents moved north-easterly and south-westerly with incoming and outgoing tides. Land lay to the southwest, sea to the northeast.

Continuing northeastwards beyond the palaeo-coast, the holomarine is represented by the Setap Shale Formation, indicating shallow inner neritic near Batu Niah and deepening northeastwards. It is likely that in the Upper Baram and Limbang areas, the Setap Shale Formation may be turbiditic, although this quality has not been described, but deeper water deposition has been suggested. The wholly marine Melinau Limestone-type locality was deposited continuously throughout the Oligocene, but elsewhere there is a widespread hiatus, indicating uplift and non-deposition, for example at Batu Gading and Upper Limbang (Figure 22).

Figure 35K indicates that the Upper Oligocene strata from beneath the Baram Delta were outer neritic and in the upper Baram region may be verging on bathyal.

A coastal fluviomarine system is shown inland SE of Bintulu. This is more than likely since the Nyalau Formation is very sandy and it is likely that a significant part of it is fluvial and involved in the northeastwards building of a delta system, as shown in Figure 35K.

VI.7.2. Lower Miocene (Aquitanian), Cycle I/II, 23 Ma

The palaeogography continues with little change (Figure 35J). Non-marine sedimentation on the west progressively gives way northeastwards to marine conditions eastwards, becoming outer neritic towards Miri. The on land stratigraphy is Nyalau Formation giving way eastwards to Setap Shale Formation. An important carbonate (the Subis Limestone) begins its growth at Batu Niah in inner neritic conditions at the eastern limit of Nyalau Formation. The early Lower Miocene carbonate platform extends offshore north-westwards seawards from Gunung Subis (Agostinelli et al., 1990). Westwards from Balingian there existed land devoid of sedimentation except within deep and actively growing half grabens. All wells west of Balingian

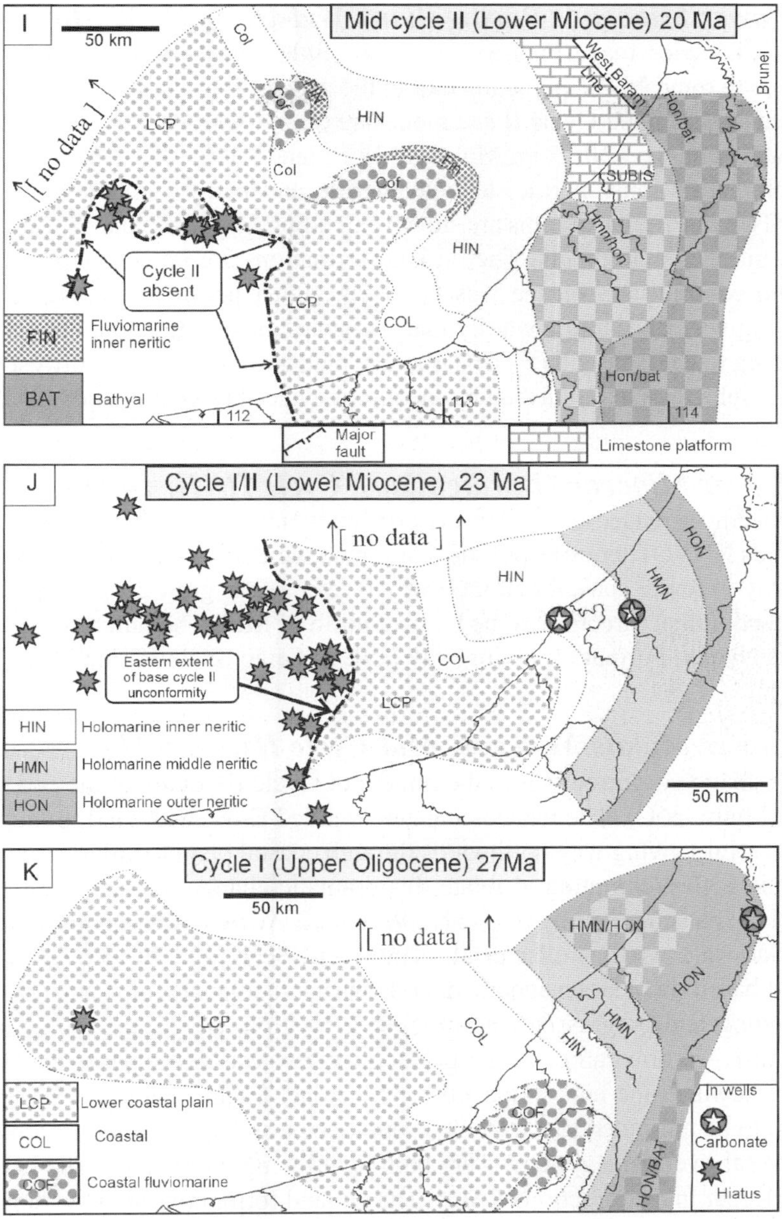

Figure 35. Palaeofacies distribution (part 1) in northeast Sarawak. Based on palaeontological well data. Compiled by Hageman (1985; unpublished). Condensed summary published by Mazlan (1999). The FIN category identification in [I] is uncertain.

encountered no strata of this age, and the prominent line showing the eastern extent of the base cycle II unconformity is well displayed. West of this line occur land surfaces known as the Penian and Sirik highs.

VI.7.3. Lower Miocene (Burdigalian) Mid-Cycle II, 20 Ma

The Shell palaeofacies map (Figure 35I) shows little change. The environmental zones continue to trend NNW, perpendicular to the present-day coastline. Land lay on the west (Penian High) and Cycle II sediments are absent. The environments of deposition gradually change ENE from Mukah to Miri: land giving way to lower coastal plain, to Holomarine inner neritic, to outer neritic and bathyal under the Miri to Brunei area. Two river delta systems are shown extending ENE from the palaeo-coastline into the marine environment. The on land formations are known as the Nyalau (ranging from coastal plain on the west to marine inner neritic in the east around Bintulu), giving way ENE to the wholly marine Setap Shale Formation, deposited in water deepening towards the ENE. The carbonate platform extending northwestwards from Gunung Subis was extinguished by the top of cycle II (Agostinelli et al., 1990).

VI.7.4. Lower Miocene (Burdigalian), Cycle II/III, 18 Ma

The same system of palaeo-environments continues unchanged (Figure 36H). The coastline trends NNW. There was non-deposition on the Penian High and westwards. Land gave way to lower coastal plain, on which river deltas grew ENE into the sea, which deepened in that direction, to be bathyal north of Miri. The outcropping strata are known as Nyalau Formation giving way ENE to Setap Shale Formation.

VI.7.5. Lower/Middle Miocene, Mid-Cycle III, 16.5 Ma

A very major change occurred before the middle of Cycle III, at about the end of the Burdigalian (Figure 36G). The palaeo-coastline trended E–W, with land in the south (mainland Sarawak) giving way northwards through coastal plain to holomarine inner neritic conditions. It is tempting to relate this dramatic change in orientation of the coastline and the palaeogography from a NNW to an E–W trend to the anti-clockwise rotation of Sarawak, which Fuller et al. (1991) and Schmidtke et al. (1990) have demonstrated based on palaeomagnetic data. However, the palaeomagnetic data suggest a continuing and progressive anti-clockwise rotation. By contrast, the change from NNW to E–W orientation of the palaeogography was geologically sudden at about 17 Ma, towards the end of the Lower Miocene. Such a geologically sudden change needs to be attributed to a tectonic event that caused uplift of the landmass of Sarawak. Several river delta systems were built up growing northwards. The sea continued to be deeper in the direction of Miri. The on land formations are now different. The major coastal to inner neritic delta, which was established NE of Batu Niah, represents the Lambir Formation, the base of which is spectacularly sharp against the underlying Setap Shale Formation. West of Batu Niah, there are no on land equivalents, for the strata were deposited off to the north. There is a major unconformity in the on land Balingian area. This was a time of strong inversion (uplift) to the south. The cause of the change of coastline from NNW (Figure 36H) to E–W (Figure 36G) may safely be attributed to uplift of the region to the south (the landmass of Sarawak), causing newly established rivers to flow directly northwards towards the sea.

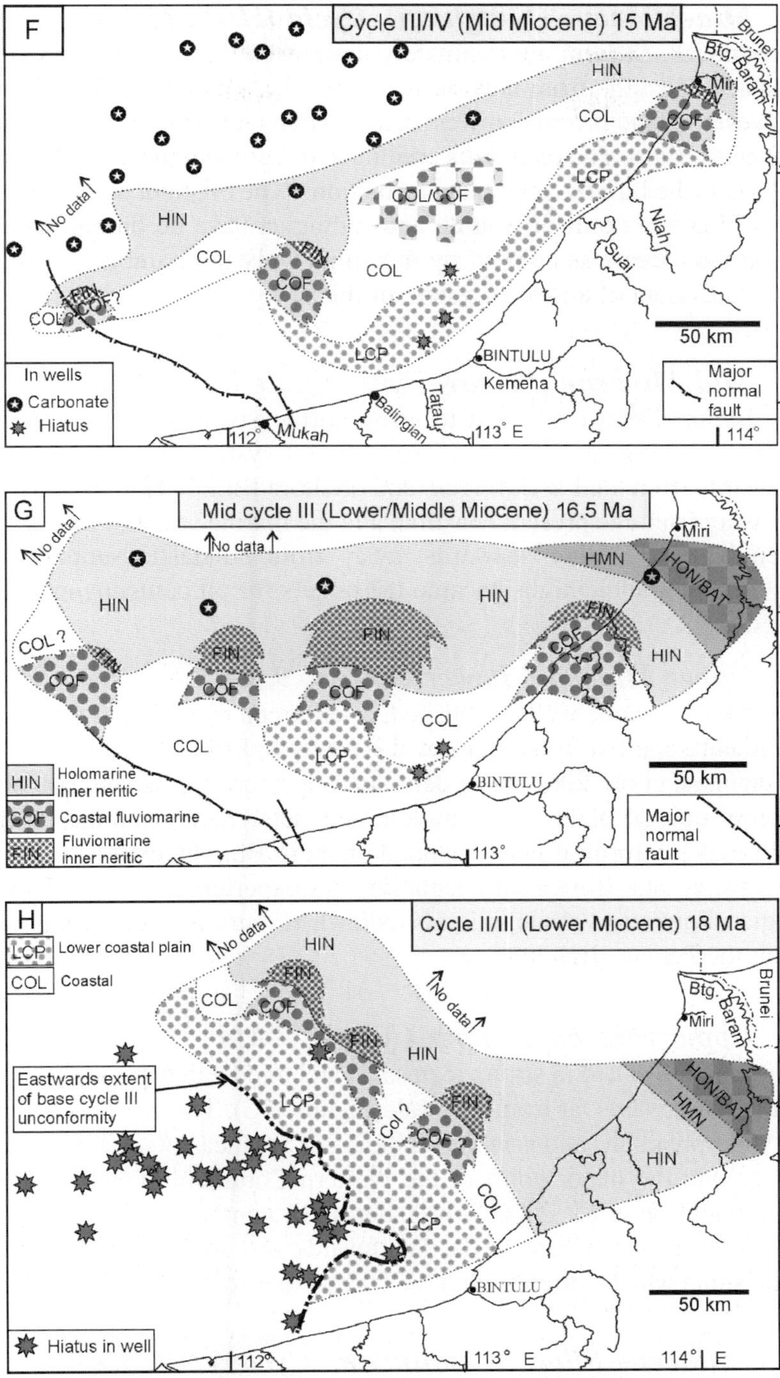

Figure 36. Palaeofacies distribution (part 2) in northeast Sarawak. Based on palaeontological well data. Compiled by Dr. Hageman of shell (unpublished). Condensed summary published by Mazlan (1999).

VI.7.6. Mid-Miocene (Langhian), Cycle III/IV, 15 Ma

Uplift of Sarawak, to form approximately the present land shape, was completed and the palaeo-coastline from now on paralleled the present-day coastline (Figure 36,F). However, a land protrusion west of Balingian and north of Mukah continued to exist (Penian High). The river delta continued to build out towards Miri, with outcrops on land as the Lambir and Tukau formations. The offshore formations beneath the South China Sea north of Bintulu and Balingian have no on land equivalents, since the on land area was uplifted by this time. Only the Lambir Hills represents an on land equivalent of a fluvial system of this age.

VI.7.7. Mid-Miocene (Serravallian), Cycle IV, 13 Ma

The early Baram Delta continued to grow northeastwards in the Lambir Hills, with water deepening towards Miri. New fluvial systems began from the coast northwestwards from land at Balingian and offshore Bintulu (Figure 37E). The sea deepened away from the present coastline and the marine strata of this age have no on-land equivalent. Farther seawards, away from siliclastic supply, carbonate build-ups on topographic highs became the targets for oil company exploration.

VI.7.8. Upper Miocene, Lower Cycle V, 9 Ma

The Baram Delta became well established, re-enforced by an impressive amount of sandy sediment supplied by erosion of the newly uplifting Western Cordillera of Sabah (Hutchison et al., 2000). The delta actively grew northwestwards in a short distance from coastal plain into a middle and outer neritic marine environment (Figure 37D). A platform of land extended offshore Bintulu, experiencing no sedimentation, but coastal fluvial sedimentation was experienced around Balingian as the Balingian Formation. A delta is also shown (Figure 37D) growing northeastwards from the Penian High land.

VI.7.9. Upper Miocene, Upper Cycle V, 7 Ma

A major inversion resulted in strong regression and middle to outer neritic conditions retreated northwestwards far from Sarawak (Figure 37C). This regression can be correlated to widespread inversions in the oil basins of Southeast Asia, e.g. Malay Basin, West Natuna and the major uplift of the Western Cordillera of Sabah (The Sabah Orogeny) (Hutchison et al., 2000). The latter uplift continued to result in excellent sediment supply to the Baram Delta. Southwards and southwestwards from Miri, there are no strata which correlate with the far offshore strata of the South China Sea.

VI.7.10. Miocene/Pliocene Boundary, Cycle V/VI, 5.5 Ma

The major unconformity continued, and marine conditions retreated far offshore Sarawak. This prominent sea level fall is illustrated in Figure 38(B, top).

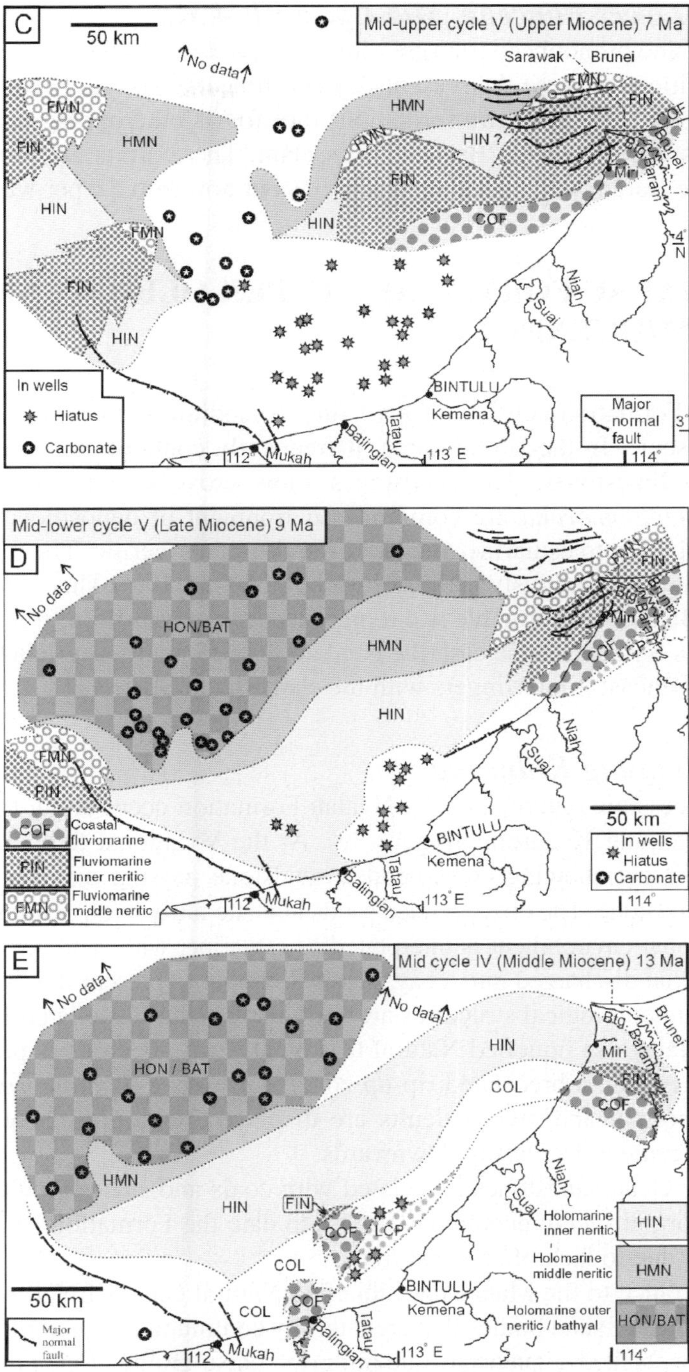

Figure 37. Palaeofacies distribution (part 3) in northeast Sarawak. Based on palaeontological well data. Compiled by Dr. Hageman of Shell (unpublished). Condensed summary published by Mazlan (1999).

VI.7.11. Lower Pliocene, Mid-Cycle VI, 4 Ma

A very rapid and pronounced transgression occurred that brought inner neritic holomarine conditions close to the present-day coastline of Sarawak (Figure 38A). The Begrih, followed by the Liang formations, are found outcropping near Balingian and northeastwards towards Brunei Darussalam. These strata are more extensive offshore. The Baram Delta continued to grow, but now into deeper water.

VI.8. NYALAU FORMATION (UPPER OLIGOCENE TO LOWER MIOCENE) [I/II]

The Nyalau Formation consists of hard fine to medium grained sandstones alternating with shale. In the Bintulu area, it frequently contains coal beds and infrequently thin limestones. The sandstones show cross-bedding and are rippled. Burrows of *Ophiomorpha* are common. Environment of deposition ranged from lower coastal plain to estuarine, shallow littoral to inner neritic. The total thickness of the formation is variously estimated to be 5000–5500 m. The contact with the underlying Buan is conformable and represents a sharp transition from argillaceous to arenaceous Nyalau. To the east, the Nyalau is conformably overlain by the Sibuti and to the northeast it interfingers with the shaly Setap Shale.

VI.8.1. Tanjung Kedurong

The most accessible outcrops of the Nyalau Formation occur along the road from Bintulu to Tanjung Kedurong (Figure 39). At the Vocational School, there is an excellent road cut showing cross-bedded sandstone passing up into interbeds of sandstone–mudstone. The cross-bedded sands indicate fairly strong currents and tides. The depositional environment is interpreted as estuarine, with sand bars formed in a river system that discharged into a bay, possibly similar to present-day Brunei Bay.

Many sedimentological structures are well displayed at Pagoda Hill, at the junction to the Malaysian liquefied Natural Gas plant. Cross-bedding is prominent and at one place is seen a brecciated rip-up zone. Burrows of *Ophiomorpha* are common. Both normal and reverse faults are displayed. One prominent thrust fault shows displacement decreasing downwards.

The fairly clean sandstones associated with coals and shales would seem to be favourable for petroleum accumulation, but to date the Formation has proved barren, possibly due to lack of seal and traps.

At the entrance to the Malaysian Liquefied Natural Gas (MLNG) plant, there is a thick sequence of sand bodies that are laterally continuous but heterogeneous. The rocks are of a stacked sequence of medium- to coarse-grained sandstones, alternating with thin, dark-grey shales. The depositional environment is coastal, mainly tide-dominated.

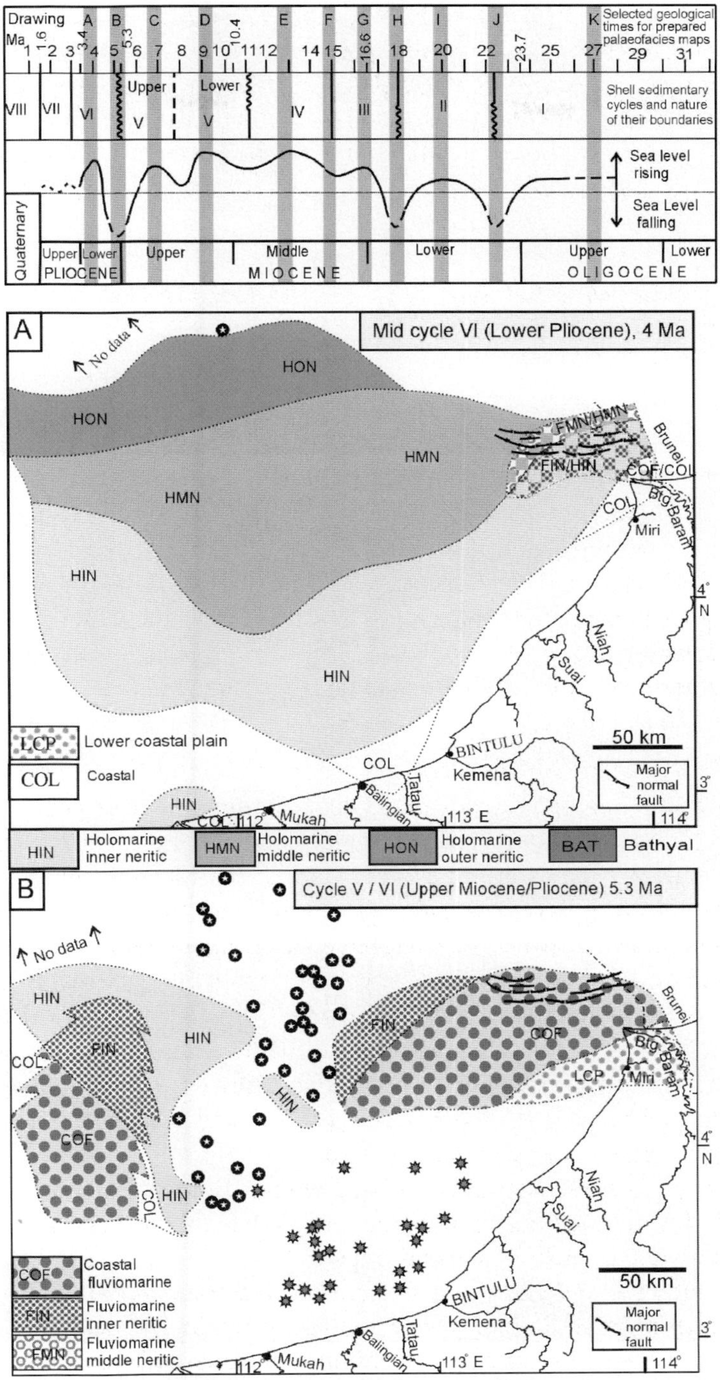

Figure 38. Palaeofacies distribution in northeast Sarawak (part 4). Based on palaeontological well data. Compiled by Dr. Hageman of Shell (unpublished). Condensed summary published by Mazlan (1999).

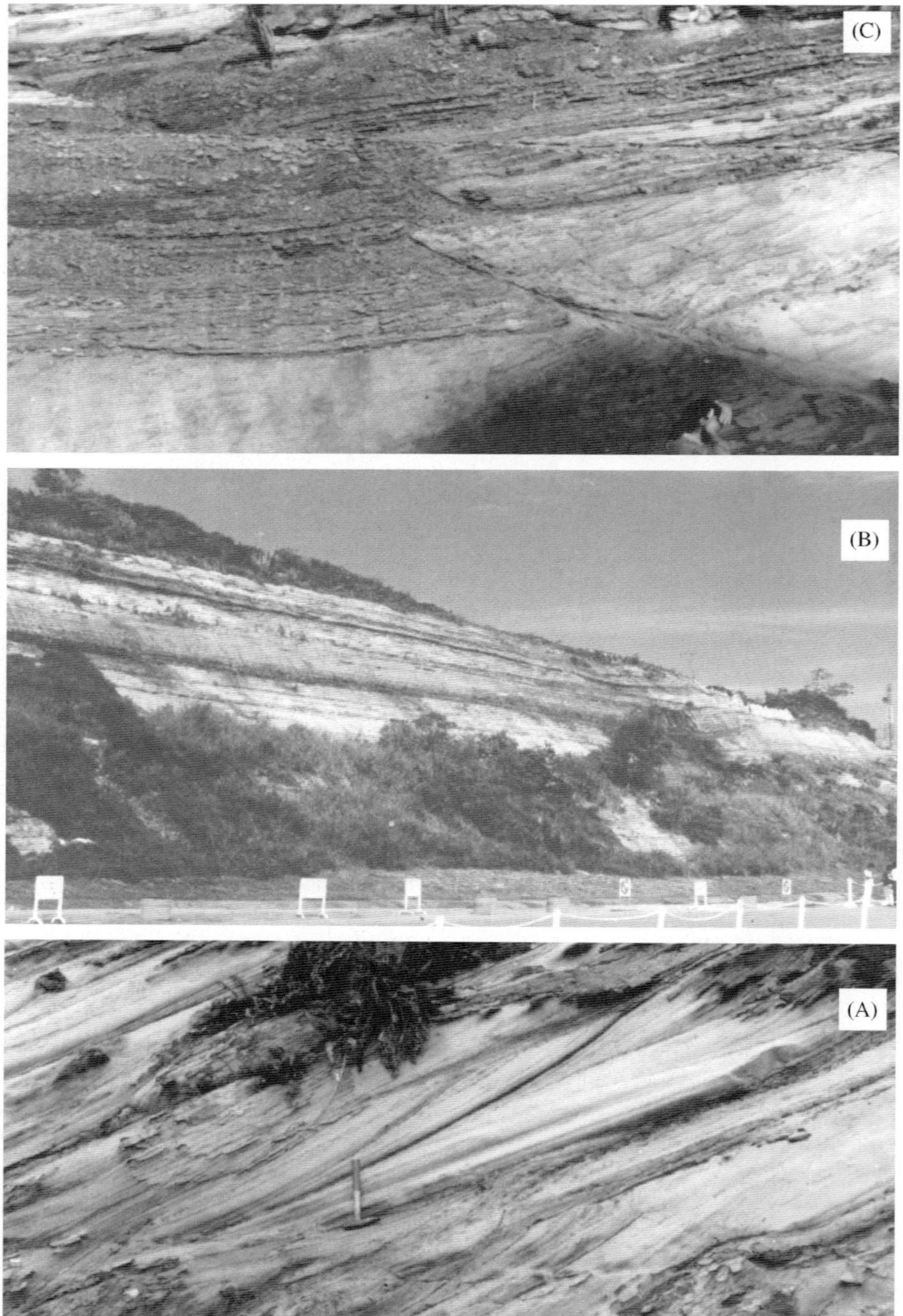

Figure 39. Outcrops of Nyalau Formation near Tanjung Kedurong, Bintulu district. (A) Thrust cross–bedding in tidal estuarine sequence. Thin sedimentary breccia beside hammer head, Locality as for C. (B) Stacked medium to coarse sandstones alternating with thin dark grey shales, entrance to MLNG plant. (C) thrust fault at Pagoda Hill, junction to the MLNG plant (After Haile and Ho, 1991).

Nyalau Formation outcrops continue intermittently as far east as Gunung Subis (Figure 40), east and northeast of which, the depositional environment gives way to holomarine conditions represented by the Setap Shale Formation.

VI.9. SUBIS LIMESTONE (TANGAP) FORMATION

The Subis Limestone is referred to as a member of the Upper Oligocene to Lower Miocene Tangap Formation by Haile (1962). It lies wholly within the Suai area. The Subis Limestone Member forms the impressive Gunung Subis of about 25 km^2 (Figures 40 and 41).

The Upper Oligocene (Te_{1-4}) to Lower Miocene (Te_5) Tangap Formation comprises calcareous shale, marl, greenish claystone and beds of purer limestone. Non-calcareous and sandy beds also occur. It interfingers with the Nyalau Formation. A general absence of sand and abundant pelagic Foraminifera indicate quiet outer neritic conditions, shallowing near the Subis Limestone Member. The thick limestone of Gunung Subis is better known, and commonly referred to as the Subis Limestone Formation.

The lower part of Gunung Subis is of algal limestone, passing upwards into coral-algal reef. The formation has been drilled into by Shell. Well Subis #2 penetrated dense impermeable limestone, interfingering with sandstones and marls of the same age, referred to the Nyalau Formation or the Tangap Formation. In Subis #2, the Subis Limestone Member is vertically 945 m thick, of which half is limestone, half of sandstone and shale assigned to the Nyalau or Tangap Formation (Figures 40 and 41).

VI.9.1. Fossils and age

The age is known to be Lower–Upper Oligocene to Lower Miocene based on a rich Foraminiferal fauna. A Te_5 (Lower Miocene: Burdigalian and Aquitanian) age is proved by the following Larger Foraminifera in the Tangap Formation (Haile, 1962): *Lepidocyclina sumatrensis* Brady; *Miogypsinoides lateralis* Hanzawa; *Miogypsinoides dehaarti* v.d. Vlerk; *Spiroclypeus orbitoideus* Douville; *Spiroclypeus tidunganensis* v.d. Vlerk and *Spiroclypeus higginsi* Cole.

The outcrop of the Subis Limestone Member of Gunung Subis is also known to be of the same Lower Miocene age from its studied abundant larger Foraminifera: *Lepidocyclina (Nephrolepidina) angulosa* Provale; *Lepidocyclina (Nephrolepidina) inflata* Provale; *Lepidocyclina (Nephrolepidina) sumatrensis* (Brady); *Lepidocyclina (Nephrolepidina)* (Scheffen); *Lepidocyclina (Nephrolepidina) flexuosa* (Rutten); *Lepidocyclina (Nephrolepidina)* (Michelotti); *Lepidocyclina (Eulepidina) dilatata* var. *tidunganensis* v.d. Vlerk; *Lepidocyclina (Eulepidina) ephippoides* Jones & Chapman; Miogypsina irregularis Michelotti; Miogypsina primitiva Tan Sin Hok; Miogypsinoides dehaarti v.d. Vlerk; *Miogypsinopides (Conomniogypsinoides) abundensis* Tobler; *Austrotrillina howchini* (Schlumberger); *Spiroclypeus* sp.; *Cycloclypeus* sp.; *Operculina* sp. and *Amphistegina* sp.

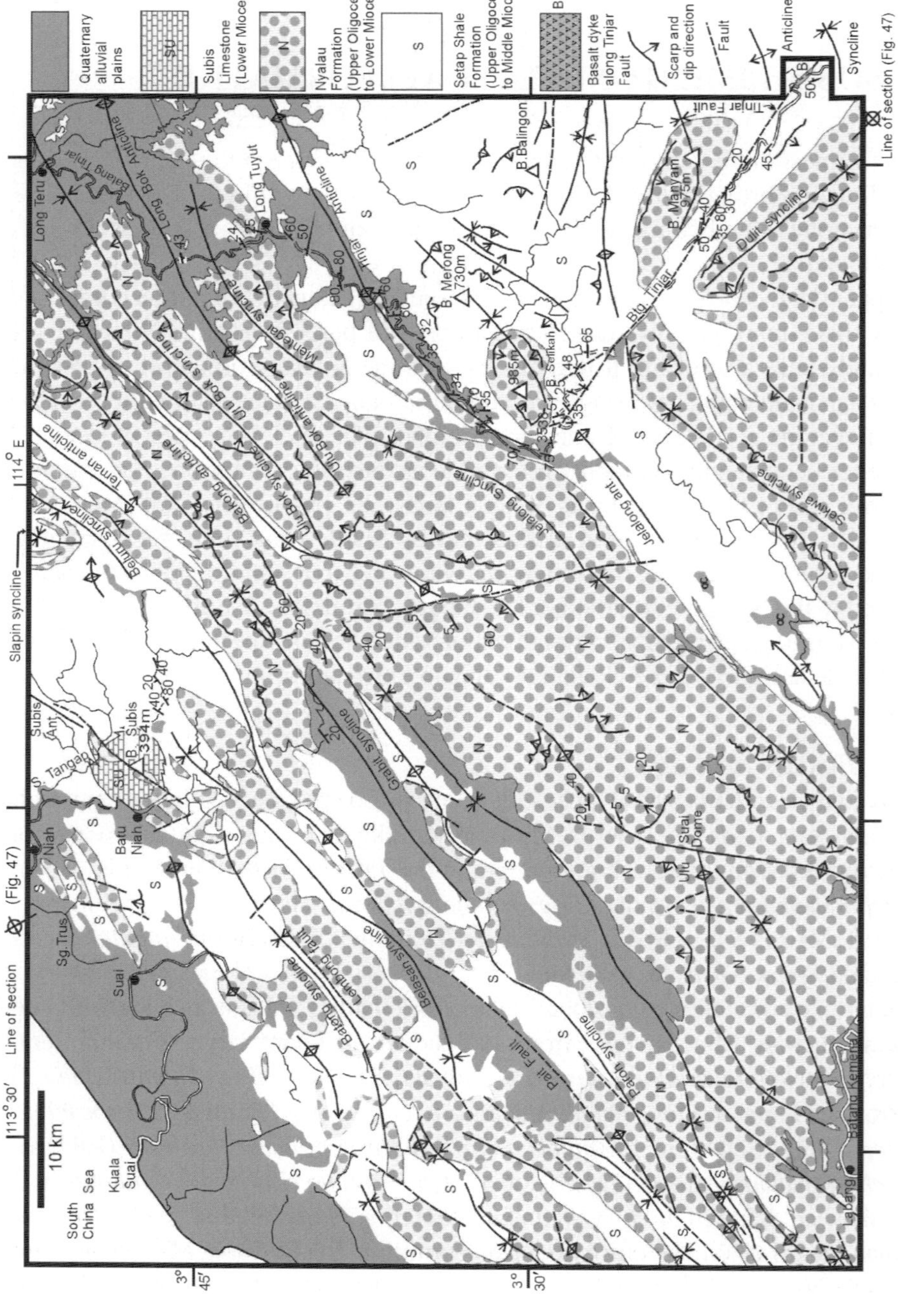

Figure 40. Geology of the Bukit Subis to Sungai Tinjar area east of Bintulu. Modified and reinterpreted after Haile (1962). With permission from the Department of Minerals and Geosciences Malaysia.

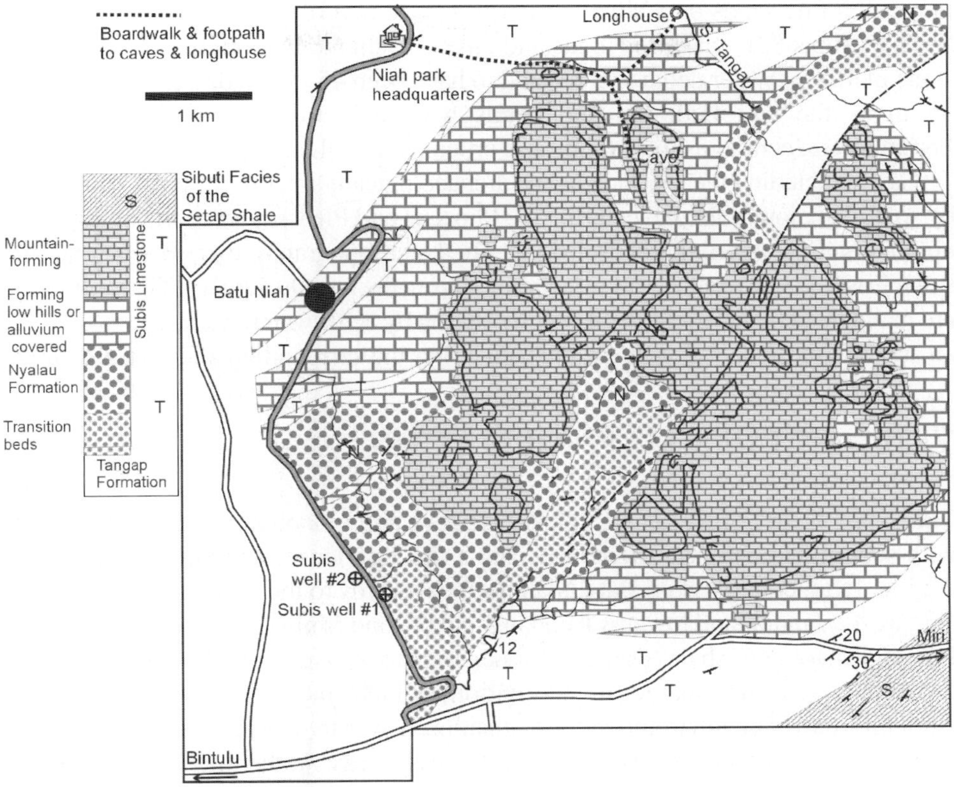

Figure 41. Geology of the Gunung Subis area near Batu Niah. From Haile (1962), updated in Haile and Ho (1991). With permission from the Department of Minerals and Geosciences Malaysia.

Corals are abundant in parts of the limestone, commonly fragmented. Algae include the following: *Lythophyllum* sp. (abundant); *Mesophyllum* sp. (abundant); *Lithoporella* sp. (abundant); *Archaeolithothamnium* sp. (less common); *Corallina* sp.; *Jania* sp.; *Halimeda* sp. (locally abundant).

The subcrop of the Subis Limestone penetrated in well Subis #2 is partly Te_{1-4} (Upper Oligocene: Chattian) and partly Te_5. Oligocene–Lower Miocene border runs through the middle to upper part of the limestone. The Upper Oligocene fauna include: *Operculina* cf. *pyramidum* Ehrenberg; *Heterostegina borneensis* v.d. Vlerk; *Neoalveolina pygmaea* Hanzawa and *Miogypsinoides ubaghsi* Tan Sin Hock. The environment of deposition of the Tangap–Subis Limestone Formation was inner to middle neritic holomarine. The onland Subis Limestone (Tangap Formation) extends NNW-wards offshore, parallel to and westwards of the West Baram Line (Agostinelli et al., 1990), as shown in the modified palaeofacies map of Shell (Figure 35).

Gunung Subis is a limestone massif that rises about 394 m above the surrounding, mainly swampy, plain (Figure 41). The reef complex accumulated during the Late Oligocene to early Miocene. It is sited at the facies boundary between the mainly sandy Nyalau Formation, to the southwest, and the shaly and marly formations (Setap Shale, Tangap, Sibuti and Tabau), to the northeast. The reef complex grew over Nyalau

Formation fine sands on a shoal area. The lowest part of the limestone is in the subsurface (Subis #2), where red algae and larger benthonic Foraminifera formed a limestone platform. Water depth is estimated to have been 45–55 m, decreasing to less than 35 m as the reef built up, and corals were able to live, together with larger Foraminifera, echinoids, bryozoans and articulate red algae.

Caves, including the spectacular and archaeologically important Niah Caves, formed by solution below water table during the Pliocene and Pleistocene, as the limestone was uplifted and the overlying beds eroded. The karst topography formed when the limestone was exposed to subaerial weathering during the late Quaternary. The vertical cliffs are maintained by undercutting through solution by acid swamp waters. Almost as large an area of limestone as forms the hills, underlies the surrounding alluvium (Figure 41).

VI.10. SETAP SHALE FORMATION

The Setap Shale Formation is a thick, extensive and monotonous succession of shale with subordinate thin sandstone beds and a few thin lenses of limestone. It occupies the low-lying country from Batu Niah northeastwards to the base of the Lambir Hills (Figure 42), and inland towards Limbang and beyond Mulu (Figure 40). The common lithologies are grey shale, grey mudstone, sandstones and a few limestones. There is a variation in the degree of induration, which increases towards the Mulu Anticlinorium. The environment of deposition is wholly marine, ranging from inner, in the west, to middle and even outer neritic eastwards. At the time of deposition, the palaeo-environments trended NNW–SSE (Figure 35). The mudstone outcrops between Batu Niah and the Lambir Hills commonly show no clearly defined bedding. However, the bedding is defined by layers of sideritic concretions, which, when broken open, commonly yield a variety of macro-fossils. The formation shows only moderate folding in the Suai area, which increases in intensity eastwards.

The Tubau Formation is a local variant of the Setap Shale; it is preferable to call it the Tubau Member. It is mainly of shales, but they are more calcareous than the regular Setap Shale. The name has been used when Shell drilled the Subis #2 well, but Leichti et al. (1960) indicated that Setap Shale was a better allocation. The main outcrop of Tubau Member is in the Kemena valley south of Suai (Kirk, 1957).

The Sibuti Formation (Lower Miocene, Te_5) consists of shale, locally calcareous with thin lenses of limestone and sandstone. The formation conformably overlies the Tangkap Formation. It is overlain by the Lambir Formation along a conformable but abrupt boundary. However, by common usage, the Sibuti Formation is included within the Setap Shale Formation.

VI.10.1. Palaeontology and age

The formation is chronologically equivalent to the Nyalau Formation, its northeastwards equivalent as water deepened (Figure 35). It is overlain abruptly by the Middle Miocene Lambir Formation at the western base of the Lambir Hills. The literature

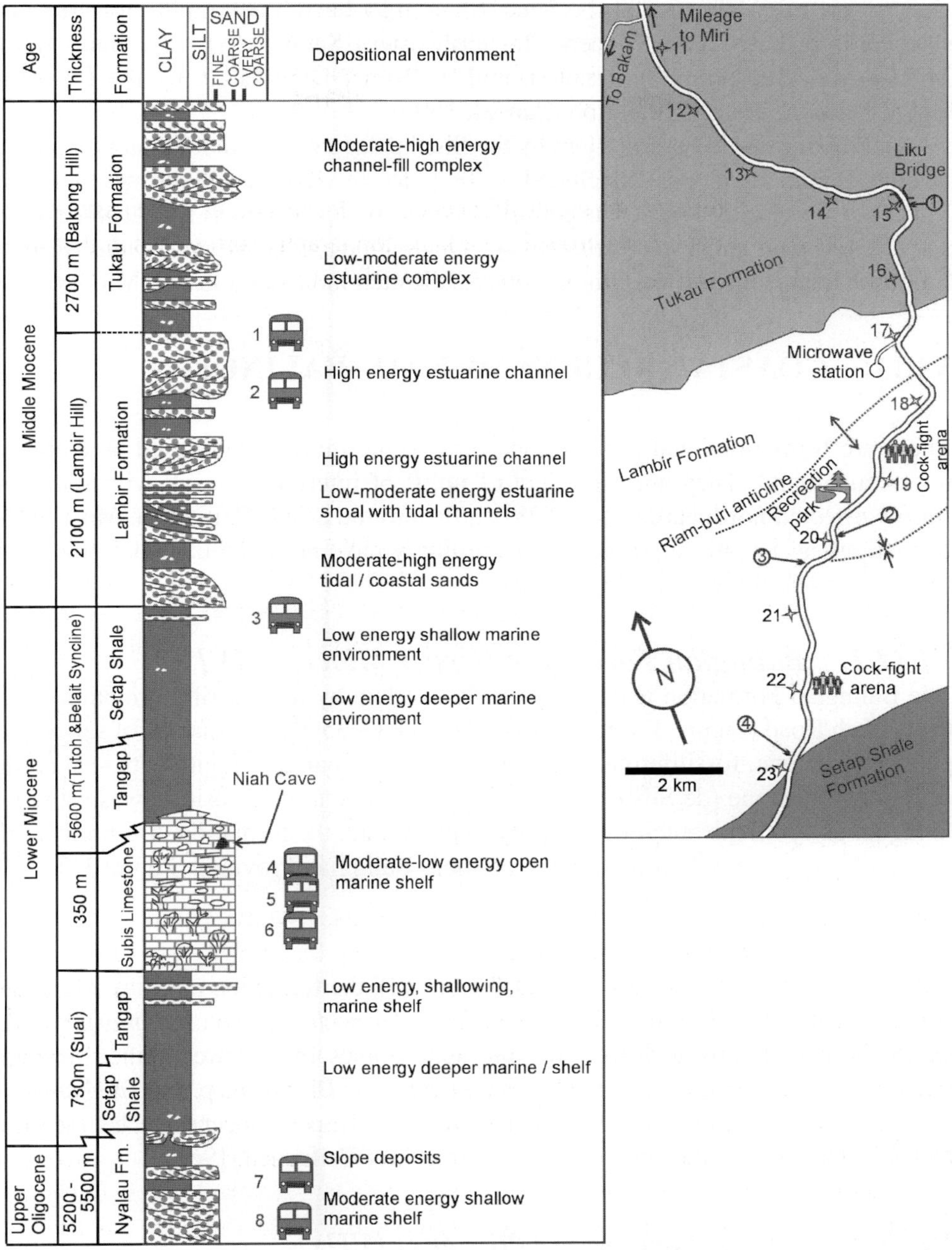

Figure 42. Left: Lithostratigraphic summary from east of Bintulu to west of Miri (after Haile and Ho, 1991). Suggested stops 1 to 7 are indicated. Right: simplified road map northwards from southern base of Lambir Hills.

(e.g., Haile, 1962) suggests an upwards extension into the Pliocene, but strictly the Setap Shale Formation should be taken as Upper Oligocene to Lower Miocene, but following Sandal (1996), the Lambir Formation may be interpreted as beginning later towards the NE, so that the Setap Shale may have a maximum upwards persistence into the Middle Miocene. In northeastern Sarawak, it is overlain by the Middle Miocene Lambir Formation, and in Brunei Darussalam by the Lower to Upper Miocene Belait Formation (Sandal, 1996).

A study of pelagic Foraminifera by the Shell palaeontological laboratory from the Middle Baram area, has concluded a Te_{1-4} to Te_5 (Upper Oligocene to Lower Miocene) age (Haile, 1962). The common forms are: *Spiroclypeus* sp.; *Lepidocyclina* sp.; *Cycloclypeus* sp.; *Miogypsinoides* sp.; *Operculina sp.; Gypsina globulus* (Reuss). From the Suai wells, *Floscullina botangensis* Rutten indicates a Lower Miocene age.

VI.11. COASTAL REGION MUKAH –BALINGIAN

There are formations that are confined only to the immediate coastal area and do not occur inland. They are a product of uplift of interior Sarawak with sediment supply directed northwards and NNW-wards into the South China Sea. Most of the formations occur only offshore and accordingly have not been named, but a small number do outcrop onland (Figure 43).

VI.11.1. Balingian Formation (Upper Miocene) [V]

The Balingian Formation is unconformably overlain by the Begrih Formation along the Mukah Road (Figure 43). It consists of a thick (≥3500 m) sequence of sandstone, pebbly sandstone, fossiliferous mudstone, abundant coal and lignite with seat earths of rootlet mudstone (de Silva, 1986b). The environment of deposition was estuarine to lagoonal. From the outcrops along the Mukah road, the formation dips northwards and has been commonly drilled in the offshore Balingian Province (Figure 33).

VI.11.1.1. Palaeontology and age

In the area adjacent to the Mukah road (Figure 43), the formation contains a brackish water small Foraminifera fauna including: *Ammodiscus* sp., *Glomospira* sp., *Haplophragmoides* sp. and *Trochammina* sp. Samples from a well drilled near the mouth of the Tatau River contained: *Ammobaculites* sp., *Discorbis* sp. and *Trochammina* sp. Ostracods were also found. The fauna indicates an Upper Miocene age, but the same fauna is also found in the younger Liang Formation (Wolfenden, 1960).

VI.11.2. Begrih Formation (Pliocene) [VI]

This formation unconformably overlies the Balingian Formation (Figure 43) and consists of a succession of laminated sandstones, fine sandstones, ortho-conglomerate, sandy conglomerate, coal and peat, clays and locally a boulder conglomerate at the base (de Silva, 1986b). The Begrih Formation is conformably overlain by the Liang Formation. The depositional environment of the Begrih is fluviatile and non-marine.

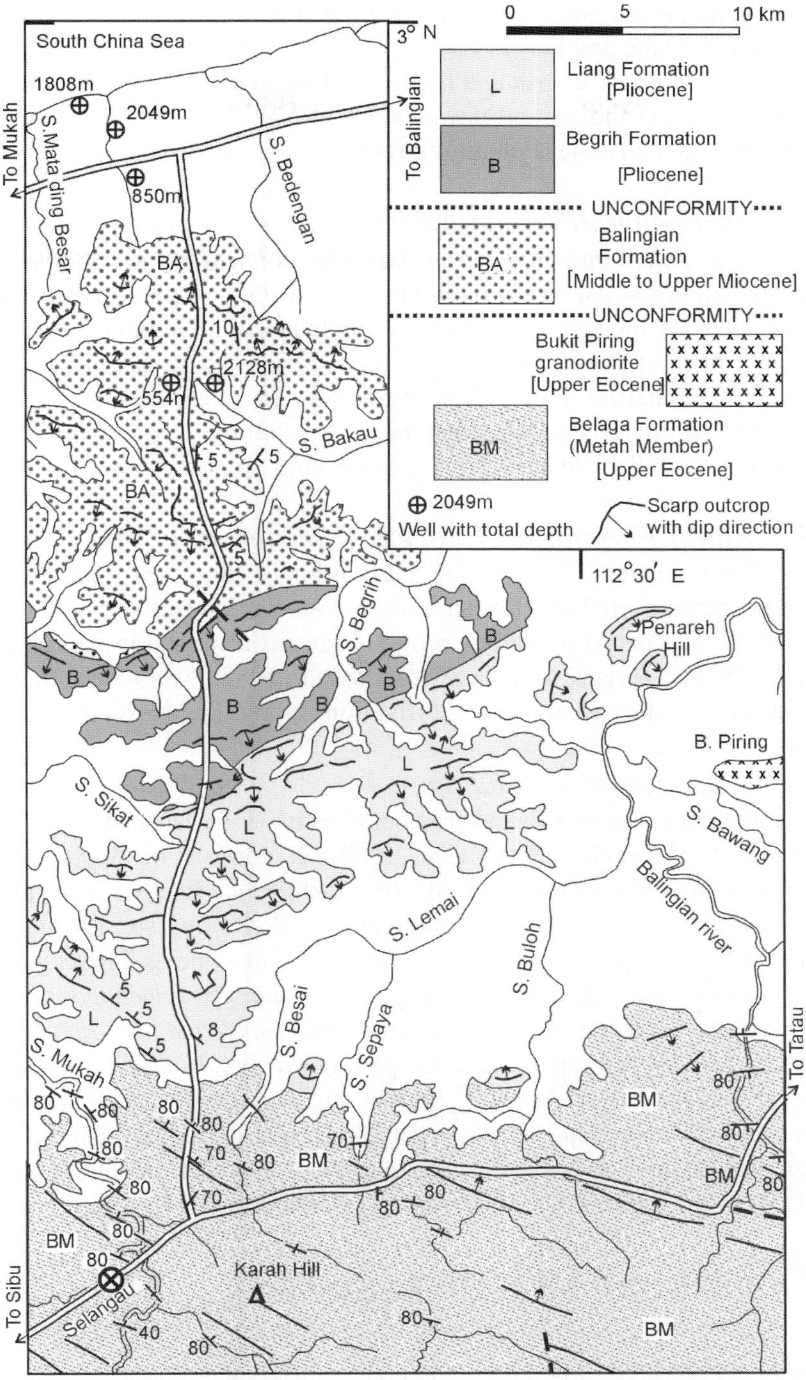

Figure 43. Areal geology along the Mukah road, showing the Upper Miocene and Pliocene formations (Balingian, Begrih and Liang), which continue offshore (after Wolfenden, 1960). With permission from the Department of Minerals and Geosciences Malaysia.

Outcrops (Figure 43) are good along the road towards Mukah (Figure 44) and have been described by Haile and Ho (1991).

The brown coals of the Begrih and Liang formations are humic and were formed in swamps identical to the present-day swamps of NW Sarawak (de Silva, 1986a). The seat earths are mudstones containing rootlets.

VI.11.2.1. Palaeontology and age

Typical Foraminifera found in outcrops are: *Ammobaculites* sp., *Bolivina* spp., *Cibicides* sp., *Elphidium* spp., *Frondicularia* sp., *Rotalia* spp., *Textularia* sp. and *Tricoculina* sp. They indicate a Lower Pliocene fauna of mixed marine and brackish-water environment (Wolfenden, 1960). A well drilled near the Balingian River yielded a littoral marine fauna. The typical forms are *Ammobaculites* spp., *Anomalina* sp., *Bolivina* sp., *Cyclammina* sp., *Elphidium* sp., *Glandulina* sp., *Quinqueloculina* sp., *Rotalia* spp., *Siphogenerinoides* sp. and *Textularia* spp.

VI.11.3. Liang Formation (Pliocene–Pleistocene) [VII]

This is a widespread succession of clay and sand, with abundant lignite and some tuff. The thickness in offshore Brunei exceeds 3000 m and offshore Balingian it exceeds 520 m. The depositional environment ranged from shallow marine in the north to coastal plain (Figure 43). De Silva (1986a,b, 1987, 1988) could find no distinction between the Begrih and Liang formations along the Mukah Road.

VI.11.3.1. Palaeontology and age

Outcrops in the area of Figure 43 yielded a poor brackish-water fauna identical with that in the Balingian Formation (Wolfenden, 1960). It consists of *Ammodiscus* sp., *Glomospira* sp., *Haplophragmoides* sp. and *Trochammina* sp. In other areas, the Liang Formation contains the typical marine Foraminifera: *Bigenerina* sp., *Elphidium* spp., *Rotalia* spp. and *Vulvulina* sp. A well drilled near the mouth of the Igan River yielded the following: *Ammobaculites* sp., *Anomalina* spp., *Bigenerina* spp., *Biloculina* sp., *Bolivina* spp., *Cancris* sp., *Cibicides* spp., *Cristellaria* spp., *Elphidium* sp., *Eponides* spp., *Gladulina* sp., *Globigerina* spp., *Gyroidina* sp., *Lagena* spp., *Nonion* spp., *Operculina* sp., *Orbulina* sp., *Quinqueloculina* spp., *Rotalia* spp., *Sigmoidella* sp., *Sigmoilina* spp., *Siphonina* sp., *Spiroloculina* sp., *Textularia* spp., *Uvigerina* sp. and *Vulvulina* sp. In addition, bryozoans, echinoid spines, ostracods, gastropods and shell fragments have been recovered. An Upper Pliocene age is interpreted (Wolfenden, 1960).

VI.12. OFFSHORE MUKAH–BALINGIAN

The area is subdivided into three distinctly different terrains by the major NNW–SSE trending West Balingian Line (Figure 45), which is a complex of closely spaced faults and comes ashore about 10 km east of Mukah, and the offshore north-westerly continuation of the Bukit Mersing Line.

Figure 44. Begrih Formation (Locality: 27.9 km along the Mukah road from the junction with the Sibu-Bintulu main road). River bar deposits overlain by abandoned channel deposits. There are three, sequences: 1 and 3 culminate in a lignite seam; 2 culminates in a paleosol (from Haile and Ho, 1991).

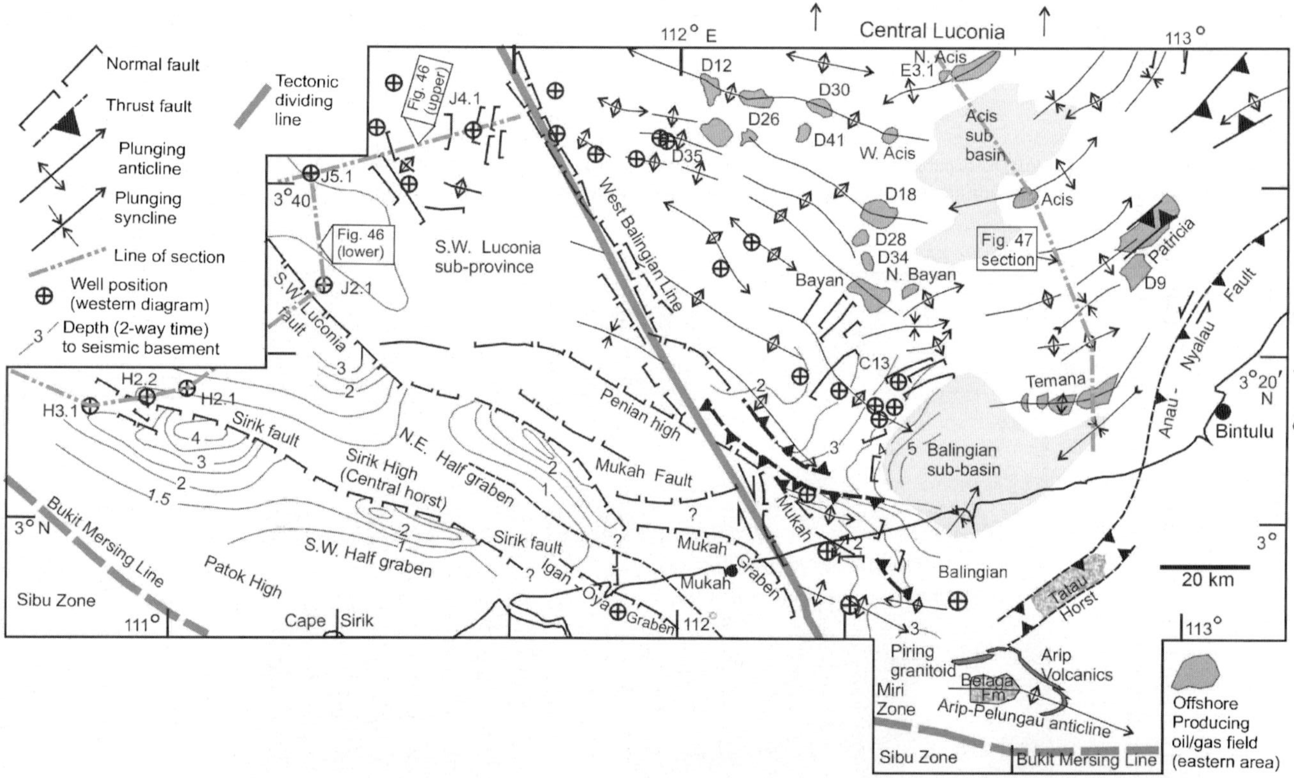

Figure 45. Structural elements of offshore Balingian and their relationship to the compressional Tatau Horst and Arip-Pelungau plunging anticline. Incorporating information from Mazlan (1999), Mazlan and Abolins (1999), Mazlan and Redzuan (1999) and Wolfenden (1960). With permission from Petronas.

VI.12.1. Southwest Sarawak offshore

A very few wells have been drilled. They indicate that the pre-Oligocene basement consists of a rhythmic sequence of low-grade metamorphic lithic feldspathic sandstone, siltstone and shale of the Belaga Formation. Overlying this unconformably is a relatively thin (<700 m) thickness of Middle to Upper Miocene shallow marine strata. The thickness of this Miocene succession increases northwestwards along strike to exceed 2 km in the Indonesian Soikang Basin. The measured geothermal gradient of the Soikang Basin averages 56°C km^{-1} (Hutchison, 1989), higher than any province farther east but characteristic of the basins, which were formed as a result of rifting of the Late Eocene extensive Sundaland landmass. The structural style is similar to that of the region continuing east of the Bukit Mersing Line as far as the West Balingian Line; characterized by NW-trending half-grabens created by normal faults (Mazlan, 1999a). The Bukit Mersing Line is therefore not of fundamental importance, representing as it does the inland north-eastern limit of outcrops of the Rajang Group (Belaga Formation) [= Sibu Zone]. From inland in the Tatau, Mulu and Batu Gading areas, as well as from subsurface information (Figure 47), the Belaga Formation continues beneath the Miri Zone, at least in substantial part.

VI.12.2. Mukah Province

Because of some unexplained aberration, Petronas refers to this province as 'Tatau'. I will more appropriately refer to it as the Mukah Province. It shares the same style with the SW Sarawak Province, but to the northeast of the Bukit Mersing Line the strata overlying the Eocene and older basement attain thicknesses of tens of kilometres in half- grabens. The eastern extent of the Mukah province is the complex fault system of the West Balingian Line, across which there is a very sharp contrast of tectonic style into the Balingian Province (Figure 45).

The province is characterized by extensional tectonics in the form of NNW-trending normal faults, most of which are downthrown to the southwest. This is also a characteristic of the Lupar Fault, which accordingly can be genetically linked with the Sirik Fault and the Mukah Fault of the Mukah Province (Figures 13 and 46) (Hutchison, 2001). The Half-graben sub-province is divided into the NE Half Graben and SW Half graben by the Sirik Fault. These half-grabens contain as much as 5 s (two-way time) of Middle to Upper Miocene siliclastic sediments (Mazlan and Redzuan, 1999). The half-grabens continue into the inland area as the Igan-Oya and Mukah grabens. They are separated by the Sirik High. The Mukah Graben is flanked to the northeast by the Penian High, which extends as far eastwards as the West Balingian Fault (Figure 45).

It is uncertain how to name the offshore formations; therefore, the sedimentary cycles of Shell Sarawak (Ho, 1978) are utilized. Cycles I and II are absent or thin over most parts of the Mukah Province and are preserved, sometimes in great thickness, within the half-grabens (Figure 35). In many cases, cycles III and IV rest directly upon Palaeocene–Eocene (= Rajang Group) basement, penetrated in J5.1 well. The master faults were active from Oligocene time through to Upper Miocene.

Thick sequences of Cycles III–and V filled the half-grabens. A Middle Miocene period of compression resulted in uplifted horsts on which carbonate build-ups developed, mostly in Cycle IV (Middle Miocene). By Upper Miocene and Lower Pliocene, an extensive sag basin developed and Cycles V and VI strata thicken northwards in response to uplift of on land Sarawak.

Depositional environments in the Mukah Province range upwards with time, from alluvial through coastal plain to open marine. The extensional faulting was initiated on a land surface, so that this region was an integral part of the Eocene Sundaland landmass, very extensive throughout Southeast Asia (Hutchison, 1992b) and added to in the Late Eocene by the Sarawak Orogeny (Hutchison, 1996a). Cycles I to IV (Oligocene to Middle Miocene) were deposited in a non-marine setting (Mahendran, 1997) within the rift-related subsiding areas. Cycle IV marked the beginning of a regional marine transgression that enabled the carbonates to develop on the horsts (Figure 46). Well H2.1 is an example of a carbonate that was developed directly on a basement horst of the Sirik High. Away from the horsts, Cycle IV was of siliclastic strata. Deltaic and shallow marine deposits typify the Cycle V and younger sequences (Mahendran, 1997). Coals within Cycles I–III may represent the hydrocarbon-generating source rocks. Substantial reserves of oil and gas have been found, with the carbonate of J4 representing the best discovery.

VI.12.3. Balingian Province

About 30% of this province is on land, where the geology has been mapped by Wolfenden (1960). The offshore area has been summarized by Mazlan and Abolins (1999). The province is bounded on the north by the Central Luconia Province and on the west by the fundamentally important Southwest Balingian Line fault-complex (Figure 45). The two major synclines, known as the Acis and Balingian sub-basins divide the province into West Balingian, characterized by NW–SE trending fold axes, and East Balingian, characterized by NE–SW trending fold axes (Figure 45). The Balingian sub-basin contains more than 6 km of sediment infill, and like the Acis sub-basin, represents almost continuous post-Eocene deposition. These two deep basins are believed to be important kitchens for oil and gas generation, which has migrated up-dip into contiguous anticlinal structures (Figure 47). The on land area east and northeast of Bintulu has been described by Haile (1962) and later summarized in the Petronas book, where it is referred to as the Tinjar province (Mohd. Idrus and Redzuan, 1999).

VI.12.4. Structure

VI.12.4.1. East Balingian

The East Balingian province (east of the Acis and Balingian sub-basins (Figure 45) is dominated by NE–SW fold axes parallel to the coastline, and this structural style continues far inland right across the Miri Zone into the Upper Baram region (Figure 40). Major shallow-dipping synclines range from 10 to 20 km wide, with much steeper

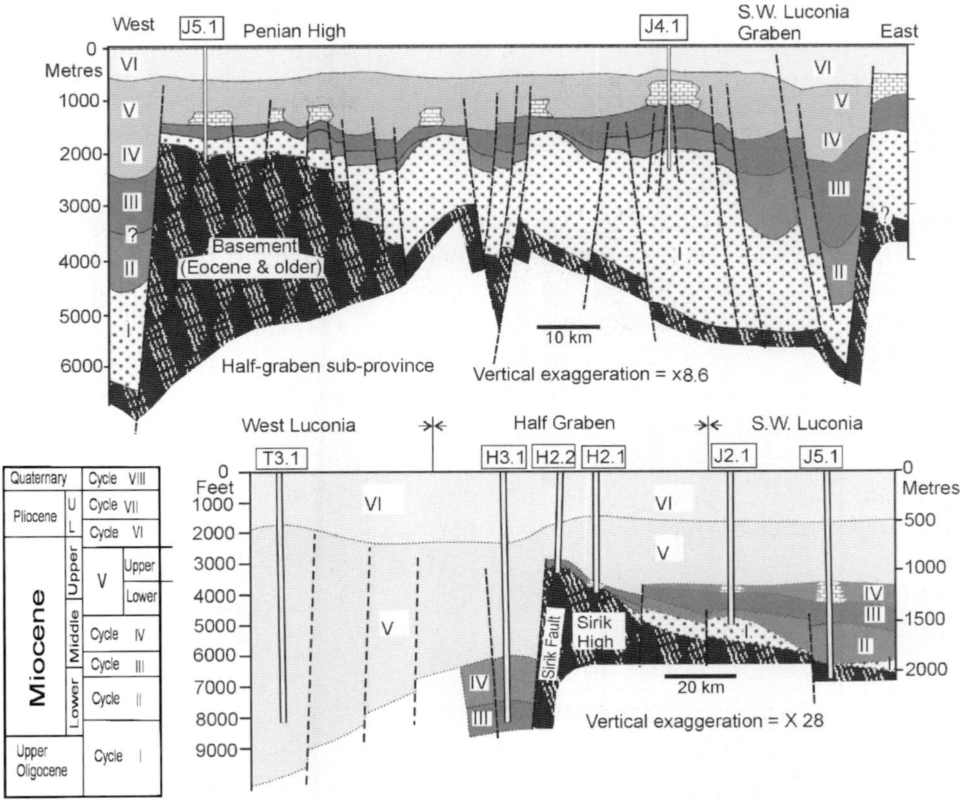

Figure 46. Cross-sections, based on seismic and well control, showing typical structures of the offshore area west of the West Balingian Fault, characterized by half-grabens. Redrawn from Mazlan (1999a) and Mazlan and Redzuan (1999). Note the different vertical exaggerations. With permission from Petronas.

and narrower anticlines (Figure 47). The compression was directed NW–SE and increased in intensity landwards towards the SE. Anticlinal structures were so much compressed as to be bounded by steeply dipping reverse faults, causing the Rajang Group basement to be squeezed up as horst-like structures. These inliers of basement are found in the Mulu Anticlinorium, Temala Anticlinorium and Tatau Horst.

Liechti et al. (1960) were the first to recognize that the Setap Shale acted as a décollement on which the sand-dominant formations were deformed differently from the underlying Rajang Group. The term they used was 'tectonique de couverture' , or superficial folding. This would now be referred to as 'thin-skinned-tectonics', and there has been an increasing awareness of the apparent relationship between wrench faulting and compression of the suprastructural strata (transpression). However, I do not feel that the mechanism proposed by Petronas staff (Mohd. Idrus and Redzuan, 1999) is sufficiently scientifically rigorous as to warrant detailing in this publication. Based on proprietary SAR imagery, they have identified faults in the area of Figure 40 that have not previously been seen. It is remarkable

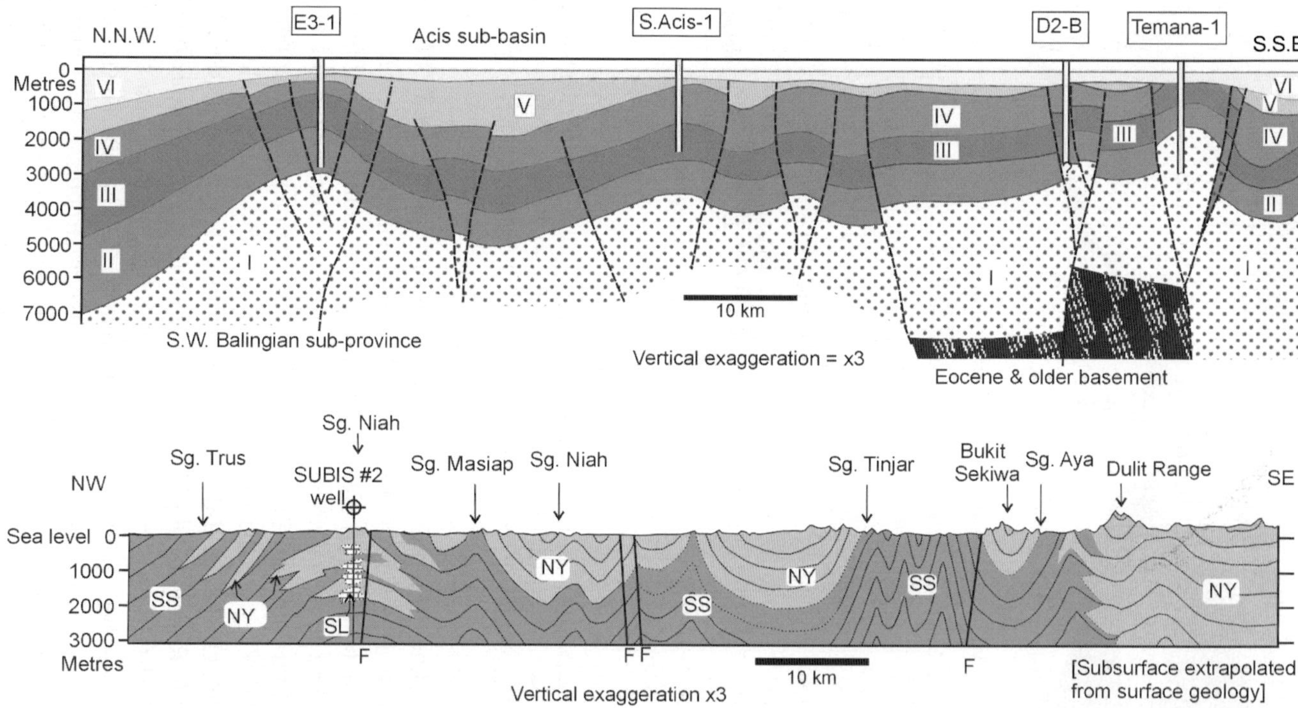

Figure 47. Cross-section across the Balingian Province. Top: offshore based on seismic with well control. The section is almost at right angles to the fold axes of the Eastern Balingian Province. Redrawn from Mazlan (1999). Lower: NW–SE cross-section onland through Batu Niah towards the Tinjar River (modified after Haile, 1962). Formations are all Upper Oligocene to Lower Miocene. SS = Setap Shale, NY = Nyalau, SL = Subis Limestone. Location in Figure 45.

that they did not record the presence of the well-known NW–SE trending Tinjar Fault lying parallel to the Dulit Range. The series of major anticlines and synclines striking NE–SW (parallel to the Grabit Syncline, Figure 40) were created by a Late Miocene maximum compression σ_1 directed along 335–155°. This is not in doubt. What remains open to scepticism is that they attribute the folding to transpression created by left-lateral wrench motion on as yet poorly authenticated NNE–SSW striking faults, which they say are minor. Minor faults are unlikely to be the driving cause of major transpressional folding.

The most spectacular structural feature of the region (Figure 40) is that the Sekiwa Syncline, which follows the regional NE–SW tectonic trend, appears to be sharply bent into a SE trend along the Dulit Range. I can do no better than quote Liechti et al. (1960) on this: ".. the Dulit strike may be conceived as a result of genuine frame folding (Rahmenfaltung). The Dulit trend is perhaps secondary, being partly the result of Neogene structural forces interfering with the pre-existing later Eocene structure." The Tinjar Fault, which is parallel to the Dulit Range, needs to be linked in some way to the Sekiwa-Dulit right-angled bend. Mohd. Idrus and Redzuan (1999) attempt this, and suggest that the 'Tinjar Fault' is dextral. Their analysis is flawed because they describe the Tinjar Fault as trending NE–SW *sic*. (p. 401). Tjia (1998) repeated the analysis of Liechti et al. (1960) using the same terminology. He suggested that the right-angled structural bend occurred in an area constrained by the strike-slip faults: Dengan fault, Tubau Line, Sekiwa Line and Tinjar Fault.

Few, if any geologists, have commented on the spectacular right-angled bend of the Tinjar River (Figure 40), but many have looked in vain for a seawards (NW-wards) continuation of the Tinjar Fault. The river flows north-westerly until it reaches the sandy Nyalau Formation Jelalong syncline, at which point it abruptly turns 90° right to become a tributary of the Baram River, taking a Z-shaped course to the sea. Rivers do not sharply turn 90° to the right unless they encounter an active right-lateral wrench fault, which has pushed the sandy Nyalau Formation ridge of the Jalalong syncline right-laterally in its path. Neither Haile (1962) nor Mohd. Idrus and Redzuan (1999) have identified a NE–SW fault bounding the rather straight margin of the Jelalong syncline ridge. Such a NE-trending fault would explain the river course, and it would have continually offset the Tinjar Fault right-laterally. Its right-laterally ultimately displaced continuation towards the sea may well be the parallel West Baram Line.

The 3D seismic data set of the Balingian Province (Figure 48) displays a clear system of en échelon high-angle Riedel shears bisecting a NE–SW orientated main fault direction (Gonguet, 2001). This direction is identical to faults within the Central Luconia province to the north. The high-angled Riedel shears are right-stepping (Figure 48), which unequivocally indicates a left-lateral (sinistral) movement. The figure also clearly shows a flower structure in vertical seismic section whereby the Riedel shears branch off upwards into multiple sub-faults at the lithological boundaries, such as the boundaries between more sandy and more shaly beds. All Riedel shears are rooted into one single vertical parent fault assumed to

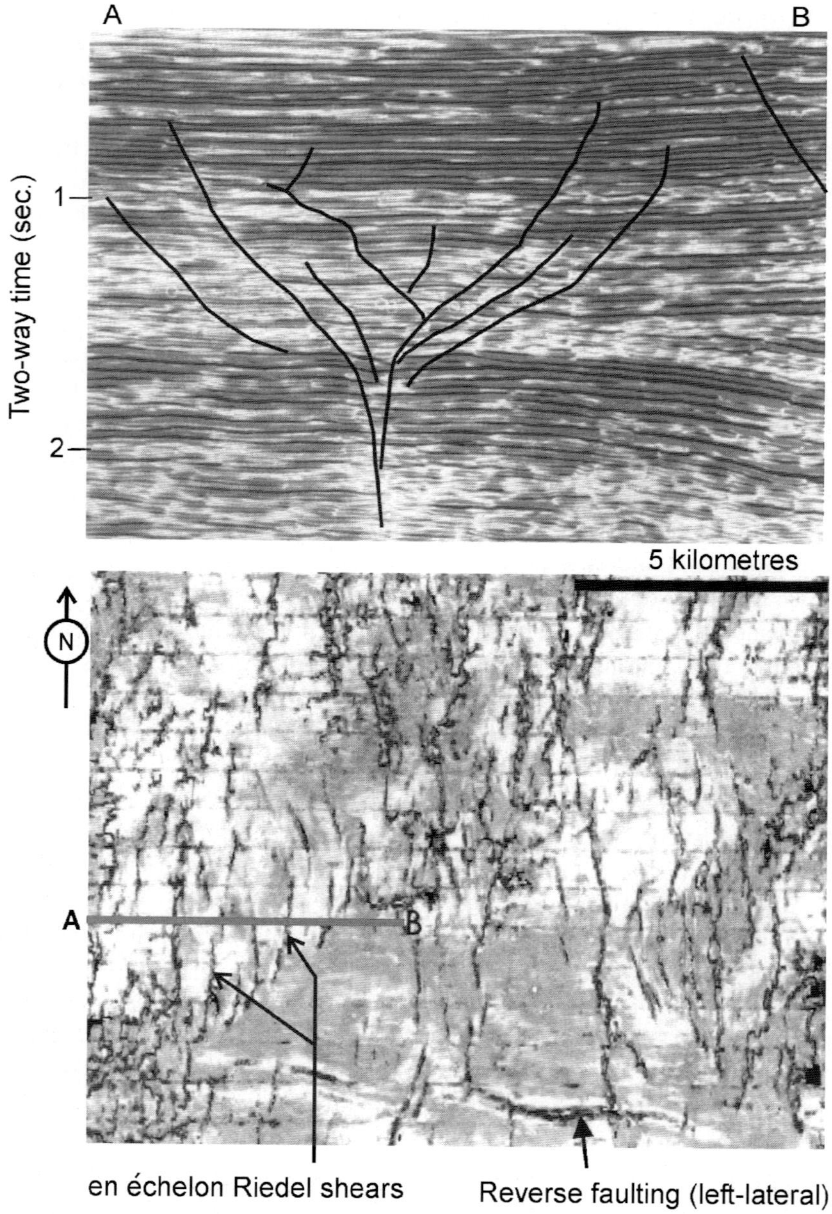

Figure 48. Flower structures in the Balingian province. Upper: E–W seismic section (about 6 km long) showing the interpreted faults. Lower: Horizontal seismic slice (horizon just below 1 s Twt). Showing en echelon N–S Riedel shears and E–W reverse fault (from Gonguet, 2001).

overlie the root basement lineament. Figure 48 also shows a WNW–ESE reverse fault system, indicating a clear compressional trend. The maximum compression σ_1, which would have caused transpressional folding, is directed NNE–SSW.

The details from the Balingian Province show a similarity with the central Luconia Province, so that a similar tectonic setting may be deduced. The wrench transpressional tectonics of this region are compatible with a change along the NW Borneo margin from Late Cretaceous–Palaeocene subduction to Late Eocene–Late Miocene collision causing oblique convergence from the north in the foredeep (Hutchison, 1996a).

VI.13. BARAM DELTA AND ITS PRECURSORS

The early development of the Baram Delta may be divided into three phases in Brunei and Limbang (Koopman, 1996). All are related to uplift and erosion of the Sibu Zone hinterland, with sediment deposition nearby towards the northwest. Phase I is called the Lower Miocene Meligan System, now occupying land south of Brunei. The sediment is interpreted to have accumulated in a foreland basin lying immediately northwest of an Early Miocene thrust front. Phase II is known in Brunei as the Middle Miocene Champion System, the sediments of which occupy much of onland Brunei and extend far offshore to the northeast and are not found in Sarawak. The sediments are interpreted as having accumulated in a sand-dominant ramp margin. Phase III is the Baram System, a delta which built out northwestwards from the Miri-Seria area in the Late Miocene to Recent.

VI.13.1. Meligan Formation

This Upper Oligocene to Lower Miocene formation is composed predominantly of massive sandstone forming the Tamabo Range and Batu Lawi, Gunung Murud Kechil in the Upper Baram district and as small occurrences in the Limbang Valley (Haile, 1962). The provenance of such a quartz-rich formation is enigmatic, for the presumed source (uplifted Rajang Group) is predominantly shaly, with only thin sandstones. The same age Nyalau Formation is very sandy and its provenance lay to the southwest (Sundaland land mass). The Shell palaeofacies maps (Figure 35) offer no suggestion, but a major fluvio-deltaic system must have built northeastwards into this region that remains poorly known because of its general inaccessibility.

The massive sandstones form prominent mountain ranges, including Batu Lawi (2027 m) and the Tamabo Range. The quartzose sandstones are coarse to medium grained. The grey and yellow-grey sandstones are interbedded with grey mudstone. There are also lenses of calcareous conglomerate that contains pebbles of radiolarian chert. However, there is a general lack of detailed descriptions.

In the Tamabo Range, the dips are regular at about 80° to the east. At Batu Lawi, the sandstones occupy a complex faulted syncline with moderate to steep dips.

VI.13.1.1. Palaeontology and age

A shale from above the lowest massive sandstone in the Tutoh headwaters yielded: *Bathysiphon* sp.; *Trochammina* sp.; *Globoquadrina altispira* (Cushman & Jarvis);

Globorotalia cf. Mayeri Cushman & Ellisor; *Cyclammina* sp.; *Globigerinoides* sp. This fauna is indicative of a Te age (Upper Oligocene to Lower Miocene) age.

VI.14. BARAM DELTA IN MALAYSIA: ITS DEVELOPMENT

In Malaysian terminology, the Baram Delta can be subdivided into two delta systems:

- Middle–Late Miocene East Baram Delta (known as the Champion Delta in Brunei)
- Late Miocene–Quaternary West Baram Delta (known as the Baram Delta in Brunei)

Hence there is a problem of terminology and the reader must distinguish whether the writing is from a Malaysian or Bruneian viewpoint. This publication follows the former. The East Baram Delta lies in Brunei and Sabah and is described later.

These two delta systems were probably supplied by separate river systems. The East Baram Delta was from the east and southeast across the Morris Fault–Jerudong Line. Uplift of the Crocker Ranges Western Cordillera of Sabah must have played an important role in providing the nearby provenance (Hutchison et al., 2000). The West Baram Delta had its sedimentary supply from the south and southwest across the West Baram Line. The main provenance for the West Baram Delta must have been uplifted and eroding very sandy outcrops (e.g. Figure 39) of the Oligocene–Lower Miocene Nyalau Formation, which ultimately were derived from a Sundaland provenance by an eastwards flowing fluvial system (Hutchison, 1996b). A Mississippi Delta model, though frequently applied by the oil industry, is inappropriate since the sediment provenance was both in the East and West deltas from actively uplifting nearby hinterlands.

VI.15. THE WEST BARAM DELTA

A very comprehensive description of the West Baram Delta, both onshore and offshore is to be found in Tan et al. (1999).

VI.15.1. Lambir Formation

In northeast Sarawak, the Baram Delta begins with a very dramatic change from Setap Shale to very sandy Lambir Formation at the southerm margin of the Lambir Hills (Figure 42). The Middle Miocene (Tf) Formation consists of sandstone, shale and some limestone. The formation occupies comparatively gentle synclines and is little changed diagenetically. Maximum thickness is around 1600 m.

The basal Lambir Formation thick-bedded sandstones have a sharp erosive base overlying medium dark-grey mudstone and calcareous mudstone (Sibuti or Setap

Shale Formation). The calcareous mudstone contains Foraminifera and crabs. The environment was shallow marine, as indicated by an abundance of gypsum. The basal sandstone is well sorted, made up of a number of sandstone cycles with hummocky cross-bedding (Figure 49). The top sandstone shows low-angle planar cross-bedding, and may be a beach deposit. The sequence therefore shows a distinct shallowing upwards transition from marine to coastal, concomitant with the change in orientation of the coastline and uplift of interior Borneo (Figure 36, change from Cycle III to IV).

VI.15.1.1. Palaeontology and age

Fossils in the Lambir area are generally of little time-stratigraphic value. A Tf_1 (Middle Miocene: Langhian) age for limestone in the Bakong area is indicated by *Flosculinella bontangensis* Rutten and a Tf_{2-3} (Middle to Upper Miocene) age for the upper part of the formation is indicated in the Bakong area by the absence of *Flosculinella botangensis*, *Austrotrillina howchini*, *Spiroclypeus* sp. and *Miogypsinoides* sp. (Wilford, 1961).

VI.15.2. Tukau Formation

The lower part of the Tukau Formation conformably overlies the Lambir Formation near Sungai Liku in the eastern Lambir Hills (Figure 42). In turn it is conformably overlain by the Liang Formation near Miri. In exploration wells Siwa #1 and #2, the Tukau Formation is underlain by the upper part of the Miri Formation (Wilford, 1961). Its base is therefore diachronous.

Much of the formation is barren of fossils, but there is a brackish-water fauna including *Ammobaculites* sp.; *Glomospira* sp.; *Haplophragmoides* sp. and *Trochammina* sp., but they have no time-stratigraphic value (Wilford, 1961). The formation is assumed to be Upper Miocene, extending into the Lower Pliocene.

The basal transition from mainly littoral to inner shelf Lambir Formation to the overlying Tukau Formation, which was deposited in a coastal plain environment, is well displayed southwest of the Sungai Liku Bridge (Figure 42). The coastal plain environment is indicated by the absence of Foraminifera except for poor brackish-water forms and the presence of thin lignite layers. The muds contain abundant amber balls.

VI.15.3. Geology of Miri Hill

Miri Hill, with good cliff exposures along the coast from Tanjung Lobang (Figure 49), was initially interpreted by the Anglo Saxon Petroleum Company as a northeast striking asymmetric anticline with a gentle northwest flank and a steep, partly overturned, southeast flank. A first glance at the block diagram (Figure 50) shows that this is a reasonable interpretation. Drilling on the Miri Hill at Miri #1 produced light oil from 138-m depth at 88 barrels per day on 22 December 1910.

Figure 49. Selected outcrops in the West Baram Delta. (A) Base of the Lambir Formation sandstones upon calcareous mudstone of the Sibuti Member of the Setap Shale Formation (locality 3 of Figure 34). (B) Miri Formation sandstones at Tanjung Lobang, Miri. The sandstones show hummocky cross-bedding. (C) Field sketch of the Airport Road outcrop, Miri. Middle Miocene Miri Formation showing coarse-grained storm beds and hummocky cross-stratified (HCS) sandstones. A and B from Haile and Ho (1991), (C) from Tan et al. (1999).

However, the success of the company was based on a wrong structural model. Simple anticlinal closure became untenable as the Miri field was further drilled. A detailed analysis only became possible when the biostratigraphic zonation,

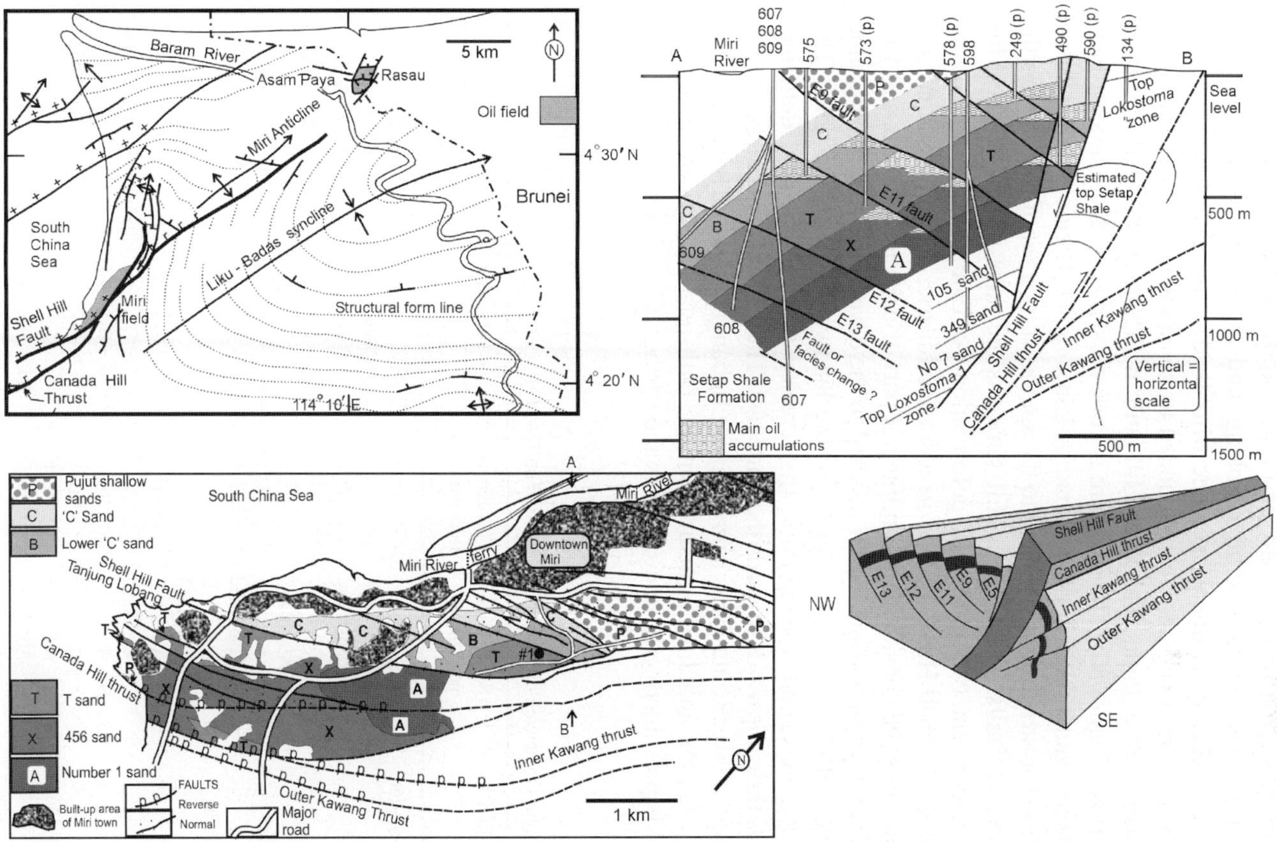

Figure 50. Simplified geological map of Miri Hill and its district. All outcrops are of Lower Cycle V Upper Miri Formation. Roads and town extent are only diagrammatic. Cross-section A–B across the Miri oil field, the first to be produced in Sarawak. (P) means projected onto the plane of the cross-section. The block diagram illustrates the relationship between the powerful Canada Hill thrust and the normal faults (after Tan et al., 1999). With permission from Petronas.

based on benthonic Foraminifera, was much later established, summarized in Table 11.

Later drilling showed that faults played a fundamental role in the trapping of oil (Figure 50). Once the palaeontologically based lithostratigraphic scheme was established (Table 11), it was found that similar looking sands, which previously had been correlated, are of radically different age. The Miri field is an asymmetric, slightly overturned, northeast trending anticline, bounded to the southeast by a set of northwest-hading steep normal faults. The Shell Hill fault, with a throw of some 750 m, is the most important (Figure 50). The structure is cut by a series of flat southeast-hading antithetic faults, and a set of northwest-hading listric faults. Their combined throw is around 300 m. The structure is bounded to the southeast by a set of merging northwest-hading reverse faults. The Canada Hill Thrust is the most powerful. The Miri Hill structure underwent two phases of deformation—a Late Miocene extension, which resulted in the Shell Hill Fault and its associated listric, and antithetic normal faults, this phase may be considered as an integral part of delta tectonics. It was followed by a Pliocene compression, which resulted in the development of the anticline and the Canada Hill Thrust, causing uplift of the landward side of the delta. The NW–SE oriented maximum compression first created the asymmetrical anticline, whose core was thrust upwards towards the northeast. These structures can be seen at low tide around the coast southwards from Tanjung Lobang. Oil would have migrated up-dip into the faulted anticlinal structure, but the fault-related traps are small and most of the oil is presumed to have escaped to the surface. Compared with the Seria

Table 11. Stratigraphic framework of the Miri Field (based on Tan et al., 1999)

Cycle	Formation	Sand Units and shales	Approx. thickness (m)	Palaeontology (established 1941)	
				Zone	Horizon
	Seria Formation			*Triloculina* 18	*Rotalia* 2
		Top *Nonion 3* sand		*Nonion*	*Cyclammina* 8
		Pujut Shallow Sand	299	3 Zone	*Bolivina* 6
		S H A L E	152		*Lagena* 9
		Upper C Sand	201		
Lower	Upper	S H A L E	46		Pre *Lagena* 9
V	Miri	Lower C Sand	192		
	Formation	S H A L E	91		
		T Sand	152		*Sigmoilina* 5
		S H A L E	30		
		456 Sand	122		Horizon 3 of *Simbatang*
		S H A L E	30		
		No. 1 Sand	61		*Bolivina* 16 *Gladulina* 2
		S H A L E	76		
		105 Sand	152	*Bolivinita* zone	Pre *Gladulina* 2
	Lower Miri Formation	Miri Shale		*Loxostoma* 1 zone	
IV	Setap Shale				

field of Brunei, the Miri field was of minor importance, the wells produced increasing amounts of water, and it was finally abandoned on 20 October 1972.

VI.15.4. Regional onland structure

The main structural styles of the onshore part of the West Baram Delta are growth faults (e.g. Shell Hill, Kuala Baram, North Lutong and North Rasau), normal antithetic and synthetic faults, and north-dipping reverse faults (e.g. Canada Hill, Kawang and Rasau thrust faults). The displacement and strike-slip movement on these reverse faults is not certain, but estimated to be some 600 m in the Miri Field (Figure 50). All these faults trend ENE–WSW.

Compressional tectonics in the Late Miocene–Pliocene resulted in numerous folds, including the Riam-Buri anticline, the Liku-Badas syncline and the Miri-Rasau anticline (Figure 50). The Liku-Badas syncline is a relatively simple unfaulted syncline, plunging to the ENE (Figure 50). Some growth faults (e.g. Shell Hill) had been reactivated as reverse faults during this compressional phase. Oil was also found trapped in the Asam Paya and Rasau structures; with the most important being Seria eastwards in Brunei.

The West Baram Line, which marks the western margin of the West Baram Delta, could not be traced onshore, but may be traced inland as the Tinjar Fault (Figure 40), displaced right-laterally westwards from it.

VI.15.4.1. Miri Formation

The Miri Formation is predominantly arenaceous, with clay and shale restricted mainly to the lower part. It has been divided into a Lower and Upper Miri Formation. The Lower part has well-defined beds of shale interbedded with sandstones; the upper part is more sandy. The outcrops around Miri Hill provide an invaluable analogue for the offshore Baram Delta, and they have been studied in great detail (Tan et al., 1999). The base of the formation is a gradual transition from the argillaceous Setap Shale to the sandy Miri Formation.

Ten different sedimentary facies have been recognized and described (Tan et al., 1999):

- *Medium-scale trough cross-bedding* The trough wavelengths vary from 0.1 to 3.0 m. The facies contains the trace fossils *Ophiomorpha nodosa*, *Ophiomorpha irregulaire*, *Palaeophycus*, *Teichichnus* and *Planolites*, all indicating a shallow marine environment. The facies was deposited in shallow sub-tidal areas of an estuary.
- *Small-scale trough cross-bedding* This is a subordinate facies containing mixed mud clasts and mud-draped cross-bedding. The environment could have been muddy tidal flats.
- *Herringbone cross-bedding* It occurs within the planar to tabular cross-bedded units. The bi-polar dip directions suggest a tidal environment as current directions reverse. There is a confirmatory presence of flat topped ripples.

- *Flaser-bedded facies* These are cross-bedded sands with numerous intercalated mud flasers. Sand and mud were continuously supplied into the shallow sub-tidal to inter-tidal environment.
- *Wavy-bedded facies* This facies is of alternating continuous wavy layers of mud draping the sandy ripples. The sand layers are 5–10 cm thick, while the mud layers are less than 2 cm.
- *Sand-clay alternation facies* Regular interbedded fine-grained thin-bedded sands and mud. The sands range from 1 to 23 cm, clays 1 to 5 cm thick. Boundaries are sharp. Load casting is common. The current action was minimal during deposition and trace fossils are rare.
- *Lenticular-bedded facies* Subordinate irregular sand bodies embedded in mud. The amount of sand does not exceed 25%. The environment was sub- to inter-tidal with low current activity.
- *Mudcrack facies and associated mudstones* This uncommon facies shows periodic exposure in between periods of deposition.
- *Hummocky cross-stratified (HCS) sandstone facies* Fine- to very fine-grained sheet sandstones displaying low-angle cross-stratification. The sandstone beds range from 15 to 90 cm thick. Bioturbation is common. The environment was shallow marine. The long low-angled undulating hummocky cross-stratification can be produced by storm waves. The bases of these sands are abrupt indicating sudden emplacement into a muddly environment. The already existing clays were actively reworked by the ingressing sands.
- *Massive coarse sandstones* Poorly sorted medium to coarse grained sandstones with no internal structures. De-watering and collapse structures are common, suggesting deposition was in a fluidized condition. These sands are thought to result from storms of typhoon strength; the transport being by turbidity current in very shallow water. But the sands may subsequently be reworked.

In summary, the Middle Miocene Miri Formation may be interpreted as a result of sedimentation in a tide-dominated estuary. An abrupt sea-level rise may be suggested by the coarse-grained sandstones and HCS sandstones sandwiched within the shallow tide-dominated deposits (Tan et al., 1999).

VI.15.4.1.1. Palaeontology and age The Lower Miri Formation, known as the Miri Shale, is dated by the Foraminifera of the *Loxostoma 1* zone. The basal sand of the Upper Miri Formation (105 sand) is identified as the *Bolivinita* zone. The overlying Upper Miri Formation sands and clays contain microfossils attributed to the *Nonion 3* zone. The unconformably overlying Seria Formation is recognized by the presence of *Triloculina 18.* The individual age-diagnostic Foraminifera are given in Table 11. The Miri Formation ranges from Middle to Upper Miocene.

VI.16. CENTRAL LUCONIA PROVINCE

The Central Luconia Province lies entirely offshore Sarawak, its geology has been described by Mohammad Yamin and Abolins (1999). It is separated abruptly from the West Baram Delta by the West Baram Line, which separates provinces of different geothermal gradient—the West Baram of average 28°C km^{-1} from Central Luconia of average 43°C km^{-1} (Hutchison, 1989). From this evidence, together with its geological stability, it is interpreted that Central Luconia is underlain by continental crust and is often referred to as a microcontinent in regional reconstructions. However, it is more likely to be an integral part of this part of Sundaland that lies beneath the South China Sea. It is separated from onland Sarawak and Bintulu by the Balingian Province, with which it shares several structural similarities.

The province is a broad and stable platform, characterized by the extensive development of Late Miocene carbonate build-ups that have developed on horsts in response to sea-level changes (Epting, 1980). More than 200 carbonate build-ups have been seismically mapped (Figure 51) and some 65 have been tested. About 20 are proven to contain commercial quantities of non-associated gas. Gas has of course leaked upwards from many of the carbonate build-ups. Production is currently in progress, and the gas piped to Tanjung Kedurong, Bintulu, for liquefaction and export.

VI.16.1. Sedimentary history

A horst and graben topography developed as part of the extensional tectonics of Sundaland, which resulted in the Oligocene to Lower Miocene rifting of the South China Sea marginal basin (Hutchison, 2004). Middle to Late Miocene carbonates developed upon the horsts. Large platform-type build-ups developed on highs, whereas pinnacle-type build-ups developed in adjacent elevated blocks within basinal areas where subsidence was more pronounced (Figure 52). The SW–NE alignment of the build-ups especially in the central and eastern parts (Figure 51) reflects basement controlled structural trends.

Biostratigraphic markers are not well preserved within the carbonates themselves. The ages of carbonate build-up are inferred from the dating of the enveloping clastic sequences (Figure 52). Carbonate deposition began in the early Miocene (late Cycle III) and continued until the present-day in deeper water areas away from siliclastic sedimentation. The peak of carbonate deposition was in Cycles IV and V. Strontium isotopic dating has been carried out to determine the carbonate stratigraphy. It indicates that the carbonates began being deposited as early as 20.5 Ma (Cycle III).

VI.16.2. Structure

Details of the fault patterns and topography of the carbonate build-ups (Figure 53) show a NE–SW composite fault system superimposed by a dense network of bisecting high-angle Riedel shears (Gonguet, 2001). The NE–SW main fault trend is a

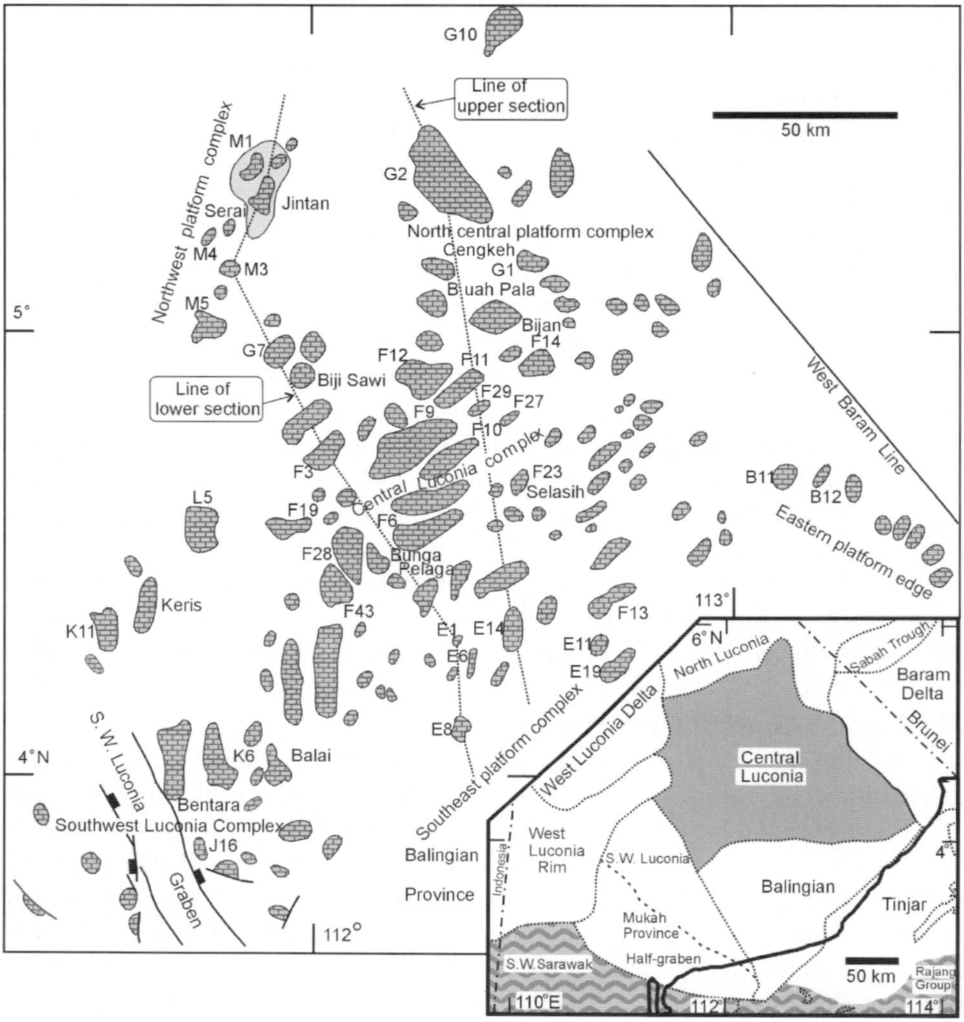

Figure 51. The Miocene carbonate build-ups of the Central Lucoinia province (from Mohammad Yamin and Abolins, 1999). With permission from Petronas.

well-known basement-related fabric that generated the typical horst and graben topography. From high quality seismic sections, the faults extend upwards from basement Eocene strata and progress upwards, but die-out upwards in a thick shale of middle to late Miocene age. Nevertheless the upwards trajectory of the faults is expressed in the overlying most recent sediments as doming and localized angular unconformities.

The left-lateral strike-slip nature of the NE–SW fault system is unambiguously proven by a pull-apart-releasing bend feature to the north where the NE–SW fault system consists of two individual faults linked by a major relay (Figure 53). The width of the pull-apart feature indicates movements of at least 300 m. Measurement of vertical

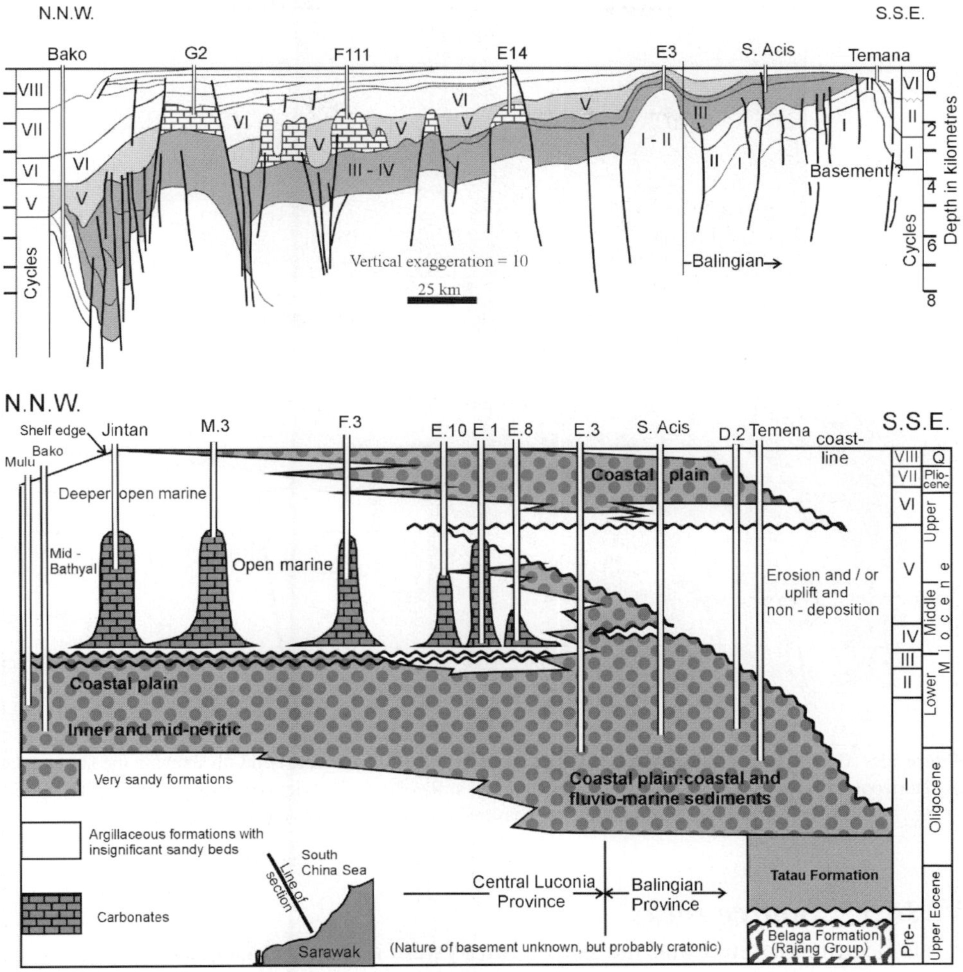

Figure 52. Central Luconia Province. Upper: Schematic NNW–SSE cross section across the Central Luconia Province and the Balingian Province. Lower: Stratigraphic NNW–SSE cross-section across the Central Luconia and Balingian provinces (from Mohammad Yamin and Abolins, 1999). With permission from Petronas.

fault throws on the largest high-angle Riedel shears (Figure 53) indicate that the main phase of fault movement took place in the Late Miocene, coeval with a known uplift event expressed as an erosional unconformity to the north (Gonguet, 2001).

The basement of the Central Luconia Province is poorly imaged on seismic, but gravity modelling indicates it to be in excess of 5 km. The clear strike-slip characters of Central Luconia indicate a similarity with the Balingian Province to the southeast and show that the topographic details of the build-ups have resulted initially from horst and graben extensional tectonics, later strongly modified by wrench or transpressional tectonics which have even resulted in pull-apart structures and significant fault offsets of the build-up topography.

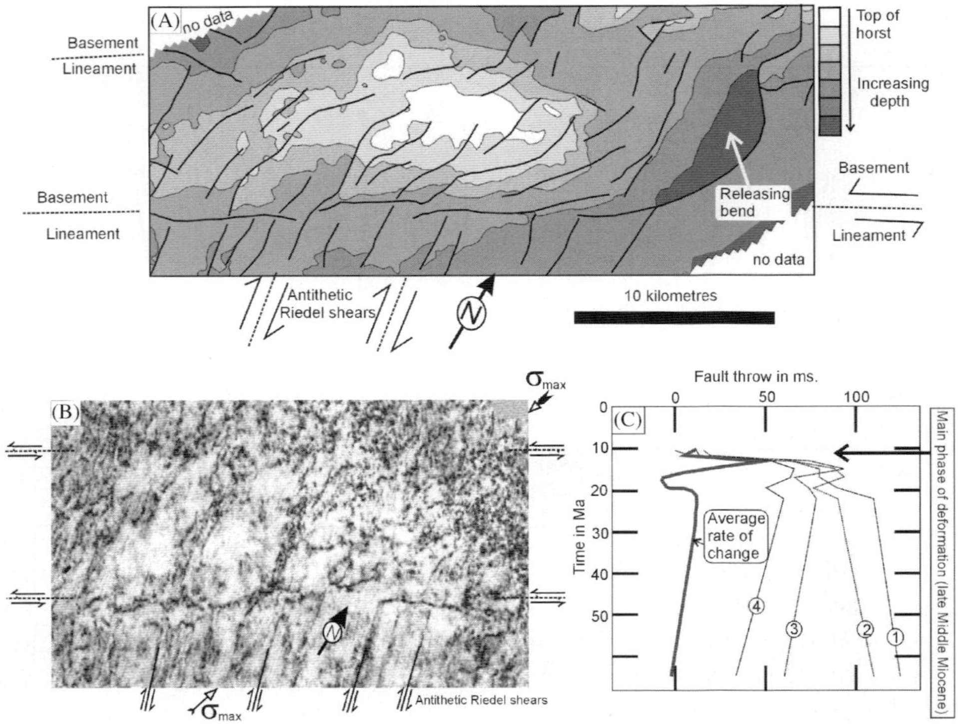

Figure 53. Central Luconia Province. (A) Depth map of a selected carbonate build-up showing the fault pattern. (B) Horizontal time-slice of the 3D seismic, showing the strike-slip fault style. (C) Vertical throw of four Reidel shears in milliseconds through time as measured on a seismic section (from Gonguet, 2001).

VI.16.3. Source rocks

It has been interpreted that the petroleum system of Central Luconia is common to that of the Balingian Province (Mohammah Yamin and Abolins, 1999). The most important source rocks are coals and carbonaceous shales deposited in a lower coastal plain setting during cycles I and II (Figure 52). Only a few wells have penetrated these cycles (because the targets were the carbonates). However, it is surprising that Cycles I and II penetrated strata contain few coals, whereas they are abundant in the contiguous Balingian Province. The geochemical analysis of the source rock material shows that the province is gas rather than oil-prone.

Chapter VII

The Passive Continental Margin

The geological history of the southern part of the South China Sea marginal basin is divided by the MMU into an earlier active rift and a post-rift episode.

VII.1. MIDDLE EOCENE – EARLY MIDDLE MIOCENE

Rifting began generally in the Middle Eocene, with local delay until the Lower Oligocene. There was abrupt cessation in the Middle Miocene at the regional break-up unconformity. The unconformity surface represents a hiatus of ~3 to 5 my. Unconformities are dated by the overlying basal draping strata, which in the Dangerous Grounds are early Middle Miocene (16 Ma), hence the name MMU. Sea-floor spreading east of 111° E began at anomaly 6b (21 to 22 Ma) in the Lower Miocene, and ceased at anomaly c5 (16 Ma) in the early Middle Miocene (Figure 1). Rising asthenosphere caused the sea-floor spreading but also caused uplift to form the break-up unconformity (Falvey, 1974). The zone of sea-floor spreading forms the abyssal plain of the South China Sea marginal basin, that is now actively subducting eastwards at the Manila Trench. All the elements of a passive margin are present (Figure 56).

VII.1.1. Crustal nature and thickness

The satellite gravity data have been recalculated into a map of depth to the Mohorovičič Discontinuity (Holt, 1998). A simplified redrawn version is given in Figure 54 (from Hutchison, 2004). The amount of crustal extension is usually quantified by a β factor, which corresponds to the ratio of crustal thickness before and after extension in uniform stretching models (McKenzie, 1978). Holt (1998) calculated β as the ratio of the thickness of the reference crust (30 km) to the basement thickness of the Sunda Shelf inferred from gravity modelling. His derived gravity and β stretching factors have shown good agreement with seismic refraction depths off the coast of China (Nissen et al., 1995). The Moho rises from an average depth of ~29 km below sea level under the Sunda Shelf to a minimum of ~16 km below the sea-floor spreading zone. Huchon et al. (2001) deduced a 6–7-km-thick oceanic crust in the centre of the SW prong of the marginal basin and a thinned continental crust on the basin edges, with the Moho depths ranging from 17 to 21 km on the northern margin, and slightly deeper in the Dangerous Grounds. Holt (1998) deduced that a smooth horizontal Moho lies beneath the Sunda Shelf (Figure 54).

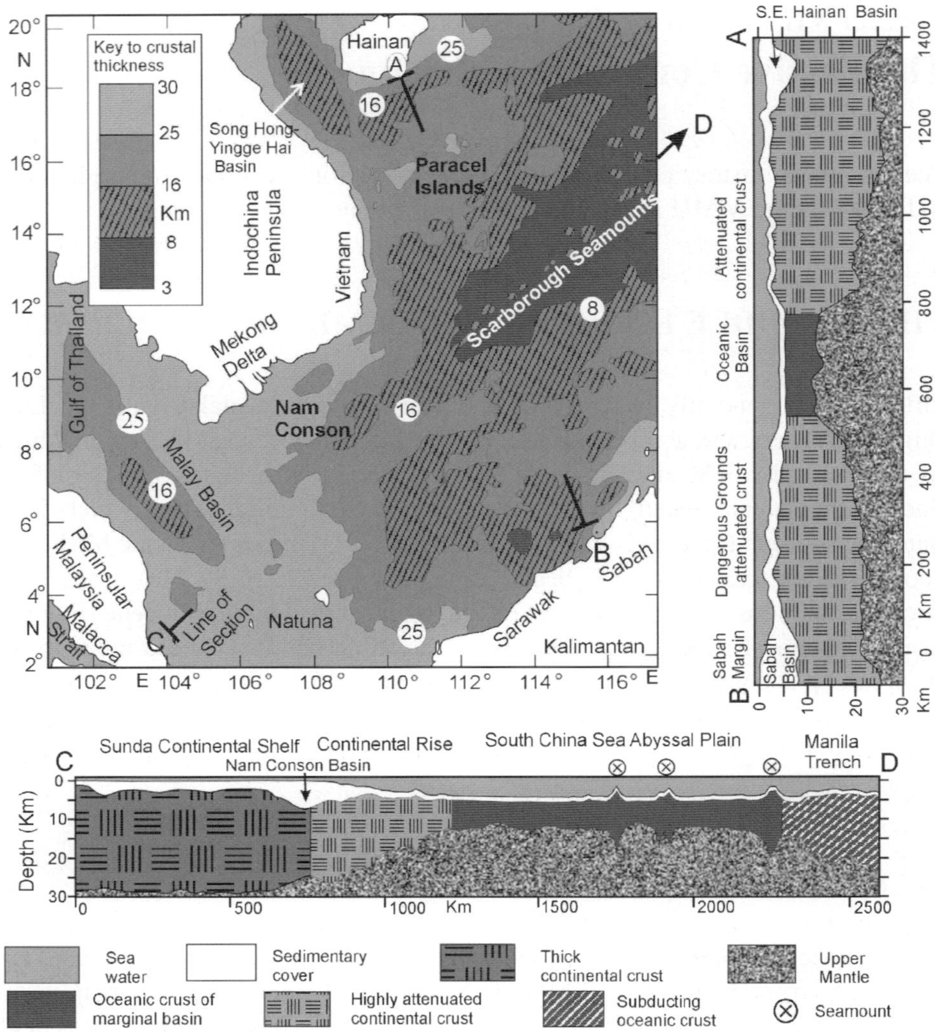

Figure 54. Simplified map of crustal thickness derived from the gravity data of Holt (1998). The two cross-sections indicate the crustal thicknesses and interpreted nature of the lithosphere (after Hutchison, 2004).

A β value of 3.0 was taken by Holt (1998) to define the outline of the zone of sea-floor spreading, since Le Pichon and Sibuet (1981) concluded that oceanic crust begins to form at β values >3.0. The Dangerous Grounds are characterized by stretching factors ranging from 1.3 to 3.0. Clift et al. (2002) quoted a maximum β factor of 1.45 for the upper crust and 1.60 for the lower crust in the Pearl River Delta, but were unable to quantify the Dangerous Grounds. Rather high values >3.5 (Mazlan et al., 1999a) characterize the central deeper parts of the Malay Basin and parts of the Baram Delta Basin. This indicates these localized regions are highly stretched. Otherwise, the Sunda Shelf has monotonous β values between 1.0 and 1.2.

VII.2. SUNDA SHELF

The Sunda Shelf is the continental shelf of East Asia. The large island of Borneo does not have its own continental shelf and western Borneo as far northeast as the West Baram Line sits within the Sunda Shelf, as do many islands such as Natuna—a region referred to as Sundaland.

Many writers have emphasized the Lupar Line, extending from the coast of western Sarawak towards Natuna, as important in regional reconstructions (Hutchison, 1996b). Its role as a plate margin ceased in the early Eocene. The related Rajang Group (Belaga Formation) was uplifted and amalgamated onto Sundaland by the end of the Eocene (the Sarawak Orogeny of Hutchison, 1996a) (Figure 55).

The associated Lupar Line lies entirely within the continental shelf, shows no bathymetric expression, and played no active role in the tectonic evolution of the present-day South China Sea.

Xia and Zhou (1993) published three regional seismic lines extending from the continental shelf across the slope onto the Dangerous Grounds. As published, their section A shows only low resolution, and my seismic sections B and C show higher resolution (Figure 56). The 200 m isobath trends SE at section A, then gradually curves to an easterly trend at section B as it approaches the West Baram Line at the G 10 promontory (Figure 59). Most seismic sections in this paper have high vertical exaggerations. The figure quoted on each figure (V.E.) applies only to the seawater layer, calculated using a seismic velocity of 1.5 km s^{-1}. Grossly exaggerated slopes result from a V.E. value of ~9x —an apparent sea-floor slope of 20° is in reality only 2°30′ and steep faults are actually of low dip.

The Sunda Shelf extends outwards from the continent approximately to the 200 m isobath. The whole shelf is characterized on the regional Bouguer gravity map of Holt (1998) by values ranging from –70 to +70 MGal. Accordingly, the crust is wholly continental, apart from old narrow sutures such as the Lupar Line, which have been amalgamated into Sundaland. A detailed analysis of the gravity data led to the conclusion that the shelf is characterized by a relatively deep Mohorovičič Discontinuity that lies between 28 and 30 km below sea level and is generally horizontal and smooth. However more than 6 km of crustal thinning has occurred in the limited area of the Malay Basin.

In common with most terrains of continental crust, the Sunda Shelf is underlain by a basement of older continental rocks and its composition may be inferred from neighbouring landmasses such as west Sarawak, Indochina and Peninsular Malaysia (Hutchison, 1989). Precambrian metamorphic rocks form the eastern Kontum Massif of Vietnam and could be expected in the shelf basement. Triassic sedimentary rocks are widespread and Mid to Upper Triassic granites form important zones related to the Indosinian Orogeny that established the general architecture of Southeast Asia. The shelf is also expected to contain older sutures, known from on land outcrops (Hutchison, 1989).

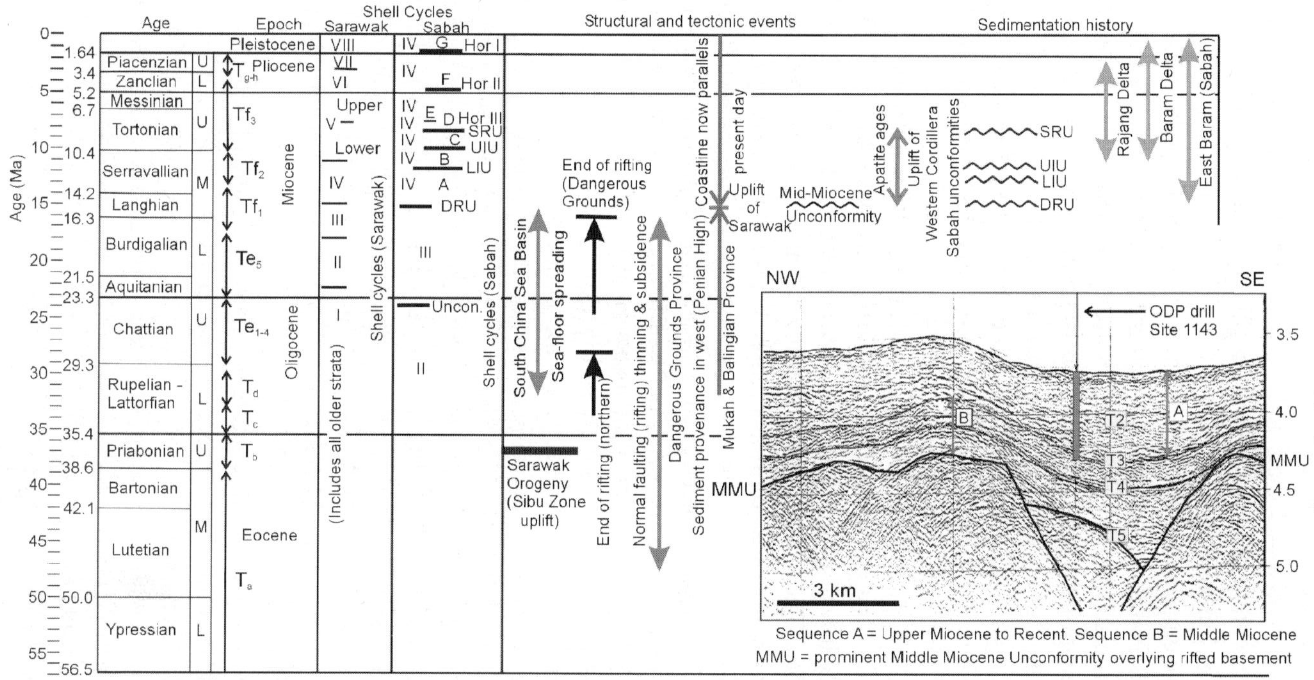

Figure 55. Important geological events in the development of offshore and coastal zone Sarawak, Brunei and Sabah. Inset: seismic section of the ODP Drill Site 1143. The 500 m cored is of Upper Miocene to Recent calcareous mudstone (from Hutchison, 2004).

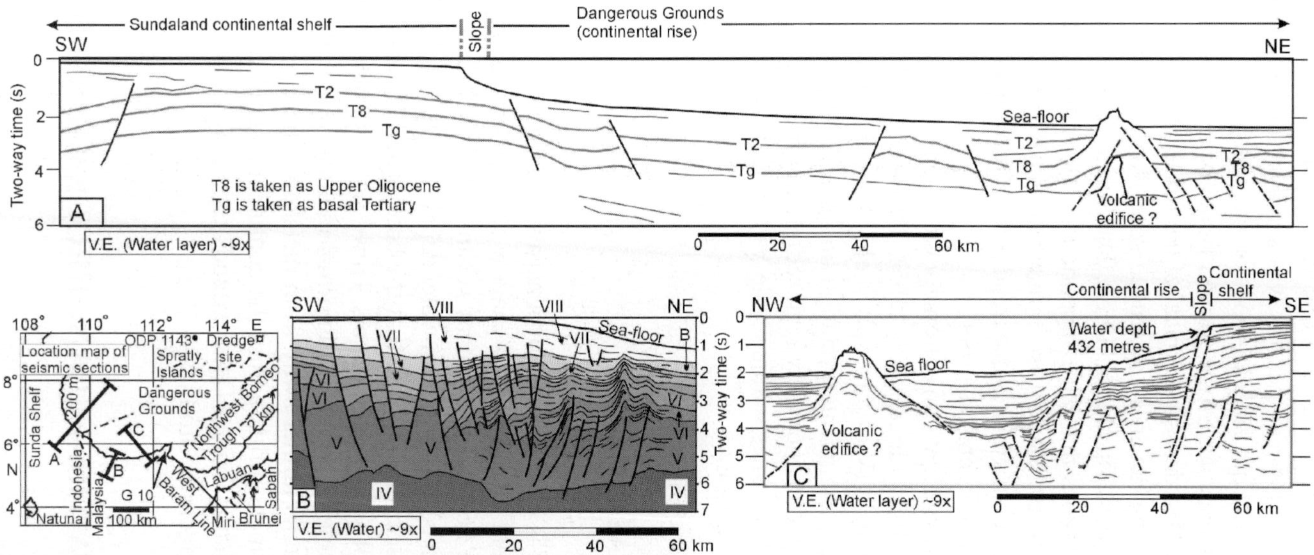

Figure 56. Interpreted seismic sections across the margin of continental Sundaland onto the strongly attenuated continental crust of the Dangerous Grounds (continental rise). (A) Regional line redrawn from Xia and Zhou (1993). (B) Detailed Petronas section of extremely thick Rajang Delta strata ranging from cycle IV of Shell, just above the Mid-Miocene break-up unconformity, to cycle VIII (Recent), as interpreted by Mohd Idrus et al. (1995). (C) A Petronas section, extending from the G10 promontory, showing a faulted continental slope. The topographic high is not a cuesta and may possibly be of igneous origin (after Hutchison, 2004).

Taylor and Hayes (1983), later followed by Clift et al. (2002), proposed that extension in the South China Sea exploited the location of a Jurassic–Cretaceous Andean-type arc. This view is based on the extensive Yenshanian granites and rhyolites of eastern China. Doubt must be expressed that this was a subduction-related volcano-plutonic arc, but it certainly was not Andean-type for the igneous rocks are overwhelmingly acidic and the granites are S-type formed from crustal anatexis, strongly mineralized in rich W–Sn–Sb–Mo deposits atypical of the Andes (Hutchison, 1996b). It is also uncertain that this Yenshanian arc extended southwards towards the landmasses bordering the Sunda Shelf (Hutchison, 1989). The Sunda Shelf is draped over by Tertiary sedimentary formations that have supported the oil industry (Hutchison, 1996b; PETRONAS, 1999). Nevertheless, these Tertiary strata have not wholly buried important belts of Upper Cretaceous granites that trend north-westwards from Sarawak through islands such as Natuna (Hutchison, 1989).

Seismic section B (Figure 56) shows an extremely thick stratigraphic post-MMU succession as presented by Mohd Idrus et al. (1995). The deepest strata are interpreted as Middle Miocene (Cycle IV of Sarawak Shell). There is no well control at this depth and the authors based their interpretation on seismic stratigraphy. If correct, then the section represents an extremely high sediment influx from the Rajang River or Mekong River. The anticlines within Cycle VI (Pliocene) bear a similarity to the toe-thrusts of the Baram Delta.

VII.3. CONTINENTAL SLOPE

The edge of the Sunda Shelf is characterized by a narrow transition zone of slightly steeper slope than the adjacent gently sloping shallow shelf and the gently sloping deep continental rise (Figure 56). Based on many seismic profiles, I have measured the foot of the Sunda Slope to be at 450–500 m water depth (Hutchison, 2002, 2004).

The majority of seismic sections indicate that the continental slope is generally unfaulted within the drape of younger sediments. However, as shown in section C (Figure 56), the slope is a zone of pronounced normal faulting, and there are more down-to-the basin normal faults farther away from the shelf. These are interpreted as a result of very high sediment influx across the continental shelf. In the area of section B (Figure 56), the major Rajang Delta extends right across the continental shelf and slope onto the Dangerous Grounds.

VII.4. CONTINENTAL RISE (DANGEROUS GROUNDS)

Water depths range from ~500 m at the foot of the continental slope to about 3.5 km at the margin of the zone of sea-floor spreading, or continent-ocean boundary of the marginal basin (Figure 57) (Hutchison, 2004). The rise, also known generally as the

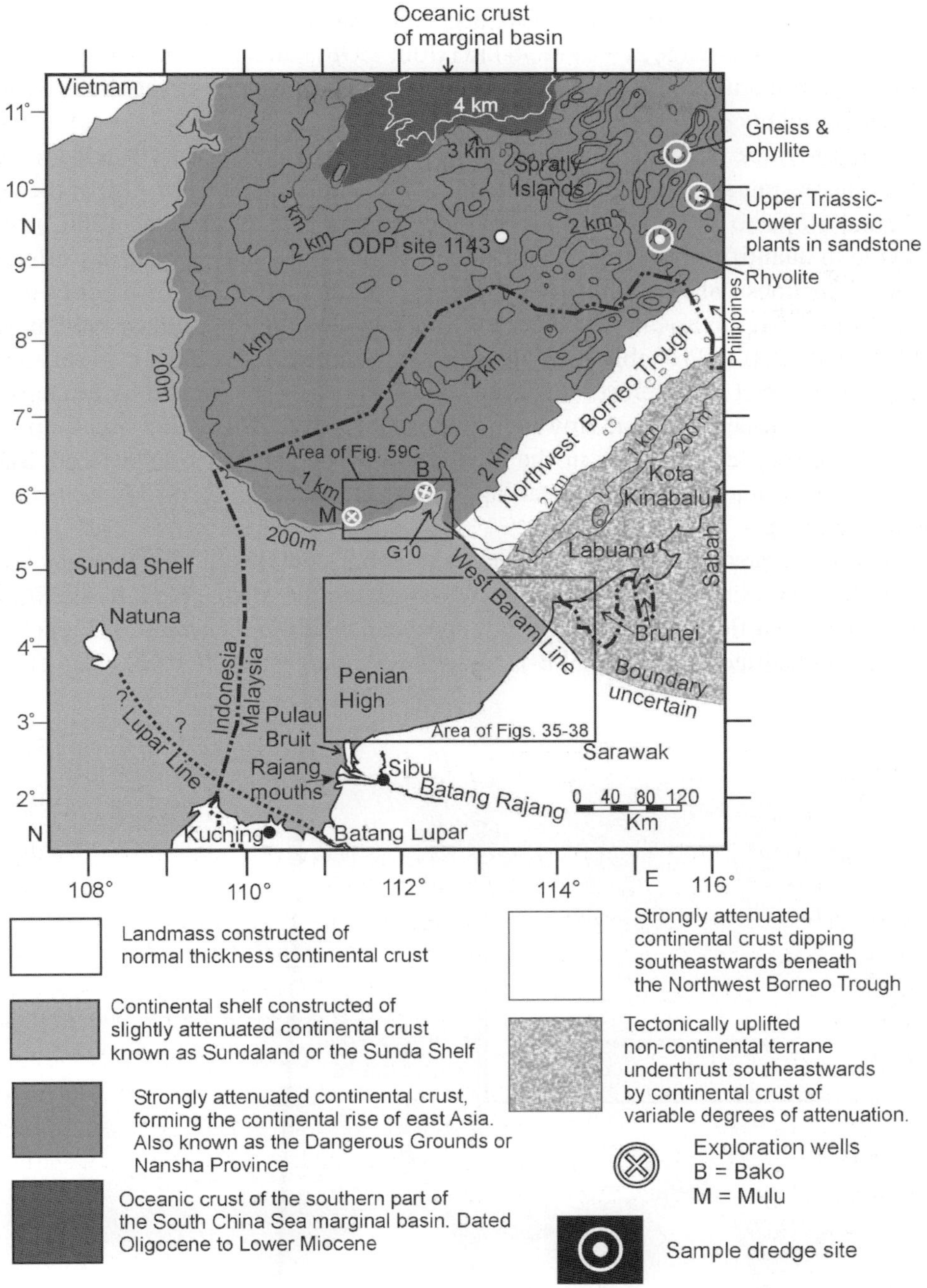

Figure 57. Major tectonic elements of the southern South China Sea and their relationship to bathymetry. The passive margin rifted transition from continent; through Sunda Shelf, slope and rise; to the abyssal plain formed by sea-floor spreading, ceases at the West Baram Line, northeast of which there is the collisional margin of Sabah.

Dangerous Grounds province, is 170–333 km wide and is composed of continental crust, thinned to a range of ~25 to ~8 km (Holt, 1998; Huchon et al., 2001). Isostatic equilibrium requires a progressive increase in depth of water with attenuation after the break-up unconformity.

The province is characterized by normal faulting in the form of half-grabens. With increasing attenuation, the stretching (β) factor ranges from ~1.2 at the continental slope to 3.0 at the continent–ocean transition. Clift et al. (2002) were unable to quantify stretching in the Dangerous Grounds. The present-day effective elastic thickness of the Dangerous Grounds continental plate (Te) is only of the order of ~8–10 km as calculated for the area between the Reed Bank and Palawan (Clift et al., 2002). Probably it was less (~ 3 km) during active rifting. A very weak continental crust is therefore suggested during break-up, facilitated by low viscosity lower continental crust. However, the end of active rifting and break-up may have happened earlier than in the Dangerous Grounds east of the Reed Bank, where contiguous sea-floor spreading started at anomaly 10 or 11, akin to the China margin.

The appearance of seismic profiles across the continental rise depends overwhelmingly on the thickness of sediments overlying the MMU. Near to the Spratly Islands, the sea-floor is characterized by spectacular half-graben cuestas (Figure 58). The area illustrated in Figure 58B is under ~1.5 km of water. Vertical exaggeration

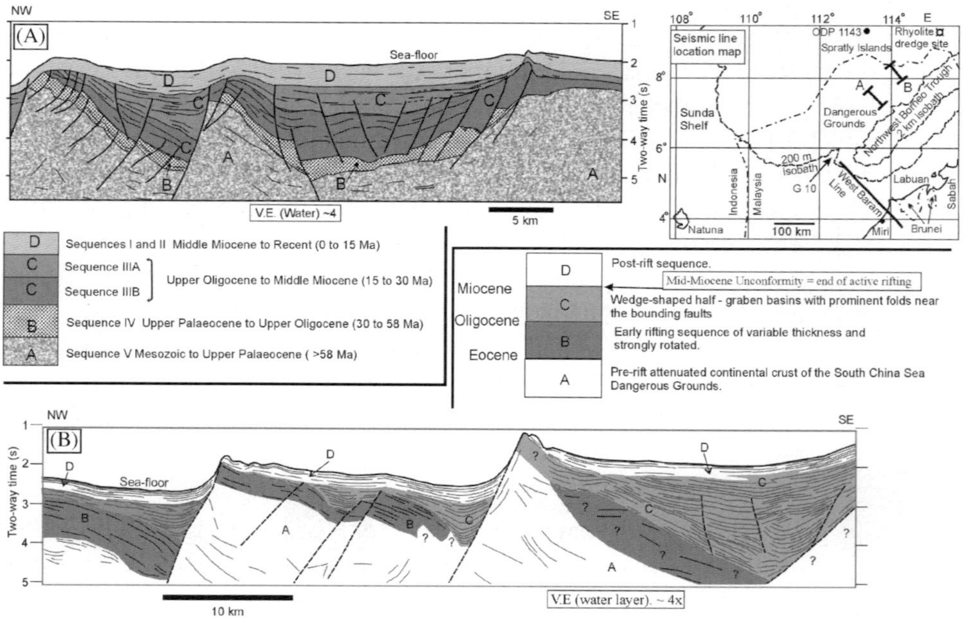

Figure 58. Interpretation of seismic lines in eastern Dangerous Grounds near the Spratly Islands. The post Mid-Miocene unconformity drape is thin and the half-graben cuestas may provide submarine outcrops of the pre-rift basement on the scarp slopes. Petronas sections A and B are interpreted respectively by Mohd Idrus et al. (1995) and by Hutchison (2004).

of the water column is ~4×, so that the scarp slope of the major cuestas appears to have a dip of 66°, but is actually only 29°. Sediment flux has been insufficient to cover the cuestas. The rift sequence is resolvable into a pre- to early-rift sequence (B) of variable thickness, and the strata are strongly rotated towards the normal faults. Overlying this is a sequence identifiable as syn-rift (C). The strata fill wedge-shaped half-graben basins between the cuestas and the well-bedded strata typically have minor folds near the bounding faults. The young post-rift sequence (D) drapes over the underlying rift sequence. Elsewhere, where it is thick enough, it completely buries the half-graben cuestas (Figure 59A).

VII.4.1. Rock formation identification

Kudrass et al. (1986) made an extensive dredge sampling programme of the Dangerous Grounds area lying between the Spratly Islands and the Reed Bank. They especially selected the slopes that support topographic highs such as banks, reefs and islands. They simply tabulated and described the large collection of rocks, but now it is possible to understand their collection by reference to the seismic sections of Figures 58. The pre-rift basement is exposed beneath the sea on the scarp slopes and has been sampled. Table 12 summarizes the range of varieties.

The dredged rocks are not unusual in the context of continental Southeast Asia. The contiguous landmasses of Vietnam, South China, Peninsular Malaysia and western Sarawak have abundant outcrops of similar Triassic strata (Hutchison, 1989). The continental terranes also contain significant belts of Triassic and Late Cretaceous granites, and localized metamorphic rocks bear witness to older sutures and deformed belts. It may, therefore, be concluded that it is the typical continental crust of Southeast Asia that has been rifted to form the basement of the shelf and continental rise of the southern South China Sea. Admittedly the sample size is small, but there is a conspicuous absence of granites that might have supported the theory of Taylor and Hayes (1983) that an Andean-type continental margin was rifting. Nevertheless, the Jurassic and Cretaceous K/Ar dates (Table 12) suggest a link to the Yenshanian tectonic events of eastern China.

Kudrass et al. (1986) have included dredge samples that date the rift-related strata of Figure 58 as Upper Oligocene to Lower Miocene. The specimens are as follows:

- Light grey-green slightly consolidated siltstone containing siliceous sponge spicules, radiolaria and planktonic Foraminifera, and Middle to Upper Oligocene nannoplankton
- Shallow-marine carbonates sampled at 23 sites. They contain Late Oligocene to Lower Miocene (Te) Foraminifera and *Nummulites*. These well-cemented shallow marine carbonates were built-up upon cuestas and show up on regional seismic records.

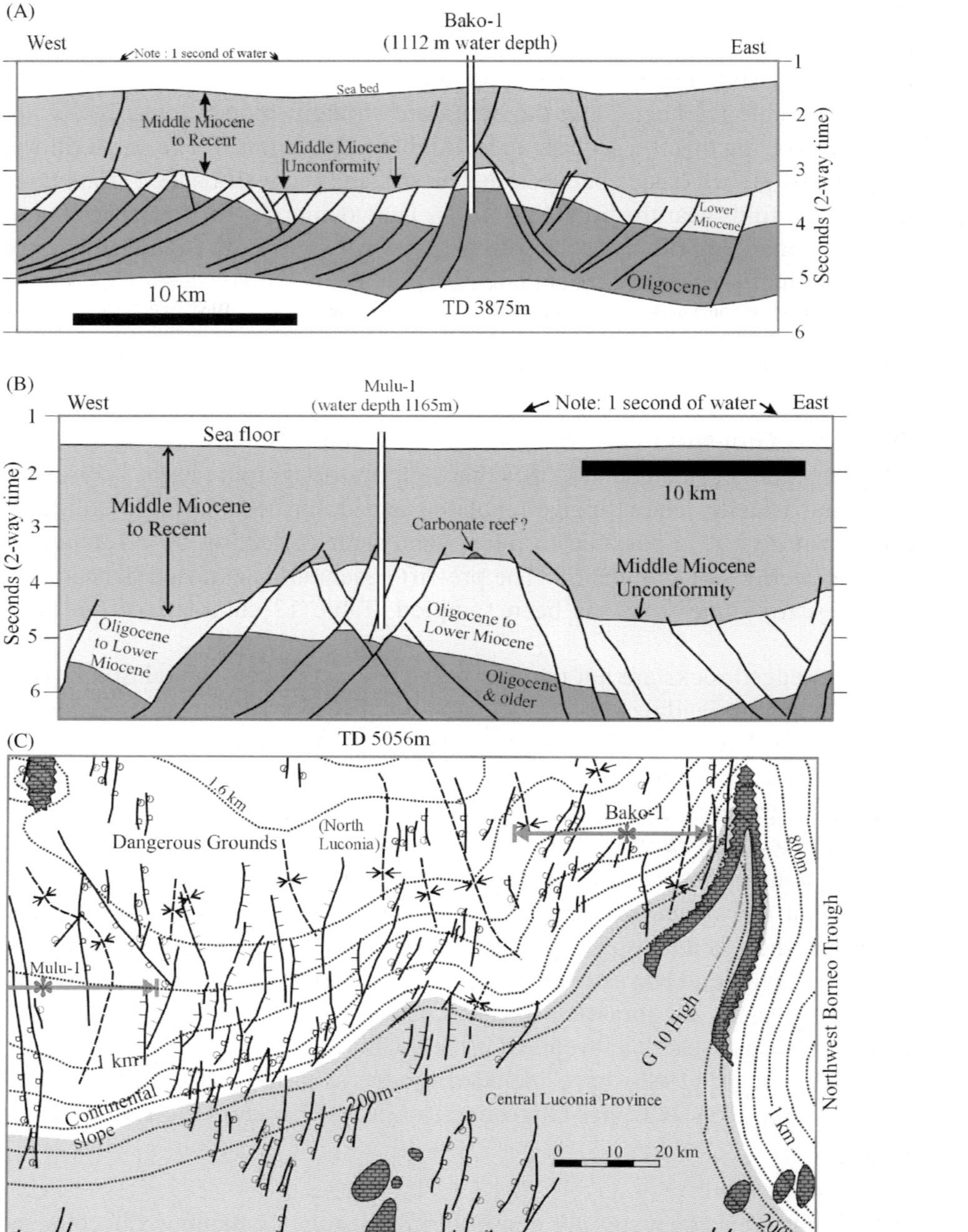

Figure 59. Western Dangerous Grounds (North Luconia) and interpreted seismic sections beneath the two deep-water wells (A, B). The chrono-stratigraphic control is imperfect, but the seismic shows a typical Dangerous Grounds signature. (C) shows the geography. All faults are normal, barbs on the down-thrown side. Redrawn from Mazlan (1999). With permission from Petronas.

Table 12. Pre-rift basement samples dredged from scarp slopes (Kudrass et al., 1986)

Sedimentary rocks	Metamorphic rocks	Plutonic rocks
Light brown-grey siltstone and sandstone containing *Clathropteris* fern leaves of Upper Triassic–Lower Jurassic age. Vitrinite reflectance: 1.0–2.5%	Biotite–muscovite–feldspar–quartz migmatitic gneiss. K/Ar date on muscovite (122 Ma) suggests Lower Cretaceous metamorphism	Boulders of dark green diorite composed of plagioclase, clinopyroxene, ilmenite and some quartz. Plagioclase and pyroxene much replaced by epidote, prehnite and chlorite. Age unknown.
Dark grey claystone containing moulds resembling Upper Triassic *Halobia* and *Daonella*	Garnet-mica schist containing sillimanite. One sample contains andalusite. K/Ar date on muscovite (113 Ma) suggests same metamorphic event	Blocks of intensely altered olivine gabbro. The olivine is almost completely replaced by aggregates of chlorite, talc and montmorillonite. Age unknown
Grey-green siltstone and fine sandstone containing Upper Palaeocene planktonic foraminifera and coccoliths. Vitrinite reflectance: 0.4%	Quartz-phyllite occurs nearby. K/Ar date on muscovite (113 Ma) indicates same Cretaceous metamorphism	
Grey-black siliceous shale with radiolarian relicts. Age unknown.	Amphibolite schist. The amphibole gave a K/Ar date of 146 Ma (Jurassic)	

Note: All data, including K/Ar dates are from Kudrass et al. (1986).

VII.5. WESTERN DANGEROUS GROUNDS

The western Dangerous Grounds, lying between the Natuna Platform and the G10 structure, is characterized by a remarkably thick (1.5–3 s TWT) post-rift sequence that completely drapes over the MMU (Figure 59).

The region is dominated by a closely spaced array of normal faults that strike N–S to NNE–SSW (Figure 59). The fault trends have been determined using a closely spaced high-quality seismic network orientated NW–SE, NE–SW and ENE–WSW (Abdul Manaf and Wong, 1995). However, 2–3° northwards, the pattern of normal faults dominantly trends northeast, approximately parallel to the magnetic anomalies of the contiguous oceanic crust (Huchon et al., 2001). Within and on the stretched continental margin, the stretching direction has been calculated to be N160° E. The change from a fault azimuth of 52° adjacent to the southwestern continent–ocean transition, to the 6° azimuth of Figure 59 has not been traced due to lack of evidence. What happens in the intervening area is unknown and a major problem of understanding exists. Obviously, the South China Sea did not have a homogeneous extensional pattern throughout, and separate sub-cells existed (Hutchison, 2004).

The draping Mid-Miocene to Recent strata are characteristically unfaulted and the faults terminate upwards at the spectacular MMU (Mazlan, 1999b). The pre-unconformity faults mostly dip and are downthrown to the west, but there are also a significant number of eastward dipping faults (Figure 60).

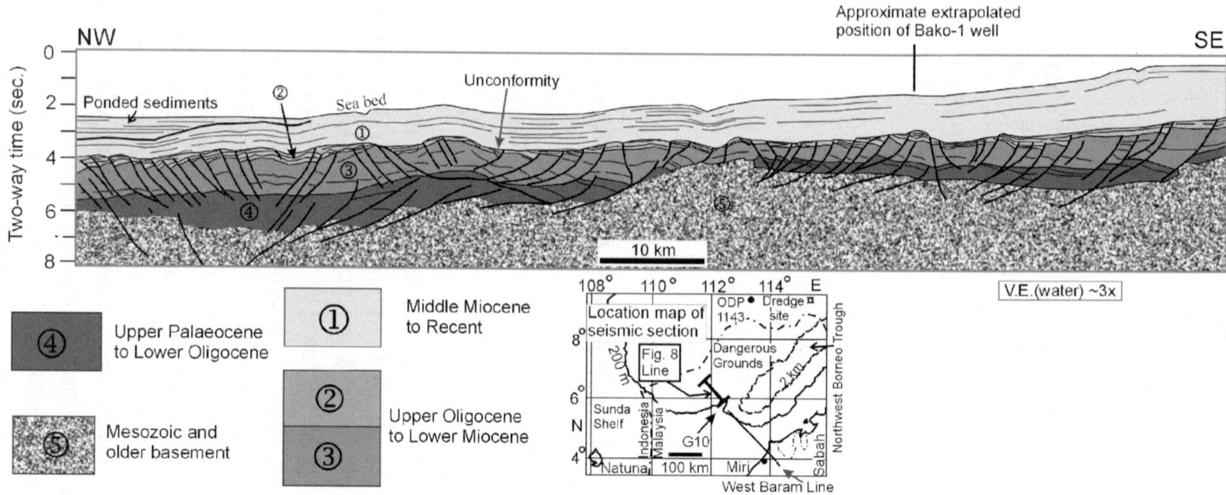

Figure 60. Interpretation using seismic stratigraphy of a section of a part of the western Dangerous Grounds province. Redrawn from Mohd Idrus et al (1995), also based on Abdul Manaf and Wong (1995). The section has some control from exploration well Bako-1 (Figure 59). With permission from Petronas.

VII.5.1. Stratigraphy

The results of a seismic stratigraphic analysis, reported by Mohd Idrus et al. (1995) and Abdul Manaf and Wong (1995), appear to be in agreement with the sparse data from the two deep-water wells—Bako-1 and Mulu-1 (Figure 60). This section has a water layer vertical exaggeration of only 3× so that the sea-floor slope is more realistic (yet 45° in reality is only ~20°).

- *Sequence 5 (Basement)* It has never been drilled, but it is reasonable to conclude that it is a composite of Mesozoic and older sedimentary and igneous rocks, with localized metamorphic belts as the dredge sampling suggests (Table 12).
- *Sequence 4 (Early rift)* This sequence is underlain and overlain by an unconformity (Figure 60). Some half-graben fill is as much as 2 s TWT thickness. The sequence frequently coarsens upwards. The lower part may be non-marine. The sequence is usually strongly rotated. The younger syn-rift deposits are characterized by strong parallel reflectors suggesting coastal plain deposits. A general age range of Upper Palaeocene to Lower Oligocene is suggested (Mohd Idrus et al., 1995).
- *Sequences 3 and 2 (Syn-rift)* The termination of sequence 2 is at the MMU. Throughout sequences 3 and 2, half-grabens were formed on a grand scale. An age of Oligocene to Lower Miocene is inferred. Sequence 3 is interpreted as shallow to open marine as the Dangerous Grounds province generally foundered after the break-up. The top of sequence 2 is a strong angular unconformity due to tilting and erosion. Coastal fluvio-marine conditions are interpreted resulting from uplift.

VII.5.2. Sea-floor edifices

Several edifices rise high from the pre-Mid-Miocene unconformity surface. Unlike the cuestas (e.g. Figure 58), they apparently have no internal seismic structure. They have little elongation and do not appear on adjacent seismic lines. Their nature is unknown but they could be Upper Cretaceous granite plutons that, unlike their country rocks, have resisted rifting. They would be analogous to islands such as Natuna. Alternatively, they could be Lower Miocene diorites or Upper Miocene adakitic plutons that are known to be abundant within the Ketungau Basin and around Kuching in western Sarawak (Prouteau et al., 2001). They do not appear to represent Pliocene basalts because they rise from the MMU surface.

When the edifices and cuestas reach into relatively shallow water, they are colonized by carbonate build-ups. Eventually, the carbonate caps formed reefs, shoals and cays known as the Spratly Islands. Since none of the islands has any outcrops other than the carbonate cover, their infrastructure remains uncertain. It was suspected that inversion may have played a role in the formation of the Spratly Islands, but inversion structures are totally absent from seismic sections such as Figure 58 across the Dangerous Grounds.

VII.6. MIDDLE MIOCENE TO RECENT STRATA

ODP Site 1143 has provided information only about the Upper Miocene to Recent post-rift strata (Figure 55) drilled as part of Leg 184 (Shipboard Scientific Party, 2000). Five hundred metres of clay and highly calcareous nannofossil ooze with Foraminifera were recovered. Kudrass et al. (1985) also dredged a large number of samples from the draping strata and young volcanic rocks (Table 13).

VII.6.1. Draping strata

Post-rift sequence 1 drapes over the unconformity surface. The age is known to be Middle Miocene to Recent and faulting is remarkably absent (Figure 61). The strata are bathyal because the Dangerous Grounds Province subsided resulting from isostatic adjustment following crustal attenuation and the break-up MMU. In this region the post-rift sequence is thick and completely drapes over the syn-rift sequences.

The post-rift strata drape over the MMU to form a fairly uniform thickness. The seafloor topography mimics, but in a subdued manner, the buried MMU (Figure 61A). Figure 60, of only 3× vertical exaggeration, shows that the draping strata are not folded. The long amplitude wave-like structure is not tectonic folding and results from the uniform deposition upon the buried rifted topography. Differential compaction may also have enhanced this effect. Figure 61A appears to suggest that the draping strata are folded. This is a misleading artefact of the large (~11×) vertical exaggeration.

The two exploration wells are non-productive (Figure 61). Micropalaeontology of Mulu-1 indicates that Oligocene to Lower Miocene strata are bathyal, whereas the transition between Lower and Middle Miocene is of coastal to inner neritic (Mazlan, 1999b). In both the exploration wells, a missing section occurs in the mid-part of the Middle Miocene and the MMU represents a hiatus of ~5 Ma (Mazlan, 1999b). The complete sequence above the unconformity is of muddy bathyal strata supplied from Sarawak by the Rajang Delta.

Table 13. Post-rift dredged and drilled formations from upper horizons (Kudrass et al., 1986; Shipboard Scientific Party, 2000)

Sedimentary formations	Volcanic rocks
500 m of core at ODP Site 1143. The well bottomed in Upper Miocene calcareous mudstones. The cored section ranges from Upper Miocene to Recent, more calcareous downwards	Porphyritic basalt; vesicular basalt containing olivine, clinopyroxene and plagioclase gave a K/Ar Pliocene date of 2.7 Ma Vesicular olivine basalt tephra surrounds lumps of Pliocene carbonaceous ooze. The basalt gave a K/Ar date of 0.42 Ma
Many sites yielded dredges of Pliocene ooze, ranging from grey-green clay to light-grey foraminiferal ooze. Coccoliths indicate a full Pliocene age range. Upper Pliocene ooze fills the outer vesicles of submarine basalt	Red and green massive dacite. Groundmass contains large plagioclase and small alkali feldspars. Secondary alteration to sericite and chlorite replace clinopyroxene. Age unknown

Note: K/Ar ages are from Kudrass et al. (1986).

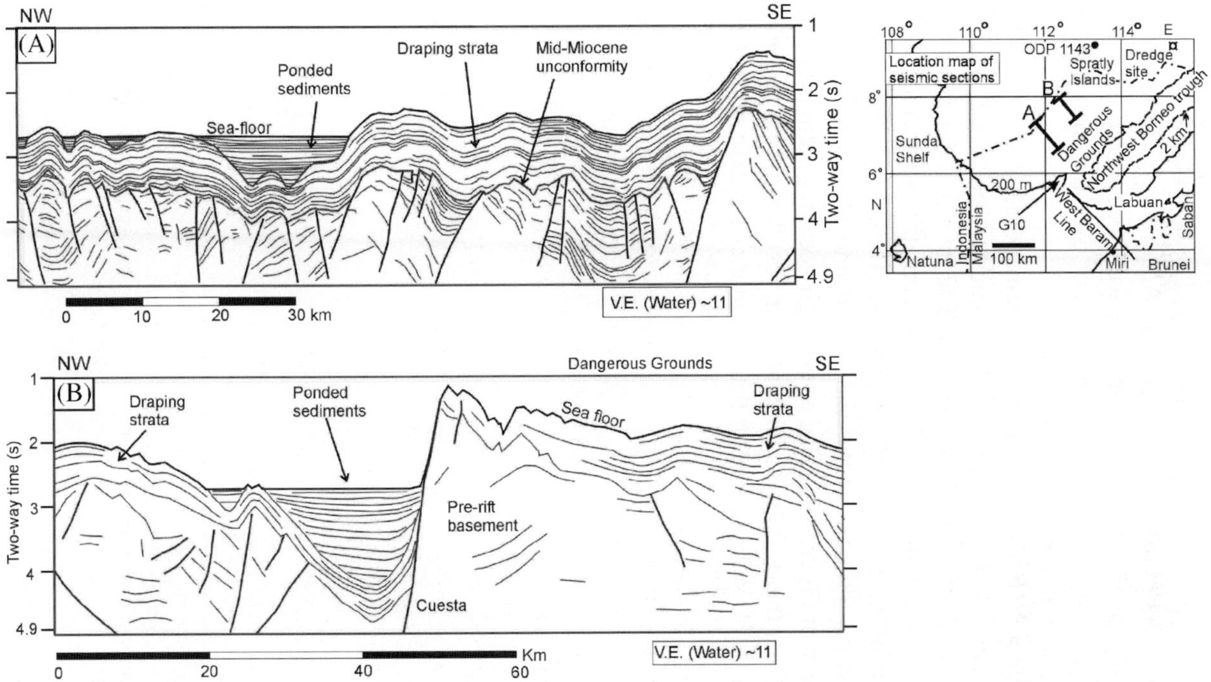

Figure 61. (A) Interpretation of a Petronas seismic section across the Dangerous Grounds north of the G10 promontory. The young ponded sediments probably represent turbidite channel 'fairways' that have cut down into the draping strata. The steepest sea-floor slope appears to be 54° because of the high vertical exaggeration. The true slope is only 7°. (B) A 20 km broad zone of young ponded sediments as wide as the Northwest Borneo Trough. The 'fairway' is bounded on one side by the sea-floor cuesta of a very large throw normal fault. The pre-rift basement is exposed on its scarp. Section included in Hinz et al. (1989).

VII.6.2. Ponded strata

Flat-lying 'ponded' strata are well displayed in Figure 61. Close inspection indicates that these were pre-existing topographic 'lows' on the top of the draping strata. The ponded strata occur in confined channels that have cut down and eroded into the top of the draping strata pile (Figure 61A). The deeper one has cut down to considerably lower levels of the draping sequence. Accordingly, these represent turbidite channels on the deep sea-floor that are now well recognized in turbidite systems. The sedimentary supply into the 'ponds' is not over the top of the adjacent ridges, but in-and-out of the line of the section.

The ponded sediments shown in Figure 61B are remarkable. One margin is formed by the impressive scarp slope of a large sea-floor cuesta. The 'pond' width is some 20 km, as wide as the Northwest Borneo Trough. The interpretation is that this wide 'fairway' has attracted many turbidite rivers that eventually meandered over the flat-lying earlier deposits of the elongate topographic low. Abdul Manaf and Wong (1995) showed blob-shaped distributions of the known Pliocene–Recent turbidites (ponded sediments) on the area of Figure 59C. Had a very high-accuracy bathymetry map been available, I believe that these blobs could have been resolved into a sea-floor meandering river system. A 3D seismic map of the sea-floor could have done the same.

Chapter VIII

Mineral, Petroleum and Coal Deposits

The commercial production of minerals in the year 1999 has been restricted entirely to industrial materials (Minerals Yearbook, 2000): Silica (quartz) sand = 274,823 t; sand and gravel = 2,941,000 t; limestone = 2,930,000 t; clay = 744,302 tonnes; building aggregate = 5,679,000 t. Exploration continues on a restricted scale, but the era of metallic mineral mining has come to an end.

Commercially viable mineral deposits were restricted to western Sarawak, west of the Lupar Line. Although, there have been traces of metallic concentrations elsewhere, they were never considered economic. As a mining province, however, western Sarawak has severely declined in importance. A summary of the mineralization is given by Hutchison (1996b).

VIII.1. NON-VOLCANIC EPITHERMAL DEPOSITS OF THE BAU DISTRICT

Stibnite, stibnite–gold and stibnite–gold–scheelite mineralization is confined to the Sundaland continental cratonic core area of SE Asia. This gold association is distinctly different from the gold–silver–telluride association of the Cainozoic volcanic arcs of the region. By contrast, it is confined to areas of continental crust characterized by high-level granite, granodiorite and diorite plutons that have provided the heat source to mobilize the metals from the old continental crustal infrastructure. Volcanic activity is lacking. There is commonly an important mercury association. The major gold-bearing quartz vein mineralization style has no volcanic association and occurs notably in meta-pelite and carbonate country rocks. The extensive Cainozoic fracture system of Sundaland, combined with the high regional geothermal gradients, have facilitated the low-temperature mobilization of the metals.

Western Sarawak is marked by important Miocene (8–23 Ma) epizonal intrusive activity. The calc-alkaline belt is marked by epithermal mineralization with gold, antimony and mercury. Only gold mining persisted later in the Bau region. Much of the gold has been from placers, but the best primary lode production has been from the Bau district. The Bau mineralization is related to a line of small high-level Tertiary dacitic to granodioritic stocks and dykes, trending NNE perpendicular to the tectonic zones, emplaced into the Jurassic–Cretaceous Bau Limestone and Pedawan Formation. The epizonal plutons provided the heat source to drive the cells of hydrothermal mineralizing solutions. The mineralization consists of complex arsenical ores containing native arsenic together with arsenical species, stibnite, pyrite and free gold. The quartz-rich veins and silicified limestone contain more free gold.

Silver is subordinate. Fluid inclusion studies indicate the temperature of formation of the mineralization to be within the range 140–250°C (Hutchison, 1983).

Sediment-hosted, or Carlin-type, Au deposits are conventionally thought to have been generated at shallow levels in a geothermal system and the Au scavenged from the host sedimentary rocks by meteoric hydrothermal fluids. However, Sillitoe and Bonham (1990) propose that the Au was contributed by magmatic hydrothermal fluids and deposited at the peripheries of base and precious metal districts, up to several kilometres from the progenitor intrusives. They classify the Bau District together with Carlin (Nevada) and Bingham (Utah), as having been formed in this way. The adakitic nature of the high-level plutons supports this conclusion.

VIII.1.1. Antimony

Antimony began to be produced from the Bau district in 1823 and since then more than 85, 000 t have been produced (Wilford, 1955). Output declined rapidly and there is no present-day production. Stibnite accompanies gold around Bau and mercury at Gading. The stibnite occurred as large granular masses and interlocking acicular crystals. The commonest gangue is quartz and calcite. Ore bodies are veins and tabular replacements in the Bau Limestone. Many mines worked eluvial ore in addition to primary veins. The main deposits were at Paku, where stibnite–quartz ore occupied faults at the Bau Limestone–shale contact. Mineralizing solutions appear to have migrated upwards through the limestone until they met the overlying impermeable shale of the Pedawan Formation, where they were impounded.

Most of the ore mined from the Bau area was from eluvial and residual deposits, concentrated as masses in solution hollows in the limestone following weathering and erosion of the shale–limestone contact. Mining of the primary veins, which continue down into the limestone, was largely unprofitable because of rapid and irregular thinning of the veins and the expense of draining the mines in this region of high rainfall.

The limestone, which forms Gunung Pangga, has been folded into a dome-like structure, subsequently broken by normal faults, along some of which dacitic dykes have been intruded (Figure 62). Hydrothermal solutions moved along the fractures to be impounded by the overlying impermeable shale of the Pedawan Formation (Wilford, 1955). The original ore bodies were probably veins or flat tabular replacement bodies in the limestone. Subsequent erosion removed the soft shale cover and weathering produced the present rugged karstic hill profile (Figure 62). During erosion, the high-density stibnite ore remained concentrated on the hill surface and settled into solution cavities that developed in the limestone, especially along the fracture system. Another favoured place for the development of deep trench-like depressions was where dacitic dykes, which had been deeply hydrothermally altered to clay, became deep and full of clay, enriched with eluvial stibnite. The mines were numerous, but individually small.

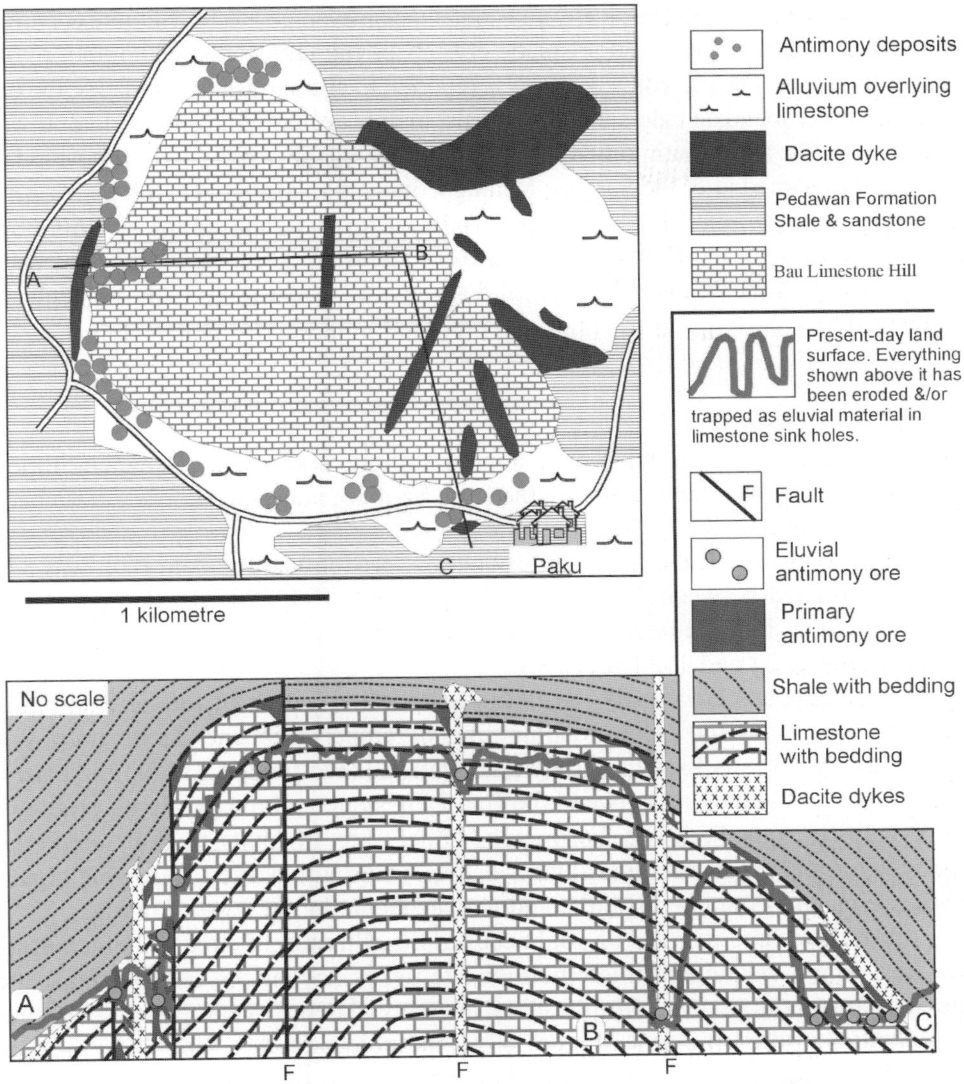

Figure 62. Non-volcanic epithermal mineralization at Bau. Antimony deposits near Paku; map and diagrammatic A–B–C cross-section. Hydrothermal alteration to clay caused accelerated weathering above faults and dykes. The released high-density stibnite accumulated eluvially within clay trapped in limestone sinkholes (after Wilford, 1955). With permission from Minerals and Geoscience Department, Malaysia.

VIII.1.1.1. Lucky Hill Mine

This antimony mine, just under 1 km south of Bau, was abandoned in 1982 due to depletion of ore reserves. The company subsequently went into producing marble slabs. The mine worked mainly calcite veins occurring along NW–WNW fracture zones within massive limestone. The main vein strikes NW and dips 50° to the SW. Its maximum strike extent was about 150 m. The veins showed rapid swelling and pinching both laterally and vertically and contained an average 5% Sb. Stibnite occurred both as massive aggregates and prismatic crystals in vugs associated with minor pyrite and arsenopyrite. The ore gold content was reported to average 5 g t^{-1}. In places, the ore also contained abundant calc-silicate minerals, mainly wollastonite and epidote. It was in this mine that a new antimony mineral, Sarabauite ($CaSb_{10}O_{10}S_6$) was discovered in 1977. It was named after the Sarabau Mining Company (Nakai et al., 1978).

Between 1970 and 1982, the mine produced an estimated 5000 t of 60–68% Sb concentrates by the flotation method from about 50,000 t crude ore. The main vein was mined by an inclined shaft and six levels, reaching an inclined depth of about 150 m.

VIII.1.2. Gold

The Bau area has been known as a gold mining district since ancient times. Intensive mining by Chinese, who came from Sambas, across the border, dates back to 1857 but flourished by 1882.

From 1864 to 1954, the Sarawak production, predominantly from Bau, was 37 t mainly extracted between 1900 and 1921 (Wilford, 1955); production peaked in 1905 and 1935, but continued until recently but on a very diminished scale. 1905 production was 2 t and 1934 was 0.9 t.

The ore deposits are mostly situated near the intersection of the ENE-trending Bau anticline with the NNE-trending line of Upper Miocene dacite porphyry, microgranodiorite stocks, sills and dykes (Figure 9) (Wolfenden, 1965). The richest deposits are at the contact between the Bau Limestone and the overlying predominantly argillaceous Pedawan Formation. Others are in fault, fracture and joint zones. Gold occurs in irregular quartz–calcite veins in silicified limestone zones. These veins also contained stibnite and native arsenic in the upper workings.

Coarse flakes and nuggets of gold occur in the alluvium of streams, which drain the areas where dacite and microgranodiorite dykes and stocks are abundant, and much of the production was from residual and eluvial deposits associated with solution cavities in the limestone (Figure 63). Fine gold particles, rarely visible even under the microscope, characterize all primary deposits of the Bau district.

Alluvial gold has been derived from weathering of sandstones and conglomerates of the Upper Eocene–Lower Oligocene Plateau Sandstone. Across the border, Sukamto et al. (1988) have shown that gold is enriched in the basal Upper Eocene sandstones of the Ketungau Basin, which unconformably overlie the Embaluh Group (Rajang Group of Sarawak). These Upper Eocene conglomerates and sandstones became a secondary source for placer gold recycled into Quaternary alluvium. The diamonds were panned last century from river gravels in the headwaters of the Sungai Sarawak Kiri, where they occurred with pebble corundum and small quantities of gold (Wilford, 1955).

Mining in Sarawak was predominantly by panning and sluicing. Trace gold is widely disseminated in the acid stocks. The deposits at the limestone–shale contacts are of auriferous silicified shale capping a mass of fine-grained quartz ore that had replaced the underlying limestone. Stibnite is common in such deposits (Wolfenden, 1965).

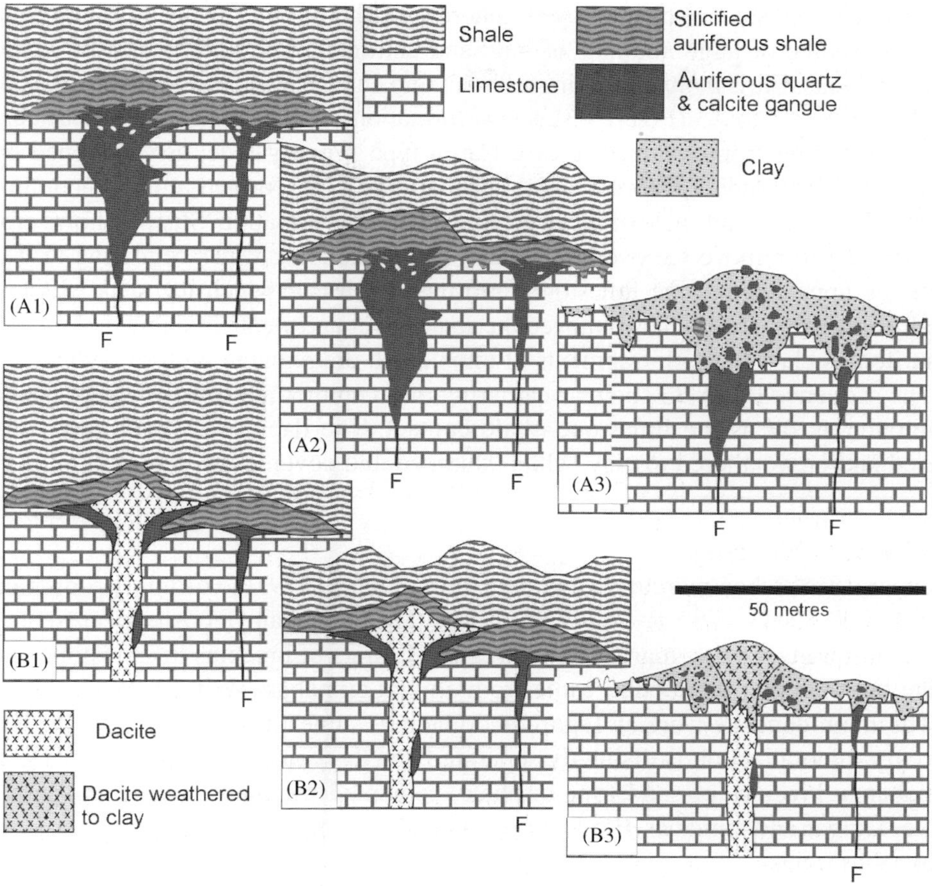

Figure 63. Non-volcanic epithermal deposits at Bau. Sequences A1–A3 and B1–B3 indicate events in the formation of secondary gold deposits related to the evolution of a regolith above the limestone hills (after Wilford, 1955). With permission from Minerals and Geoscience Department, Malaysia.

The primary ore occurs in the limestone as: quartz–calcite veins, quartz ore with calc-silicate minerals and within the contact aureole of intrusions or in quartz veins within the intrusions.

Residual deposits are of weathered ore in Au-bearing clay (Figure 63) and as cave-filling alluvium (Pimm, 1967b). A large lens-like body of disseminated Au mineralization, associated with arsenopyrite and pyrite, was discovered in shale cut by minor dykes at Jugan. Later mining in the Bau area was by opencast pits.

Arsenical gold ore characterizes many deposits in the Krokong area of Bau. They contain native arsenic, arsenopyrite, realgar and orpiment (Pimm, 1967b). A representative bulk analysis of such ore contains: Au 12.0 g t^{-1}, Ag 1.68 g t^{-1}, As 8.09%, Sb 1.91%, Fe 1.33% and Cu 0.00173%.

Coarse gold is absent and the gravity methods of recovery unsuitable. Cyanidation was used to remove the undesirable sulphides. The old workings therefore present an unsolved environmental problem.

Residual deposits are formed by weathering of the primary deposits. They have been the source of large amounts of gold and antimony. They occur mostly on the limestone flats. Eluvial ore occurs in sinkholes and channels and has also been mined. The sequence A1 to A3 (Figure 63) shows formation of a secondary auriferous clay deposit resulting from weathering of a Carlin-type primary lode, as envisioned by Wilford (1955). Stage **1** shows a typical primary ore body formed at the junction of the Bau Limestone and shale of the overlying Pedawan Formation. Stage **2** shows erosion starting to remove the overlying shale, and percolating water beginning to dissolve the upper part of the limestone, into the cavities of which the shale slumps. Finally at stage **3,** erosion has removed the shale cover and weathered much of the ore to auriferous clay in which fragments of unoxidized ore remain (Wilford, 1955).

The sequence B1 to B3 shows weathering of an epithermal deposit where dacite penetrated as far as the limestone–shale contact. The dacite was hydrothermally altered and/or weathered to clay, which slumped and covered the gold ore.

VIII.1.3. Mercury

Between 1868, when mercury was first exported, and 1949, the official production was 22,000 flasks (760 t) (Wilford, 1955). The rich Bau deposits occurred as cinnabar–quartz–pyrite–marcasite and realgar fillings of breccias in sandstone and shale, but the great production came from eluvial cinnabar (Wilford, 1955).

The main deposits were at Tegora and Gading, some 11 km south of Bau. The mercury mineralization is distinctly separate from the gold–antimony, occurring as cinnabar in breccias of sandstone or shale, cemented by quartz, pyrite and lesser barite, realgar and cinnabar. Production was mainly from eluvial ore and later from primary breccias.

The breccia localities occur close to dacite and microgranodiorite stocks. The most abundant ore mineral is cinnabar, occurring mostly as crystalline encrustations, rarely massive and along the joints of silicified sandstone and shale. The main gangue is quartz, pyrite, calcite, stibnite and talc. Barite and fluorite also occur and the breccias contain traces of gold.

The breccias are lens-shaped with overlying and contiguous eluvial and alluvial deposits. The dacite and microgranodiorite provided the geothermal heat engine for the gold–antimony–mercury mineralization, which must be regarded as one system. Temperature of deposition resulted in a separation of the individual components. The mercury deposition represents the lowest temperature of all the mineralization in the Bau district.

VIII.2. DIAMONDS

Diamonds were panned last century from the river gravels in the headwaters of the Sungai Sarawak Kiri. They are interpreted to have been derived from Upper Eocene sandstones and conglomerates of the Plateau Sandstone (and Kayan Sandstone), a source similar to that proven for occurrences in the Barito-Meratus area of Kalimantan (Hutchison, 1996b). The diamonds occur with pebble corundum and small quantities of placer gold (Wilford, 1955). The ultimate primary source of the diamonds would have been kimberlite pipes outcropping on the Sundaland landmass, the nearby part of which has foundered beneath the South China Sea near Mukah. We do not know the nature of this foundered landmass, but detrital diamonds would not have had far to travel in the fluvial system on this nearby landmass, which was an integral part of Asia at the time of deposition of the Plateau Sandstone.

VIII.3. PETROLEUM

The Miri onshore field of the Baram Delta province produced 80×10^6 barrels of oil until it was shut down in 1972. The first well Miri #1 was drilled in 1910. As of 1 January 1998, a total of 237 wildcat and 202 appraisal wells have been drilled in what is referred to as the 'Sarawak Basin', although actually it is not a basin at all (Figure 64). The most prolific oil basin is the West Baram Delta (Table 14), which is centred in Brunei, and extends continuously through the Miri area as far westwards as the West Baram Line.

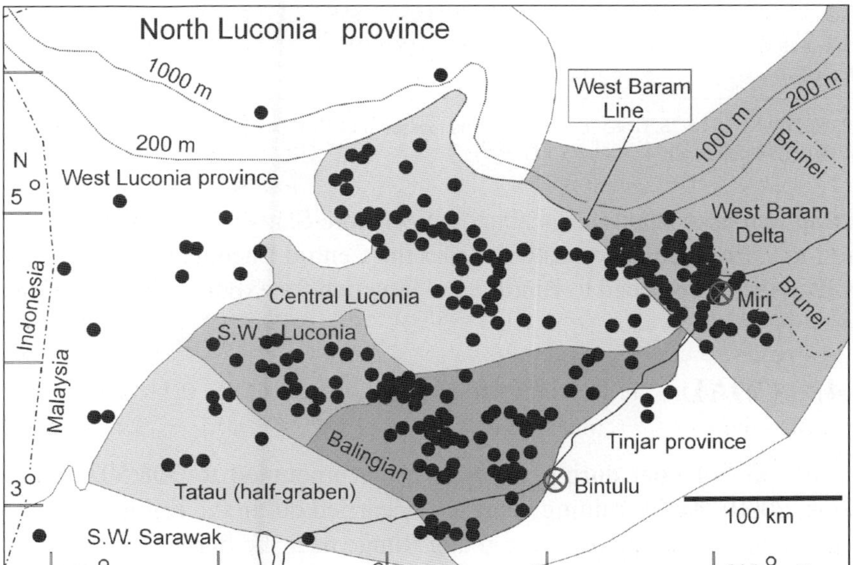

Figure 64. Sarawak well-density map, also showing the divisions into provinces (after Othman Ali and Salahuddin, 1999). The term 'Tatau' is inappropriate because the town lies within the Balingian province. It is referred to in this book as 'Mukah'. With permission from Petronas.

Table 14. Discovered oil resources in barrels, as at 1 January 1998 (Othman Ali & Salahuddin, 1999)

Basin province	Oil initially in-place	Estimated ultimate recovery	Production to 1 January1998	Reserves as at 1 January 1998
West Baram Delta	3364×10^6	1450×10^6	982×10^6	459×10^6
Balingian	2003×10^6	580×10^6	218×10^6	370×10^6
Central Luconia	901×10^6	63×10^6	nil	64×10^6
Mukah Tatau (SW Luconia)	32×10^6	8×10^6	nil	8×10^6
Sarawak total	6.3×10^9	2.1×10^9	1.2×10^9	0.9×10^9

Table 15. Discovered natural gas resources in standard cubic feet, as at 1 January 1998

Basin province	Gas initially in-place	Estimated ultimate recovery	Production to 1 January 1998	Reserves as at 1 January 1998
Gas associated with oil				
West Baram Delta	$10\,515 \times 10^9$	$6\,694 \times 10^9$	$1\,520 \times 10^9$	$5\,183 \times 10^9$
Balingian	$6\,029 \times 10^9$	$3\,295 \times 10^9$	192×10^9	$3\,101 \times 10^9$
Central Luconia	$43\,462 \times 10^9$	$34\,936 \times 10^9$	$6\,288 \times 10^9$	$28\,622 \times 10^9$
Tatau (SW Luconia)	$7\,010 \times 10^9$	$5\,230 \times 10^9$	nil	$5\,183 \times 10^9$
West Luconia	$2\,313 \times 10^9$	$1\,464 \times 10^9$	nil	$1\,462 \times 10^9$
Tatau (half-graben)	771×10^9	680×10^9	nil	709×10^9
Sarawak total	70.1×10^{12}	52.3×10^{12}	8.0×10^{12}	44.3×10^{12}
Non-associated gas				
West Baram Delta	$4\,399 \times 10^9$	$2\,814 \times 10^9$	nil	$2\,842 \times 10^9$
Balingian	$3\,544 \times 10^9$	$1\,782 \times 10^9$	nil	$1\,786 \times 10^9$
Central Luconia	$43\,137 \times 10^9$	$34\,987 \times 10^9$	$6\,300 \times 10^9$	$28\,704 \times 10^9$
Tatau (SW Luconia)	$6\,721 \times 10^9$	$5\,112 \times 10^9$	nil	$5\,116 \times 10^9$
West Luconia	$2\,322 \times 10^9$	$1\,454 \times 10^9$	nil	$1\,462 \times 10^9$
Tatau (half-graben)	978×10^9	750×10^9	nil	690×10^9
Sarawak total	61.1×10^{12}	46.9×10^{12}	6.3×10^{12}	40.6×10^{12}

1 ft^3 = 0.028,316,85 m^3 = 28.31685 l. Standard ft^3 is measured at 70°F (= 21.11°C) under a pressure of 1 atm (1 atm = 101.325 kiloPascals, 1 Pascal = 1 N m^{-2})

Natural non-associated gas is abundant offshore Sarawak, reservoired in carbonate build-ups. Production comes only from the Central Luconia province (Table 15), from which the gas is piped to Bintulu and liquefied for export at Tanjung Kedurong.

VIII.4. COAL

The production of coal during the year 1999 amounted to 308,502 t (Minerals Yearbook, 2000). Active mining is presently carried out in the Kapit Region (Merit-Pila). The Bau district and the Klingkang Range near the border are under study.

Several lignitic coal-bearing basins have been identified and some have supported past mining activity, while others have subsequently been developed. The Silantek Formation in the Klingkang Range, some 4 km north of the Kalimantan border, contains several coal seams within a sequence of siltstone, shale and sandstone (Haile, 1954). One coal seam has a thickness of 1.9 m. The formation is locally intruded by Miocene stocks and dykes, which have locally metamorphosed the coal. Fifty Mt of coal are estimated on the Sarawak side of the border. The coal-bearing Upper Eocene Silantek of the Klingkang Range continues beneath the Plateau Sandstone into Kalimantan, where it dominates the Ketungau Basin (Zeijlmans van Emmichoven, 1939).

The Balingian coal area occupies the coastal plain between Mukah and the Balingian River (Wolfenden, 1960). The coal-bearing formations are the Miocene Balingian and Pliocene Begrih-Liang formations. Lignite seams occur within the Miocene strata, and three seams of 1.5–5 m thick occur in the Begrih-Liang Formation, both investigated by drilling. Reserves are estimated to be of the order of 100 Mt of lignite grade.

The Bintulu coal deposit occurs in Miocene paralic strata (Kho, 1966). About 2 Mt of bituminous coal are regarded as mineable, but less than half will be accessible by open pit.

VIII.4.1. The Sadong colliery

This formerly mined deposit occupies the west end of Gunung Ngili, close to the Sadong River (Figure 13). The mine started in the 1850s and Mr. Coulson became the Borneo Company resident engineer and attempted without success to reopen the mine that closed because it was losing money. In 1855, the famous naturalist Alfred Russel Wallace visited the site and stayed several months with Mr Coulson.

By 1874, however, coal was being worked. Buffaloes were initially used, but a railway with steam locomotives was built to transport the coal to the wharf on the river. Between 1874 and 1931, about 1 Mt of coal were mined, some of which was used locally, and 862,502 t were sold. The peak annual production was 30,894 t in 1898. The mine closed after 1931 because of loss of the local market, closing of most of the mines at Bau, and the change from coal to oil in ships and power stations.

The coal was mined from two seams in the Silantek Formation: the Top and the Main, which lay about 76 m vertically apart. The Top seam varied from 46 to 61 cm of bright, clean looking, friable coal, underlain by soft shale. The Main seam was duller and had a roof and floor of shale. It was composed of two parts, separated by a thin shale horizon. The clean coal thickness averaged 0.91 m. All the coal was sub-bituminous (Haile, 1954).

VIII.4.2. Merit-Pila coal mine

The basin is located 75 km upstream from Kapit in the Rajang River basin (Kirk, 1957; Johari and Mohamad, 1994). It is the biggest coal-bearing basin in the country, and covers 160 km^2 , elongated in an E–W direction. The coal is contained within this the largest outlier of the Upper Oligocene to Miocene Nyalau Formation, which unconformably overlies the Rajang Group orogenic belt of the Sibu Zone. The Miocene rocks include conglomerate, sandstone, mudstone and coal seams. Two or three seams range in thickness from 1.5 to 3.8 m and have commercial potential. An estimated 300 Mt have been reported from drilling on the western side of the basin and 50–60 Mt may be mined by open pit, now in operation. The rank is high-grade lignite.

Wan Hasiah Abdullah (1997) showed that coal samples from the Stapang field of Merit-Pila lie within the range 0.36–0.41% R_0. There is virtually no difference between outcrop and core samples.

VIII.5. BAUXITE

Mining began in the Sematan district early in 1958 and ceased in 1965. The total production of washed bauxite ore was 1.617×10^6 t. Since 1959, bauxite was Sarawak's most valuable mineral commodity and peak production of 2.80×10^5 t was attained in 1961. The deposits all occurred in the Sematan district of western Sarawak, and have been described by Wolfenden and Haile (1963). The bauxite is a residual deposit formed by *in situ* weathering of basic and intermediate igneous rocks. Four types of parent igneous rocks have been identified: pyroxene andesite (e.g. at Munggu Belian); gabbro (e.g. Bukit Gebong) and greenstone (e.g. small low-grade deposits at Tanjung Serabang) and medium-grained amphibolite (e.g. small low-grade deposits in the Gunung Puting area). The bauxite shows the residual texture of the underlying parent rock, and contains core boulders of unweathered parent rock. The boulders have a skin of bauxitized material, and the transition from parent to bauxite is sharp. A thick bed of kaolinitic clay separates the bauxite from the underlying fresh rock (Wolfenden and Haile, 1963).

VIII.5.1. Munggu Belian deposit

High-alumina bauxite was formed from andesite at Munggu Belian, Sematan, where the hills rise about 24 m above the surrounding alluvial plain. The hill contained 1.69 Mt of washed bauxite prior to mining with an estimated 228,000 t that could be mined below the alluvium.

The bauxite on the hills forms a bed about 3 m thick. It occurs as pink to brown nodules (0.6–30 cm, average 2.5 cm across) scattered in yellowish clay composed of gibbsite and kaolinite. The bauxite nodules contain 70–80% gibbsite and the Al_2O_3 averages 51–54 wt%. Massive soft kaolinitic clay with spheroidal weathering underlies the bauxite and may be as thick as 18 m. It contains a few core boulders of the underlying parent andesite.

VIII.5.2. Bukit Gebong

Bauxite has been formed from gabbro on the steep-sided Bukit Gebong (334 m). The deposit was estimated to contain a mineable 1.5 Mt of washed bauxite with an average composition of 49.4% alumina (Wolfenden and Haile, 1963).

The bauxite formed a bed averaging 2 m thick. The bauxite is friable and reddish-brown, contains boulders of all sizes up to 12 m diameter of gabbro, which would hinder mining. The core of the boulders is fresh, with a thin skin of bauxite. The bauxite ranged from 45–53% Al_2O_3 and contains 64–78 % gibbsite. A thick bed of kaolinite, goethite and some gibbsite underlies the bauxite. The parent rock is an olivine gabbro, composed of labradorite, pyroxene and olivine.

VIII.5.3. Other deposits

Eight other small bauxite deposits occur in the Samatan area. Most are low-grade, none are economic. At Munggu Bajo Bergantong, bauxite was formed from gabbro forming a low hill. At Gunung Pandan, bauxite was formed from gabbro. At Tanjung Serabang, low-grade bauxite was formed from greenstone. At Gunung Angus, bauxite was formed from sheared gabbro. At Gunung Tamin Tungku, bauxite and laterite were formed over basic igneous rocks. At Bukit Batu, bauxite was formed over metamorphosed intrusions of dolerite. In the Lower Samunsam Valley, low-grade bauxite and laterite were formed from greenstone. In the Gunung Putting area, bauxite and laterite were formed over amphibolite.

VIII.6. GLASS SANDS

Surficial white quartz sands of extremely high purity form thin veneers and dunes along the coastal plain, especially between Bintulu and Tanjong Kidurong. They have been mined and exported for optical glass. They commonly have an extremely sharp base upon underlying peat swamp material, so they are definitely transported. Wind along the coastal zone seems to have had a role in depositing and purifying these surficial deposits.

Chapter IX

Tectonic Elements and Models

The tectonic evolution of Sarawak cannot be analyzed without recognition of the geology of the whole of Borneo. The detailed map of Tate (2001), meticulously compiled from Indonesian, Malaysian and Brunei sources, is invaluable in this respect. Equally needed are the first order offshore structural elements. Oil company seismic data are naturally aimed at finding oil. Regional research seismic lines exist, but they are not in the public domain. SEASAT gravity maps derived from satellite altimetry, such as that of Sandwell and Smith (2000), are of immense importance. Unfortunately only the low-resolution SEASAT data are in the public domain (Figure 65). Higher resolution gravity maps do exist, and offshore Sarawak data are under the control of Shell. However, they add nothing of importance to our knowledge of the first-order tectonic elements.

IX.1. GRAVITY DATA RELATED TO STRUCTURE

The gravity measured over the Balingian Province (Figure 65) shows a characteristic signature. There are no steep gradients, the contours are widely spaced, and values are low, generally less than 50 $\mu m\ s^{-1}$, suggesting a thick blanket of Tertiary low-density sediments. Outboard of the Balingian Province lies the Central Luconia Province, also without steep gradients, but the gravity readings are within the range 100–250 $\mu m\ s^{-1}$, suggesting that the blanket of Tertiary low-density sediments is considerably less and that a basement of higher density lies closer to the surface.

The Tatau compressive horst is well shown by the gravity data (Figure 65), displaying values as high as 450 $\mu m\ s^{-1}$, combined with very high gradients on either side, indicating that a high-density basement has been uplifted to the surface by steeply dipping faults. The closely spaced gravity contours extend north-eastwards parallel to the coast, a steeply dipping structure known as the Anau–Nyalau Fault.

The western extent of the low-gradient terrains of Balingian and Central Luconia may be identified as the powerful NNW–SSE-trending West Balingian Fault, separating terrains of distinctively different character (Figure 65). West of this fault lies the Mukah Province, characterized by NW–SE-trending master faults downthrown to the SW. The graben and fault trends are clearly displayed by the gravity contours. Higher values up to 300 indicate where the basement almost reaches the surface, and lower values less than 150 $\mu m\ s^{-1}$ overlie the thick half-graben fills of Oligocene and Miocene strata. Southeast of the Anau–Nyalau Fault, high gravity values as high as 400 $\mu m\ s^{-1}$ are common, in keeping with the regional geology in which high-density basement rocks such as the Mulu and Kelalan formations outcrop as inliers from

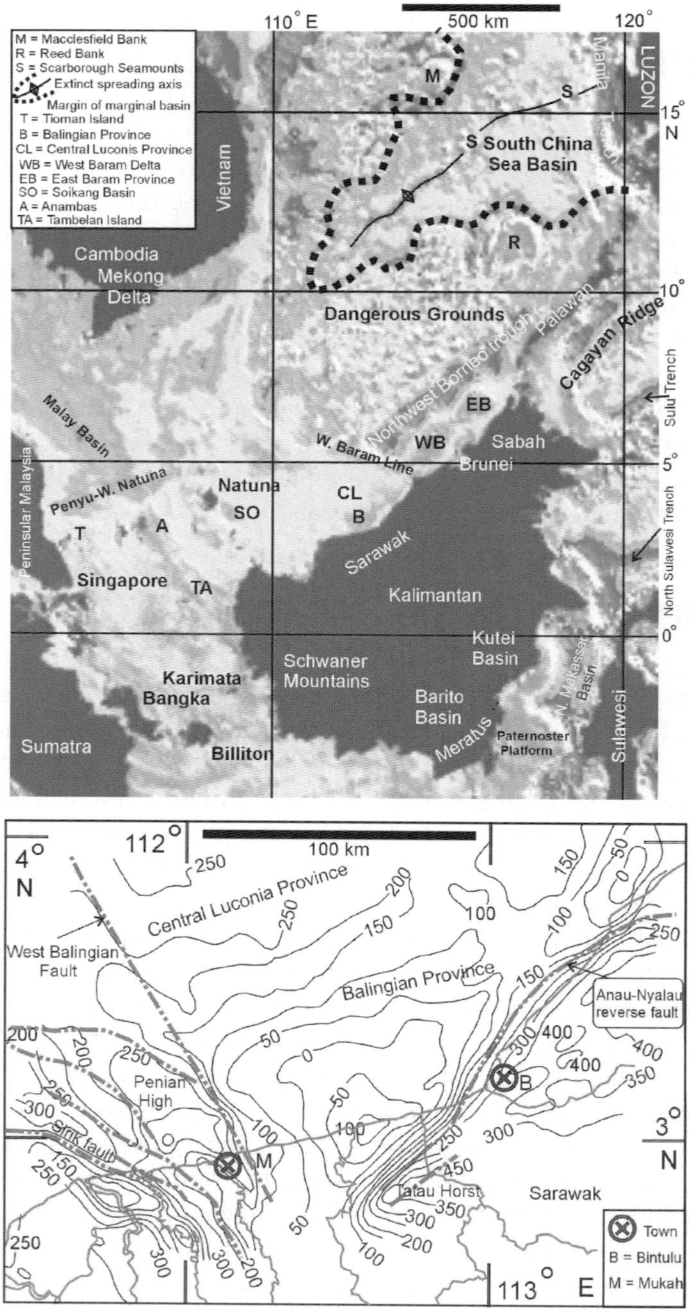

Figure 65. Top: SEASAT marine gravity anomaly map of selected part of Southeast Asia. Download from Oxford University Earth Sciences Department web site http://www.earth.ox.ac.uk/~geodesy/downloads.html. Lower: Detail of gravity measurements of the coast of Sarawak (taken from Hutchison, 1991). Units of measurement are: μms^{-1} . The faults shown are those highlighted by the gravity data.

beneath the Nyalau Formation. Thus the gravity map of this selected area may be readily interpreted in terms of the first-order structure in an elegant way.

IX.2. SEASAT GRAVITY FROM SEA SURFACE ALTIMETRY

Several SEASAT images are available on the Internet. I have selected the Southeast Asia portion posted by Oxford University Department of Earth Sciences (Figure 65) obtained from site www.earth.ox.ac.uk/~geodesy/downloads.html shown in shades of grey. Higher resolution SEASAT imagery is available, but not in the public domain. Such imagery adjacent to Sarawak is under the custody of Shell, who kindly allowed me to see the higher resolution version as well as offshore magnetic surveys, but they add nothing of significance to the SEASAT image shown in Figure 65.

IX.2.1. Tectonic trends from West Borneo

At first appearance, without the aid of SEASAT, the important trend is ENE from north Johore in Peninsular Malaysia, eastwards through Pulau Tioman, Anambas and thence to Natuna. But this trend is spurious, resulting from the younger rifting of the Penyu and West Natuna basins, and the important Miocene inversion of that area.

The real trends are SSE through Singapore, through Bangka, thence swinging eastwards towards the SW tip of Borneo, as pointed out by Hutchison (1993), based on an analysis of the structural geology of the so-called 'tin islands'. The arc is dominated by well-dated Mid- to Late-Triassic granites (Cobbing et al., 1992).

The most prominent feature of western Kalimantan is the Schwaner Mountains, dominated by Cretaceous granites and volcanic rocks. In the east, the trend is due W, gradually curving northwestwards to link with the Anambas islands along a strong broad SEASAT lineament where low-density granitoid basement lies beneath shallow seas of the Sunda Shelf (Figure 65). Radiometric dating of the northern Schwaner Mountains granitoids has been summarized by Hutchison (1996a) from various sources. The plutonic complex has been dated by whole rock K:Ar to range from 104 to 123 Ma and biotite and hornblende separates gave K:Ar dates ranging from 75 to 115 Ma. The southern region has yielded whole rock ages of 86–91 Ma and mineral separates have given 78–86 Ma. The volcanic rocks of the southern region have yielded whole rock ages of 65–75 Ma. The ages indicate entirely Cretaceous magmatism, extending over a wide range from Barremian to Maastrichtian. However, this wide range is unlikely to be wholly accurate and future radiometric work is needed to refine it. The plutonic arc of the northern Schwaner Mountains extends to the Anambas Islands, an arc on which the granite of Tambelan Island has been K:Ar dated at 84 Ma using biotite (Table 5). Another broad parallel arc trends due south through Natuna Island then gradually curves towards the SE into Sarawak along the estuary of the Rajang River (Figure 66).

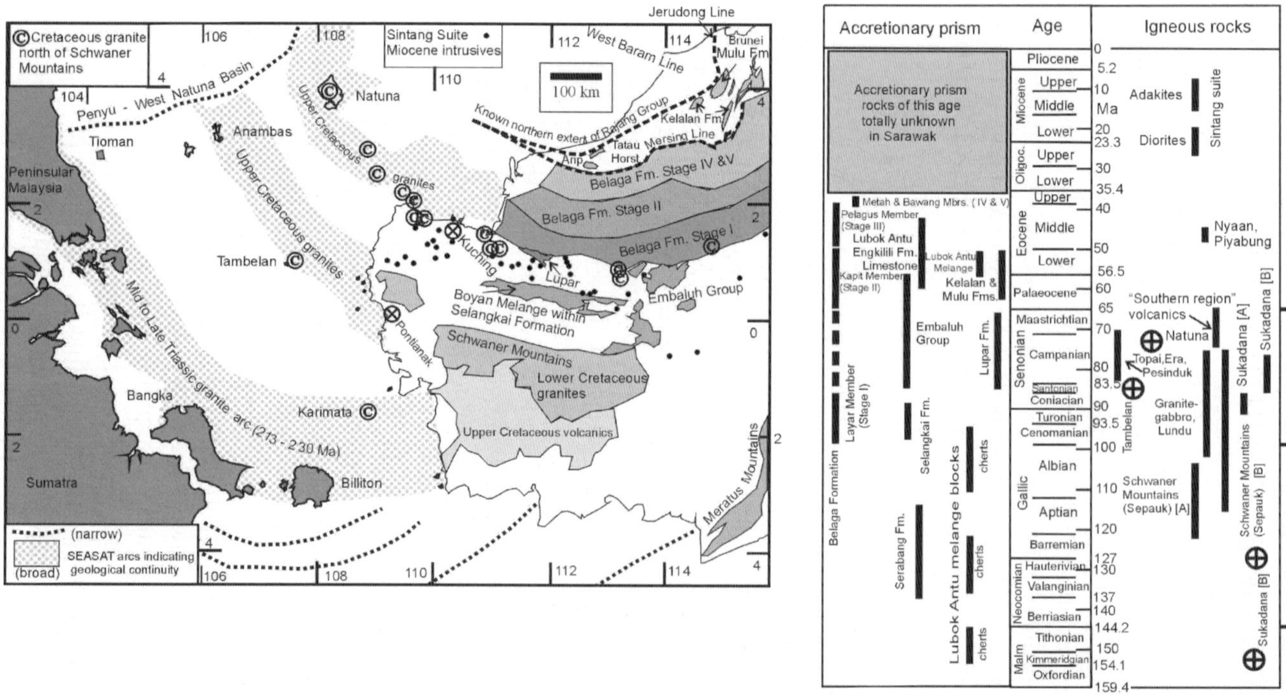

Figure 66. Extrapolation seawards of accretionary prism and igneous arcs from Sarawak and Kalimantan (based on SEASAT) and table showing the known palaeontological and radiometric ages. Future research may refine the ages.

The SEASAT gravity of the arc through Natuna is dominated by Cretaceous granitoids, K:Ar dated 73 Ma in Natuna (Table 5). The arc continues through Tanjung Datu to the Lundu area, thence to Tinteng Bedil and eastwards into Kalimantan. Consistent K:Ar biotite ages within the Upper Cretaceous age 76–79 Ma (Table 5) confirm the continuity of this arc. Shaded areas within this broad arc (Figure 57) represent less elevated basement. The most prominent represents the large Soikang Basin and probably other smaller half-graben basins can be 'seen' on the imagery. The broad arc through Natuna not only includes Late Cretaceous granitoids but also accretionary prisms comprising the Sibu Zone: the palaeontologically dated Sejingkat (115–137 Ma) and Belaga Formation (95–45 Ma). If the Lupar Fault is possibly 'recognizable' on Figure 57, then it could be represented by the southern and SW margin of the low-density arc. It would then continue west of the Lupar estuary, parallel and close to the coast, then swinging NW and finally N between Natuna and Anambas, through several small anomalies.

IX.2.2. Tectonic elements NE from Tanjung Sirik

North-eastwards along the coast from Cape Sirik towards Miri, the SEASAT imagery (Figure 65) clearly shows the moderate-gravity Balingian Province, indicating a relatively thin sedimentary sequence overlying a relatively dense basement. Outboard of the Balingian Province, the SEASAT imagery prominently shows the Central Luconia Province. The SEASAT colouring of Central Luconia is misleading, for the detailed gravity contours (Figure 65, lower) indicate that higher gravity results from elevated high-density basement (SEASAT appears to suggest low density).

The NW–SE trending West Baram Line is clearly seen, terminating the south-westwards extension of the high-density Northwest Borneo Trough. Outboard of the Northwest Borneo Trough is the very distinctive patch work of the Dangerous Grounds (including the Spratly Islands), where numerous horsts of the attenuated continental crust have supported a large number of carbonate build-ups and present-day pinnacle reefs.

Landward of the Northwest Borneo Trough, two very prominent low-density circular areas, which have amalgamated to a figure-of-eight (Figure 65) represent the great thickness of low-density Neogene sediments in the West Baram Delta and East Baram Province. Northeast of the Dangerous Grounds lies the large and prominent Reed Bank and to its west are other smaller carbonate banks, which delineate the southern margin of the Oligocene to Lower Miocene South China Sea marginal basin.

The northwest margin of the South China Sea basin likewise has small carbonate banks, as well as the large Macclesfield Bank (Figure 65). The fossil spreading axis of the marginal basin, which de-activated in the Early Miocene (Briais et al., 1993), is very clear on SEASAT and towards the northeast it passes through several clearly 'seen' basaltic seamounts (The Scarborough Seamounts). The active high-density Manila Trench is very prominent, representing eastwards present-day subduction beneath Luzon. Younger sedimentary cover in the South China Sea Basin makes the pixel colours misleading in terms of gravity.

IX.2.3. Tectonic elements extending from eastern Borneo

The extinct volcanic arc named the Cagayan Ridge is obvious and links the Philippines to Sandakan. The dark area that parallels it along its NW side is not understood. It is not a trough, but is a great thickness of Crocker Formation with younger superimposed Upper Miocene basins (Hutchison, 1996). The presently active Sulu Trench is obvious, lying along the NW margin of the Sulu Archipelago.

Further south, the North Sulawesi Trench is distinct. South of it is the North Makassar Basin, with the Mahakam Delta extending into it from the Kutei Basin. The Paternoster Platform continues with similar low-density shallow basement linking Kalimantan with Java, Billiton, Bangka and Sumatra (Sundaland). The Karimunjava Arch links SW Borneo with SE Sumatra in a gentle curve, and the parallel Billiton Basin is recognized by small light grey pixels to the south of Billiton Island. The extensive medium grey platform, which extends continuously from Tioman, Anambas and Natuna, southwards through Billiton, then wrapping around southern Borneo as far NE as the SW margin of the Kutei Basin (Paternoster Fault) represents the Sundaland Peninsula which was a continuous landmass at the end of the Cretaceous, upon which Tertiary basins began their sedimentary history with continental and lacustrine sedimentation (Hutchison, 1992b, 1996a). The darker areas north of Anambas and Natuna were also part of this Sundaland Landmass, but have foundered more as a result of greater crustal attenuation.

IX.3. TECTONIC MODELS

Several tectonic models have been proposed for Sarawak, not all equally successful and some must be categorized as bizarre. The pre-Plate Tectonic models include the outstanding work of van Bemmelen (1949), who recognized and named great arcuate belts or zones that passed through The Malay Peninsula and Borneo. Without our present–day knowledge of the South China Sea (Figures 61 and 62), van Bemmelen (1970) was remarkably perceptive in drawing in the various tectonic zones. Haile (1969) made a great achievement by recognizing that all the geosynclinal elements of orogenic zones (with the exception of a foreland), as documented by Aubouin (1965), could be identified in Sarawak. This clearly showed that Sarawak is tectonically similar to other orogenic belts of the world. Hutchison (2001) later showed that the foreland, or stable continental block towards which tectonic deformation youngs, is present and shallowly buried beneath and offshore Mukah to the west of the West Balingian Line. It is an integral part of the Sundaland foreland required by the geosynclinal theory.

IX.3.1. Important features for modelling

IX.3.1.1. Old oceanic lithosphere (Proto South China Sea)

Ophiolitic igneous rocks have been described from the Serabang, Sejingkat and Lupar Formations. In the latter, they are collectively known as the Pakong Mafic

Complex. Collectively they are interpreted as the basic crust of a former basin, either an integral part of an ocean, or more likely of a marginal basin. The basalts preserve excellent pillow structures. Although not radiometrically dated, their age may be inferred from identified radiolaria in associated ribbon cherts, interpreted as the first sediments directly deposited upon the pillow basalts of the deep basin. The chert clasts within the Lubok Antu Mélange fall into three groups (Figure 66):

- Kimmeridgian–Tithonian (Upper Jurassic)
- Valanginian–Barremian (Lower Cretaceous)
- Albian–Cenomanian (Lower–Upper Cretaceous boundary)

It is common to find such a range of chert ages, but in this case the total age range of 60 Ma (from 154 to 94 Ma ago) is rather excessive for the active duration of a marginal basin. An ocean is more likely. In the geosynclinal terminology of Aubouin (1965), this was the sea or basin that was filled by a thick flysch sequence, collectively termed the Rajang Group, which dominated the Sibu Zone and the northern part of West Sarawak. Following Brondijk (1964), Haile (1969) suggested that this flysch-filled basin floored by pillow basalts, overlain initially by chert, should be called the 'Danau Sea'. However, the term has not been widely accepted, and 'Proto South China Sea' has come into wide usage.

IX.3.1.2. Support for an arc-trench convergent margin

The Schwaner Mountains are dominated by Lower to Upper Cretaceous I-type plutonic and volcanic rocks (Williams et al., 1988; Bladon et al., 1989; Haile et al., 1977). Because they lie equidistant from The Rajang Group flysch belt and the Meratus Mountains (Figure 66), Hamilton (1979) suggested that the volcano-plutonic arc may have been related to a trench in the neighbourhood of the Lupar Line, or alternatively within the Meratus Mountains. Of the two alternatives, the curvature of the Meratus Mountains favours a former Benioff Zone dipping northwestwards beneath the Schwaner Mountains. If we are to use the Lupar Line, then the curvature would require a bent or segmented Benioff Zone to enable it to dip beneath the Schwaner Mountains.

The tabulation of radiometric ages of igneous rocks, which could possibly be subduction-related (Figure 66), in conjunction with flysch formation deposition, that may be interpreted as accretionary prism material, indicates that subduction must have ceased in the Palaeocene, about 63 Ma ago. After that, the flysch strata must be interpreted as deposited in a basin which was not subducting. There is no problem in accepting the Serabang Formation as accretionary prism. It contains all the necessary elements, or foliated mudstones, mélange zones and ophiolite. The Lupar Formation, including the Pakong Mafic Complex, may also be accepted as accretionary prism material (Figure 66). The early trench would have been located within the Serabang Formation outcrop. Later, it would have migrated to the Lupar Formation outcrop. This interpretation has been carefully formulated by Hutchison (1996a) and by Moss (1998). The gap between the accretionary prism and the Schwaner Mountains is the natural place for shallow water fore-arc basin formation, and the Cenomanian to Turonian Selangkai Formation is ideally interpreted as occupying a fore-arc basin. The Boyan Mélange is constructed of broken and disrupted Selangkai Formation (Moss, 1998) and hence it can have no plate tectonic significance other than that of a younger major shear zone, which separates the Ketungau and Mandai basins.

The Upper Cretaceous granite–gabbro-hybrid arc through Tanjung Datu, Lundu and Tinteng Bedil, has intruded the Lower Cretaceous accretionary prism within the Serabang Formation outcrop. Granites of this age also continue eastwards into Kalimantan (Figure 67). They are interpreted as products of the end-stage of subduction, as the formerly subducting Proto South China Sea slab no longer subducts, but falls away with a steeping trajectory into the mantle under its own weight, a process colloquially known as 'slab roll-back' (Figure 67).

IX.3.1.3. Post-subduction Proto South China Sea basin in collision

From mid-Palaeocene time (~64 Ma) there is no further subduction-related igneous activity, either within the Schwaner Mountains, or within the Late Cretaceous arc through Lundu and Tinteng Bedil. Therefore, the Palaeocene and younger flysch strata of the Rajang Group do not represent accretionary prism material, but sedimentary fill of the Proto South China Sea Basin which is in a collisional tectonic situation, compressed between Southern Sundaland and Northern Sundaland, composed of the Mukah-Central Luconia province (Figure 67). Moss (1998) therefore interprets the Rajang-Embaluh Group as representing deposition as a large submarine fan, fed by the proto Mekong River upon a substrate of already deformed oceanic crust. The basin was closed and its strata deformed into a large anticlinorium by the Middle Eocene before extrusion of the Nyaan Volcanics and formation of the overlying Tertiary basins, such as the Ketungau.

The absence of subduction and accretion during deposition of the later part of the Embaluh Group and Belaga Formation (Kapit, Pelagus, Metah and Bawang Members) and Mulu and Kelalan Formations, is consistent with the absence of features characteristic of accretionary tectonics. Unlike the Rajang Group, the Embaluh Group is not metamorphosed. It is a turbidite sequence which youngs southwards, whereas the Belaga Formation is metamorphosed and youngs northwards. Low grade metamorphism of the Belaga Formation needs to be attributed to collision by the Mukah-Central Luconia part of northern Sundaland. Fission track dating of zircons from the Embaluh Group (Moss, 1998) indicate a Triassic provenance, widespread in Sundaland as a result of the Indosinian Orogeny.

IX.3.1.4. Rocks of the Lupar Line in need of reinterpretation

Whereas the Lupar Formation (and its contained Pakong Mafic Complex) are interpreted as part of accretionary prism tectonics, this interpretation cannot be extended to include the Lubok Antu Mélange, whose matrix is palaeontologically dated Lower Eocene (Figure 67). The limestones of the Engkilili Formation, which occurs

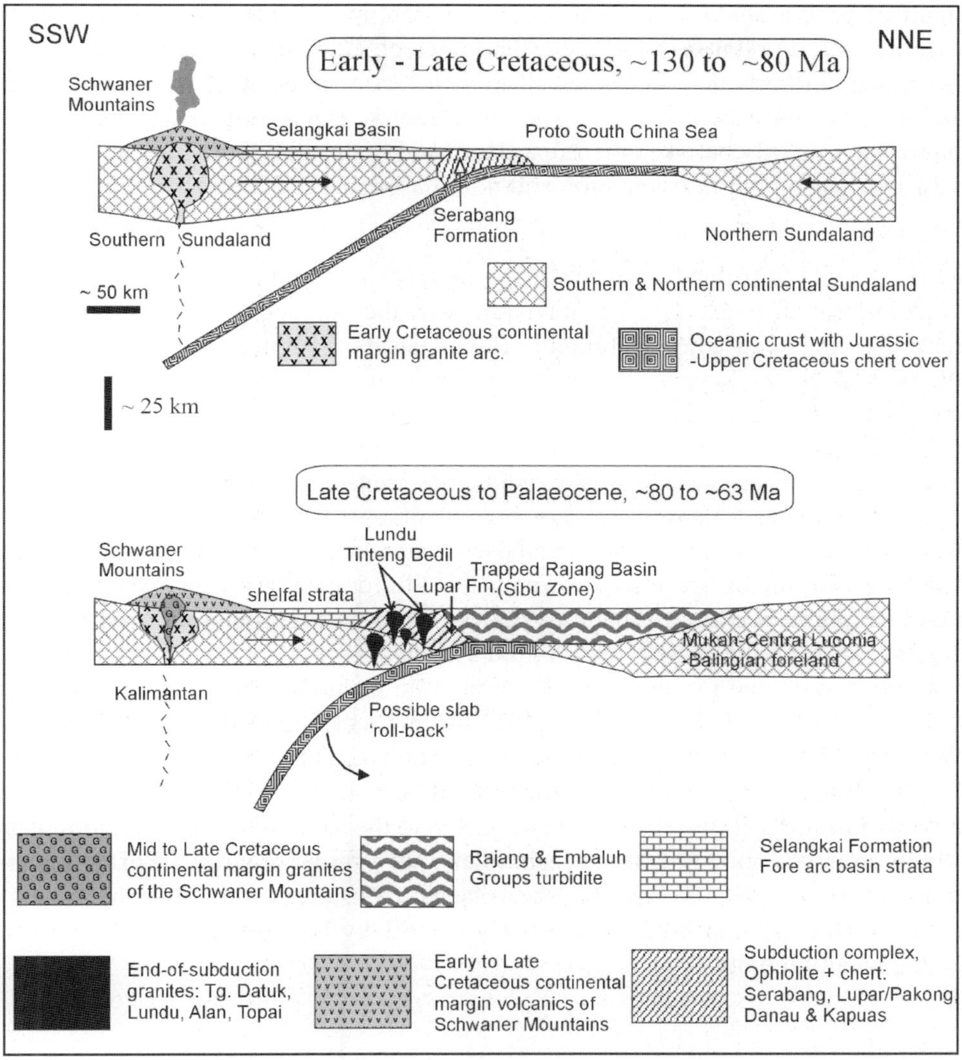

Figure 67. NNE–SSW diagrammatic cross-section to suggest the plate-tectonic model for Early Cretaceous to Middle Eocene convergent tectonics (modified after Moss, 1998). (by permission of the Geological Society of London)

within the outcrop of the Lubok Antu Mélange, indicates that by Early Eocene times, this region was not part of the Proto South China Sea. The limestones should be contrasted with the deep water and much older cherts, which are also found as clasts within the Lubok Antu Mélange. The mélange has, therefore, incorporated clasts of unrelated rock types. The Engkilili Formation limestones are post-subduction and unrelated to the Proto South China Sea, whereas the cherts are pre-subduction, but could not have been incorporated into the mélange by a subduction or accretionary prism process. The Lubok Antu Mélange is accordingly interpreted as a fault of deep-seated importance which has sampled the basement geology. It is probably entirely related to the Lupar Fault, which is the master normal fault controlling the northerly margin of the half-graben Ketungau Basin, downthrown to the south to contain the Silantek Formation in Sarawak. Hutchison (2001) drew attention to the similarity between the Lupar Fault and faults within the Mukah Province to the northeast, both in orientation and non-marine basin fill.

IX.3.1.5. Sintang suite high-level diorites and adakites

No plate-tectonic scenario can be envisioned for the Sintang Suite, which has preferentially intruded the Ketungau Basin strata (Figure 66). The Upper Oligocene–Lower Miocene diorites have normal calc-alkaline characteristics, but no convergent plate margin has been discovered in or offshore Sarawak of this age.

The Middle to Upper Miocene adakites are interpreted to have resulted from partial melting of an oceanic slab flake, which has continued to reside in the underlying Upper Mantle. Although they are calc-alkaline, they are characteristically low in potassium and are interpreted as modern analogues of Archean greenstone belt granitoids. Examples are also known in the Philippines, for example the recent dacitic eruption of Mount Pinatubo. Both greenstone belt granitoids and modern adakites, for example at Bau, have a gold association.

There is no value in attempting to model the Cainozoic volcanic belt through Usun Apau and the Hose Mountains. The volcanic rocks have not been dated radiometrically, and their chemical signature unknown. There is no trench of Late Miocene–Pliocene age with which they might be connected. Accordingly they are probably not subduction-related. A relationship to the suitably positioned Northwest Borneo Trough (Figure 66) would seem likely, but a future programme of radiometric dating would have to show that the volcanic rocks are Lower Miocene, as suggested by Hutchison (1989). Such a study is unlikely because the plateaus remain geographically remote.

Part B

Sabah

Chapter X

Introduction

The first landmark in the beginning of understanding of the geology of Sabah was the publication of Reinhard and Wenk (1951). In 1948, the Shell Oil Company informed the Colonial Office that it had arranged for Professor Max Reinhard and Dr Eduard Wenk of Basel University to compile a comprehensive report on the geology of North Borneo. Both had previous field experience in the country. It was published as the first bulletin of the newly created Geological Survey Department. All the fieldwork on which it was based was completed before 1942. The manuscript was mostly written in German and translated by staff of Royal Dutch Shell in The Hague.

X.1. CHERT–SPILITE FORMATION

Molengraaf (1902), from his mapping of the lake region of Kalimantan, had coined the term *Danau Formation* for a characteristic association predominantly of diabase and chert. Reinhard and Wenk (1951) followed the terminology, now firmly interpreted as an ophiolite association. Fitch (1955) used the term Chert–Spilite Formation for what formerly was known as the Danau Formation. Wilson (1961) and Kirk (1962) ensured that the term Chert–Spilite Formation became widely accepted in the stratigraphic nomenclature of Sabah for a characteristic association of spilite, pillow basalt and ribbon chert. However the ophiolitic nature, clearly recognized in the term Danau Formation, became lost within the usage Chert–Spilite Formation. A major problem also arose because other apparently unrelated rocks were included and the Chert–Spilite Formation became something of a *miscellanea*.

X.2. ULTRABASIC AND BASIC ROCKS

Large regions of Sabah are characterized by ultrabasic and basic plutonic rocks, clearly recognized as ophiolitic (Kirk, 1968), but their link with the Chert–Spilite Formation, following the wisdom of his ophiolitic 'Trinity' (Steinmann, 1906), was not made in Sabah until later. The basic plutonic rocks are commonly layered and the ultrabasic rocks, when not completely serpentinized, demonstrate a similar fabric.

X.3. CRYSTALLINE SCHISTS

Extensive outcrops of ophiolitic ultrabasic and basic rocks, commonly layered, have been metamorphosed to amphibolite and gneiss. Their main mafic mineral is hornblende, which has partially to completely replaced the original pyroxene. Reinhard and Wenk (1951) used the term Crystalline Schists for these metamorphosed ophiolitic rocks. Of course, they were writing for the Shell Oil Company and the term was carefully chosen to indicate that such a terrain had zero oil potential. They frequently wrote that it was difficult, if not impossible to map with certainty crystalline schist or ophiolite: "Great difficulties and uncertainties were encountered in the attempt to distinguish between the occurrences of crystalline schists and those of the ophiolite of the Danau Formation…certain types of rocks are as likely to appear in the one as in the other category" (Reinhard and Wenk, 1951, p. 99). This problem persists until today, for it is impossible in the field to make the distinction whether an ophiolite is metamorphosed or unmetamorphosed, a problem commonly encountered elsewhere in the world.

The same authors summarized the problem and expressed a hope, which remains unrealised: "Once a better knowledge is obtained, it has to be demonstrated whether the adopted three-fold subdivision into

crystalline schists and associated intrusives
slate Formation
Danau Formation

is to be maintained. The delimitation between these three formations is uncertain, and another subdivision would perhaps give more satisfactory results" (Reinhard and Wenk, 1951, p. 105).

Reinhard and Wenk (1951), however, were of the opinion that "the gneisses, amphibolites and metamorphic slates and the associated quartz-dioritic rocks of the Dent Peninsula are older than the Danau Formation." But their opinion was not held with conviction, for "However, similar metamorphic rocks occasionally occur closely associated with ophiolites, and it must be left an open question as to whether these are metamorphic varieties of Danau ophiolites or older crystalline schists." (Reinhard and Wenk, 1951, p. 64).

X.4. CRYSTALLINE BASEMENT

Within this atmosphere of great uncertainty, Leong (1974) elevated the crystalline schists to the status of crystalline basement and concluded from the available radiometric dating that this basement was Lower Triassic and older. Yet, the doubts of Reinhard and Wenk (1951) remain valid. The same area (Leong, 1974) had been previously mapped by Fitch (1955), who had interpreted the crystalline schists of the Segama area as banded dioritic and acid igneous rocks, genetically associated and contemporaneous with the ultramafic bodies. Leong (1974) subdivided the same rocks into a crystalline basement and a separate group of banded gabbros and associated ultramafic rocks.

X.5. CALC-ALKALINE GRANITOIDS

Within the crystalline basement of the Segama area, there also occur several bodies of calc-alkaline granitoids not of ophiolitic affinity. Their relationship with the ophiolite is poorly defined. Hutchison et al. (2001) have agreed with Leong (1998) that these indicate that older continental crust must underlie and support the ophiolite, for these granitoids do not have their origin in ophiolitic lithosphere.

X.6. MODERN INTERPRETATION OF OPHIOLITE

Following the recognition that ophiolites represent uplifted ancient oceanic lithosphere, not of the major oceans but rather of smaller marginal basins as documented by Coleman (1977), Hutchison (1975) published the next landmark paper genetically linking together the various ophiolitic components of Sabah. Although most outcrops have been strongly imbricated, there are a few well-studied areas that demonstrate a typical ophiolite stratigraphy, from mantle peridotite, through gabbro layer, to overlying pillow basalt or spilite, overlain by ribbon cherts.

In the next landmark study, Hutchison (1978) showed that amphibolite gneissic rocks of the Darvel Bay islands, which had been mapped as 'crystalline basement', are all meta-ophiolite of igneous parentage. The metamorphism is universally incomplete and patchy, as is common in many ophiolite complexes. The igneous pyroxenes have generally been replaced by amphibole clusters that show a prominent lineation parallel to the fold axes, but hornblende relicts persist. The plagioclase in any one rock is of bimodal composition—relict labradorite and metamorphic oligoclase. In no single specimen did the assemblage represent total metamorphic equilibrium. This casts doubt upon the possibility that the crystalline schists can be separated from the ophiolite suite.

The stratigraphic problems of Sabah remain and may never be completely resolved. Tectonic imbrication and deep weathering are major obstacles. Existing maps, based upon outdated concepts, cannot be reinterpreted without remapping, but the outcrops would prove to be beyond the capability of all but the most experienced geologists.

Chapter XI

Geomorphology

Sabah exhibits a spectacular range of topographic terrains, from mangrove coastal plains to Mount Kinabalu, which rises to 4101 m elevation. The physiography of the country has been outlined by Collenette (1963) and Wilford (1968).

XI.1. WESTERN LOWLANDS

The Western Lowlands occupy a narrow strip of land bordering the South China Sea, broadest at Klias and narrowest at Kota Kinabalu and Tuaran (Figure 68). This terrain has been subdivided by Collenette (1963) into the Crocker Foothills, the Crocker Plains, the Klias Hills and the Western Islands. The division between the Crocker Foothills and the Western Cordillera is arbitrarily set at ~300 m elevation.

Wilford (1968) lists marine terraces of 24 m at Klias and at ~30 m at Sipitang in the Brunei Bay area. Terraces of ~9 m elevation occur on Labuan, Klias, Beaufort and Sipitang. Low-level terraces of ~3–5 m occur on Labuan, Klias and Sipitang. Marine caves, 3.5 m above sea level on Burong Island, have now all been quarried away. Shells and wood radiocarbon ages of 22,450 and 2840 a have been obtained on Klias. Terraces at 4.5, 9, 15 and 21 m are recorded near Kota Belud.

XI.2. WESTERN CORDILLERA

This is the mountainous core of Sabah, on average 80 km wide (Collenette, 1963). It is an amalgamation of mountain ranges with the following geographical features:

Crocker Range: The arbitrary boundary with the Crocker Foothills is taken at 300 m. The elevation generally lies within the range 1220–1830 m, rising to 4101 m at Mount Kinabalu in the north (N) and to Mount Lamaku in the south (S), about 2440 m elevation.
Trusmadi Range: The central peaks exceed 2440 m.
Witti Range: The central part is over 1520 m, decreasing to 600 m in the north.
Meligan Range.
Intermontane plains of Tenom, Keningau, Tambunan, Sook-Dalit, Nabawan, Patau and Ranau.

The Western Cordillera has been dramatically uplifting and eroding mainly in the Upper Miocene (Hutchison et al., 2000), and there is no reason to suspect that

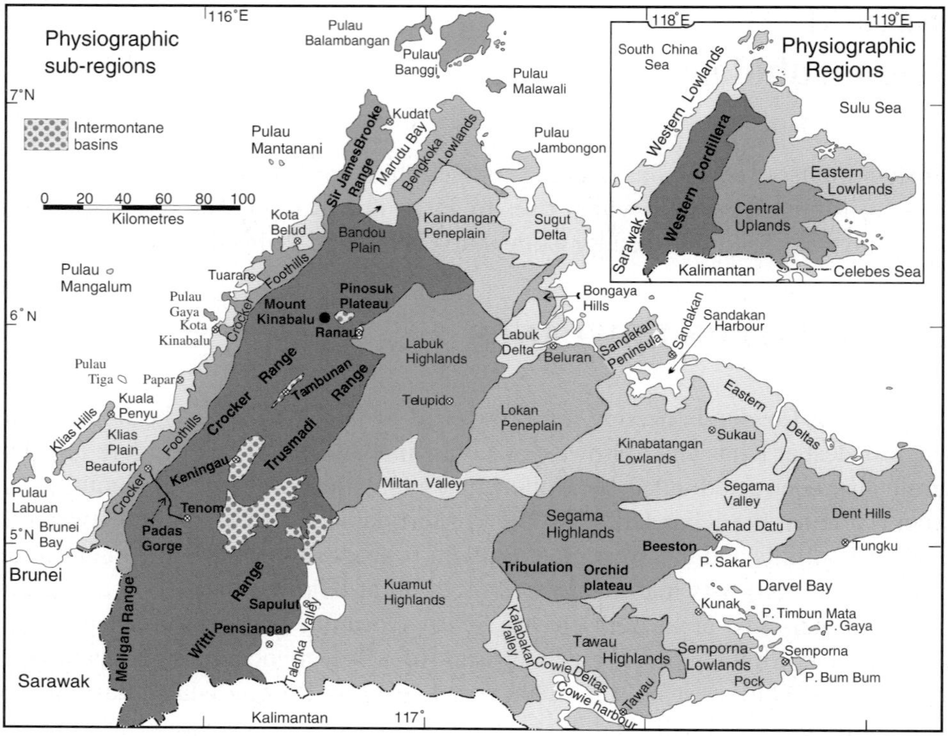

Figure 68. Physiographic regions of Sabah (after Collenette, 1963, Wilford, 1968). With permission from Minerals and Geoscience Department, Malaysia.

the process has come to an end. The preferred mechanism for uplift is isostatic rebound, discussed later in this monograph.

Faults played an important role in the uplift, and an integral part of this process is that certain areas became down-faulted to produce intermontane basins, which received alluvial infillings. Terraces at three or four main levels occur at Tambunan, Keningau, Tenom and Sook-Dalit (Wilford, 1968). The surfaces of the higher terraces dip east-south-easterly (ESE).

XI.2.1. Mount Kinabalu glaciation

Mount Kinabalu, of elevation 4101 m, was capped by an ice field at least once during the Pleistocene, covering some 5.4 km^2 of the summit area (Koopmans and Stauffer, 1967). A few precipitous peaks projected through the ice. Glacier action formed a smooth and gently sloping summit area above 3600–3800 m elevation. This level coincided with the snow line. At least two valley glaciers developed, one descended Low's Gully, where remnants of a terminal moraine exist at elevation 2840 m. A second one moved over an icefall above the Panar Labah Mountain and formed a small glacier tongue in the upper Sungai Kolopis, with a terminal moraine at altitude 3230 m.

XI.2.1.1. Pinosuk Gravels

The Pinosuk Gravels occur to the south and west (W) of Mount Kinabalu, the main occurrence being in the Pinosuk Plateau, which is of tilloid covering 65 km^2. The plateau falls gradually from 1680 m over a distance of 8 km to the Liwagu Valley (760 m elevation). Outliers extend as far a Ranau (550 m). Maximum thickness near Kampung Pinosuk is 140 m (Jacobson, 1970).

Wood samples from several places within the gravels gave ages beyond the carbon method (>39,900 a). Wood from an overlying soil gave 7980±100 a. The tilloids are therefore Pleistocene. There are two distinct lithologies.

The *lower unit* occurs in the Mesilau and Mantaki valleys and appears to be of fluvial origin. It is of unconsolidated, poorly sorted sandy gravel, with clasts ranging up to boulder size. Clasts >2 mm make up more that 70% of the formation. The clasts are angular to sub-angular and of locally derived sedimentary and ultrabasic rocks.

The *upper unit* is much more widespread and of poorly sorted clayey to sandy boulder gravel. Clasts also make up >70% of the formation. The clasts are of varied lithologies and include rounded boulders of granodiorite derived from Mount Kinabalu. This unit is probably a mudflow derived from reworking of a glacial moraine (Jacobson, 1970). One striated granodiorite boulder was found downstream in the Sungai Mesilau. The main road, east (E) of Kundasang, transects the Pinosuk Gravels. Typical granitoids occurring as clasts are of porphyritic biotite–hornblende monzodiorite, with phenocrysts of 2–15 mm set in a 0.5–2.0 mm matrix of quartz–feldspar–hornblende–biotite. Such lithologies occur commonly as boulders in the river at Ranau beneath the bridge over the main road.

XI.3. CENTRAL UPLANDS

The highland districts are Labuk Highlands, Milian Valley, Kuamut Highlands, Segama Highlands, Kalabakan Valley and Tawau Highlands (Collenette, 1963). Among these, the Kuamut Highlands are mostly above 300 m and Mount Rara is about 1520 m elevation. The Segama Highlands have summits rarely exceeding 900 m. They contain a notable flat area known as the Orchid Plateau. The Tawau Highlands are dominated by the volcanic forms of Mount Magdalena and Mount Wullersdorf. However the volcanic edifices are not as young as previously held; now known to be Miocene. Wilford (1968), however, notes that dacitic pyroclastic flows near Tawau contain charcoal dated 27,000 a, and soils contain volcanic bombs.

XI.4. EASTERN LOWLANDS

The eastern low plains may be subdivided into the following subdistricts: Northern Islands, Kaindangan and Lokan Plains, the Deltas of Kinabatangan, Segama, Sugut and Labuk rivers, the Bongaya Hills, Sandakan Peninsula, Kinabatangan Lowlands, Segama Valley, Dent Hills and the Semporna Lowlands (Collenette, 1963). The major rivers all flow to the Sulu Sea. The Segama River is thought to have once reached the sea at Lahad Datu, and captured due to tilting of the country as a result of Miocene rifting in the Sulu Sea marginal Basin.

Chapter XII

Introduction to the Stratigraphy

The Western Cordillera is composed predominantly of the Crocker Formation, a flysch of great but undefined thickness, whose structure has yet to be satisfactorily resolved. The lack of marker beds and the fact that most specimens have yielded the so-called "flysch-Foraminifera", which are arenaceous and not age-diagnostic, have hindered progress. Nevertheless, a fauna of definite Palaeocene to Eocene age is confirmed for the area south of Keningau and a definite age of Oligocene to Lower Miocene for the Temburong facies of the Tenom and Beaufort area. They represent the anchor points of knowledge.

Figure 69 is a somewhat simplified geological map, based on Leong (1999), which in turn is based on Tongkul (1994, 1997). The map shows lineations obtained from radar imagery, but close scrutiny shows that some selected lineations are imaginary rather than real.

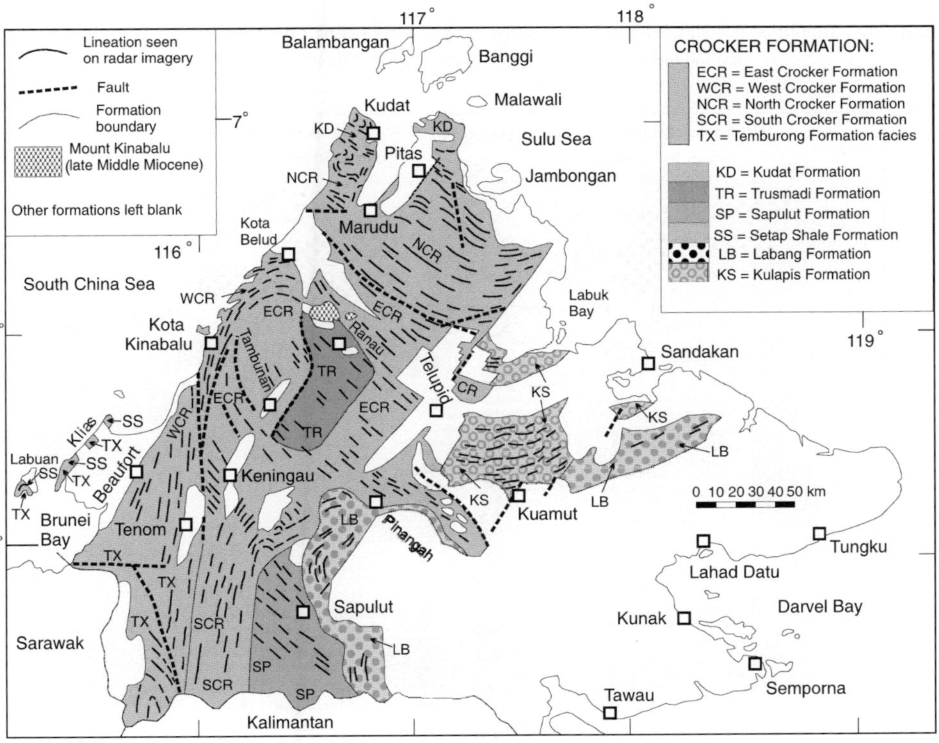

Figure 69. Simplified partial geological map of Sabah, largely based on Leong (1999). Blank areas are of formations excluded for clarity from this compilation. They are shown on the complementary map of Figure 70.

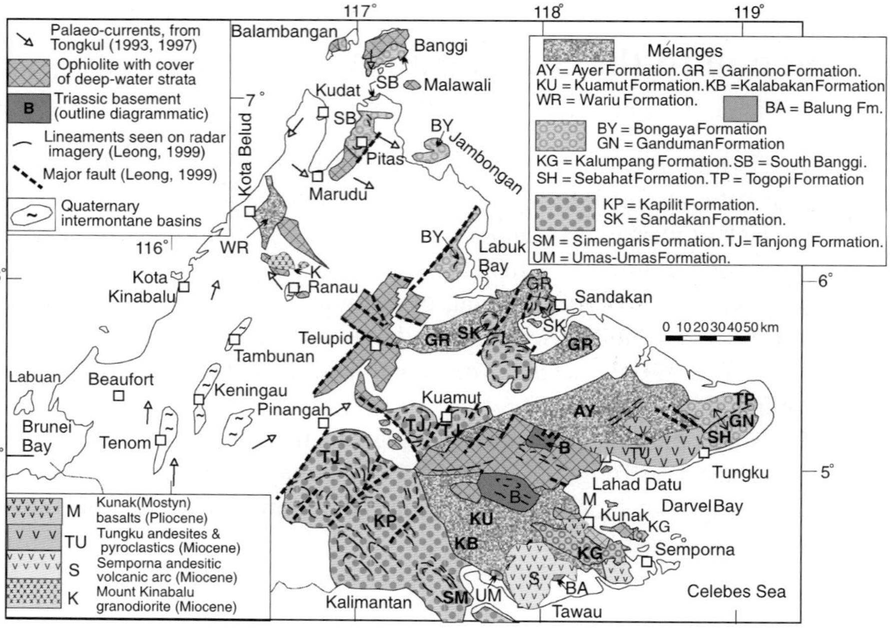

Figure 70. Simplified partial geological map of Sabah, based on Leong (1999). Blank areas are of formations excluded for clarity from this complication. See also Figure 69.

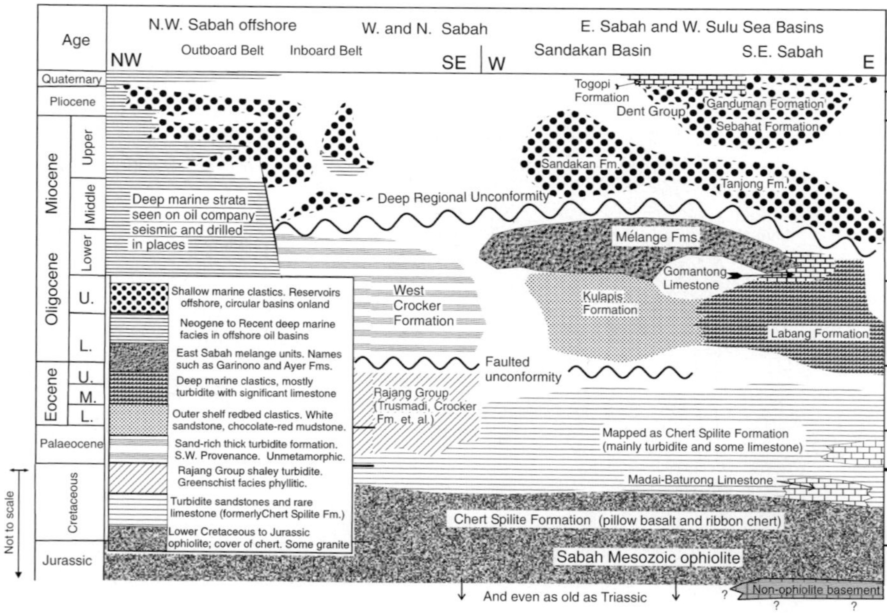

Figure 71. Schematic stratigraphy of Sabah (modified after Hutchison et al., 2000; Clennell, 1996; Noad, 1998). The end Eocene unconformity has been attributed to the Sarawak Orogeny and the Mid- to Late Miocene to the Sabah Orogeny (Hutchison, 1996). Extent of the non-ophiolite basement poorly defined and its contact with the ophiolite not investigated. (by permission of the Geological Society of London)

The continuation of the geology into the Eastern Lowlands is given in Figure 70. This map also includes the basement ophiolite and non-ophiolite rocks, whose radiometric ages indicate may be as old as Triassic. A particular feature of this region is the "circular basins", formed by the Tanjong Formation. They post-date the south-east (SE). Sulu Sea rifting, and overlie extensive syn-rift mélange formations.

A simplified stratigraphic diagram across the country is given in Figure 71. Not all formations are included, and more names may be found in Figures 69 and 70. Figure 70 emphasises that the mélange formations subdivide the stratigraphy of Sabah into pre-rift and post-rift. The rifting is a universal time of unconformity, except in the deep-water areas offshore in the South China Sea.

Chapter XIII
Sulu Sea Marginal Basin

It is from knowledge of the Sulu Sea that the Neogene stratigraphy and tectonics of on-land Sabah become possible. This was first brought to light by Hutchison (1992a) because the fabric of the marginal basin trends south-westerly towards Sabah (Figure 72).

XIII.1. NORTH-WEST SULU SEA

This province has a thick crust. The seismic basement is of highly deformed layered rocks (Figure 73) and the most attractive interpretation is that it represents a continuation of the Crocker and Trusmadi formations of Sabah (Rangin et al., 1990;

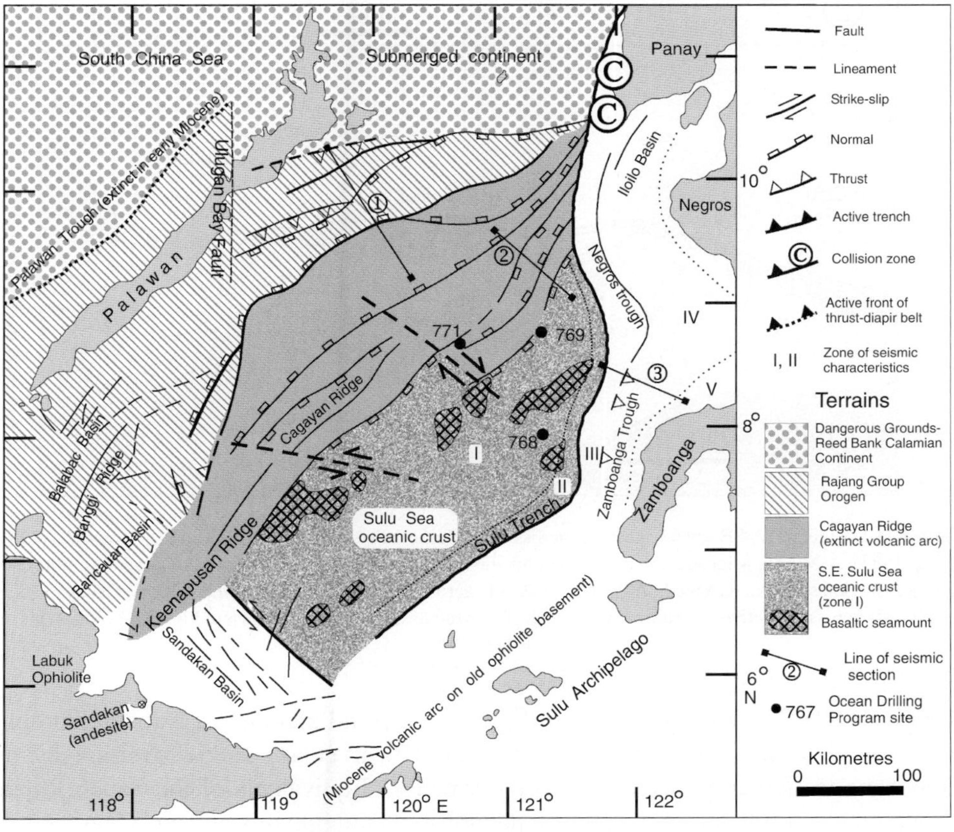

Figure 72. Structural and geological features of the Sulu Sea (after Hinz et al., 1991; Rangin, 1989).

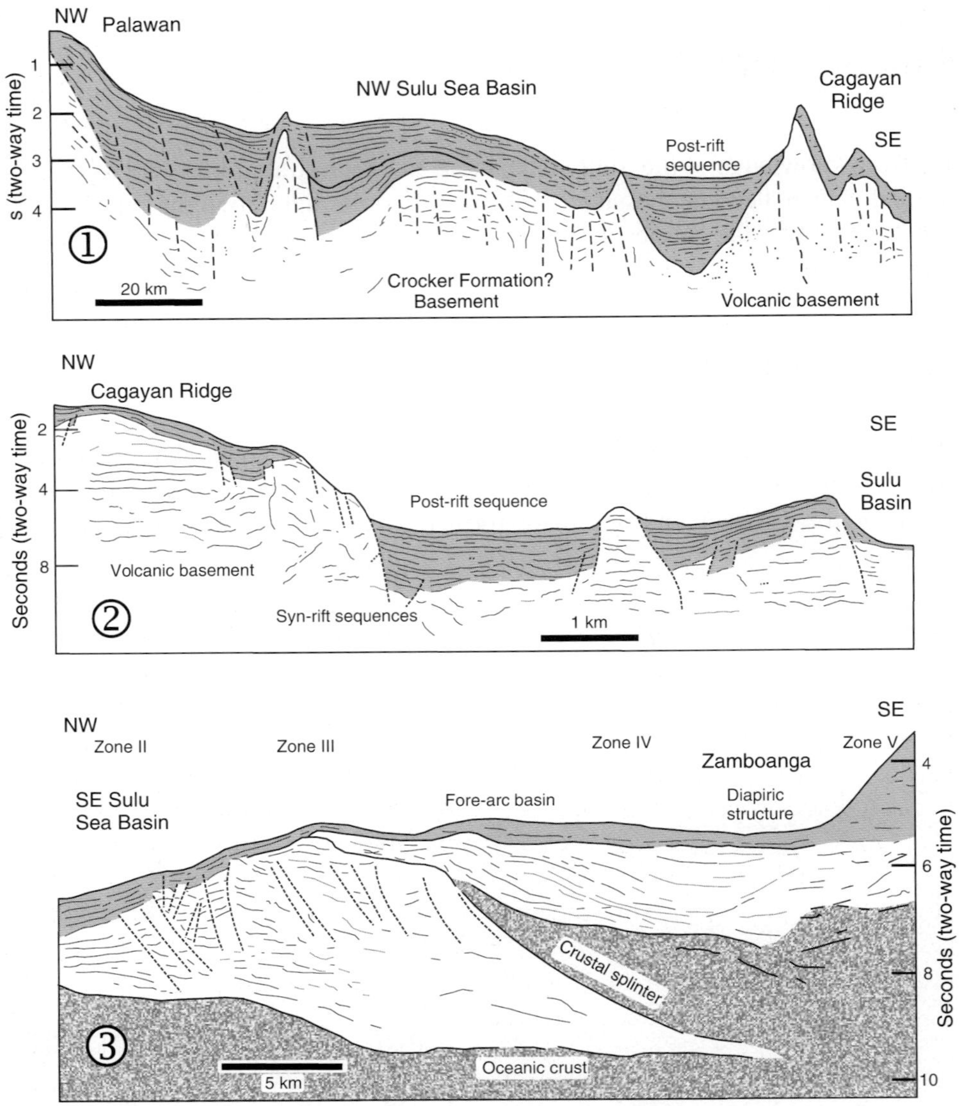

Figure 73. Three NW–SE seismic sections across the Sulu Sea. See Figure 72 for locations. 1 = NW Sulu Sea (Hinz, 1983). A Middle Miocene age is assumed for the unconformity between 2.5 and 3 s (two-way time). The sediment thickness above it exceeds 7000 m. 2 = Profile across the rifted SE flank of the volcanic Cagayan Ridge. 3 = Profile across the active immature Sulu Trench. The structural zones are as shown in Figure 72 (after Hinz and Block, 1990).

Rangin 1989). The basement terrain (Figure 73) would include the Neocomian ophiolite complete with Barremian ribbon chert (Basir et al., 1985), overlain by the Oligocene to Lower Miocene Crocker Formation. Equivalents of the Sabah Rajang Group on Central Palawan have given nannofossils ranging in age from Maastrichtian to Upper Oligocene (Rangin et al., 1990).

A pronounced unconformity, shown to be early Middle Miocene in the drilled Balabac Basin (Beddoes, 1976), separates the basement from the overlying relatively undeformed Neogene strata, which commonly attain a thickness of 7000 m in the main depocentre (Figure 73). It is therefore very likely that the deformed Rajang Group that now underlies the north-west(NW) Sulu Sea, formed a landmass continuous with Sabah (the Sabah–Palawan Orogenic Belt) before the early Middle Miocene unconformity. This landmass was fringed along its south-east (SE) cordilleran-type margin by the Cagayan volcanic arc. Early Middle Miocene rifting affected the whole Sulu Sea and eastern Sabah, causing extensive subsidence, the greatest being centred in the SE Sulu Sea marginal basin that experienced sea-floor spreading.

XIII.2. CAGAYAN RIDGE

Andesite, containing phenocrysts of plagioclase, clinopyroxene and olivine, was dredged from the Meander Reef on the Cagayan Ridge (Kudrass et al., 1986). Its K:Ar age of 14.7 Ma (Langhian; Middle Miocene) is consistent with Lower to Middle Miocene carbonate sediments dredged from another seamount on the Cagayan Ridge. The basement at OPD site 769 (Figures 72 and 74) is of massive unstratified basaltic-andesite, devoid of admixed sediment. Ocean Drilling Program (OPD) site 771 bottomed in late Lower to early Middle Miocene massive, structureless volcaniclastic lapilli-stone underlain and capped by basaltic lava flows (Silver et al., 1989a, b). The intercalations of flows and pyroclastic strata suggest close proximity to volcanic vents that are likely to have extended above sea level and to have vented subaerially.

ODP sites 769 and 771, on the flank of the Cagayan volcanic ridge, penetrated a thin sedimentary section devoid of turbidites (Figure 74). The volcanic and pyroclastic rocks are immediately overlain by brown claystone containing late Lower to early Middle Miocene radiolaria. This age therefore represents cessation of volcanism of the Cagayan Ridge as well as spreading of the SE Sulu Sea Basin. Rifting of the volcanic arc must have resulted in its rapid foundering. The basaltic to andesitic tuffs in OPD site769 may correlate with the tuffs underlying the onshore Sandakan Formation beneath the mosque on the coastline, which contain Oligocene fossils (Rangin et al., 1990). The andesite tuff (94SB31A), however, has yielded an Upper Cretaceous K:Ar age of 76 Ma (Table 24). The same rock yielded an apatite fission track age of 33.9 ± 7.7 Ma (Rupelian; U. Oligocene) (Figure 77) (Hutchison et al., 2000), consistent with the content of Oligocene fossils. The apatite crystals are interpreted to have fallen directly into the tuff from a nearby volcanic explosion. The K:Ar plagioclase age (Table 24) may support the interpretation of Lee (1970) that this outcrop is of mélange containing clasts of ophiolitic basement. I am inclined to support the view that the 76 Ma old age is unreliable (the sample contains extremely low potassium), and I favour an Oligocene interpretation.

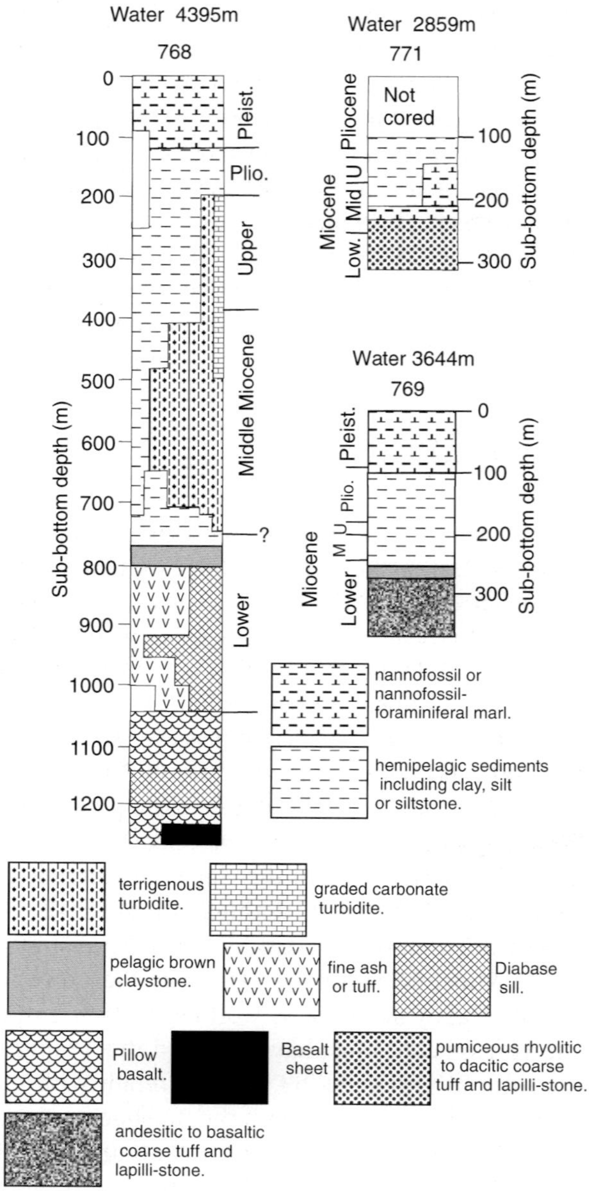

Figure 74. ODP sites in the Sulu Sea (after Silver et al., 1989a, b).

XIII.3. SULU – ZAMBOANGA – NEGROS SUBDUCTION SYSTEM

The SE Sulu Sea has been shown by Hinz and Block (1990) to be divisible into five structural zones (Figure 72 and 73). Zone I forms the major part of the deep SE Sulu Sea Basin. Heat flow is relatively high, ranging up to 4.74 HFU. Its igneous basement is relatively smooth and has been drilled at ODP site 768 (Figure 72 and 74). Its seismic signature is typical of oceanic crust, overlain by a sedimentary layer varying in thickness from 1 to 2 s (two-way time). It is dominated by a chain of basement highs trending 60°, from which basaltic rocks have been dredged. Hinz and Block (1990) report that there is no evidence for the existence of any magnetic lineations, and certainly not of the anomalies 17–20 erroneously reported by Lee and McCabe (1986), further disproved by the drill data (Figure 74).

The basement at ODP site 768 is of pillow basalt and sheet flows, intruded by dolerite sills (Figure 74). The basalt has trace element geochemistry signatures transitional between mid-ocean ridge basalts (MORB) and island-arc tholeiite.

The SE Sulu Sea seems to have originated as an intra-arc basin in the late Lower to early Middle Miocene, as determined by radiolarian biostratigraphy in sediments overlying the volcanic basement (Figure 74). Initial deposition of volcanic brown clay was followed shortly thereafter by rapidly deposited coarse mass-flow pyroclastic rhyolitic and dacitic tuffs composed of devitrified glass shards and pumice. Slow accumulation rates (9 m Ma^{-1}) marked the early part of the Middle Miocene, but in the late Middle Miocene (10.5 Ma) very rapidly deposited (300 m Ma^{-1}) continentally derived turbidites inundated the basin, interpreted as derived from erosion of the rapidly uplifting Western Cordillera of Sabah. Similar turbidites also inundated the Celebes Basin (Silver et al., 1989a, b). Volcanic ash layers appeared at 6 Ma (Figure 74).

Zone II represents the 4550 m deep trench of the active subduction zone. The trench, however, does not extend as far as Sabah (Figure 72). The E to SE-descending oceanic basement dips 10° and is overlain by 0.5–1.8 s (two-way time) of sediments, suggesting that subduction may have ceased. Heat flow lies between 1.87 and 2.37 HFU.

Zone III is of a sedimentary sequence characterized by imbricate thrust sheets. The faults dip SE and E, and curve downwards to join a major décollement surface that lies just above the descending igneous oceanic crust (Figure 73).

An up-thrust slab of oceanic crust supports a sedimentary apron 2.4 s (two-way time) thick, which resembles a "fore-arc basin" (Zone IV). Such a small-scale immature subduction system, however, could not possibly have supported a volcanic arc. The sediments of the basin have a gentle 8° dip towards the east. The basin continues northwards to be represented by the Iloilo Basin of central Panay, which contains Middle Miocene limestones, Middle to Upper Miocene coarse clastics, and basalt flows. Heat flow values range from 1.07 to 1.69 HFU. The basin has been drilled, indicating a very low geothermal gradient of 21°C km^{-1} (Bureau of Mines, 1976; Hutchison, 1996b).

Zone V forms the western slope of Zamboanga and Negros. A sedimentary apron, ranging from a few hundred metres to 2.8 s (two-way time), overlies the basement complex of the Sulu Archipelago and Zamboanga. The Sulu Archipelago islands consist of Cretaceous ophiolite (e.g. Tawi–Tawi) and Neogene volcanic rocks (e.g. Jolo), and similar rocks build the Semporna and southern Dent Peninsulas of Sabah, lying on the same geological trend.

The seismic profile across zones II–V (Figure 73) is remarkably similar to that of a normal arc-trench system, but on a miniature scale. The distance from the trench to Zamboanga Peninsula and Jolo Island is little more than 100 km and no Benioff zone has been documented. The miniature scale makes it virtually impossible for subduction of SE Sulu Sea oceanic lithosphere deep enough to have created a volcanic arc.

The borehole televiewer showed that the short axis of the elliptical drill hole at OPD site 768 is oriented north-east (NE), representing the present maximum stress direction of the crust (Silver et al., 1989a). This is consistent with north-easterly subduction beneath the active Sulu–Negros–Zamboanga trench system.

XIII.4. RELATIONSHIP TO SABAH STRATIGRAPHY

The tectonic events of eastern Sabah clearly reflect the tectonic history of the Sulu Sea (Hutchison, 1992a). The initial rifting and sea-floor spreading to give pillow basalts occurred in the Lower Miocene (Figure 74). Only Middle Miocene and younger formations of Sabah show coherent structures. The eastern Sabah formations, which pre-date the initiation of the SE Sulu Sea, are all imbricated and show no coherent structure. Only locally coherent areas can be analysed structurally, but there is no broad scale continuity. Faults abound in the pre-Lower Miocene formations.

The formations that separate the younger formations of coherent stratigraphy from the older imbricated formations are mélanges. They have been given various names: Garinono, Ayer and Kuamut. Clennell (1991, 1992) has made a comprehensive study of the mélanges and shown that they were formed in the uppermost

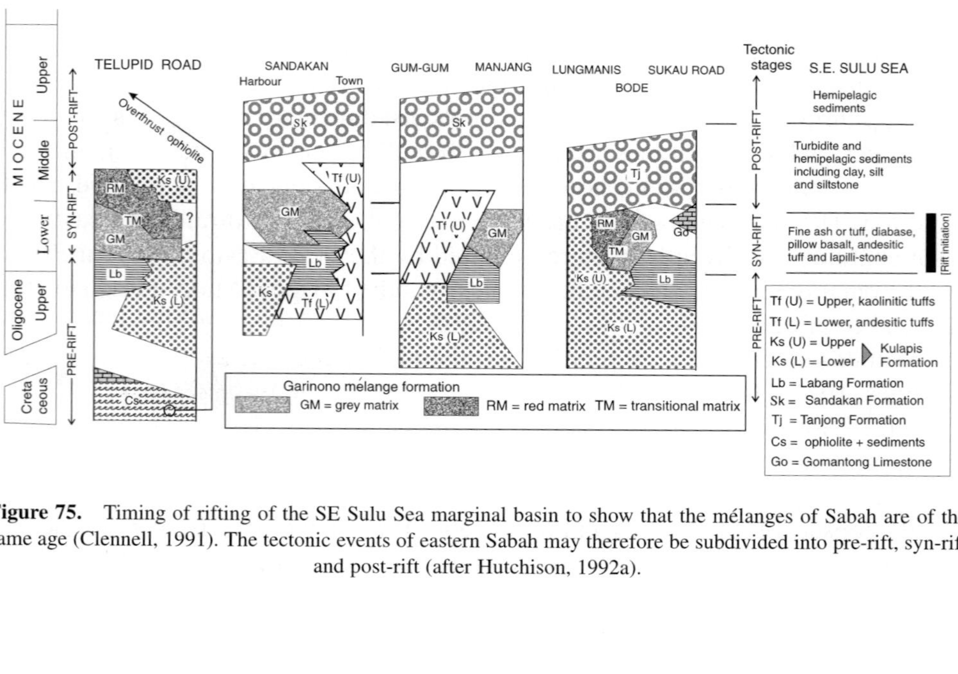

Figure 75. Timing of rifting of the SE Sulu Sea marginal basin to show that the mélanges of Sabah are of the same age (Clennell, 1991). The tectonic events of eastern Sabah may therefore be subdivided into pre-rift, syn-rift and post-rift (after Hutchison, 1992a).

Lower Miocene to lowermost Middle Miocene. Hutchison (1992a) therefore subdivided the stratigraphy of eastern Sabah into three major divisions (Figure 75):

Post-rift. These are formations such as Tanjong and Sandakan, which form structurally coherent basins, often saucer-shaped or 'circular'. They are Middle Miocene or younger.

Syn-rift. These are the marine mélange formations of uppermost Lower Miocene to lowermost Middle Miocene age. They were formed as a result of active rifting that gave rise to the SE Sulu Sea marginal basin that trends south-west (SW) into Sabah. There is no indication that sea-floor spreading extended into Sabah, but the rifting certainly did and eastern Sabah was an integral part of the SE Sulu Sea. The syn-rift stratigraphy of the SE Sulu Sea includes tuffs and lapilli-stone, indicating that an active volcanic arc (Cagayan Ridge) was being rifted to form the SE marginal basin. There are tuffs on-land associated with the mélanges (Figure 75).

Pre-rift. The imbricated Lower Miocene and older formations, including the Cretaceous and older basement. All pre-rift formations were imbricated and incorporated as clasts within the syn-rift mélanges

Chapter XIV

The Ophiolitic Basement

To form any understanding of the nature and age of the basement, the traditional stratigraphic terminology needs to be replaced by the modern terminology of ophiolite stratigraphy. The obvious starting point is with the ribbon cherts (formerly of the Chert–Spilite Formation). Ribbon cherts were the first deep-marine sediments to be deposited upon ocean or marginal basin floor where there had been sea-floor spreading (Coleman, 1977). Ribbon cherts of Cretaceous age are abundant throughout the world, and are well represented in Sabah. In oceanic seismic stratigraphy, these cherts represent the sedimentary layer (Layer 1) of a complete ophiolite.

XIV.1. RIBBON CHERTS

Well-bedded cherts, commonly complexly contorted and folded, outcrop in all areas of basement ophiolite: Banggi Island, around Kudat, in the Labuk Highlands near Telupid and especially in the Segama Highlands.

XIV.1.2. Lithology

The most common colour is orange, red, brown, but light green and grey are also known (Leong, 1974; Kirk, 1962). Most outcrops are strongly brecciated, with veinlets of quartz and chalcedony occupying the fractures. Radiolaria are common, frequently recrystallized and deformed. Bedded chert is commonly folded and intensely jointed. Structural measurements on chert outcrops indicate no analytical pattern, and this is because the ophiolite is imbricated and the chert commonly occurs as randomly oriented clasts within the mud-matrix Miocene mélange formations.

XIV.1.3. Age

The cherts commonly contain *radiolaria*, but all older palaeontological identifications, contained in the memoirs, may now be discarded as misleading for they were based on thin-section identifications, now known to be woefully inaccurate. We must recognize as valid only those identifications of three-dimensional *radiolaria*, which have been dissolved out of the chert by hydrofluoric acid.

The first such determinations were made by W. R. Riedel on specimens from the Upper Segama area and reported by Leong (1977). He reported the radiolaria given in Table 16 from specimen J7250 and J7989. This assemblage indicates a Lower Cretaceous Valanginian to Barremian age (~127–137 Ma).

Table 16. Radiolaria identified in the districts of Kudat, Telupid, Lahad Datu and Segama

Radiolarian species and genera	Kudat 1	Telupid 2	Lahad Datu 3	Segama 4
Acaeniotyle diaphorogona (Foreman)		X		
Acaeniotyle umbilicata (Rüst)		X		
?*Acaeniotyle umbilicata*				X
Alievium sp.,		X		
Alievium sp. (Pessagno)			X	
Archaeodictyomitra sp.	X	X	X	
Archaeodictyomitra brouweri (Tan Sin Lok)		X		
Archaeodictyomitra pseudoscalaris (Tan Sin Hok)	X	X		
Archaeodictyomitra lacrimula (Foreman)	X			
Archaeodictyomitra sp. cf., A. sliteri (Pessagno)			X	
Archaeodictyomitra sp. cf. A. vulgaris (Pessagno)			X	
Archaeodictyomitra vulgaris (Pessagno)		X		
Archaeodictyomitra puga (Schaar)	X			
Artostrobium urna				X
Conosphaera tuberosa (Tan Sin Hok)	X	X		
Crucella sp.			X	
?*Dictyomitra boesii*				X
Dictyomitra duodecimcostata				X
Eucyrtis micropora (Squinabol)		X		
Hagiastrum sp.		X		
Hagiastrum euganeum (Squinabol)	X			
Hemicryptocapsa pseudopilula (Tan Sin Hok)	X	X		
Higumastra sp.		X		
Histiastrum aster (Lipman)	X			
Lithomitra pseudopingius (Tan Sin Hok)		X		
Obesacapsula somphedia (Foreman),	X			
Orbiculiforma sp.		X		
Orbiculiforma cf. perampala (Rüst)	X			
Paronaella sp.	X	X		
Parvicingula boessi (Parona)	X			
Patellula planoconvexa (Pessagno)	X			
?*Patulibrachium*			X	
Pseudodictyomitra sp. (Pessagno)			X	
Pseudodictyomitra carpatica (Loznyak)		X	X	
Pseudodistyomitra leptonica (Foreman)	X			
Pseudodictyomitra lilyae (Tan Sin Hok)		X		
Saturnalis sp.				X
Sethocapsa cribrata		X		
Sethocapsa orca (Foreman)	X			
Siphocampium davidi (Schaaf)		X		
Sithocapsa cribrata (Hinde)		X		
Spongocapsula palmerae (Pessagno)	X			
Spaerostylus lanceola				X
Sphaerostylus lanceola (Parona)			X	
Spongodiscus sp.		X		
Spongodiscus renillaeformis (Cambell and Clark)	X			
Spongodruppa coccos (Rüst)	X			
Staurosphaera septemporata (Parona)		X		
Staurosphaera septemporata				X
Stichomitra asymbatos (Foreman)		X		

Contniued

Table 16. (*Continued*)

Radiolarian species and genera	Kudat 1	Telupid 2	Lahad Datu 3	Segama 4
Stichomitra communis (Squinabol)			X	
Sethocapsa orca (Foreman)	X			
Stichocapsa pseudodecora (Tan Sin Hok)	X			
Thanarla conica (Aliev)			X	
Thanarla pulchra (Squinabol)		X		
Ultranopora sp.		X		
Xitus alieri (Foreman)		X		
Xitus spicularius (Aliev)		X	X	

Data Source: 1, Basir et al. (1985); 2, Basir (1992); 3, Aitchison (1994) and 4, Leong (1977).
Note: X, means present.

A more comprehensive study of bedded chert from the Ruku-Ruku Valley, Telupid, was made by Basir (1992). The chert overlies pillow basalt and extensive serpentinized peridotite. The identified radiolaria are given in Table 16. This fauna indicates a Lower Cretaceous age of Late Valanginian to Barremian (127–135 Ma).

Basir et al. (1985) have also identified and described Lower Cretaceous radiolaria from cherts in the Kudat area (Table 16). The chert lies upon pillow basalt along the Sin Sin-Bak Bak road. Fossiliferous outcrops occur at Tanjung Bangau and Bukit Pengaraban. Some of the species are short lived and give good stratigraphic control. The age indicated is Lower Cretaceous (Barremian to early Albian). Lower Cretaceous (pre-Albian) radiolaria were extracted and identified by Aitchison (1994) from a large block of ribbon chert in the Ayer Formation mélange in a quarry on the main road between Lahad Datu and the Segama Bridge.

The radiolaria data are consistent. There are no modern determinations on extracted fossils younger than Albian or older than Upper Valanginian, so it must be concluded that the first ocean or marginal basin sediments were laid down at this time in water deeper than the carbonate compensation depth.

XIV.2. AGE OF THE UNDERLYING IGNEOUS OPHIOLITIC BASEMENT

Since the Upper Valanginian to Barremian (Gallic) ribbon cherts represent Layer 1 of the ophiolite, it follows that the pillow basalts and spilites, which represent Layer 2 of an ophiolite sequence, must be Upper Valanginian (Neocomian) or older ($\gtrsim$135 Ma). There are several outcrop localities, such as Ruku-Ruku Valley, Telupid, and Segaliud Estate, where the cherts can be observed in stratigraphic relationship to pillow basalts, with the pillow tops younging in the direction of the overlying chert. The gabbros and layered gabbro that represent Layer 3 of an ophiolite, must likewise have an age $\gtrsim$135 Ma. Table 17 is a listing of all known radiometric ages of the ophiolitic rocks.

Table 17. Potassium–argon radiometric ages of Sabah ophiolite and related rocks

Specimen	Source	Rock type	Locality	Mineral	K (wt.%)	Age with error (Ma)	Geological age
94SB87	2	Amphibolite	Silumpat Island	Amphibole	0.15	131±6	Hauterivian Neocomian
J5500A	3	Gneissic amphibolite	Pulau Tanna, Darvel Bay	Hornblende	0.185	87±2.5	Coniacian U. Cretaceous
J1060	1	Amphibolite gneiss	Pulau Silumpat, Darvel Bay	Hornblende	0.15	101±5	Albian L. Cretaceous
J1166	1	Epidote amphibolite	Pulau Adal, Darvel Bay	Whole-rock	0.03	140±20	Berriasian Neocomian
EKc	6	Garnet amphibolite	Tungku River, Dent Peninsula	Hornblende	0.03	75.6±21.3	Campanian U. Cretaceous
JD1	6	Amphibolite	Danum Valley	Hornblende	0.262	127±5	Barremian, L. Cretaceous
LD6	6	Isotropic gabbro	Lahad Datu, Darvel Bay	Hornblende	0.22	164±7	Callovian M. Jurassic
KS2	6	Metagabbro	Kg. Silam, Darvel Bay	Hornblende	0.053	217±17	Norian. U. Triassic
S14a	6	Isotropic gabbro	Pulau Silumpat, Darvel Bay	Hornblende	0.12	179±11	Aalenian M. Jurassic
94SB38A	2	Gabbro	Segaliud oil palm estate	Plagioclase	0.11	81.7±4.3	Campanian U. Cretaceous
S28-2	4	Large gabbro block or slice	Bole River, Ulu Segama	Whole-rock[a]	0.16	137.54±6.88	Valanginian Neocomian
94SB39A	2	Microgabbro	Segaliud oil palm estate	Plagioclase	0.12	52.0±3.5	Ypressian L. Eocene
S87-40	4	Gabbro in mélange	Mount Silam	Whole-rock[a]	0.33	33.41±1.67	Rupelian L. Oligocene
The following determinations are on non-ophiolitic granitoids associated with the Segama Valley ophiolites							
NB11714	1	Tonalite	Litog Klikog Kiri, Segama	Biotite	6.80	150±6	Tithonian U. Jurassic
NB10852	1	Hornfels	Litog Klikog Kiri, Segama	Biotite	7.29	160±8	Callovian M. Jurassic
94SB86	2	Two-mica granite	Bole River, Segama	Muscovite	8.6	156±3	Oxfordian U. Jurassic
				Biotite	7.0	105±2	Albian L. Cretaceous
				Plagioclase	1.5	99.5±2.6	Albian L. Cretaceous
J5712	3	Granite	Lower Telewas Valley	Biotite	6.91	120±1.5	Aptian L. Cretaceous
J5698B	3	Tonalite	Kawag Gibong River, Segama	Biotite	6.08	210±3	Norian U. Triassic

Data Source: 1, Kirk (1968); 2, Swauger et al. (1995), Graves et al. (2000); 3, Leong (1974); 4, Rangin et al. (1990); 5, Omang and Barber (1996) and 6, Omang (1993).
* whole rock wt.% quoted in K_2O.

All are by potassium–argon and the accuracy is unfortunately limited because of the very low K concentrations in these rocks. The dates obtained for amphibolite (94SB87) from Silumpat Island and gabbro (S28-2) from Bole River, Ulu Segama, are 131 (Hauterivian) and 137 Ma (Valanginian), respectively. Both are of rocks assigned to the gabbroic layer (Layer 3) and their dates indicate an appropriate Neocomian, or earliest Cretaceous age. The appropriate ages appear to be independent of metamorphic condition. However, meta-gabbros from the Silumpat Island display the problem of attaining the true age. Obtained K:Ar ages from extracted amphibole range from 101 Ma (J1060), through 131 Ma (94SB87) to 179 Ma (S14a). Clearly all cannot be the true age. The age 101 Ma may be rejected as too young since it post-dates the cherts, but there is no way at present to ascertain if the Neocomian age (131 Ma) or the much older Middle Jurassic age (179 Ma) is correct. The oldest age determination so far is of amphibole separated from meta-gabbro from Silam (KS2), which gave an age of 217 Ma (Norian; Upper Jurassic).

The epidote amphibole (J1166) from Pulau Adal, which has been interpreted as a meta-basalt of Layer 2 by Hutchison (1978), also gives an appropriate Neocomian age of 140 Ma (Berriasian). The ultrabasic rocks, which underlie the gabbros, do not lend themselves to radiometric dating, and a similar age must be assumed since they are differentiated cumulates genetically related to the basic rocks of the lower crust. The younger ages in Table 17 may be regarded as spuriously young and to represent the results of tectonic and recrystallization events and/or loss of argon during tectonic events and recrystallization.

Our present state of knowledge, therefore, indicates that the Sabah ophiolite, including the overlying chert, is Neocomian (Lower Cretaceous) irrespective of whether the age determinations were on metamorphosed or unmetamorphosed basic igneous rocks (Graves et al., 2000). The meaning of the older determined ages cannot be resolved with our present understanding.

XIV.2.1. Fission track dating

Twenty apatite crystals from meta-gabbro (95SB38A) from a large block in Garinono Formation mélange in the Segaliud oil palm estate (Figure 76) (Hutchison et al., 2000), gave fission track data as summarized in Figure 77. Unfortunately no zircon crystals were recovered (Swauger et al., 1995). The gabbro is massive, altered and medium-grained, composed of altered plagioclase and pyroxene, most of which has been altered to chlorite. The data are ambiguous leading to no clear-cut interpretation. This is an igneous rock; hence the apatite ages would be expected to form a compact normal cluster (all of one age). The interpretation is that the rock and its contained apatite crystals have been reheated after igneous crystallization, so that the fission track ages have been annealed, but not completely. Averaging fission track data has no meaning for such a wide spread. The four measurements between 160 and 170 Ma are extremely interesting and may be true relicts of the igneous parentage with no subsequent annealing. This might indicate that the gabbro initially crystallized in the Middle Jurassic (Dogger). There is no reason to eliminate these four Jurassic dates. Two conclusions are possible: not all of the ophiolite is Neocomian, and some of the ophiolite basement, including this metagabbro, may be older (at least Jurassic).

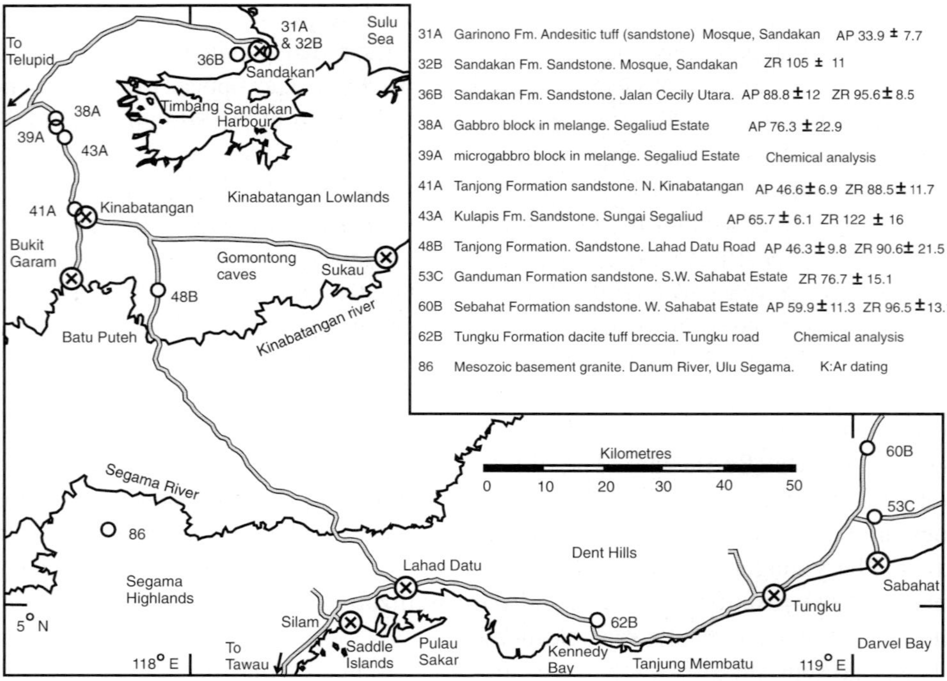

Figure 76. Locality map of samples collected for fission-track dating for the Eastern Lowlands of Sabah (after Swauger et al., 1995).

The gabbro block may have been somewhat heated by the mud matrix of the Garinono mélange. Clennell (1991) showed that the mélange mud matrix was not particularly hot (< 100°C). However, apatite is extremely sensitive and experiences partial annealing at as low as 60°C and complete annealing at 110°C (Gallagher et al., (1998). This specimen has definitely been annealed, but not completely and Jurassic-aged crystals remain.

XIV.3. STRATIGRAPHY OF THE OPHIOLITE

It is only on the islands of Darvel Bay that the ophiolite structure (Hutchison and Dhonau, 1969) and stratigraphy (Hutchison and Dhonau, 1971) have been successfully unravelled. Inland exposures are limited to rivers and, unlike the Zambales Complex of the Philippines, where only grass grows over the ultrabasic rocks and trees over the gabbroic rocks, there are no grasslands in Sabah and dense forest covers both gabbros and ultrabasic rocks. Vegetation gives no aid to mapping, and inland outcrops cannot be structurally mapped adequately.

The detailed geology of Pulau Sakar (Figure 78) is of a steeply dipping succession of ultrabasic rocks, overlain by banded (layered) gabbroic rocks, in turn overlain by fine grained amphibolitized basalt. No bedded chert occurs on the island, but there are

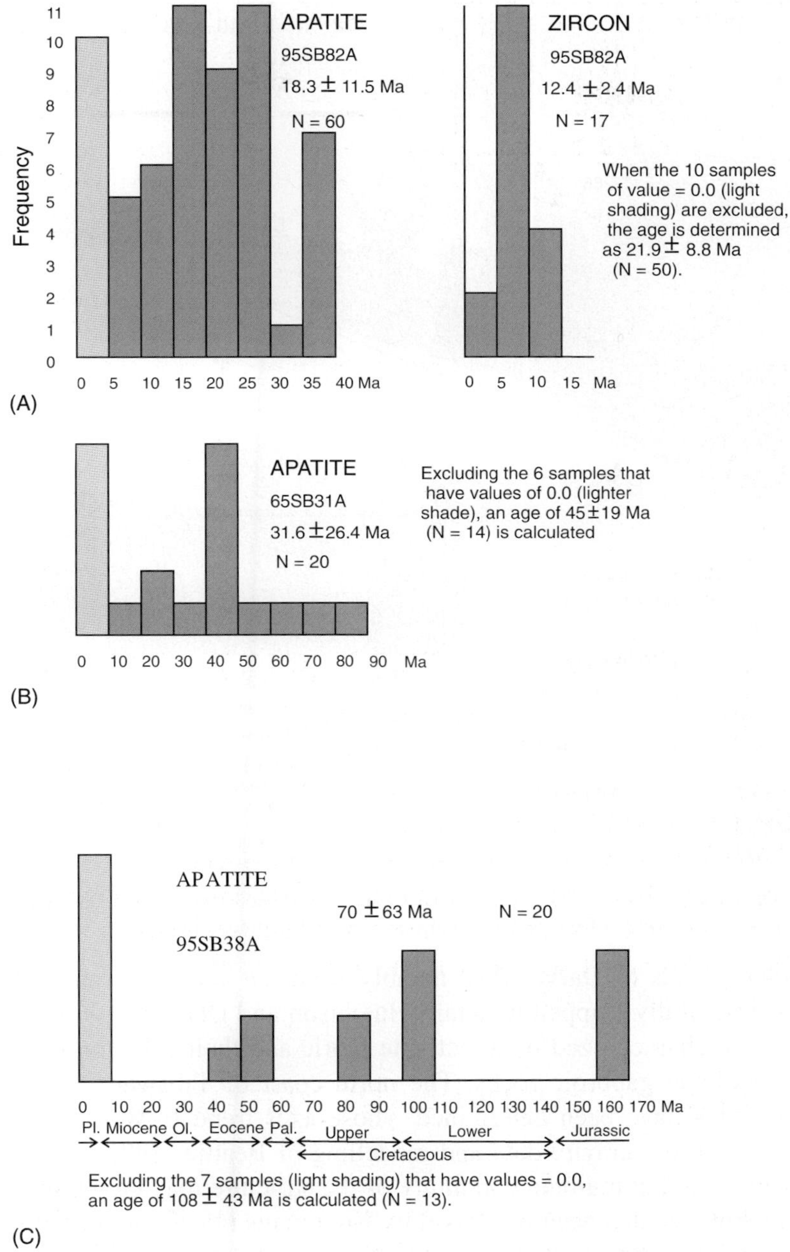

Figure 77. Fission-track histograms of igneous rocks from the Eastern Lowlands. (A) from the Semporna volcanic arc, Tawau; (B) andesitic tuff from the Sandakan Mosque; (C) a gabbro block in mélange at Segaliud.

nearby outcrops to the west of Lahad Datu. Thus we have all the elements of ophiolite stratigraphy in correct mutual position on Sakar Island. The south coast is remarkably straight and most of the ultrabasic rocks are serpentinized, so that a major fault or shear zone is inferred, along which the ophiolite has been up-faulted.

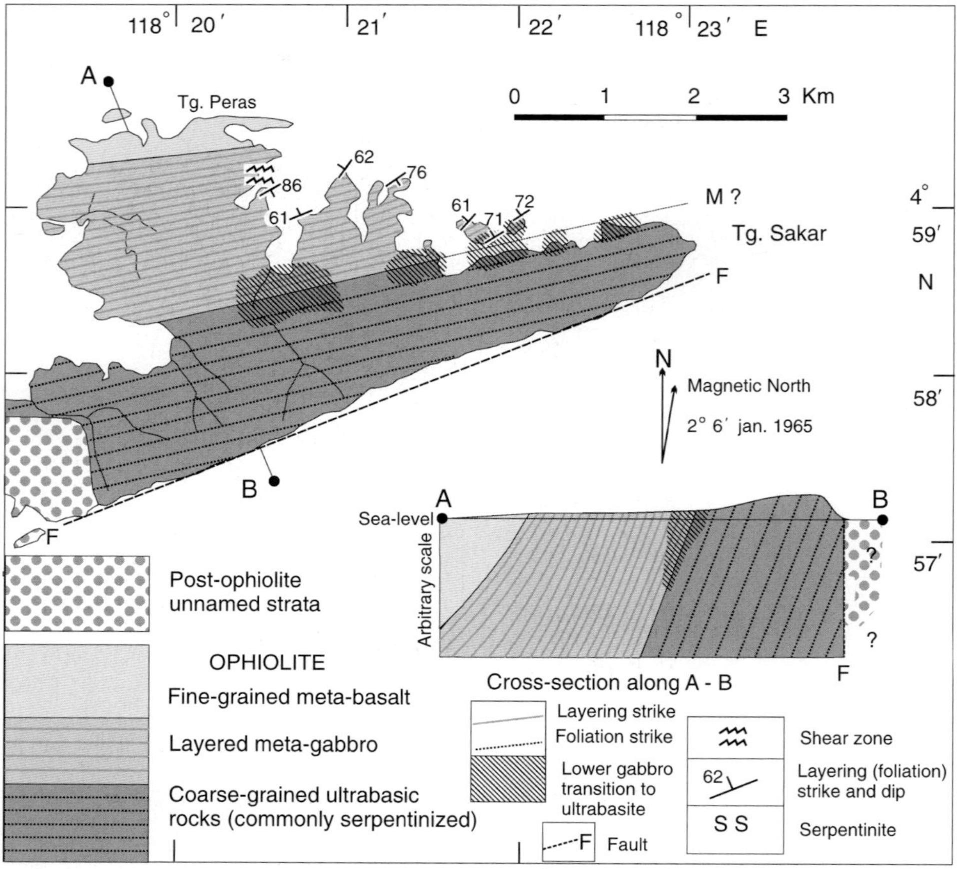

Figure 78. Ophiolitic geology of Sakar Island, near Lahad Datu (after Hutchison and Dhonau, 1971). With permission from Minerals and Geoscience Department, Malaysia.

The outer islands of Darvel Bay, notably Bohayan and Tabawan (Figure 79), have been structurally mapped in detail (Hutchison and Dhonau, 1969). The ultrabasic rocks are characterized by a tectonite fabric and their foliation is parallel to the layering of the gabbroic rocks. The north coast of Tabawan island is fault-controlled. Folds have been determined whose axes trend E–W. This geological trend agrees with the gravity data and modelling of Beattie (1986) and Ryall and Beattie (1996), so that regional continuity of the structure may be interpreted. The E–W fold trend has also been described by Koopmans (1967) in northern Timbun Mata. The folding of the gabbro is readily discerned because it is of outcrop scale layers of grey plagioclase-rich and dark green/black layers rich in amphibole with pyroxene relicts. A prominent mineral lineation is seen on all layering surfaces. It is represented by elongated aggregates of actinolitic amphibole. The lineation is always parallel to the fold axes, whose plunge is low to sub-horizontal (Hutchison and Dhonau, 1969). The crystallization of the actinolitic amphibole is therefore synchronous with the ophiolite folding, of broad open style (Figure 79). However, as shown later, the metabasites preserve much of their original igneous fabric and

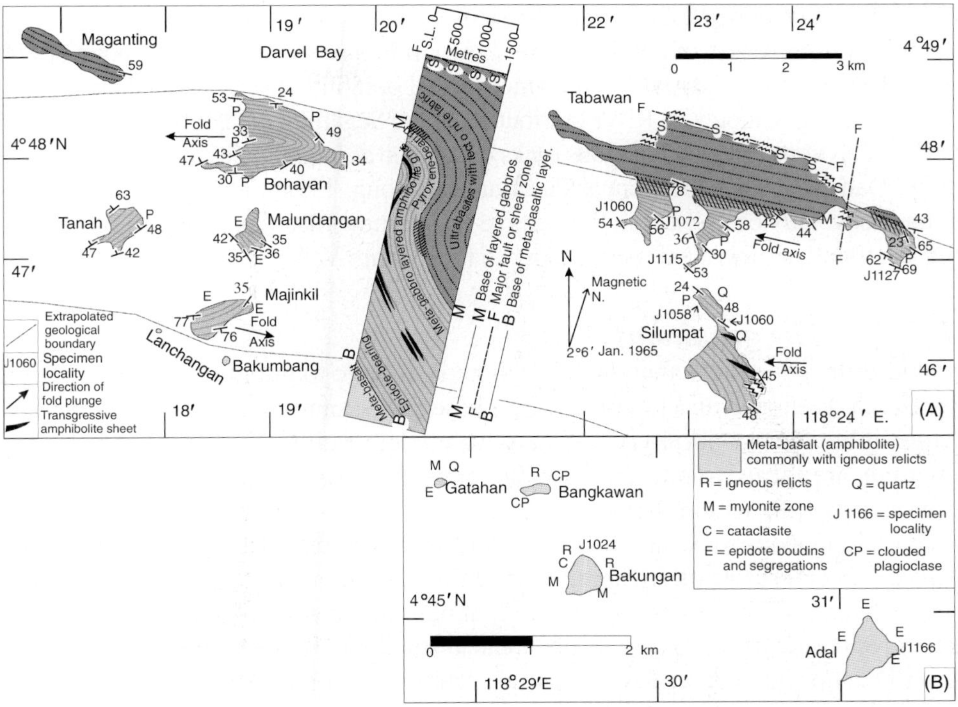

Figure 79. Geology of the main islands of Darvel Bay (after Hutchison and Dhonau, 1971). Legend also in Figure 78 (A) large central islands, (B) far outer islands, composed entirely of meta-basalt. Note that (B) lies some distance towards the SE. With permission from Minerals and Geoscience Department, Malaysia.

commonly contain igneous relicts. The metamorphism is retrogressive and may have been sub-sea-floor before uplift. The common partial metamorphic condition, with igneous relicts, presents a problem for interpretation of the radiometric dating.

Although there is structural continuity of outcrops in the limited area of Darvel Bay, the ophiolitic rocks are generally faulted and imbricated, with little regional continuity. The ultrabasic rocks are usually up-faulted and serpentinized. Their relationship to other rocks is usually faulted. Only in isolated inland area are the stratigraphic relationships undisturbed, such as near Telupid and in the Segaliud estate, where ribbon cherts conformably overlie pillow basalts.

XIV.4. NON-OPHIOLITIC BASEMENT

There are scattered occurrences of granitoids within the ophiolitic terrain of the Segama Valley that need to be interpreted as non-ophiolitic basement for the following indisputable reasons. They are normal calc-alkaline granitoids not of the plagiogranite suite that is to be expected genetically associated with ophiolite. They have given the oldest radiometric dates of Sabah (Table 17), pre-dating the Neocomian ophiolite and yielding Triassic and Jurassic ages.

XIV.4.1. Geographical distribution

The known occurrences of granitoids are shown in Figure 80, taken from Leong (1974, 1998). They always occur within metamorphosed ophiolitic basalt and gabbro terrain, and in the Litog Klikog Kiri River (a tributary of the Segama), the granite is surrounded by a hornfels zone. The acid igneous rocks occur also as large bodies, notably in Dismal Gorge, Danum Gorge, and Sungai Purut areas (Leong, 1998). There may be other as yet undiscovered occurrences in this remote region, now made more accessible by logging roads and an upgraded road leading to the Danum Valley biology field station.

XIV.4.2. The granitoids

Granodiorite is the most abundant, but there is a range to granite and to more basic types such as diorite and tonalite. Late phase veins and dykes are of granite pegmatite, aplite, quartz porphyry and massive epidote-rich rocks (Leong, 1974). The granitoids are intrusive into amphibolite, amphibole schist and gneiss, which also occur as xenoliths within them.

The acid igneous rocks are invariably deformed, brecciated and sheared to various degrees into cataclasites showing mortar or flaser structures. The granodiorites show a crude foliation. The largest outcrop of granodiorite stretches from Ulu Purut to the Segama River at Dismal Gorge. There are also smaller outcrops forming gorges in the Lower Danum and Segama River. Granite porphyry occurs in Dismal Gorge. At Eagle Gorge the granite contains xenoliths of the layered gabbro, which Fitch (1955) called diorite. The granodiorites of Ulu Bole have a crude gneissic fabric.

Abundant granite, granodiorite and tonalite occur in the Kawag, Litog Klikog Kiri and Babayas (Babias) rivers (Figure 80). Granodiorite also occurs to the east in the Lower Telewas area, Tempadong Gorge, Upah and Mawan valleys. In the Bilin (Batu Beling) area, thick granitic veins cut amphibolite, and coarse granite contains xenoliths of spilite.

XIV.4.2.1. Chemistry

A selection of the available whole-rock analyses is given in Table 18. These granitoids are clearly not of the ophiolite suite and contain significant concentrations of K_2O, implying that they have been derived by partial melting of an underlying continental basement, which does not outcrop, but may be held responsible for supporting the dense ophiolitic rocks that make the majority of outcrops. All available analyses are plotted on a K_2O versus SiO_2 diagram (Figure 81) that clearly indicates the calc-alkaline non-ophiolitic signature of the granitoids. The samples c, g and j belong to the low-K tholeiite series. Accordingly they are plagiogranites and genetically related to the ophiolitic gabbros.

XIV.4.3. Contact aureole rocks

A contact aureole of hornfels has been mapped by Kirk (1964), as shown in Figure 80. The aureole in the Litog Klikog Kiri valley ranges from about 300 m to over

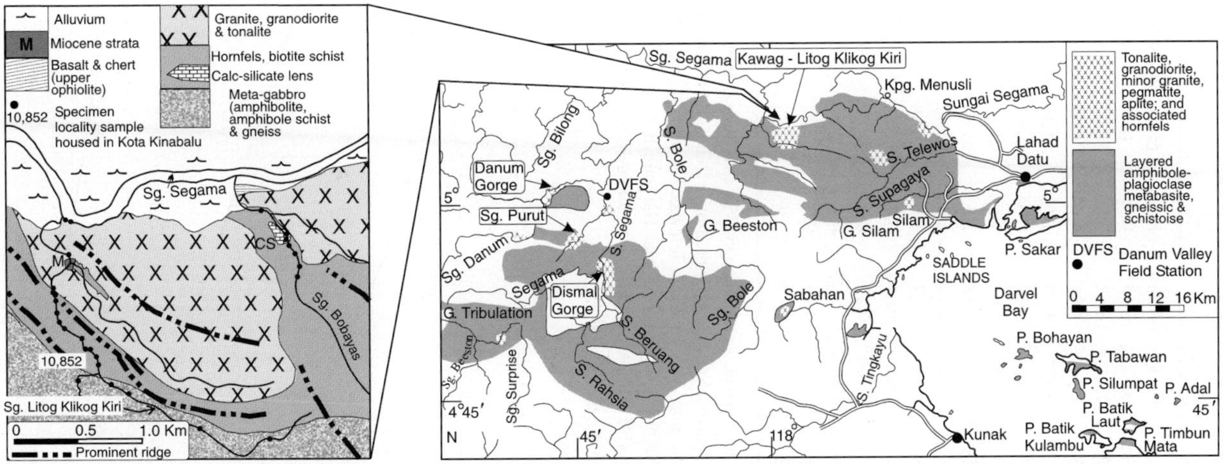

Figure 80. Distribution of metamorphic ophiolitic basalt and layered gabbro and associated granitoids of the Upper Segama basement. Redrawn from Leong (1974, 1998). Associated ultrabasic rocks are not shown. The left-hand map was drawn by Kirk (1964). It shows the zone of hornfels surrounding the basement granitoid,which has given the oldest radiometric age of Sabah. With permission from Minerals and Geoscience Department, Malaysia.

Table 18. Wt.% chemical analyses of granitoids occurring predominantly within Segama Valley ophiolite terrain

Sample	a	b	c	d	e	f	g	h	i	j
Source	1	2	1	1	1	2	1	1	1	1
SiO_2	55.10	56.34	56.8	59.3	66.03	68.34	68.56	69.1	71.1	72.7
TiO_2	0.94	0.68	0.59	0.57	0.53	0.36	0.43	0.25	0.17	0.11
Al_2O_3	19.71	17.88	18.3	17.1	16.63	8.86	16.91	15.5	15.0	14.9
Fe_2O_3	2.76	2.69	2.85	2.30	1.51	8.07	0.54	0.60	0.54	0.34
FeO	2.35	3.99	3.75	3.25	1.77	1.65	1.10	1.26	1.27	0.89
MnO	0.14	0.16	0.14	0.12	0.10	0.04	0.07	0.05	0.06	0.04
MgO	3.39	3.85	3.10	2.80	1.63	1.56	1.06	0.49	0.63	0.39
CaO	5.15	6.76	7.45	5.55	4.48	3.32	1.66	3.03	2.75	2.50
Na_2O	5.14	4.19	4.45	4.85	4.40	4.54	6.47	5.39	4.75	4.95
K_2O	1.23	1.36	0.55	1.27	1.27	1.49	0.58	2.00	2.25	1.36
H_2O+	3.14	0.93	1.24	2.05	1.07	0.73	1.48	1.48	1.04	0.48
H_2O-	0.12	0.65	0.16	0.26	0.16	0.42	0.32	0.24	0.20	0.28
CO_2	0.95	0.20	0.05	0.15		0.12	0.60	0.57	0.31	0.61
P_2O_5	0.36	0.18	0.19	0.21	0.19	0.08	0.07	0.05	0.08	0.05
Total	100.48	100.44	99.62	99.78	99.77	99.93	99.85	100.01	100.14	99.60

a = diorite (Nh95), South coast, Kudat Harbour. b = tonalite (J5833), Ulu Sungai Purut, Upper Segama Valley. c = tonalite (NB11661B), Sungai Upah, Segama Valley. d = granodiorite (NB10855), Sungai Batu Beling, Segama Valley. e = granodiorite (L103), Sungai Segama, west of Kuala Telewas. f = granodiorite (J5911). Sungai Segama, south of Dismal Gorge. g = tonalite (Nh99), South coast, Kudat Harbour. h = granodiorite (NB8732), Sungai Babias, Segama Valley. i = granodiorite (NB1-851), Sungai Litog Klikog Kiri, Segama Valley. j = granodiorite (NB11655), North of upper Sungai Telewas, Segama Valley.
Data Source: 1, Kirk (1968); 2, Leong (1974).

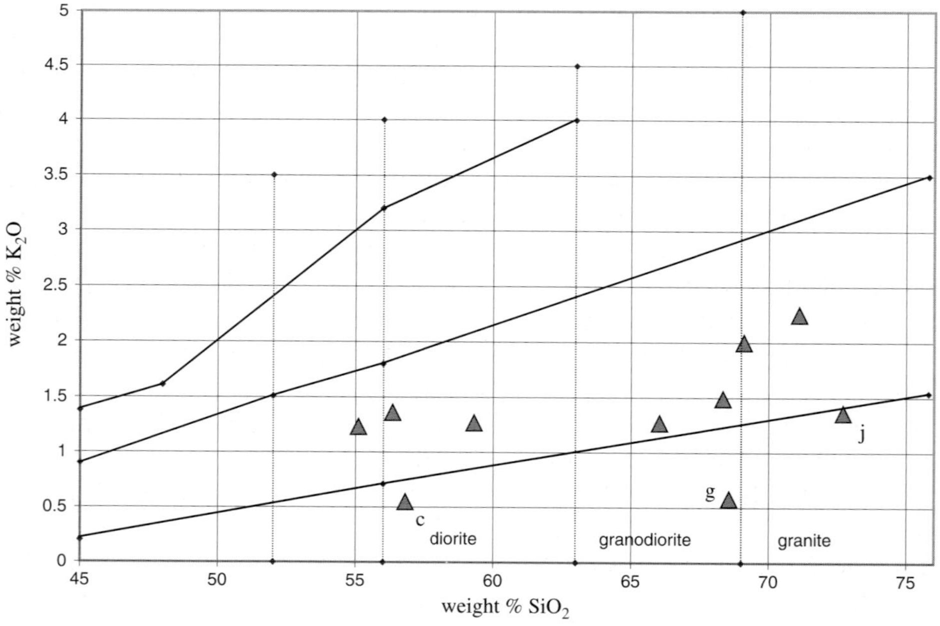

Figure 81. Calc-alkaline granitoids within ophiolite terrains. With permission from the Department of Minerals and Geosciences Malaysia.

1 km in width. Many of the rocks have hornfels texture, but some are schistose and there are relict meta-volcanic textures (Leong, 1974). Both the schistose and non-schistose hornfels of the Litog Klikog Kiri area contain plagioclase and quartz. The commonest mafic mineral is biotite, sometimes accompanied by amphibole and rarely by garnet and epidote. Some hornfels outcrops are crudely banded with alternating amphibole-rich and plagioclase-rich layers. This is a common property of metagabbros of Darvel Bay where there is no contact aureole and no granitoids.

The contact aureole in the Babayas Valley contains outcrops of metamorphosed limestone (calc-silicate rock) (Figure 80). They contain garnet (grossularite), calcite, pyroxene and wollastonite and form layers in amphibole hornfels and hornblende and biotite schist. Schistose rocks predominate. The common minerals in the non-calcareous rocks are plagioclase, quartz, amphibole, garnet and sillimanite. Andalusite (chiastolite) occurs in garnet mica schist in a tributary of the Sungai Kawag.

XIV.4.3.1. The problem of the contact aureole rocks

Biotite "is commonly present and in most hornfelses constitutes the only mafic mineral" (Leong, 1974). Indeed it was biotite (containing 7.29% K) that gave the 160 ± 8 Ma Jurassic age from the Litog Klikog Kiri River (Figure 80, Table 17). By contrast the metabasalts and metagabbros of the ophiolitic rocks of the Segama Valley and Darvel Bay area do not contain biotite because their protoliths are low to deficient in potassium. The country rocks that surround the granitoids may not therefore represent contact metamorphosed ophiolite, but a separate formation, richer in potassium that pre-dates the Neocomian ophiolite. The lens of calc-silicate along Sungai Bobayas (Figure 80) also supports this conclusion, for limestones cannot occur within the basaltic and gabbroic layers of an ophiolite, but only as overlying sediments. This older formation into which the granitoids intrude has been called the "crystalline basement", but the term has been extended too comprehensively to include most of the basaltic and gabbroic part of the Neocomian ophiolite, but wrongly excluded the ultrabasic rocks, which were regarded as unrelated. There may be structural differences between the crystalline basement rocks (that include the granitoids) and the younger Neocomian ophiolite, which has generally been subjected to retrogressive metamorphism giving amphibolite assemblages with common igneous relicts, but devoid of biotite because of a deficiency in potassium.

It appears necessary to regard the generally cataclastic granitoids (granodiorite, tonalite and granite) and their envelopes of biotite-bearing metamorphic rocks and including calc-silicates, which have been mapped as contact aureoles of the granites, as tectonic windows of Jurassic and Triassic basement underlying and supporting the Neocomian ophiolite. The cataclastic fabric of the granitoids may have been imposed near the sole of the ophiolite overthrust mass. Field reinvestigation will be necessary to address this possible interpretation, which logically follows from the older radiometric dates of the granitoids and their envelope.

XIV.5. OPHIOLITIC ULTRABASIC ROCKS

The main ultrabasic rock type is serpentinized peridotite, mainly harzburgite, with less abundant lherzolite (Kirk, 1968; Omang, 1995). Dunite is rare. The large ultrabasic bodies invariably have steeply dipping, intensely brecciated and serpentinized faulted margins. Their emplacement was tectonic and cold, since there are no thermal effects on adjacent rocks. The largest ultrabasic massifs are in the Labuk Highlands, notably at Mount Tavai and the Bidu-Bidu Hills. Quite large massifs occur in the Segama Valley, the largest at Mount Silam. There are scattered bodies SE of Mount Kinabalu, and in the north-eastern islands of Balambangan, Banggi and Malawali.

Foliation and banding is common, and parallel to that of contiguous gabbros, if the rocks are still in mutual stratigraphic position. Porphyroblastic and neoblastic textures are interpreted as a result of plastic deformation in the upper mantle. The lower gabbro layer also contains layers that themselves are ultrabasic. In Figures 78 and 79, the gradational contact between the gabbroic and ultrabasic rocks has been interpreted as the petrological Mohorovičić Discontinuity. If not the actual boundary between the crust and upper mantle, it is a dramatic and clearly mappable geological boundary.

XIV.5.1. Segama Valley and Darvel Bay

The major faults that have facilitated the uplift of the major serpentinized mantle peridotite bodies trend E–W. The largest massif is Mount Silam, and ultrabasic bodies extend eastwards along Pulau Sakar (Figure 78) and Tabawan (Figure79).

XIV.5.1.1. Mount Silam

The E–W trending ultrabasic massif is about 32 km long and up to 6 km wide, forming Mount Silam (890 m) and Mount Beeston (863 m). Spilites occur along the southern margin and gabbros along the northern. The contacts are steep, rather as in Pulau Sakar. Peridotite, mainly harzburgite, is the predominant rock. Dunite forms lenticles 1 m wide. Swarms of gabbro dykes cut serpentinite in the Hitam and Kamut Valleys. The dykes parallel the shear structures of the serpentinite. This is an example of the sheeted dyke complex, so characteristic of ophiolite complexes but rarely seen in Sabah. The region was mapped in detail by Bailey (1963) in his unsuccessful search for economic chromite deposits. Most are too thin and lenticular to be economically mined. The economic chromite deposits of the Zambales occur well below the Mohorovičić Discontinuity. The Sabah outcrops accordingly are too high in the ophiolite sequence to be economically viable.

XIV.5.1.2. Darvel Bay

The ultrabasic rocks, where not serpentinized, include peridotite, pyroxenite and dunite, commonly bearing small amounts of bytownite or anorthite (Hutchison and Dhonau, 1971). All are coarse-grained (1–3 mm). Olivine is commonly $Fo_{90.4}$, *ortho*-pyroxene $En_{82-90}Fs_{0-8}Wo_{1-10}$, *clino*-pyroxene $En_{69}Fs_3Wo_{28}$ (Omang, 1995). The pyroxene is generally diopside, less commonly enstatite or hypersthene. All peridotites contain forsterite, commonly serpentinized. The most widespread accessory mineral is chromite. Its occurrences and physical properties have been detailed by Hutchison (1972). A range of X_{Mg} 0.59–0.62 and X_{Cr} 0.38–0.39 was determined by Omang (1995), typical of Alpine-type chromites. .

The ultrabasites display typical cumulate textures. They consist of commonly rounded crystals of even coarse grain, the interstices being filled by finer-grained material. Usually they are without structure, but near the contact with gabbros they display a crude foliation parallel to the gabbro layering. Typically there is a transition zone between the ultrabasites and gabbro.

Using the analysed chemistry of the mineral phases, Omang (1995) deduced a mantle formation temperature of 1030–1100°C and a pressure of 11–16 kbar. Such a *P–T* combination sits exactly on the Island-Arc Geotherm.

XIV.5.2. Labuk Highlands

Large ultrabasic massifs occur in this area, associated with minor gabbroic rocks. They form a total of 900 km^2 of mountainous country rising to over 1200 m elevation in the Labuk and middle Kinabatangan valleys. The main outcrops are at Mount Meliau (250 km^2), Mount Tavai (280 km^2), Bidu-Bidu Hills (140 km^2), Mount Tingka (116 km^2) and Gunung Rara and adjacent areas (90 km^2). The outcrops form a N–S trending belt. Individually they are separated by faulted strips of pillow basalt and sedimentary rocks.

XIV.5.2.1. Meliau Range

This range of steep hills, rising to Mount Meliau 1336 m, is predominantly of ultrabasic rock. Pillow basalts occur on its southern side, with Crocker Formation on the north. The bulk of the massif is of enstatite peridotite grading into lherzolite. Dunite forms lenticular areas. Serpentinization is most intense along major faults and shear zones. Banding generally strikes N–S, parallel to zones of brecciation and shearing (Kirk, 1968).

XIV.5.2.2. Bidu-Bidu Hills

The geology has been described by Newton-Smith (1967). The main massif is of serpentinized peridotite containing various proportions of ortho- and clinopyroxene. The peridotite is commonly foliated, containing numerous fault and breccia zones. The foliation results from sub-parallel alignment of pyroxene crystals and has a markedly NE trend in the northern sector and NW in the southern sector.

Uplift of the ophiolite The Bidu-Bidu Hills is the only ultrabasic massif from which evidence of ophiolite uplift and erosion has been presented by Newton-Smith (1967) and Hutchison and Tungah Surat (1991). The most accessible locality is at Kuala Dudor in the Rumidi Estate. There is a NW-trending ridge 1.5 km long, 0.4 km wide, of serpentinite conglomerate. The surface is gossanized around the abandoned manager's house. The conglomerate contains blocks up to 1.2 m diameter of ultrabasic rock, spilite and feldspar–tremolite schist embedded in fine-grained serpentinite matrix. The conglomerate is interbedded with grit beds up to 3.8 cm thick, containing angular serpentinite clasts up to 2 mm size interbedded with red mudstone. One sample contains pelagic Foraminifera that unfortunately are not age-diagnostic. Other grains are of chrome spinel and picotite. The new road leading off milestone 69 from Sandakan, leading northwards to the Leadstar mining prospect and the Rumidi Estate, has provided good outcrops of serpentinite conglomerate and sandstone (Hutchison and Tungah Surat,1991). The sandstone contains fresh angular pyroxene and angular to sub-angular serpentinite grains that appear to have been transported as single olivine crystals before being serpentinized. The original porosity of the serpentinite sandstone has been cemented by calcite. The relationships between the serpentinite conglomerate and sandstone are sedimentologically and structurally complex, and Hutchison and Tungah Surat (1991) suggest these rocks resulted from submarine mass flow.

It is not possible to define when the ophiolite was uplifted and subjected to erosion. Newton-Smith (1967) ascribed the serpentinite conglomerates and sandstones to the "Chert–Spilite Formation", but contiguous outcrops of his Kamansi Beds might also be the host. Sandstones of the Kamansi Beds contain grains of quartz, sodic plagioclase, chert, magnetite, olivine, zircon and epidote. There was therefore a nearby ophiolite provenance.

The Foraminifera, identified from mudstones of these formations, are listed in Table 19. The Shell palaeontologists, who identified the Foraminifera, concluded that the fauna was not age-specific, and characteristic of flysch of Late Cretaceous to Eocene or Oligocene age. From this, we might conclude that the Bidu-Bidu Hills ophiolite was uplifted and eroding in early Tertiary time.

Along the main realigned Sandakan road 5 km SE of Telupid, slices of ophiolite, dipping 40–60°, have been thrust from the south over Miocene Garinono Mélange (Clennell, 1992). The deep road cut still remains intact. Grey matrix Garinono Mélange is overlain by massive serpentinite of the Telupid ophiolite complex (Figure 82). Slices of mélange and serpentinite are interleaved *schuppen* fashion. Some show down-dip slickensides, dipping south irregularly but generally steep 40–60°. The geometry suggests that the ophiolite was thrust over the

Table 19. Foraminifera in the Chert–Spilite Formation and Kamansi Beds (Newton-Smith, 1967)

Species of Foraminifera	Chert–Spilite Formation	Kamansi Beds
Ammodiscus sp.	X	
Bathysi*phon* sp.	X	X
Cyclammina sp.	X	
Glomospira sp.	X	X
Haplophragmoides sp.	X	X
Haplophragmoides walteri (Grzybowski, 1898)	X	
Sigmoilina sp.	X	
Trochammina sp.	X	X
Trochamminoides sp.	X	X
Trochamminoides subcoronata (Rzehak and Grzybowski, 1896)	X	

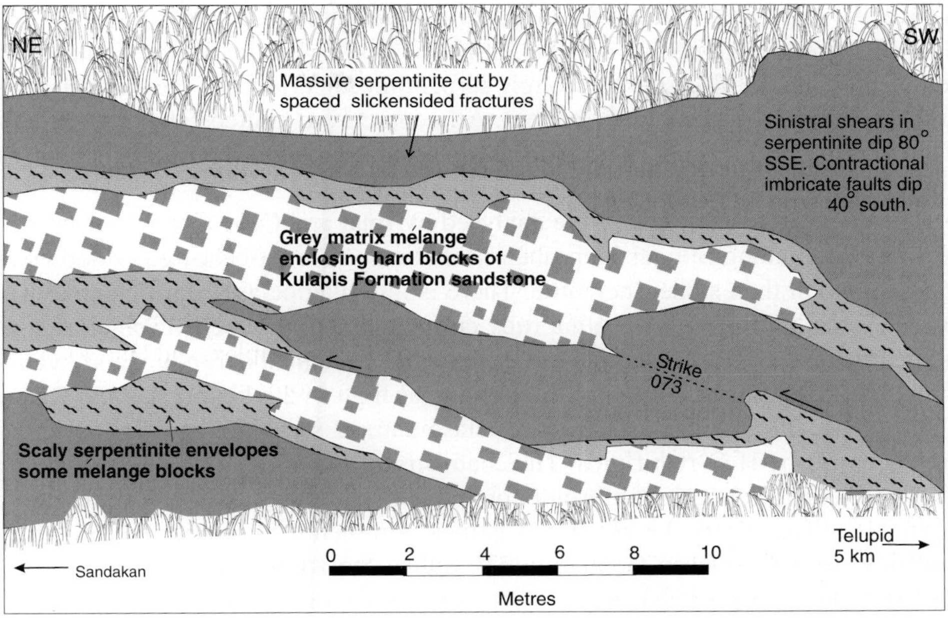

Figure 82. Imbricated base of the Labuk Highlands ultrabasic ophiolite exposed in a road cutting 5 km east of Telupid. The ophiolite is inferred to have been thrust over the Garinono Mélange from the south. The road cut continues to persist. Drawing by Clennell (1992).

mélange from the south (Clennell, 1992). The overthrusting of the ophiolite post-dates the Garinono Mélange and therefore occurred in post-Middle Miocene time.

XIV.5.2.3. Mount Tavai

The Mount Tavai ultrabasic massif was mapped by Hancock (1964) while prospecting for commercial chromite. The massif is predominantly of coarse-grained harzburgite showing various degrees of serpentinization. Much of it has a crude gneissic fabric. Gabbro forms lenticular bodies. Wide zones of breccia characterize the contacts. Frequently mylonitized serpentinites are prominent. They indicate that the ultrabasic body was tectonically emplaced.

Mineral banding results in lenticular structures; the main rock types being dunite, feldspathic peridotite and pyroxenite within the harzburgite. The bands vary in width up to 100 m, are invariably steeply dipping and usually parallel to the peridotite gneissic foliation.

XIV.5.2.4. Mount Tingka

The main rock is strongly serpentinized peridotite. A large arcuate fault controls the outcrop shape. Basalt and chert are commonly in faulted contact with the ultrabasic rocks. The Tingka massif probably connects at depth with Gunung Raya south of the Kinabatangan River.

XIV.5.3. Mount Kinabalu–Mount Tambuyukon area

Several large and numerous smaller ultrabasic bodies are in faulted contact with sedimentary strata. They are of strongly serpentinized peridotite. There are also several associations of hornblende gabbro. All contacts are faulted. The most accessible outcrops are road cuts on the northern side of the road to the Poring hot springs only 1–2 km from the Ranau–Telupid road junction.

The varieties of peridotite were analysed by Imai and Ozawa (1991) from samples collected along Sungai Mensabah near Ranau. Garnet-bearing peridotites are more common than spinel lherzolite. There is rare plagioclase-bearing spinel lherzolite. Some hornblendites contain ruby corundum, strongly zoned spinel, pyrope-rich garnet and zoisite. A sample of hornblendite contained corundum and kyanite. Such rocks represent an equilibrium pressure as high as 20 kbar (Imai and Ozawa, 1991). The ultrabasic rocks were metasomatized on their upwards tectonic emplacement by H_2O-rich fluids. The chemistry of the mineral phases of the ultramafic rocks requires an equilibrium combination of 850–950°C and a pressure of 7–15 kbar. The garnet lherzolite and garnet–hornblende peridotite *P–T* combination lies along the Ocean Geotherm. The spinel lherzolite and plagioclase–spinel lherzolite lies along the Island-Arc Geotherm. Thus the ophiolite originated in an oceanic environment, possibly near or in the region of an island arc.

XIV.5.4. Northern islands

Malawali Island is constructed mainly from ultrabasic rocks. The most common rocks are serpentinized harzburgite, with rare lherzolite. The rocks are cut by E–W shear zones, in which there are angular to rounded boulders within sheared serpentinite.

Banggi Island is strongly imbricated and the serpentinized ultrabasic rocks are associated with gabbro, many outcrops being in fault relationship.

XIV.5.5. Chemistry

A selection of chemical analyses from Kirk (1968) is given in Table 20. All available analyses are plotted on a K_2O versus SiO_2 diagram (Figure 83). Clearly, the majority of rocks are ultrabasic cumulates, and do not represent magmatic liquids. However, a few rocks named ultrabasites have been misnamed and fall into the basaltic and basaltic-andesite fields. There are many occurrences of melanocratic gabbro rich in pyroxene, which may be named ultrabasic during field mapping. There is indeed a gradation between the ultrabasic and the gabbroic layers, for example on Tabawan and Sakar islands. All are naturally deficient in potassium.

XIV.6. THE GABBRO ZONE

The traditional arbitrary division into "basic igneous rocks" and "crystalline basement" cannot be upheld. The overwhelming majority of rocks assigned to the "crystalline

Table 20. Wt.% chemical analyses of selected ophiolitic ultrabasic rocks of Sabah

Sample	a	b	c	d	e	f	g	h
Source	1	1	1	1	1	1	1	1
SiO_2	37.2	38.14	40.1	41.75	43.06	33.2	44.6	53.3
TiO_2	0.01	0.04	0.01	0.12	0.02	Tr.	0.45	0.14
Al_2O_3	0.63	0.47	1.24	3.73	2.15	0.17	12.9	2.80
Fe_2O_3	3.90	4.64	3.10	2.34	2.40	5.85	0.75	2.40
FeO	3.40	2.59	4.55	5.54	4.15	1.65	6.15	5.95
MnO	0.10	0.14	0.12	0.16	0.12	0.09	0.12	0.16
MgO	37.9	38.47	37.9	37.24	33.5	39.6	19.4	28.2
CaO	0.90	tr.	1.73	2.90	3.95	0.43	11.1	6.40
Na_2O	0.08	0.50	0.06	0.38	0.13	0.08	1.15	0.19
K_2O	tr.	0.29	0.01	0.18	0.01	tr.	0.01	0.02
H_2O+	13.5	13.27	9.75	4.58	7.90	15.8	2.35	0.49
H_2O-	1.12	0.84	0.57	0.22	0.71	1.62	0.31	0.15
CO_2	0.72	0.23	0.59	-	0.64	1.02	0.06	0.06
P_2O_5	0.01	0.23	0.01	0.03	0.01	0.01	0.02	0.02
Cr_2O_3	0.42	0.26	0.28	0.31	0.56	0.32	0.32	—
NiO	0.34	0.28	0.23	0.28	0.13	0.39	0.11	—
Total	100.23	100.39	100.25	99.76	100.0	100.23	99.80	100.30

Data Source: 1, Kirk (1968).
a = serpentinized harzburgite (NB10381), Sungai Porog, Labuk Valley. b = serpentinized harzburgite (NB5419), Pulau Malawali, Darvel Bay. c = Lherzolite (NB10441A), Sungai Diwata, near Lahad Datu. d = harzburgite (NB6103), Pulau Maganting, Darvel Bay. e = tremolite peridotite (NB 10441B), Sungai Diwata, near Lahad Datu. f = dunite (NB10382), Sungai Porog, Labuk Valley. g = Garnet pyroxenite (NB10360), Sungai Meliau, near Telupid, Labuk Valley. h = pyroxenite (NB11693), Mount Beeston, near Lahad Datu.

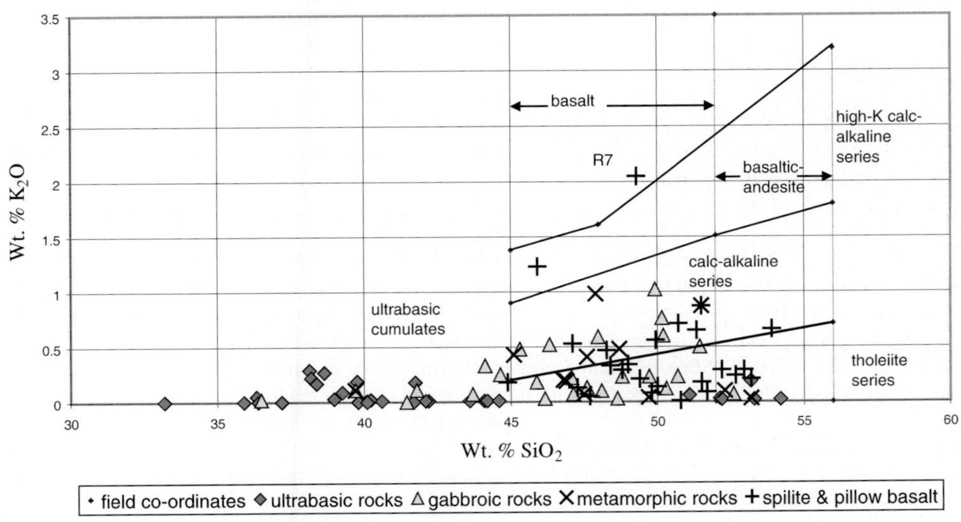

Figure 83. K_2O versus SiO_2 diagram for Sabah ophiolite rocks.

basement" are either meta-gabbro or meta-basalt. All meta-gabbros and meta-basalts contain abundant igneous relict crystals, and the retrogressive metamorphism has never gone to completion. The main criterion for assigning a rock to the category "crystalline basement" is a foliation, which in the case of meta-gabbros, is a layering or banding of igneous origin. The replacement of pyroxene by amphibole is also a usual criterion for assigning a rock to the "crystalline basement", but igneous relicts invariably persist. On the other hand, few of the rocks assigned to the category "basic intrusive rocks" by Kirk (1968) are without retrogressive metamorphism. In Table 21, NB372 from Danum River appears to be entirely of igneous mineralogy, mainly plagioclase, clinopyroxene and olivine.

Elsewhere in the world, gabbros of the ophiolite suite are frequently characterized by rather patchy metamorphism. Ophiolites were formed at a spreading axis of a marginal or oceanic basin, and sub-sea-floor metamorphism has been documented. Dredged rocks from the Mid-Atlantic Ridge can be ascribed to the greenschist and amphibolite facies of metamorphism while other specimens are unmetamorphosed (Aumento et al., 1971; Miyashiro et al., 1971). Such rocks correlate well with outcrops in Sabah; some ascribed to the "crystalline basement" others to "basic intrusive rocks". Both can be expected in an ophiolite as a result of mineralogical and textural changes

Table 21. Wt.% chemical analyses of ophiolitic gabbro, trondhjemite and dolerite rocks from Sabah

Sample	a	b	c	d	e	f	g	h	i	j
Source	2	1	1	1	1	2	1	4#	1	4#
SiO_2	36.5	41.8	43.7	45.31	47.08	48.1	50.2	51.42	52.6	67.27
TiO_2	0.76	0.85	1.04	3.16	0.22	0.76	0.64	1.55	0.73	0.37
Al_2O_3	15.0	17.8	19.3	16.53	21.46	14.6	14.6	15.32	15.6	16.60
Fe_2O_3	0.82	4.15	7.00	8.33	0.86	1.85	2.10	13.94*	1.83	2.89*
FeO	4.0	5.80	6.45	6.26	3.69	6.0	6.60	*	6.40	*
MnO	0.10	0.16	0.17	0.11	0.12	0.16	0.13	0.23	0.14	0.044
MgO	27.6	7.50	6.35	5.44	9.00	8.40	8.90	4.64	7.53	1.06
CaO	4.22	11.2	13.5	8.96	14.02	10.40	9.70	8.18	10.5	3.52
Na_2O	0.58	2.25	1.75	2.97	1.81	2.96	3.20	4.10	2.15	7.68
K_2O	0.02	0.11	0.07	0.48	0.22	0.10	0.60	0.20	0.07	0.044
H_2O+	6.90	6.80	0.67	2.71	1.20	5.30	3.00	2.60**	1.94	0.73**
H_2O-	0.10	1.43	0.20	0.16	0.29	1.30	0.29	**	0.16	**
CO_2	<0.01	0.22	0.06	—	0.19	0.08	0.20	—	0.24	—
P_2O_5	0.02	0.01	0.24	tr.	0.01	0.08	0.05	0.21	0.05	0.149
Total	98.6	100.8	100.50	100.42	100.23	99.60	100.21	99.79	99.94	99.62

Data Source: 1, Kirk (1968); 2, Swauger et al. (1995); 4, Omang and Barber (1996).

Notes: a = gabbro dyke (94SB69A). Silam quarry. b = olivine gabbro (NB10418), Sungai Diwata, near Lahad Datu. c = gabbro (NB11695), Mount Beeston, Silam, near Lahad Datu. d = gabbro (J202). Sungai Subahan, near Lahad Datu. e = olivine gabbro (NB372), Sungai Danum, Segama Valley. f = gabbro (94SB38A), Segaliud palm estate. g = dolerite (NB10892), sungai Kinarom, near Kota Belud. h = metadolerite (PT3a), Pulau Saddle, Darvel Bay, i = hornblende dolerite (NB10323), Sungai Moilu, Labuk Valley. j = trondhjemite (plagiogranite) (PS12c), north shore Pulau Sakar.

analysis given on volatile-free basis.* Total iron given as Fe_2O_3 ** water not determined, but figure given is 'Loss on Ignition'.

known to occur at a spreading axis. The most attractive hypothesis of Spooner and Fyfe (1973) suggests that seawater brine circulates through the hot crustal rocks. Gass and Smewing (1973) show that thermal gradients, and hence metamorphic isograds, are locally extremely high and unmetamorphosed basic rocks may occur close to metabasites, as in the Troodos Massif of Cyprus. If circulating brine is held responsible, then metamorphic reactions will be sluggish to impossible within the dry gabbro layer. In Sabah metabasites, igneous relicts abound. Omang and Barber (1996) offered an alternative hypothesis that the ophiolites of the Segama Valley were deformed at high temperature, but low pressure, along a major transform fault which offset the spreading axis. They further speculated that the environment was above a subduction zone.

XIV.6.1. Chemistry

A selection of chemical analyses of metabasic rocks, which have been allocated to the "crystalline basement" (Kirk, 1968) is given in Table 22. Chemically they are identical to unmetamorphosed ophiolitic rocks (Table 21), and when both are plotted on a K_2O versus SiO_2 diagram (Figure 83), they occupy an identical field. There is no justification therefore for continuing to consider these two groups of rocks as being genetically different.

Table 22. Wt.% chemical analyses of metamorphic ophiolitic rocks (formerly called *Crystalline Basement*)

Sample	a	b	c	d	e	f	g	h	i	j	k
Source	1	2	1	1	1	1	1	1	1	1	1
SiO_2	43.1	46.8	46.9	47.5	47.6	47.9	48.7	49.7	50.1	51.5	52.3
TiO_2	0.08	0.13	0.77	0.51	1.54	0.79	0.90	0.49	1.1	1.04	1.44
Al_2O_3	19.4	20.3	16.4	13.0	17.6	18.3	16.3	14.0	13.9	16.4	16.5
Fe_2O_3	1.05	0.99	3.64	1.62	4.25	2.09	2.90	1.70	2.3	3.39	5.78
FeO	2.64	2.7	7.01	9.90	5.59	4.46	5.40	5.75	8.7	6.01	5.21
MnO	0.10	0.08	0.23	0.20	0.18	0.17	0.17	0.14		0.15	0.13
MgO	9.68	7.86	8.74	7.50	6.24	7.07	9.60	10.8	4.6	5.54	2.18
CaO	19.30	12.70	12.42	12.7	9.60	9.68	10.9	14.3	12.6	8.59	12.36
Na_2O	0.69	2.60	2.49	2.95	4.44	3.69	3.40	2.05	4.5	3.91	1.61
K_2O	0.59	0.20	0.19	0.07	0.41	0.98	0.48	0.04	0.5	0.87	0.10
H_2O+	3.16	3.90	0.81	2.05	1.76	3.64	0.77	0.61	0.7	1.67	1.38
H_2O-	0.20	0.30	0.12	0.09	0.16	0.41	0.14	0.11	0.1	0.25	0.33
CO_2	0.07	<0.01	0.01	0.59	0.25	0.50				0.47	0.06
P_2O_5	0.02	0.02	0.14	0.13	0.14	0.08	0.07	00.03	0.2	0.19	0.40
total	100.08	98.60	99.87	98.81	99.76	99.76	99.73	99.72	99.3	99.98	99.78

a = gabbro gneiss (J1215), Pulau Enam, Darvel Bay. b = metagabbro (94SB-72A) KTS road, NW Silam. c = amphibolite gneiss (J1060), Pulau Silumpat, Darvel Bay. d = amphibolite gneiss (NB10450), Silam Harbour, Darvel Bay. e = epidote amphibolite (J1166), Pulau Adal, Darvel Bay. f = amphibolite (NB8671), Sungai Litog Klikog Kiri, Segama Valley. g = amphibolite schist (NB6364), Islet near Pulau Tanna, Darvel Bay. h = amphibolite gneiss (NB6180), Pulau Bohihan, Darvel Bay. i = amphibolite (T59), Sungai Meliau, near Telupid, Labuk Valley. j = amphibolite (NB8756), near Kuala Agob, Segama Valley. k: epidote schist (J1278), islet SW of Pulau Sakar, Darvel Bay. *Data Source*: 1, Kirk (1968) and 2, Swauger et al. (1995).

The great majority of the analysed rocks are chemically gabbro, but some have <45 wt% SiO_2 and represent ultrabasic cumulates within the gabbro zone (Figure 83). When plotted on a standard AFM diagram (Figure 84) the ophiolitic suite (metagabbros, metadolerite and amphibolite) of the Lahad Datu area plot within the field of actual samples of oceanic gabbros of Kirst (1976), and they are tholeiitic, as to be expected.

The metabasites from the Lahad Datu district were analysed for a range of minor and trace elements (Omang and Barber, 1996). The titanium–zirconium–yttrium discriminant plot of Pearce and Cann (1973) shows that the Sabah metamorphic ophiolite rocks fall within the fields of MORB and IAT (Figure 84). The obvious interpretation is that the rocks have oceanic and island-arc tholeiite character. They therefore represent the uplifted (obducted) crust of a marginal basin.

XIV.6.2. Petrography

The gabbro zone is composed of both isotropic as well as layered gabbroic rocks, doleritic and basaltic dykes as well as some plagiogranite. Labradorite and pyroxene (both diopside–augite and hypersthene) are the main constituents. Olivine occurs in some varieties and is commonly serpentinized. The gabbros may be equigranular and medium-grained. But even the isotropic olivine gabbro is crudely foliated (Yan, 1979). In several localities the gabbros are cut by dykes and sills of dolerite, more appropriately named epidiorite since their mafic minerals are predominantly altered to amphibole. More commonly, however, the gabbros are layered and consist of a complexity of original mineralogy as well as retrogressive metamorphic phases. There is no clear large-scale separation of unmetamorphosed and metamorphosed gabbro. They are closely associated in the field, for example in the Bole River area (Yan, 1979).

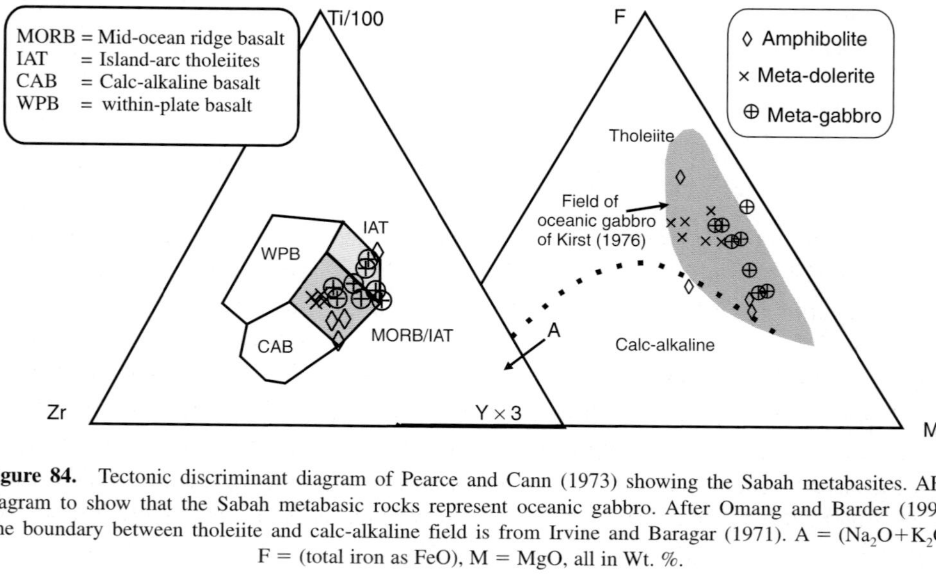

Figure 84. Tectonic discriminant diagram of Pearce and Cann (1973) showing the Sabah metabasites. AFM diagram to show that the Sabah metabasic rocks represent oceanic gabbro. After Omang and Barder (1996). The boundary between tholeiite and calc-alkaline field is from Irvine and Baragar (1971). A = (Na_2O+K_2O), F = (total iron as FeO), M = MgO, all in Wt. %.

Cumulate texture is very common in both the gabbro and meta-gabbros. The cumulus phases are of somewhat rounded plagioclase and olivine, surrounded and enclosed by large intercumulus dark pleochroic green–brown hornblende of igneous origin. The mafic minerals such as pyroxenes and hornblende are commonly partly altered to lighter coloured actinolitic hornblende or actinolite. Cumulate rocks within the gabbro layer do not represent magmatic liquids, and their analysis may be ultrabasic (Figure 83).

Throughout the gabbro and meta-gabbros zone occur bodies of trondhjemite (plagiogranite), which is characteristic of the gabbro layer ophiolite suite. Plagiogranite is distinctly different from the calc-alkaline granites described above. In the Lower Bole River (~5°07′N; 117°53′E) is a large trondhjemite body about 1 km wide, which shows no clear intrusive contacts with the host gabbro. There is neither contact metamorphism nor a chilled margin (Yan, 1979). The rock is leucogranite, crudely foliated, medium-grained, containing mostly quartz and plagioclase and minor amounts of biotite, muscovite and epidote. Plagioclase, of grain size 3 mm, forms 65% and quartz 30% of the rock. There are also leucogranite dyke rocks, commonly deformed into boudins within amphibolite gneiss. Mortar structure is common.

Typical meta-gabbros, such as found on Silumpat and Bohayan islands, are medium- to coarse-grained gneisses, white and dark green, normally well lineated by the orientation of amphibole crystal aggregates, and commonly banded into plagioclase- and mafic-rich layers (Hutchison and Dhonau, 1971). The amphibolite gneisses are associated with concordant or slightly discordant sheets of dark epidiorite (amphibolite). They commonly include randomly oriented xenoliths of metagabbros. Where gabbro zones on the same islands contain more pyroxene and less amphibole, they are less layered, show only crude banding and have a totally igneous aspect.

The gneissic meta-gabbros of Darvel Bay are identical to those samples from the Verma transform fault zone of the equatorial Mid-Atlantic Ridge (Honnorez et al., 1984). A popular hypothesis, therefore, is that amphibolites and gneissic meta-gabbros result from sub-sea-floor metamorphism related to fluids circulation along an active transform fault zone.

XIV.6.3. Mineralogy

Most of the mineral phases have been analysed by Hutchison (1978) and a summary is given in Figure 85.

Plagioclase (partly igneous and partly metamorphic) occurs as a mosaic of polygonal grains. The lamellar twinning is difficult to use for composition determination for it is pericline and other complex varieties. Immersion methods suggested an unbelievably wide range from An_4 to An_{84} (Hutchison and Dhonau, 1971). Later microprobe work by Hutchison (1978) showed that the plagioclase of igneous paragenesis is of labradorite An_{58}–An_{65} with little compositional zoning. Labradorite persists in the same rock together with oligoclase or andesine that has resulted from incomplete metamorphism. The metabasites do not represent an equilibrium metamorphic assemblage (Figure 85). Although the gabbros are low in potassium, it is not uncommon for oligoclase to contain significant K, with a formula of $Ab_{73}An_{22}Or_5$. In thin section the

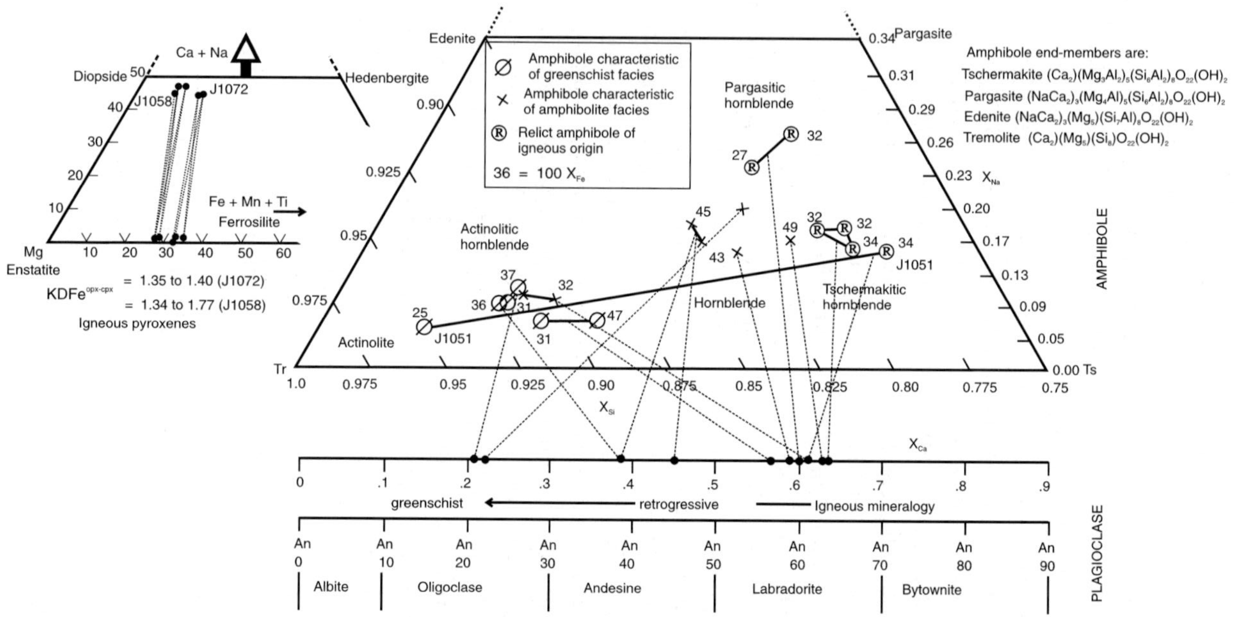

Figure 85. Amphibole analyses plotted relative to four end-members, based on X_{Na} = Na/(Na+Ca) and X_{Si} = Si/(Si+Al) in the structural formulae. Tie lines connect coexisting phases. The plagioclase plots are based on X_{Ca} = Ca/(Ca+Na+K). The X_{Fe} values indicate how much Fe replaces Mg in the structural formula. J1051 contains dark brown igneous relicts rimmed by colourless tremolite resulting from retrogressive metamorphism. The analysed coexisting orth- and clinopyroxenes are plotted on the Ca–Mg–Fe triangle. The partitioning or distribution coefficients are given.

plagioclase crystals show a distinctive cream coloured clouding, probably due to oriented inclusions of spinel (Poldervaart and Gilkey, 1954).

Pyroxene (wholly igneous) relicts are augite with a composition ranging from $Wo_{43}En_{44}Fs_{13}$ to $Wo_{45}En_{39}Fs_{17}$. Coexisting hypersthene has a composition ranging from $Wo_1En_{71}Fs_{28}$ to $Wo_2En_{65}Fs_{33}$ (Figure 85). These pyroxenes are in igneous equilibrium with partitioning or distribution coefficients K_D $Fe^{opx-cpx}$ ranging from 1.34 to 1.77. The orthopyroxene is conspicuously pleochroic from α salmon pink, β yellow and γ pale green. $2V\alpha = 71°$. The diopsidic augite is also pleochroic, though less intensely, $2V\alpha = 58°$ and γ^$Z = 48°$

Ilmenite (partly igneous) is present with an analysed formula $Fe_{(1.95)}Ti_{(2.02)}O_6$. It coexists with magnetite of analysed composition $Fe_{(1.0)}Ti_{(0.01)}O_4$.

Pargasitic or tschermakitic hornblende (wholly igneous) relicts occur in several metabasites (Hutchison, 1978). They are plotted in Figure 85 and coexist with labradorite as an equilibrium assemblage. The crystals are commonly clouded with fine speckles of iron oxide. The igneous hornblendes contain more than 2.0% (by wt) TiO_2, whereas metamorphic recrystallization reduces this amount to 0.4% in amphibolite facies and to 0.05% in greenschist facies. The titanium is therefore released and ilmenite is formed during retrogressive metamorphism. The K_D Fe $_{amphibole-clinopyroxene}$ in the same rocks is predominantly within the range 1.00 – 1.50, so that the pargasitic hornblende shares an igneous parentage with the pyroxenes.

Actinolitic hornblende. As a result of retrogressive metamorphism, the mafic minerals are replaced by amphiboles whose compositions move towards the tremolite end member. The amphibolite facies amphiboles lie within the common hornblende field (Figure 85) and have higher iron contents (X_{Fe} 0.43 – 0.49) than either the igneous or the greenschist facies compositions. The actinolitic amphibole is pleochroic from green to pale yellowish green. $2V\alpha = 77°$ and γ^$Z = 15°$.

Epidote occurs commonly in meta-basites, which have been downgraded to greenschist facies, throughout the rocks as well as segregated into boudins. A typical analysis gives the formula $(Ca_{2.06}\ Na_{0.02})_{2.08}(Al_{2.08}\ Fe_{0.73}\ Mn_{0.01}\ Ti_{0.01})_{2.83}\ Si_{3.03}\ O_{12}$, which is close to the iron-rich end member.

Olivine, of forsterite composition, may constitute up to 10% of some gabbros and 15% in troctolite. It is commonly serpentinized, and does not persist when the rock is metamorphosed. There is commonly a preferred orientation fabric of the olivine, plagioclase and pyroxene (Leong, 1974).

Quartz is usually a rare commodity, but occurs in accessory amounts (of grain size 1–5 mm) in some gabbro specimens, but not in the cumulates.

XIV.7. THE BASALT ZONE

A thick basalt layer overlies the gabbro zone and is itself overlain by the Lower Cretaceous ribbon cherts. Much of the ophiolite is imbricated, but stratigraphic relationships are well displayed in the islands of Darvel Bay (Figures 78 and 79).

Traditionally the basalt, which is sometimes pillowed but otherwise isotropic, is closely associated in the field with the ribbon chert. This led Fitch (1955) to establish the term "Chert–Spilite Formation", which was useful for field mapping. The problem arose when areas devoid of basalt (spilite) and of chert were also allocated to the Formation. I have taken out such younger strata from the "chert–spilite association". They are described below (Section XV) as part of the Rajang Group.

XIV.7.1. Chemistry

A selection of chemical analyses of rocks from the basalt zone is given in Table 23. When plotted on a K_2O versus SiO_2 diagram, they are seen to be predominantly basaltic with a slight spread into the basaltic–andesite field (Figure 83). The majority are tholeiitic, but there is some spread into low K_2O calc-alkaline. The very few samples with K_2O values >1.0% may be disregarded as atypical and the analyses are suspect.

A wide spectrum of trace elements was analysed in samples of basalt (Swauger et al., 1995). The chondrite-normalized rare-earth analyses are summarized in Figure 86. The horizontal to concave patterns of elements 57–72, with relatively low rare-earth element(REE) contents (<10–20 times chondrite) are typical of MORB. The patterns of ophiolitic basalts and gabbros are closely similar, but that of the Silam metagabbro shows depletion in the whole range, except of europium.

Table 23. Wt.% chemical analyses of selected spilites and pillow basalts (formerly Chert–Spilite Formation)

Sample	a	b	c	d	e	f	g	h	i
Source	1	1	2	1	1	2	1	1	3
SiO_2	44.89	47.11	48.4	49.0	49.4	50.8	51.7	53.2	58.39
TiO_2	4.06	2.01	1.28	1.20	1.44	1.08	1.28	1.01	0.62
Al_2O_3	15.80	15.51	14.8	15.3	15.4	14.3	15.3	14.7	14.63
Fe_2O_3	1.40	4.02	4.54	3.43	4.00	3.57	2.85	2.90	7.32*
FeO	9.10	6.05	4.5	4.92	4.80	4.5	6.15	5.90	
MnO	0.08	0.17	0.30	0.15	0.18	0.11	0.18	0.14	0.12
MgO	4.76	5.05	6.60	6.35	7.35	5.93	6.6	5.55	6.42
CaO	5.44	5.92	10.00	10.5	10.2	12.0	7.05	8.65	10.06
Na_2O	5.53	7.34	2.83	3.25	3.20	1.83	4.9	3.70	2.36
K_2O	0.18	0.53	0.33	0.34	0.21	<0.01	0.09	0.21	0.06
H_2O+	4.94	4.81	2.70	4.83	2.35	4.40	2.75	3.35	
H_2O-	0.52	1.07	1.40	0.58	1.32	0.40	0.83	0.57	
CO_2			0.08	0.12	0.07	<0.01	0.04	0.03	
P_2O_5	1.10		0.12	0.11	0.15	0.06	0.11	0.10	0.06
Total	97.80	99.59	98.10	100.8	100.07	99.50	99.83	100.37	100.04

Data Source: 1, = Kirk (1968);. 2, Swauger et al. (1995) and 3, Omang and Sanudin (1995).

a = spilite (R26), Sungai Brantian. b = spilite (H59), Pulau Balambangan, near Kudat. c = pillow basalt (94SB-40A), Segaliud Estate, off Kinabatangan road. d= spilite (NB8890), Sungai Bongkulat, Kinabatangan Valley. e = spilite (NB10331), Sungai Labuk, near Telupid. f = pillow basalt (94SB-19), Wonod River, near Telupid. g = spilite (NB8877), Sungai Kinabatangan, near Kuala Karamuak. h = spilite (NB10311), Sungai Mengkadait, Labuk Valley. i = TP pillowed tholeiitic basalt, Wonod River, Telupid.

* Total iron expressed as Fe_2O_3. Loss on ignition 3.96%.

From a plot of chromium versus yttrium contents, and Ti/100 versus Zr versus Yttrium x3, Omang and Sanudin (1995) concluded that the Telupid pillow basalt lies on the border between the fields of MORB and island-arc tholeiite.

XIV.7.2. Petrology

Pillow basalts occur at many localities, for example on the Wonod River west of Telupid, in the Segaliud Quarry and on Timbun Mata Island (Koopmans, 1967), but

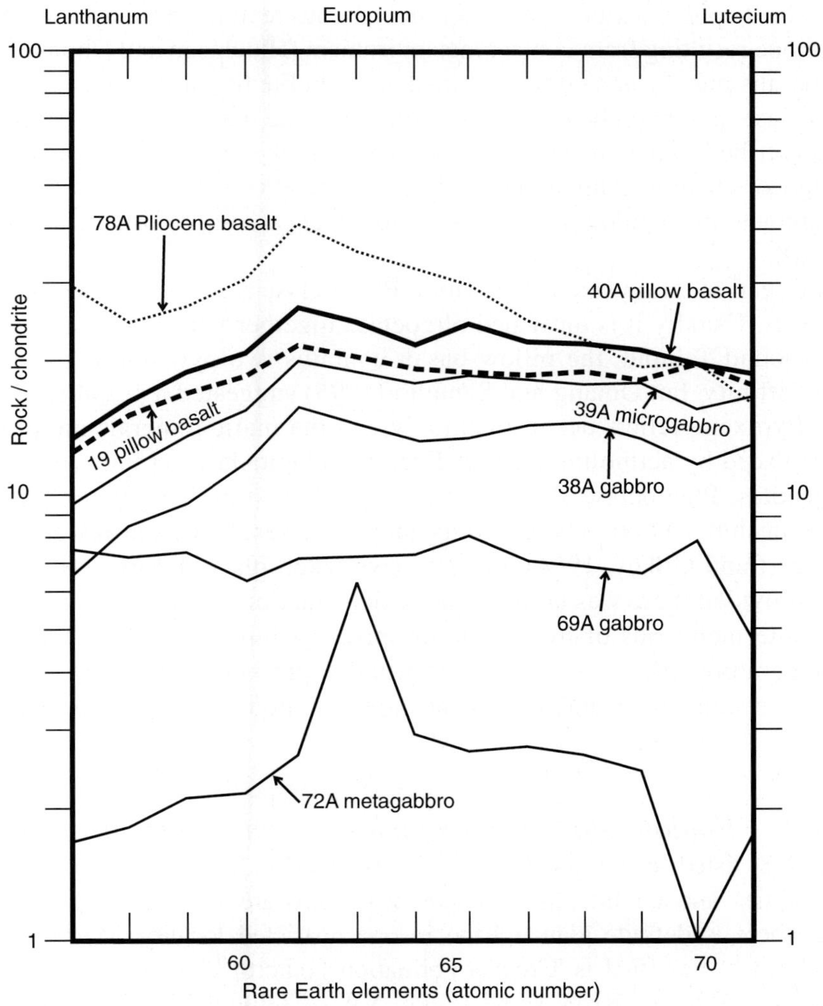

Figure 86. Chondrite normalised rare-earth plot of Sabah Lower Cretaceous ophiolite and Mostyn Estate Pliocene basalt. From Swauger et al. (1995). Ophiolite: 19, pillow basalt, Wonod River, Telupid. 38A, gabbro, Segaliud estate. 39A, microgabbro, Segaliud estate. 40A, pillow basalt, Segaliud estate. 69A, gabbro, Silam quarry. 72A, metagabbro, KTS road, NW Silam Non-Ophiolite: 78A, vesicular basalt, Mostyn Estate.

the great majority of outcrops are of isotropic featureless basalt. Some exhibit a foliation. The name "spilite" given by Kirk (1968) is superfluous. They are predominantly basalt. Outcrops around Telupid and in the Segaliud quarry show that bedded chert was deposited directly upon the upper surface of pillowed basalt. The contacts are generally near-vertical, and stratigraphic relations cannot be traced far because of faults. The pillowed basalt is well exposed beneath the bridge where the main Telupid road crosses the Wonod River, and also at 85th kilometer on the main road (Omang and Sanudin, 1995).

These fine-grained dark green to black rocks have been named "hornfels" both because of their isotropic field character as well as thin-section fabric. Many have a fine-grained ophitic texture. This is a relict igneous texture and should not be confused for one resulting from contact metamorphism (Hutchison and Dhonau, 1971).

The basalts may be named fine-grained amphibolite because they are composed of plagioclase and amphibole, usually actinolitic. The metamorphism is retrogressive and ascribed to the greenschist facies. Every rock contains igneous relicts and a relict igneous texture. Although there is no compositional banding, on Adal Island it is common to find epidote-rich bands, up to 50 cm across, deformed into boudins (Figure 79).

In thin section, grain size is 0.1–1mm. Plagioclase forms a mosaic with larger phenocrysts. Usually it is untwined oligoclase together with labradorite igneous relicts. Around Telupid, the pillow basalt contains two pyroxenes suggesting a tholeiitic affinity, but Omang and Sanudin (1995) suggested a possible boninitic affinity. Pyroxene relicts occur commonly, but the mafic minerals are predominantly replaced by actinolite needles. Even the plagioclase crystals contain actinolite needles. Plagioclase showing relict ophitic texture is commonly clouded and pink in thin section. The clouding may have resulted from metamorphism (Poldervaart and Gilkey, 1954), or may have taken place during cooling of the magma, provided there was no penetrative deformation. Clouding usually results from minute inclusions of iron oxide or spinel (Whitney, 1972) resulting from high-temperature diffusion. Epidote is usually present, and quartz present in only a few specimens. Cataclastic zones are common and some may be named mylonite.

XIV.7.2.1. Epidote-glaucophane facies metamorphism near Telupid

Leong (1978) listed several localities of "blueschists" in Sabah. I have examined all these localities and confidently conclude that there are no blueschists in Sabah. However there is definite glaucophane in one restricted locality. Metabasalts and metasandstones (classified as 'Crocker Formation') outcrop along the Sungai Lividoi, a tributary of the Labuk River, some 13 km NW of Telupid. This area was mapped by Johnston and Walls (1974) but no detailed bulletin has been published. I have studied the samples and identified perfectly euhedral porphyroblasts both of glaucophane and piedmontite in a quartz-rich metasandstone or metachert. This must be the most attractive Malaysian rock in thin section, because of the blue–violet pleochroism of glaucophane and the yellow–pink pleochroism of piedmontite, set in a mosaic of normal quartz. The contiguous outcrops of metabasalts (fine-grained amphibolite) also contain small amounts of glaucophane and piedmontite. Mineralogists of the U.S. Geological Survey also studied the samples at Menlo Park. A confirmatory report of the mineralogy was returned to the Geological Survey, Kota Kinabalu. These rocks have not been sheared and the metamorphism to high-pressure low-temperature epidote-glaucophane facies took place under static conditions of strong loading without shearing. The manganese epidote, piedmontite, is to be expected because the ocean floor is a region of manganese enrichment.

These rocks are not "blueschists" for the small porphyroblasts cannot be discerned nor suspected in outcrop and are seen only in thin section. The general conclusion, made by Hutchison et al. (2000), is that the static metamorphism occurred under 7–8 kbar at a low geothermal gradient. This knowledge allows the interpretation that the Sabah Trench was located in the neighbourhood of Telupid. The trench was the depocentre that became infilled with a great thickness of Trusmadi and Crocker Formation sediments. These sediments are therefore not strictly "accretionary prism" and the trench was not located at the NW Borneo Trough. The tectonic models are given and discussed later. Nevertheless the Labuk Highlands area around Telupid must have been dramatically inverted and exhumed to their present outcrop position from a depth of around 20 km.

Chapter XV

Eastern Rajang Group (Gallic to Eocene)

The marine strata that overlie the Lower Cretaceous chert consist of sandstone–shale turbidite and calcareous rocks. They represent an infilling phase of the marine basin by terrestrially derived turbidites and localized carbonates as the water shallowed above the carbonate compensation depth. These strata were included under the all-encompassing term Chert–Spilite Formation. The stratigraphy of this group has not been resolved because of lack of continuity between outcrops. Koopmans (1967) noted the problem thus: "well stratified layers occur interbedded with highly cataclastic, disturbed, and contorted layers". This, of course, is a universal character of flysch sequences. Fossil assemblages indicate ages ranging from Lower Cretaceous to Eocene; many localities are Upper Cretaceous (Figure 87). This group of strata overlies the Lower Cretaceous ribbon chert, but should not be grouped with it. Therefore there is no value in persisting to include so many different entities in the term "Chert–Spilite Formation". Localized carbonates could well be interpreted as having formed in shallower water overlying seamounts (Lee, 2003). Limestone localities have been more studied because of their palaeontological yield. In the absence of an appropriate name, I have assigned them to the Rajang Group, which occurs in Sarawak and western-central Sabah, and with which they are, at least in part, chronologically equivalent.

The older publications used an Asian Tertiary letter classification that has since been revised. The revised version of Adams and Wilford (1972) is used in this book and older ages have been updated:

Tg-h = Upper Pliocene and younger
Tf_3 = Upper Miocene–Lower Pliocene (Tortonian, Messinian, Zanclian)
Tf_2 = Middle Miocene (Serravallian)
Tf_1 = Lower-Middle Miocene (Upper Burdigalian + Langhian)
Te_5 = Lower Miocene (Aquitanian and Burdigalian)
$Te_{1\text{-}4}$ = Upper Oligocene (Chattian)
Td = Lower Oligocene (Rupelian)
Tc = Lower Oligocene (Lattorfian)
Tb = Upper Eocene (Priabonian)
Ta = Palaeocene + Lower and Middle Eocene

XV.1. AGOB-DABALAN AREA

Thin beds of tuffaceous limestone and calcilutite occur in the Agob-Dabalan area of the Segama River, 32 km NW of Lahad Datu.

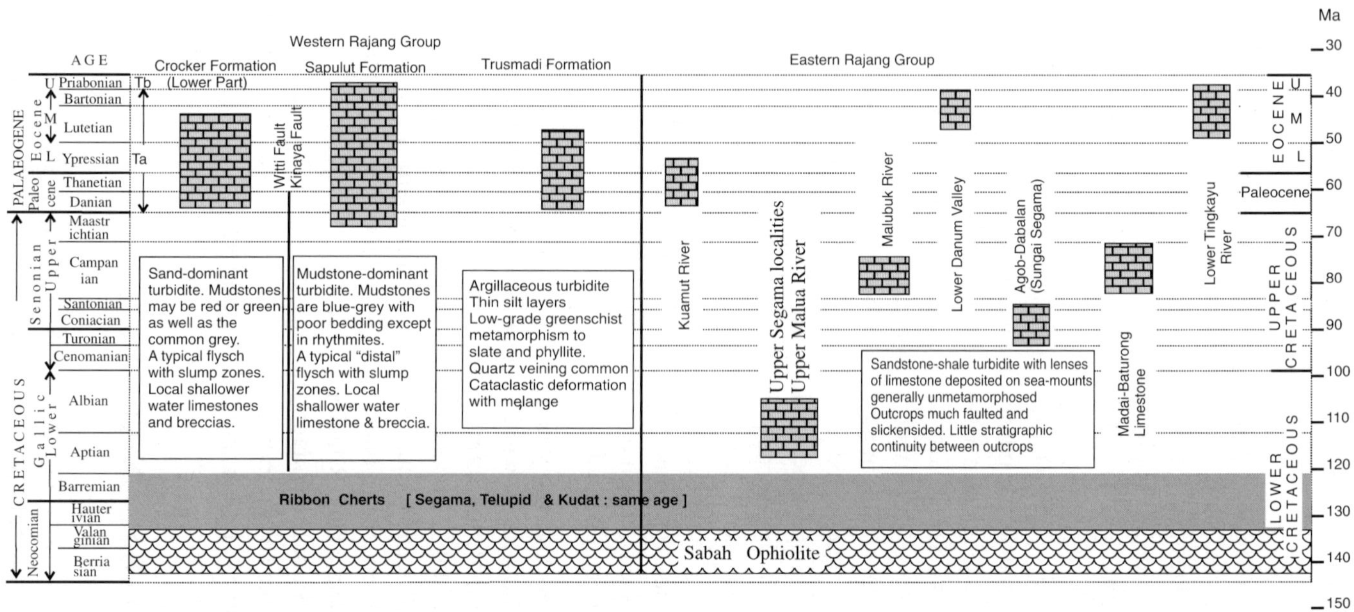

Figure 87. Chronological chart of lenticular limestones within the predominantly turbiditic Rajang Group of southern Sabah. The Eastern Rajang Group was formerly included within the now obsolete term "Chert–Spilite Formation". Based on palaeontological determinations documented in text.

XV.1.1. Age

The following Foraminifera have been tabulated by Leong (1974): *Globigerina cf. cretacea* d'Orbigny, *Globotruncana* cf. *arca* Cushman, *Globotruncana* cf. *bulloides* (Volger), *Globotruncana concavata* (Brotzen), *G* cf. *concavata* (Brotzen), *G cf. linneiana* (d'Orbigny), *Globotruncana* cf. *renzi* Gandolfi, *Globotruncana* cf. *schneegansi* Sigal, *Globotruncana tricarinata* (Quereau), *Globotruncana* sp. Indet., *Gumbelina sp*., *Hedbergella* cf. *delrioensis* (Carsey), *Hedbergella* cf. *planispira* (Tappan), *Hedbergelloa* sp., *Hetrohelix globulosa* (Ehrenberg), *Hetrohelix* sp. Indet., *Heterohelicidae* indet., *Praeglobotruncana* (*Hedbergella*) spp., *Rotaliidae* indet. and *Textularidae* Indet.

The above fauna is entirely of Upper Cretaceous age, with a range from Turonian to the lower part of the Santonian. Thin beds of limestone in the lower Sungai Bole yielded the Cretaceous *Praeglobotruncana* (*Hedbergella*) spp.

XV.2. MADAI

Gunung Madai, 13 km NW of Kunak, is a notable landmark, lying close by the main road from Lahad Datu. It is well seen from the Sungai Tingkayu Bridge and readily accessed by a side road. At the foot of the karstic hill is a small village of wooden houses, in which the bird's nest collectors live during the collection season. Leong (1974) called the rock the Madai–Baturong Limestone. There is another small limestone hill, Supad Batu, lying 2 km SE of Gunung Madai. However it contains neither Foraminifera nor algae, but is assumed to be a continuation of the Madai–Baturong Limestone.

The main part of the Madai Limestone is detrital and all fossils broken and abraded. Therefore reworking of older fossils makes it impossible to give a precise age for the Madai limestone. The most fossiliferous specimens are of limestone breccia (Adams and Kirk, 1962). The breccia is adjacent to massive chert and altered basalt and greywacke in the vicinity along River Madai. However no regular structure can be discerned. The nearby presence of chert indicates that the limestone, which indicates shallowing of the sea to above the carbonate compensation depth, must overlie the chert, which was deposited below the carbonate compensation depth (CCD).

Lee (2003) concluded that both the Madai and the Baturong limestones were deposited on seamounts occurring within the Chert–Spilite Formation deep sea. This conclusion is consistent with the extreme purity of the limestones with a lack of terrigenous detrital input, and their shallow water contrast to that of the surrounding cherts and turbidites of the Chert–Spilite Formation.

XV.2.1. Age

Algae are the main fossils, and the following have been identified by C. G. Adams and G. F. Elliot (Leong, 1974) and detailed in Adams and Kirk (1962). Fontaine and Ho (1989) have added some additional: *Acicularia* sp., *Bacinella* sp., *Cayeuxia* sp.,

Cayeuxia piai Frollo, *Cayeuxia moldavica*, *Cayeuxia* cf. *kurdistanensis* Elliot, *Cayeuxia* cf. *jurassica* var. *lanquinei* Pfender, *Cypeina* sp., *Cylindroporella* sp., *Girvanella* sp., *Lithocodium* cf. *aggregatum* Elliot, *Lithophyllum torinosuensis* Endo, *Marinella lugeoni* Pfender, *Munieria* sp., *Neomeris* possibly *Neomeris pfenderaea* Konishi and Epis, *Nipponophycus ramosus* Yabe and Toryama, ?*Parachaetetes* sp., *Pseudoepimostopora jurassica* Endo, *Pycnoporidium lobatum* Yabe and Toryama, *Tripoerella* sp., *Solenopora* cf. *jurassica* var. *lanquinei* Pfender, *Solenopora* sp.,*Stenoporidium* sp., *Stenoporidium* cf. *chaetetiformis* Yabe and Toryama, *Suppliluliumaella polyreme* Elliot, *Thaumatoporella parvovesisulifera* (Raineri) Pia. No algae were found at Supad Batu.

Foraminifera are locally abundant and have been identified by Adams and Kirk (1962). The pelagic forams are: *Cuneolina* sp., *Dicyclina* sp., *Globotruncana concavata* (Brotzen), *G. elevata* cf. *stuartiformis* Dalbiez, *G.* aff. *fornicata* Plummer, *G.* cf. *renzi* Gandolfi, *G. Tricarinata* (Quereau), *Orbitolinidae* sp., *Praeglobotruncana* (*Hedbergella*) sp., *?Praeglobotruncana delrioensis* Plummer, *Quinqueloculina* sp. *Hetrohelix* cf. *costulata* Cushman and *Hetrohelix* spp.

Adams and Kirk (1962) place more emphasis on the Foraminifera, pointing out that the algae are stratigraphically long ranging, and they conclude that the Madai and Baturong Limestone are Upper Cretaceous, no older than Campanian.

Mollusc fossils, which are locally abundant and suggest littoral conditions, have been found but they indicate only a Mesozoic age. The species identified by N. J. Morris and C. P. Nuttall (Leong, 1974) is *Pleisioptygmatis* of the family *Nerineidae.* The latter ranges from Upper Jurassic to Upper Cretaceous. Fontaine and Ho (1989) discovered molluscs in isolated groups, commonly fragmentary, but rare. They are *Caprinidae*, which ranges from Valanginian, important in the Barremian, peaking in the Cenomanian, and became extinct by end Cretaceous. It was never seen in Bau, Sarawak. Corals, abundant at Bau, are extremely rare and only debris have been observed. Sponges are absent.

XV.3. BATU BATURONG

Batu Baturong is another limestone hill, lying 27 km due west of Kunak, but can be reached by road and track.

XV.3.1. Age

The following fossils are recorded by Leong (1974) and Fontaine and Ho (1989)—Foraminifera: *Cuneolina* sp. and *Dictyconous* sp., problematic microfossil: *Hensonella* cf. *cylindrical* Elliot, and the algae: *?Arabicodium* sp., *Girvanella* sp., *Lithocodium aggregatum* Elliot *Cayeuxia* sp., and *Pycnoporidium* cf. *sinuosum* Johns and Kan.

A general Lower Cretaceous age is interpreted, similar to that of Gunung Madai, but Adams and Kirk (1962) indicate that the Foraminifera suggest a general Upper Cretaceous, or Senonian age.

XV.4. OTHER LIMESTONE LOCALITIES IN THE SEGAMA AREA

Brecciated limestone, partly oolitic, occurs in many areas throughout the Segama Highlands, Segama Valley and along the Kuamut River (Leong, 1974). These restricted occurrences are similar to the Madai–Baturong Limestone. Some have inclusions of volcanic rocks. Most are devoid of fossils. Fossiliferous localities also occur on the large island of Timbun Mata.

XV.4.1. Age

The sparsely fossiliferous localities have yielded the following Foraminifera (Leong, 1974): *Orbitolina* sp., *Hedbergella* sp., *Orbitolina lenticularis* (Blumenbach) *sensu* Hofker. The algae *Solenopora sp. and ?Cayeuxia piai* also occur. This assemblage indicates a Lower Cretaceous age, most probably Aptian–Albian according to C. G. Adams.

Limestone breccia, which contains clasts of ribbon chert, occurs along the Kuamut River, 15 km south-south-westwards of Kuamut. The rocks contain an age-diagnostic fauna, identified by Keij (1963) as follows: *Nummulites* sp., *Discocyclina* sp., *Alveolina* sp. *and Kathina* sp. In addition to the Foraminifera, the limestone breccia contains the distinctive *Distichoplax biserialis* (Dietrich), whose worldwide distribution is confined to the Palaeocene to Lower Eocene. It was formerly thought to be an alga, but is the creeping stems of *Rhabdopleura*, belonging to the *Pterobranchia* of the sub-phylum *Hemichordata* (Keij, 1963). This distinctive fossil also occurs in the Trusmadi and Sapulut formations, all of the Rajang Group.

XV.5. LOWER TINGKAYU RIVER

Thin lenses of Foraminiferal limestone and marl occur in the Lower Tingkayu River NE of Gunung Madai (Kirk, 1962). The river also yields outcrops of bedded chert, volcanic breccia and conglomeratic limestone, but relationships between these rocks is obscure. Near where the main Lahad Datu–Kunak road crosses the Tingkayu River are extensive road cuts of sandstone–shale turbidite showing well-developed sole marks. The outcrop is extensively affected by normal faults with prominent slickenside zones. This extremely imbricated nature of pre-Miocene rocks (pre-Sulu Sea rift) of this region explains how the regional structure cannot be determined and how rocks of different type come into outcrop juxtaposition.

XV.5.1. Age

The calcareous localities of the Lower Tingkayu River (near Gunung Madai) contain the following larger Foraminifera (Kirk, 1962; Leong, 1974):

Aktinocyclina sp., *Assilina* sp., *Asterocyclina* sp., *Discocyclina* sp., *Globorotalia* cf. *pseudobulloides* (Plummer), G. cf. *pusilla* Bolli, G. cf. velascoensis (Cushman), G. cf. *abundocamerata* Bolli, *Lithamnium* sp., *Miliola* sp., *Nummilites javanis* (Verbeek), *Operculina* sp., *Pellatispira* sp., *Globigerina* sp., *Alveolina* sp., *Lithothamniun* sp., *Miliola* sp. and *Nummilites baguelansis* (Verbeek). This assemblage is Middle Eocene, probably extending to the Upper Eocene.

Chapter XVI
Rajang Group (Western)

The Rajang Group extends from Sarawak north-eastwards along the NW Borneo Trend. In Sabah it is known as the Sapulut, Trusmadi and Crocker (undifferentiated) formations (Figure 87). These three formations are mutually contiguous in the Keningau district of SW Sabah. The equivalent, across the border in Kalimantan, of the Rajang Group is the Embaluh Group (Moss, 1998).

XVI.1. TRUSMADI FORMATION

The Trusmadi Formation occupies a broad belt of hilly to mountainous country, approximately 40 km wide, stretching from the southern end of the Trusmadi Mountains north-north-east (NNE) to Sungai Liwagu. They are described in the early memoir of Collenette (1958). The main body of the Trusmadi Formation is bounded by N–S faults through Tenompok and Kundasan in the Kinabalu area (Jacobson, 1970). The outcrop extends 65 km to the Keningau area. The formation is well exposed SSW of Ranau, where the road to Tambunan crosses the Sungai Kenipir, a tributary of the Liwagu. The main road near Kundasan, close to the entrance to the Kinabalu National Park, also offers good exposures. The Trusmadi Formation also underlies the Klias Peninsula, locally brought to the surface by structural complexity.

It is frequently stated that the Trusmadi Formation is always in fault contact with the Crocker Formation. However, Collenette (1965) has stated that one may be traced into the other along strike. There is a difference in lithology, but the change in metamorphic grade could possibly be related to depth of burial. Collenette (1965) felt that the faults that bound the formation were active during deposition.

As a generalization, the Trusmadi Formation lithologically resembles the Sapulut Formation, but has been subjected to low greenschist facies metamorphism. These two formations contain identical age fossils. The interpretation, therefore, is justified that the Trusmadi Formation has been more deeply buried within a sedimentary depocentre by a greater thickness of Crocker Formation.

XVI.1.1. Lithology

The most distinctive features are

- Dark argillaceous rocks predominate
- Low-grade greenschist facies metamorphism to slate and phyllite and moderate to strong deformation
- Quartz veining of the rocks is very characteristic.

Phyllites and slates predominate. The predominantly dark grey-argillites are interbedded with thin sandstone and siltstone beds. There are localized thin limestone lenses. In the Keningau district limestone micro-breccias are enveloped in phyllite. Cataclastic zones are common, and around Kundasan there are outcrops of mélange. The highly cataclastic zones have traditionally been ascribed to the Wariau Formation, which occupies a N–S zone through Kota Belud.

The argillaceous beds are up to 30 m thick. They are dark grey, commonly sheared and phyllitic. Slaty cleavage occurs in folded zones. The mineralogy is quartz, muscovite, chlorite, opaque minerals and abundant carbon, and the rocks may be described as low-grade greenschist facies. Quartz veining is very characteristic especially where there are tight folds.

The formation is generally of argillaceous turbidite. The argillite beds vary from < 2 cm to 0.6 m in thickness, mostly in the range 7–15 cm. The typical flysch sequence is laminated with small-scale cross beds, graded bedding and sole marks, and quartz veining is universal (Jacobson, 1970). There are a few massive sandstones, but always fine- grained (0.1 mm). Isolated spilitic volcanic rocks outcrop within the Trusmadi Formation body, indicating that the formation was deposited upon oceanic crust. They are metabasaltic and contain epidote and relicts of pyroxene and plagioclase. Everywhere the Trusmadi is faulted against other rock types, but is intruded by granitoids of Mount Kinabalu in the Mamut copper mine.

XVI.1.2. Age

The best age-diagnostic fossil assemblages are documented by Keij (1963). The fossils occur in limestone lenses. Limestone breccia, 24 km SSE of Tambunan, has yielded the Foraminifera: *Alveolina* sp., *Linderina* sp., *Opertorbitolites* sp., *Discocyclina* sp., *Nummulites* sp., *and Asterocyclina* sp. This assemblage occurs together with the very distinctive *Distichoplax biserialis* (Dietrich). It is the creeping stems of *Rhabdopleura*, a *Pterobranchia* of the sub-phylum *Hemichordata* (Keij, 1963). Its known occurrence worldwide is Palaeocene to Lower Eocene. 24 km east of Apin-Apin are several boulders of limestone breccia that have yielded the same assemblage.

Foraminifera were obtained from shale in the Kinabalu area (Jacobson, 1970) but they are not age specific: *Bathysiphan* sp., *Cyclammina* sp., *Gaudryina* sp., *Globigerina* sp., *Haplophragmoides* sp., *Rotalia* sp., *Trochammina* sp. and *T. renzi* nom. nov.

Muddy calcarenite on Burong Island, north of Klias Peninsula, has also yielded *Distichoplax biserialis* (Dietrich), together with *Discocyclina sp.*, *Nummulites sp.*, *Operculina sp.* and *Rotaliidae sp.*, together indicating a Palaeocene–Lower Eocene age (Keij, 1963).

Collenette (1958) found limestone along the Sungai Liwagu, SE of Ranau, which has yielded Middle Eocene Foraminifera, but no faunal list is given. He does, however, indicate that the most commonly occurring genera are: *Cyclammina* sp., *Bathysiphon* sp., *Haplophragmoides* sp., *Trochammina* sp. *and Trochamminoides* sp. Other less common genera are: *Ammodiscus* sp., *Discocyclina* sp., *Globorotalia* sp., *Glomospira* sp., *Lepidocyclina* sp. and *Nummulites* sp.

The Trusmadi Formation extends as far south-westwards as the Keningau district, where Collenette (1965) listed the following Foraminifera from limestone and limestone breccia occurrences: *Actinocyclina* sp., *Alveolina* sp., *Assilina* sp., *Alieolina* (*Flosculina*) sp., *Discyclina* sp., *Globigerina* sp., *Heterostegina* sp., *Miscellanea* (*Ranikothalia*) sp., *Nummulites* sp., *Operculina* sp., *Operculinella* sp., *Opertorbitolites* sp. and *Somalina* sp. There are also long-ranging Foraminifera, which are not age-specific, and occur commonly in flysch strata: *Bathysiphon* sp., *Cyclammina* sp., *Haplophragmoides* sp. and *Trochammina* sp. The Foraminifera assemblage indicates a generally Lower Eocene age, but the common occurrence of *Distichoplax* indicates that the assemblage extends from the Palaeocene to the Lower Eocene (Collenette, 1965). This means that the three formations that are contiguous in the Keningau district of SW Sabah are the same age. They have been given different formations names, but they contain the same fossils and therefore represent facies differences.

XVI.2. SAPULUT FORMATION

This is a predominantly argillaceous formation occurring within the Logungan Valley. Its outcrop extends northwards from the Kalimantan border, through Pensiangan, as far as the Milian Valley, and covers ~6,500 km^2 of country in SW Sabah. The mapping and description was carried out by Collenette (1965). The formation is interpreted as having been deposited in deep-marine conditions in which there were localized shallower areas and unstable zones with submarine slumping.

XVI.2.1. Lithology

The type locality is in the Sapulut River between the Paya Rapids and Kuala Sablangan. The oldest part of the Formation, known to be Upper Cretaceous, is predominantly argillaceous. The thickness estimate is 1500–3000 m. Predominantly it is poorly bedded blue–black mudstone with rare beds (<1 m thickness) of chert-conglomerate and bedded sandstone. There is a 90 m thick sandstone bed at the Atub Rapids. The Palaeocene to Lower Eocene section, estimated to be up to 1200 m thick, is predominantly of unbedded mudstone, in places highly disturbed. There are thin beds of argillaceous limestone. Several conglomerate beds contain blocks and cobbles of limestone, chert, sandstone and quartzite in a mudstone matrix that contains Foraminifera. The Upper Eocene zone is also predominantly of mudstone, but there are thick beds of chert conglomerate and sandstone at Bukit Saap. The general aspect is that the Sapulut Formation is of mudstone-dominant turbidite with several incursions of sandstone, and periods of instability causing slumping of blocks and clasts into a generally muddy environment (conglomerate). Some sandstone beds have flute casts which demonstrate that part of the section is inverted (Collenette, 1965). The Upper Eocene section is overlain by the Labang Formation.

In the Lower Logungan Valley, the shale is fissile, and occurs with siltstone beds (15 cm to 1 m thick) of strongly contorted rhythmite flysch. Blocks of siltstone occur within the mudstone matrix and the sequence has been interpreted as a typical turbidite flysch. Collenette (1965) comments that earlier reports have referred to the fissile shales as "slate". He disagrees with this, pointing out that there are no phyllites and no slaty cleavage. Locally within the Sapulut Gorge are thick lithic-quartz sandstones up to 6 m thick with ripple marks.

The Upper Talankai Valley, from Sapulut to Punan Batu, is also characterized by mudstone and shale, with thick lithic sandstones in places. Punan Batu is a light-grey Foraminiferal limestone with steep cliffs. Its bedding is obscure. The basal beds are argillaceous and the top is of cross-bedded sandstone.

In the Labu and Pingas Valleys the mudstone contains beds of lithic sandstone up to 30 m thick. Quartz veins occur indicating the effects of low-grade metamorphism. Lithic sandstones occur within the shales in the Milian Valley, but limestone lenses are rare.

The overall structure of the Sapulut Formation is of steep anticlines and synclines trending N to NW.

XVI.2.2. Age

Upper Cretaceous: Collenette(1965) states that fossils of this age are rare and he gives few details, except to state that *Globotruncana* spp. have been discovered in the type locality. He also states that Upper Cretaceous fossils have been reworked into Eocene strata

Palaeocene to Lower Eocene: The described fauna is as follows: *Globigerina dissimilis* Cushman & Bermudez, *Globigerina* spp., *Globigerinoides orbitormis mexicana* spp., *Globorotalia aragonensis* Nuttall, *Globorotalia centralis* Cushman and Bermudez, *Globorotalia crassata* (Cushman), *Globorotalia aff. Globigeriniformis* van Bellen, *Globorotalia lehneri* Cushman and Jarvis, *Globorotalia spinulosa* Cushman, *Globorotalia wilcoxensis* Cushman and Ponton, *Globorotalia* spp., *Hantkeniana* cf. alabamensis Cushman and *Hastigerina micra* (Cole).

Lower to Middle Eocene (late Ta to early Tb): The following Foraminifera occur in the Punan Batu limestone: *Aktinocyclina sp.*, *Assilina* sp., *Discoclina* sp., *Gypsina* sp., *Nummulites* sp. and *Operculina* sp. Limestone in the Sapulut River, 62 – 64 km upstream from Sapulut contains the algal-like organism *Distiichoplax biserialis* (Dietrich), which is confined to the range Palaeocene to Lower Eocene (Keij, 1963). It occurs together with *Gypsina* sp., *Discocyclina* sp., *Operculina* sp. and *Archaeolithothamnium* sp.

Upper Eocene (Tb): The following fauna has been identified from the Sapulut River (Collenette, 1965): *Cibrohantkenina bermudezi* Thalman, *Globigerina dissimilis* Cushman and Bermudez, *Globigerina* cf. *increbescens* Bandy, *Globorotalia* cf. *cerro-azulensis* (Cole) and *Globotruncana* spp.

XVI.3. CROCKER FORMATION (BASAL PART)

In the Sapulut–Keningau district of SW Sabah, the oldest part of the Crocker Formation is separated from the Sapulut Formation by the NNE-trending major Witti and Kinaya faults (Collenette, 1965). The Crocker Formation is distinguished from the Sapulut Formation on lithological grounds. It is sand-dominant, whereas the Sapulut is mudstone-dominant. This part of the Crocker Formation contains the same-age fossils as the Sapulut and hence they are one and the same, representing only a major facies change. Both formations are turbiditic; the Crocker occurred where there was a major influx of sand into the same basin.

XVI.3.1. Lithology

The Crocker Formation is of massive as well as thinly bedded lithic sandstones and siltstones interbedded with grey, red, green and black mudstone, with rare beds of conglomerate and limestone.

In the Baiayo River, near Keningau, the formation is predominantly of mudstone altered to sub-phyllite (Collenette, 1965). There is a tectonic breccia composed of blocks of lithic sandstone that contains quartz veins, enclosed in a mudstone matrix. Conglomerate contains clasts of quartz-veined siltstone. Intraformational conglomerate contains clasts of chert and spilite. Limestone, calcilutite and limestone microbreccia occur in the Baiayo River. About 5 km north is the Lian Cave, of microbrecciated limestone containing Foraminifera.

The Crocker Formation strata dip steeply and strike north to NNE. Collenette (1965) states that this part of the formation is equivalent to the Trusmadi and upper Sapulut Formation.

XVI.3.2. Age

In the Keningau district, limestones at Pingas, Rompun and in the Baiayo River have yielded a Palaeocene to Lower Eocene assemblage: *Aktinocyclina* sp., *Alveolina* sp., *Alveolina* (*Flosculina*) sp., *Assilina* sp., *Discocyclina* sp., *Heterostegina* sp., *Miscellanea* (*Ranikothalia*) sp., *Nummulites* sp., *Operculina* sp., *Opertorbitolites* cf. *douvillei* Nuttall, *Opertobitolites sp*. and *Somalina* sp.

Keij (1963) reported the Palaeocene to Lower Eocene *Distichoplax biserialis* (Dietrich) from limestone breccia boulders in the Baiayo River near Keningau.

Crocker Formation shales characteristically also contain long duration flysch fauna of little stratigraphic value: *Ammodiscus* sp., *Bathysiphon* sp., *Cyclammina* sp., *Haphlophragmoides* sp., *Trochammina* sp. and *Trochamminoides* sp.

Palynomorphs were obtained from a specimen of grey mudstone, interbedded with thick sandstones and mapped as Crocker Formation (94SB20A), on the main road west of Telupid (Swauger et al., 1995). It contains the following Palaeocene–Eocene flora: *Spinozonocolpites baculatus*, *S. echinatus*, *Triporopollenites*, *Leiotriietes adriennsis*,

Echinatisporis levidensis, *Echistephanoporites obscurus*, *Triorites multipora*, *Cleistosphaeridium* and *Pediastrum.*

XVI.3.3. Kudat Peninsula

The southern Kudat Peninsula is occupied by flysch allocated to the Crocker Formation, which overlies an ophiolite basement. It is an alternation of sandstones and mudstones showing deep-water turbiditic characteristics (Tongkul, 1994). The formation has undergone intense deformation producing a fold-thrust belt, oriented NW–SE across the Kudat and Bengkoka peninsulas (Tongkul, in press).

XVI.3.3.1. Palaeontology and age

Stephens (1956) carried out the original mapping and listed a collection of Foraminifera. Not all are age-diagnostic, but an age of Lower to Upper Eocene has been assigned for the fauna: *Bathysiphon* sp., *Bolivina* sp., *Camerina* sp., *C. thalicus* L.M. Davies, *Cyclammina* spp., *Discocyclina* sp., *D. (Discocyclina)* sp., *D. (Asterocyclina)* sp., *Gaudryina* sp., *Globigerina* sp., *Glomospira* sp., *Halkyardia* cf. *minima* (Libeus), *Haplophragmoides* sp., *Textularia* sp., Trochammina sp., and *Trochamminoides* sp.

Chapter XVII

Kinabatangan Group

A hiatus occurs in the Sapulut area, and the section from the youngest part of the Sapulut Formation (Upper Eocene) to the Upper Oligocene is absent. The underlying Sapulut Formation is unconformably overlain by the Labang Formation, which Collenette (1965) has designated as beginning the Kinabatangan Group. However, sedimentary conditions below and above the hiatus are generally the same, deep marine predominantly turbiditic, with shoaling carbonate areas. Further east, in the Dent Peninsula, the base of the Labang Formation is nowhere exposed. Widespread turbiditic sedimentation does continue from the underlying Rajang Group, so it is unlikely that there is a universal unconformity.

XVII.1. LABANG FORMATION

The Labang Formation was first named from the Sapulut River in SW Sabah (Collenette, 1965), since then widely mapped in the Dent Peninsula (Haile and Wong, 1965) and in the Kudat Peninsula in the far NE. These widely separated localities indicate that the formation was formerly widely deposited and much eroded. The common environment of deposition was deep marine, with characteristic turbidite deposits, with local shoals upon which there were carbonate buildups. Tectonic disturbance caused resedimentation from the topographic highs in the form of limestone breccias. The southerly extension of the formation is transitional to the Kalumpang and Kuamut formations. In Sekong Bay of Sandakan Harbour, the Labang Formation interfingers with the Kulapis Formation. The Labang Formation is overlain by the Gomantong Limestone along the Sukau Road. The Labang Formation is also overlain by the Tanjong Formation in apparent structural conformity, but there is an extremely strong contrast across the contact in the Bukit Garam circular basin. The Upper Labang Formation is of turbidite, and the Lower Tanjong is of extremely shallow water deposition.

XVII.1.1. Lithology

The Labang Formation outcrop in the Sapulut River district is 11 km long by 8 km broad. It sits unconformably upon the Sapulut Formation. The basin structure is generally synclinal, with dips varying from 20–90°, on average 30–40°. But some outcrops are overturned (Collenette, 1965). The sequence is of sandstone, siltstone, grey–blue mudstone, and localized limestones that occur more frequently than in the underlying Sapulut Formation. The sandstones have a typical turbidite character. The thickest limestones form Batu Pinahas, Batu Punggul and Batu Timbang.

The limestones show slump bedding. There are two very thick sandstones, separated by contorted mudstones. One forms the Paya Rapids.

Clennell (1992) made a comprehensive study of the eastern Telupid–Sandakan and Sukau road area of the Dent Peninsula,. The formation is predominantly of calcareous, thinly bedded (5–30 cm thickness) greywackes, sublitharenite, siltstone and mudstones. Carbonate build-ups and limestone breccias and thin detrital limestones occur (Figure 88). Much of the formation has Bouma-graded turbidite sequences. The sandstones and siltstones are commonly calcite-cemented. The general aspect of the formation is that it was deposited in deep marine conditions and represents a typical flysch together with slump folds. Shoal areas received carbonate sedimentation, with limestone breccias slumping off the shoal areas.

Chert–pebble conglomerate occurs at Batu Puteh Estate and near Bukit Gomantong (Haile and Wong, 1965), indicating that the ophiolite basement was uplifted and eroding by Oligocene time.

Based on a detailed study of outcrops along the Kinabatangan–Sukau road, Noad (1998) subdivided the Labang Formation into three distinct lithofacies that do not occur together and hence represent regional facies differences:

- *Sandy lithofacies association.* It is composed of interbedded grey mudstone, up to 200 cm but generally 10 cm thickness, interbedded with fine-grained structureless sandstones usually of thickness 1–7 m. The sequence also includes thinly laminated siltstone beds.
- *Transitional lithofacies association.* It is composed of medium-to fine-grained sandstone of thickness 35–240 cm thickness, interbedded with thin grey muddy siltstone beds.
- *Muddy lithofacies association.* It comprises thin grey fine-grained sandstones interbedded with thick featureless mudstone beds that contain benthic Foraminifera.

Noad (1998) interpreted the environment of deposition of the Labang Formation as middle to lower shoreface, with upper slope (gully fill) turbidites and distal turbidites (middle shelf). A wide range of water depth is interpreted. The basin contained deep-water passing laterally into shelfal deposition.

The *Napu Sandstones* (Haile and Wong, 1965) outcrop on a ridge east of Sukau. The sandstones are grey colour and finely laminated. A few thin shale beds are interbedded with the sandstones. A conglomeratic zone contains clasts of chert and siltstone. A conglomeratic mudstone contains large blocks of limestone and calcareous sandstone. The problem with the Napu Sandstones is that the sequence contain a Palaeocene to Eocene (Tab) fauna: *Nummulites* sp., *Discocyclina* sp., *Alveolina* sp. and *Asterocyclina* sp. Haile and Wong (1965) regard this fauna to be reworked, and have ascribed the Napu Sandstones to the Labang Formation. The dip is generally westwards, however, and the Napu Sandstones could represent a lower, or Eocene part of a similar sequence underlying the Labang Formation.

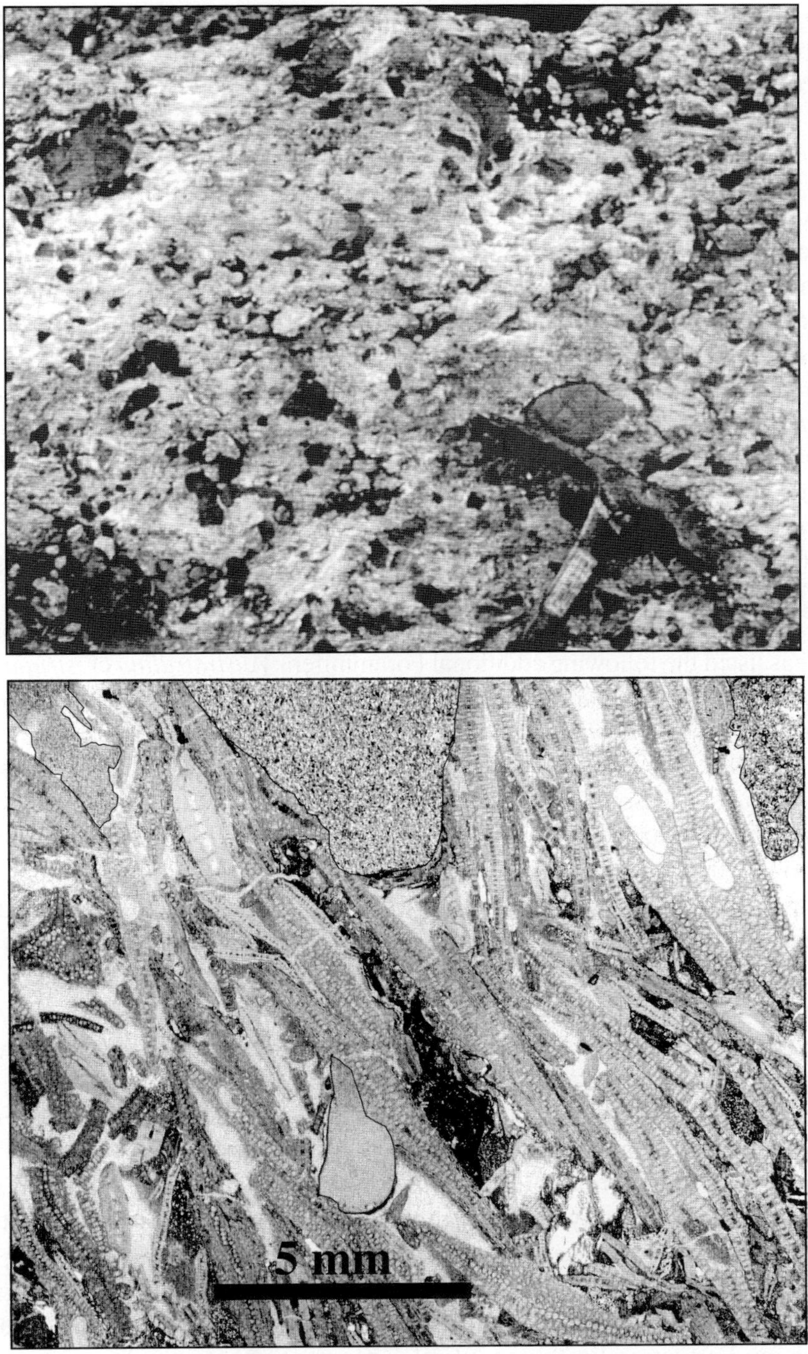

Figure 88. Labang Formation, Sukau Road near Gunung Gomantong. Top: Mudstone–limestone conglomerate (breccia), interpreted as slump deposits sourced from carbonate build-ups. Lower: Bioclastic limestone breccia, composed of bivalve shell fragments and large forams (1–2mm diameter) and ostracods in a calcite cement together with subangular clasts of fine-grained quartz litharenite and calcareous micrite (outlined).

The *Tambang Beds* (Newton-Smith, 1967) represent a fault-bounded outlier of the Labang Formation in the Bidu-Bidu Hills area. They are of steeply dipping grey sandstone, calcareous sandstone, calcarenite and grey shale.

XVII.1.2. Age

An excellent Lower to Upper Oligocene (Te_{1-4}, but including Tcd) (Lattorfian, Rupelian and Chattian) Foraminifera fauna has been collected and identified (Collenette, 1965; Haile and Wong, 1965): *Alveolina* sp., *Amphistegina* spp., *Austrotrillina howchini* (Schlumberger), *Cycloclypeus* sp., *Globigerina* spp., *Globigerina binaiensis* Koch, *Globigerina ciperoensis* Bolli, *Globigerina dissimilis* Cushman and Bermudez, *Globigerina cf. increbescens* Bandy, *Globigerinoides* spp., *Globoquadrina* sp., *Globoquadrina venezuelana (Hedberg)*, *Globorotalia mayeri* Cushman and Ellisor, *Gypsina* sp., *Heterostegina* sp., *Heterostegina* cf. *borneensis* van der Vlerk, *Lepidocyclina* spp., *Lepidocyclina* (*Eulepidina*) sp., *Lepidocyclina* (*Nephrolepidina*) spp., ?*Miogypsinoides* sp., *Neoalveolina* sp., *Neoalveolina pygmaea* (Hanzawa), *Operculina* sp., *Operculinoides* sp., *Sphaerogypsina globulus* (Reuss) and *Spiroclypeus* sp.

The Tambang Beds contain several of the above Foraminifera, and Newton-Smith (1967) has listed the following additional Foraminifera: *Austrotrillina* cf. *striata*, *Borelis* sp., *Lepidocyclina* (*Eulepidina*) cf. *ephippioides* (Jones and Parker), *Lepidocyclina* (*?Eulepidina*) sp. and *Lepidocyclina* (*Nephrolepisina*) cf. *parva* (Oppenoorth).

The strata also contain reworked Upper Eocene Foraminifera, such as *Aktinocyclina* sp., *Discocylina* sp., *Fabiana* sp., *Nummulites* sp. and *Pellatispira* sp. The limestones contain the coral *Cyphastraea gemmulifera* Gerth, which is known to extend into the Lower Miocene. Many rocks contain the trace fossil *Palaeodictyon*

Clennell (1992) reported palynomorphs from the Sukau and Telupid road sections that are Oligocene to Lower Miocene: *Acostichum*, *Anacolasa*, *Canthium*, *Casuarina*, *Crudia* sp., *Florschuetzia trilobata*, *Graminae*, *Laevigatosporites*, *Magnastriates howardii*, *Meyeripollis*, *Myrtaceae*, *Polypodiisporites*, *Spinozonocolpites echinatus* and *Timonius* type. There are also palynomorphs which are less age-specific such as*: Polypodiisporites usmensis*.

Rangin et al. (1990) reported Upper Oligocene calcareous nannofossils from the Batu Puteh Limestone and from the Sukau road: *Helicosphaera recta*, *Cyclicargolithus abisectus*, *Sphenolithus ciperoensis and Sphenolithus distensus*. All of these indicate wholly marine, probably bathyal conditions. The limestone hill on the main Sukau road, west of Bukit Gomantong, yielded the following nannofossils: *Cyclocargalithus floradinus*, *Coccoiithus pelagicus*, *Reticulofenestra* sp. and *Sphenolithus* sp. (Swauger et al.,1995). The age is early Lower Miocene.

Lambang Formation limestone NE of Gomantong contains: *Amphistegina radiata* (Fichtel and Moll), *Gypsina globula* (Reuss), *Lepidocyclina* (*Eulepidina*) *Formosa* (Schlumberger), *Lepidocyclina* (*Eulepidina*) *planate* (Oppenoorth) and *Heterostegina borneënsis* van der Vlera (Hashimoto and Matsumaru, 1981). This fauna indicates an Upper Oligocene (Te_{1-4}) age, considered to be near the top of the formation. The palaeontological data indicate that the Labang Formation extends over the whole Oligocene and into the early Lower Miocene.

The following list of nannofossils from the Labang Formation of SW Sabah indicates an Uppermost Eocene to an Upper Oligocene age: *Chiasmolithus altus*, *Chiasmolithus solotus*, *Coccolithus* sp., *Coccolithus miopelagicus*, *Coccolithus pelagicus*, *Coronocyclus nitescens*, *Cribrocentrum reticulatium*, *Cyclicargolithus abisectus*, *Cyclicargolithus floridanus*, *Dictyococcites bisecta*, *Discoaster* sp., *Discoaster barbadiensis*, *Discoaster deflandrei*, *Discoaster ladoensis*, *Discoaster multiradiatus*, *Discoaster tani*, *Discoaster saipanensis*, *Ericsonia formosa*, *Helicospheara compacta*, *Helicospheara intermedia*, *Marthasterites tribrachiatus*, *Micrantholithus flos*, *Pontosphaera* sp., *Reticulofenestra umbilica*, *Sphenolithus* sp., *Sphenolithus ciperoensis*, *Sphenolithus dissimilis*, *Sphenolithis distentus*, *Sphenolithus microformis*, *Sphenolithis moriformis*, *Sphenolithus praedistentus*, *Sphenolithus pseudoradians*, and *Zygrhablithus bijugatus*.

XVII.1.3. Gomantong Limestone

This is a thick feature-forming limestone of Lower Miocene age, famous for its birds'nests (Haile and Wong, 1965). The hill is readily accessed from the Sukau road. It is a synclinal formation, with dips of 5–25°, sitting unconformably upon Labang Formation. It may be massive or blocky and rubbly. Thickness at Bukit Gomantong is approximately 300 m. Several limestones occur along the Sukau road forming a SSW–NNE trend extending 50 km. Limestone also occurs as far south as Batu Puteh near the Kinabatangan River.

XVII.1.3.1. Lithology

The lithofacies suggest a range of environments from open marine marls to shore-attached reefal deposits (Noad, 1998, 2001). The following lithofacies have been studied by Noad (1998):

- *Green mudstone.* Thin beds of poorly cemented dark green mudstone are interbedded with limestone. Thickness rarely exceeds 10 cm and they pinch out laterally.
- *Labang clast conglomerate.* Clasts are embedded in a green mudstone matrix. The clasts are subangular to well rounded and individually may reach 50 cm thickness. Mudstone clasts are the most common in the matrix-supported conglomerate, commonly making up to 30% of the rock.
- *Labang Formation clasts in limestone matrix.* The limestone is well cemented and contains up to 30% clasts of siltstone to mudstone. Clasts range up to 10 cm diameter. Many have a coralline-algal overgrowth. The matrix is micritic with a siliciclastic component. Sorting of the clasts is poor.

- *Calcarenite with siliclastic component.* The calcarenites are made of mixed carbonate detritus and siliclastic silt-grade material.
- *Foraminifera-dominated facies. Lepidocyclina* packstones make up at least 60% of the Gomantong Limestone (Noad, 2001). The Foraminifera are commonly imbricated and usually sub-complete. Individual fossils can reach 10 cm in diameter, but more typically are ~4 cm. The matrix is micritic to calcarenite composition. Other fossil components are coralline algae, bivalves, gastropods, bryozoans and worm tubes. Isolated corals may occur. There are also rare Foraminiferal grainstones.
- *Coral-dominated lithofacies.* At least 39 species of corals have been identified and they are autochthonous to the build-ups. Varieties are coral bafflestone, massive coral framestone, mat-like coral framestone and coral bindstone (Noad, 1998). There are also coral floatstone and coral rudstone containing allochthonous coral clasts.
- *Coralline algae-dominated lithofacies.* The varieties are rhodolithic packstone, coralline-algal packstone and coralline-algal grainstone.
- *Comminuted bioclastic packstone (calcarenite).* This facies is of well sorted finely comminuted skeletal debris, forming 50 cm beds.

Noad (2001) compared the Gomantong Limestone with limestones of the Luconia Province of offshore NW Sarawak, but it is not obvious that the geological setting is similar.

XVII.1.3.2. Palaeontology and age

There is an abundance of *Lepidocyclina* (*Nephrolepidina*) spp., including *Nephrolepidina sumatrensis*, *N. inflata* and *N. borneensis* groups. *Lepidocyclina perornata* Douvillé is also common. Other Foraminifera are: *Eulepidina* sp., *Miogypsinoides dehaartii* van der Vlerk, *Miogypsina* spp., *Cycloclypeus eidae* Tan, *Flosculinella globulosa* (Rutten) and *"Orbitolites"* cf. *vandervlerki* (de Neve) (Haile and Wong, 1965). In addition, Hashimoto and Matsumaru (1981) also identified *Acervulina inhaerens* (Schultz) and *Miniacina miniacea* (Pallas). The age is Lower Miocene (Aquitanian to Burdigalian) (Te5).

Noad (1998) published a comprehensive analysis of nannofossils and large benthonic Foraminifera showing an age range from Upper Oligocene Te4 to Lower Miocene Te_5. Some Foraminifera were also extracted from interbedded mudstones:

- Nannofossils: *D. deflandrei*, *C. pelagicus*, *Helicospaera* cf. *perch nielsenae*, *H. euphratis*, *Clausicoccus* sp., *Umbilicospaera* cf. *jafari*, *Coronocyclus nitescens*, *Triquetrorhabulus carinatus*, *Spenolithus moriformis*, *S. ciperoensis* and *Cyclicargolithus floridanus*. The nannofossils indicate an uppermost Oligocene age.
- Large benthonic Foraminifera: *Amphisorus*, *Amphistegina*, *A. asmanensis*, *Amphistegina*, *Austrotrillina striata*, *Borelis pygmaeus*, *Carpenteria*,

Cibicidella, *Clavulina*, *Cycloclypeus*, *E. bandjirraensis*, *E. ephippoides*, *Gypsinid*, *Heterostegina* (*V*) *borneensis*, *H* (*V*) *assilinoides*, *Lepidocyclina* sp., *Lepidocyclina brouweri L. davidpricei*, *L* (*N*) *gibbosa*, *L* sp. *aff. nipponica*, *Lepidocyclina* (*N*) *oneatensis*, *Lepidocyclina* (N) *parva*, *L.* (*N*) *soebandi*, *L.* (*N*) *sipoerensis*, *L.* (*N*) *sondaica*, *L. sumatrensis*, *L.* (*N*) *verbeeki*, *L. volucris*, *Massilina*, *M. borodinensis*, *M. dehaarti*, *M. tani*, *Neorotalia*, *Operculina*, *Operculinella* sp., *Planogypsina*, *Planorbulinella larvata*, *Planostegina*, *Rotalia* sp., *Sorites*, *Sigmoidella* sp., *Sphaerogypsina*, and *Victoriella*.

The large benthonic Foraminifera indicate an age range of Chattian, Aquitanian and Burdigalian (Noad, 1998).

XVII.1.3.3. Occurrences in SW Sabah

Collenette (1965) reported steeply dipping limestone outcrops at Batu Punggul and Bukit Timbang. Balaguru (2001) mapped small outcrops of calcareous sandstone and shallow water Foraminifera-rich bioclastic packstones along the Tibow–Keningau road and in the Kalabakan river, and a tributary of the Kuamut river. These beds are of limited exposure and only 0.5–1 m thick and they occur at the boundary between the Labang and the Tanjong Formations (Balaguru, 2001), hence are equivalent to the Gomantong Limestone of the Sukau road. The following rich Lower Miocene (Burdigalian, Te – Tf_1) fauna was identified: *Amphisorus*, *Amphistegina*, *Austrotrillina howchini*, *Carpentaria*, *Cycloclypeus*, *Dasyclads*, *Dentoglobigerina altispira*, *Flosculinella bontangensis*, *Flosculinella reicheli*, *Globigerinoides*, *Globogerinoides triloba*, *Globoquadrina dehiscens*, *Gypsina*, *Heterostegina*, *Heterostegina* (*Vlerkina*) sp., *Lepidocyclina*, *Lepidocyclina brouweri*, *Lepidocyclina* (*Nephrolepidina*) *brouweri*, *Lepidocyclina ferreroi*, *Lepidocyclina isolepidinoides*, *Lepidocyclina nipponica*, *Lepidocyclina oneatensis*, *Lepidocyclina* (*Nephrolepidina*) *oneatensis*, *Lepidocyclina parva*, *Lepidocyclina praedelicata*, *Lepidocyclina rutteni*, *Lepidocyclina* (*Nephrolepidina*) *rutteni*, *Lepidocyclina stratifera*, *Lepidocyclina sumatrensis*, *Lepidocyclina* (*Nephrolepidina*) *sumatrensis*, *Lepidocyclina* (*Nephrolepidina*) *verrucosa*, *Miliolids*, *Miogypsina*, *Miogypsina digitata*, *Miogypsina inflata*, *Miogypsina sabahensis*, *Miogypsina tani*, *Miogypsinoides*, *Miogypsinoides dehaarti*, *Operculina*, *Operculinella*, *Planorbulinella larvata*, *Quinqueloculina*, *Rodophytes* and *Spiroclypeus*.

This abundant fauna indicates a Lower Miocene (Burdigalian) age and confirms the equivalence to the Gomantong Limestone, representing the upper part of the Labang Formation.

XVII.2. KULAPIS FORMATION

This distinctive formation is a facies variant of the Labang Formation but their relationship has not been seen in outcrop. The Kulapis is a wholly marine redbed formation, readily mapped because both the sandstones and shales are red. The sandstones are commonly pink coloured in outcrop and the shales chocolate-brown. The formation is extensive, from the Telupid area on the west, through Beluran to the Sandakan area and southwards towards Kinabatangan.

XVII.2.1. Lithology

The main lithologies are greywacke, sublitharenite, calcareous lithic arenite and laminated mudstones. All have a red colour varying from pink sandstones to chocolate-brown shales. Calcitic concretions are common in the thicker sandstone beds. The sandstones are quartz-rich and contain plagioclase and chert grains. The mudstones are rich in illite, smectite and chlorite (Clennell, 1992). There are massive sandstone beds up to 10 m thickness. Some thinner sands show normal grading, flute marks and groove marks. Siltstones show ripples. Sole marks are rare in Kulapis Formation sandstones. The thick sands usually contain intraclasts of mudstone. Slump folds indicate deep water, but younger parts of the formation indicate shoaling of the sea.

This wholly marine redbed formation is very distinctive in outcrop. Sometimes the shales are chocolate-brown and the sands pale reddish-pink. The red colour probably indicates an iron-rich source, perhaps nearby uplifted ophiolite. By contrast, the same age Labang Formation is not red and probably its area of deposition lay farther from the iron-rich provenance.

To the east of the Bidu-Bidu Hills, the gently to steeply dipping Kulapis Formation strata are of pale greyish-red sandstones interbedded with chocolate-brown mudstones. The beds are commonly 0.3–15 m thick but most commonly are 0.3–1.0 m thick (Newton-Smith, 1967). The sandstones are commonly calcareous and there are concretions in the thicker sandstones. A distinctive feature of the chocolate-brown mudstones is that they are usually extremely fissile, and broken into biconvex flakes with conchoidal fracture and polished surfaces, resulting from dewatering. By contrast, the interbedded sandstone beds remain intact in outcrop.

Noad (1998) proposed a type locality at km 35 on the Telupid–Sandakan road, where all four lithofacies outcrop.

- Thick massive pink coarse-grained sandstone grading up to fine-grained sandstone. The base commonly exhibits load casts, but may have grooves or flutes. The interpretation is of high density turbidity currents.
- Thick beds of featureless red mudstone. The beds are commonly contorted. They are low-energy deposits in a lower fan or basin plain.
- Thin laterally persistent dark-red sandstones, 20–200 cm thick, with grooved and fluted bases, grading upwards into finely laminated siltstone then into mudstone. They are interpreted as sandy turbidites deposited during waning turbulence.
- Thin interbeds of very fine-grained red sandstone and thicker mudstone. They are interpreted as distal turbidites.

XVII.2.2. Relationship with the Garinono Mélange

The Kulapis Formation forms many outcrops along the main Labuk road from Telupid to Sandakan. Angular pink sandstone blocks of the Kulapis Formation are the most common clast within the mudstone matrix of the mélange. Some road cuts show a gradual transition from normally faulted Kulapis Formation to mélange containing isolated pink angular sandstone blocks. This is particularly well shown at the junction between the Labuk and Kinabatangan road, where there is a recently constructed housing estate. It is clear from such outcrops that the Kulapis Formation was undergoing pervasive extensional (normal) block faulting and was contributing these blocks into the adjacent Garinono Mélange basin. The timing of this event is the rift event of the SE Sulu Sea marginal basin.

XVII.2.3. Age of the Kulapis Formation

The Foraminifera of the Kulapis Formation are arenaceous and undiagnostic benthonic of no stratigraphic value, for they range typically from Upper Cretaceous through Oligocene (Newton-Smith, 1967). However, Fitch (1958) stated that shales in the Bode area near Sandakan contain an Oligocene fauna (post-Tb, pre-Te_5). He stated that the identifications were by Shell, but details were not available.

Clennell (1992) tabulated an extensive flora and fauna from red shales of the Kulapis Formation, which allow the conclusion that it seems likely to extend from Uppermost Eocene, through Oligocene, to Middle Miocene:

Palynomorphs: *Ainipollenites verus*, *Alangium*, *Anacolosa*, *Antidesma*, *Avicenia*, *Barringtonia*, *Brownlowia*, *Canthium type*, *Casuarina*, *Ceophalomappa*, *Crudia*, *Dactylocladus*, *Ephedra*, *Eugeissona insignis*, *Florschuetzia semilobata*, *F. trilobata*, *Graminae*, *Inaperturopollenites*, *Laevigatosporites*, *Longetia*, *Lycopodium cemum*, *Margocolporites vanwijhei*, *Myrtaceae*, *Polypodiisporites*, *Pometia*, *Pteris type*, *Retitricoporites*, *Rhizophora*, *Shorea albida*, *Spinozonocolpites echinatus*, *Stemonurus*, *Timonius* and *Zonocostites*. This palynomorph flora all represents wholly marine to brackish water conditions with a range from Oligocene to Middle Miocene (Clennell, 1992). Swauger et al. (1995) also obtained palynomorphs from red and grey mudstones, interbedded with buff coloured sandstones, along the road to Rumidi Estate and east of Telupid: *Dicoipopollis borneensis*, *Spiriferites*, *Mytaceidites*, *Tetracoiporopollenites espotoidea*, *Salixporienites*, *Spinozonocolpites echinatus*, *Pallabrevitricolpites*, *Polypodiscoisporites regularis* and *Florschuetzia trilobata*. This flora represents an Oligocene to Middle Miocene range.

Calcareous nannofossils: Clennell (1992) tabulated the following, representing a Mid-to Upper Oligocene age: *Cyclicargolithes floridans*, *C. abisectus*, *Discoaster tanii*, *Helicosphaera compacta*, *H. recta*, *Reticulotenestra gelida*, *R. scissura*, *Sphenolithus predistentus*, *S. celsus* and *S. trilobosus*. However, the Rumidi Estate road locality also yielded the nannofossils: *Discoaster neohamirtus* and *Coccolithus pelagius* representing a late Middle Miocene age (Swauger et al., 1995).

Therefore it appears that the range of deposition of the Kulapis Formation ranged from uppermost Eocene, through Oligocene, to the Middle Miocene.

XVII.3. KALUMPANG FORMATION

This formation is the continuation south-eastwards into the Semporna Peninsula and forms the eastern part of Timbun Mata Island in Darvel Bay (Kirk, 1962). It may extend as far as Sebatik Island on the Indonesian border, where it is known as the Sebatik Sandstone–Shale Member. There must be doubt about its inclusion within the Kalumpang Formation. The outcrops around Mount Wullersdorf have been described by Lim (1981).

XVII.3.1. Lithology

The formation is strongly folded and imbricated. A characteristic is that the sedimentary rocks are interbedded with volcanics and pyroclastic rocks (Kirk, 1962). Predominantly it is of shale and mudstone, associated with sandstone, conglomerate, limestone and rare chert. Dark green tuff is common. There are scattered pyroclastic rocks and volcanic rocks ranging from andesite to dacite. The conglomerate is of pebbles of radiolarian chert and siltstone. Sandstone and dark red chert occur in the Umas Umas Valley. Dacitic rocks occur at Mount Mostyn. Eastern Timbun Mata Island is of feldspathic tuff and keratophyre

The Sebatik Sandstone–Shale Member is carbonaceous and contains coaly shale and impure lignite. By contrast with the main Kalumpang Formation, it is distinctly of shallow water facies, and Kirk (1962) interpreted it as occurring on the margin of the main Kalumpang Basin.

The Sipit Limestone Member forms four hills near the coast across from Timbun Mata Island. It consists of very pure limestone. Limestone forms seven small N–S trending hills at km 50 from Tawau along the Kunak Road (Figure 114). It overlies tuff and contains corals, algae and echinoids in addition to Foraminifera, which allow a very precise Middle Miocene (Tf1–Tf2) age (Lim, 1981).

XVII.3.2. Palaeontology and age

Kirk (1962) and Lim (1981) recorded a large fauna of Foraminifera from shales, limestones and tuffs of the Kalumpang Formation, as follows: *Acervulina* sp., *Alveolinella globulosa* Rutten, *Amphistegina* sp., *Asterigerina spp.*, *Austrotrillina howchini* Schlumberger, *Bolivina* spp., *Borelis* sp., *Bulimina* sp., *Carpentaria* sp., *Cibicides* spp., *Cristellaria* spp., *Cycloclypeus* sp., *Cycloclypeus* (*Katacycloclypeus*) *annulatus* Martin, *Dentalia* sp., *Discorbis* sp., *Ehrenbergina* spp., *Frondicularia* spp., *Flosculinella* sp., *Gaudryina* spp., *Globigerina* sp., *Globigerina binaiensis* Koch, *Globigerina binaiensis* Koch var., *Globigerina* cf. *ciperoensis* Bolli, *Globigerina dissimilis* Cushman and Bermudez var., *Globigerina* cf. *increbescens* Bandy, *Globigerina subcretacea* Chapman, *Globigerinatella insueta* Cushman and Stainforth, *Globigerinoides* sp., *Globigerinoides* cf. *trilobus* (Reuss), *Globigerinoides* sp. (*Gr. rubra* group), *Globigerinoides* sp. (*Gr. triloba* group), *Globigerinoides glomerosa* Blow, *Globigerinoides rubra* d'Orbigny var., *Globigerinoides sacculiferus* Brady, *Globigerinoides subquadratus* Bronniman, *Globoquadrina* sp., *Globoquadrina altispira* Cushman and Jarvis, *Globoquadrina dehiscens* Chapman Parr and Collins, *Globoquadrina venezuelana* Hedberg, *Globoquadrina venezuelana* Hedberg var., *Globorotalia* sp., *Globorotalia centralis* Cushman and Bermudez, *Globorotalia cerroozulensis* Cole, *Globorotalia crassata* Cushman, *Globorotalia* cf. *fohsi arrisanensis* LeRoy, *Globorotalia mayeri* Cushman and Ellisor, *Globorotalia seitula* Brady var., *Globorotalia siakensis* (LeRoy), *Globorotalia spinulosa* Cushman, *Globorotalia wilcoxensis* Cushman and Ponton, *Glomospira* spp., *Gypsina* sp., *Gypsina globulus* Reuss, *Gyroidina* sp., *Halkyordia* sp., *Haplophragmoides* sp., *Heterostegina* sp., *Lepidocyclina* sp., *Lepidocyclina* (*Nephrolepidina*) sp., *Lepidocyclina* (*Nephroplepidina*) cf. *sumatrensis* Brady, *Lepidocyclina* (*Eulepidina*) sp., *Lepidocyclina* (*isolepidina type*), *Lepidocyclina ferreroi Provale*, *Marginospora*, *Miogypsina* sp., *Miogypsinoides* sp., *Neoalveolina* sp., *Nodosaria* sp., *Nummulites* sp. (reticulate type), *Nodosaria* spp., *Operculina* sp., *Operculinoides* sp., *Operculinella* sp., *Orbulina* sp., *Orbulina bilobata* (d,Orbigny), *Orbulina universa* d'Orbigny, *Orbulina suturalis* Bronniman, *Planarbulina* sp., *Planarbulinella* sp., *Psammosiphonella* spp., *Pullienia* spp., *Quinqueloculina* sp., *Rotalia* sp., *Sorites*, *Sphaeroidinella multiloba* LeRoy, *(?)Spiroclypeus* sp., *Textularia* spp., *Trochammina* sp., and *Uvigerina* sp. This fauna indicates an age range of Lower Oligocene to Middle Miocene (Td–Tf). The Kalumpang Formation of the Wullersdorf region is limited to a Lower to Middle Miocene age (Lim, 1981).

XVII.4. BALUNG FORMATION

This gently dipping formation appears as windows beneath the volcanic rocks of the Wullersdorf area near Tawau (Lim, 1981). Age is Middle to Upper Miocene. The abundance of tuffs indicates the presence of active volcanism. It is thought that the Balung Formation overlies the Kalumpang Formation because its dips are usually gentle.

XVII.4.1. Lithology

Grey volcanic ash, mudstone, lithic and crystal tuffs, and rare coaly beds are characterized by abundant plant remains. The volcanic ash forms beds a few centimeters to 1 m thick. It contains abundant plant material and spores and pollen. Medium-to fine-grained tuff forms beds a few centimetres to 30 cm thick. The common crystals are quartz, plagioclase and rounded rock fragments. Some tuffs contain lapilli.

XVII.4.2. Palaeontology and age

The Formation is dated by age-diagnostic sporomorphs. The age is Tf_{2-3} (Middle to Upper Miocene) and clearly overlaps in age the volcanic rocks since tuffs and ash are common. The Foraminifera are not age-diagnostic (Lim, 1981). The following pollen and spores have been identified: *Alangium* sp., *Alnus* sp., *Amylotheca* sp., *Anacolosa* sp., *Arenga* sp., *Avicennia* sp., *Barringtonia* sp., *Blumeodendron* sp., *Brownlowia* sp., *Calamus* sp., *Camptostemon* sp., *Canthium* sp., *Casuarina* sp., *Cephalomappa* sp., *Ceratopteris* sp., *Crudia* sp., *Cyclophorus* sp., *Dacrydium* sp., *Dactylocladus* sp., *Dipterocarpus* sp., *Duria* sp., *Ephedra* sp., *Eugeissona* sp., *Eugeissona insignis*, *Eugeissona minor*, *Ficus* sp., *Gonystylus* sp., *Gramineae* sp., *Hibicus* sp., *Longetia* sp., *Lycopodium* sp.,, *Lycopodium cernum*, *Lycopodium phlegmaria*, *Myrtaceae* sp., *Nenga* sp., *Palmae* sp., *Pices* sp., *Pinus* sp., *Podocarpus* sp., *Podocarpus polystachyus*, *Pometia* sp., *Rhizophora* sp., *Sonneratia* sp., *Stenochlaena* sp., *Stenochlaena aerolaris*, *Stemonurus* sp., and *Tsuga* sp.

Lim (1981) listed the following Foraminifera that however are considered to be non age-diagnostic: *Ammobaculites* sp., *Ammodiscoides* sp., *Ammodiscus* sp., *Ammodiscus grzybowski* (Emillani), *Ammonia* sp., *Angulogerina angulosa*, *Bathysiphon* sp., *Bolivina* sp., *Cibicides* sp., *Cribostomoides* sp., *Cyclammina* sp., *Cyclammina cancellata* (Brady), *Cyclammina amplectens* (Grzybowski), *Discorbis* sp., *Eggerella* sp., *Eggerella bradyi* (Cushman), *Elphidium* sp., *Elphidium koeboeense LeRoy*, *Gaudryina* sp., *Glandulina* sp., *Globigerina* sp., *Glomospira* sp., *Haplophragmoides* sp., *Haplophragmoides walteri* (Grzybowski), *Hormosina* sp., *Kalomopsis grzybowski* (Dylanzanka), *Karrerielia* sp., *Lagena* sp., *Len cullna* sp., *Loxostomum karrerianum var. carinatum* (Millet), *Nonion japonicum* (Asano), *Psammosphaera placenta* (Grzybowski), *Psammosiphonella* sp., Psammosiphonella carapitana (Hedburg), *Psammosiphonella cylindrical* (Glaessner), *Psammosiphonella irregularis* (LeRoy), *Quinqueloculina* sp., *Recurvoides deformis* (Andreas), *Reophax* sp., *Reusella* sp., *Rotalia* sp., *Sigmoilopsis schlumbergeri* (Silvestri), *Spiroculina* sp., *Spirosigmoilinella* sp., *Strebius ketienzeis* (Ishozaki), *Textularia* sp., *Trochammina* sp., *Trochammina renzi* (Renz), *Trochammina renzi* Brouwer, *Trochamminoides* sp., *Trochamminoides subcoronata* (Rzehak and Grzybowski), *Usbekistania* sp., and *Valvulina* sp.

XVII.5. PALAEOGEOGRAPHY FOR END OLIGOCENE, BEGINNING MIOCENE

Tjia (1988) proposed a “Kinabalu Suture Zone” that meanders through Sabah from north to south, drawn to include the ophiolitic rocks. The implication of this proposal is that everything east of this “suture” is unrelated to west Sabah. More specifically it means that the East Crocker, Labang and Kulapis are unrelated to the West Crocker. Tjia (1999) then proposed a scenario that the "East Sabah Terrane" originated between Taiwan and Luzon in the Oligocene and mysteriously migrated southwards to "dock" with greater Borneo in Miocene time (Tjia, 1999, p. 606). Such a fantasy ignores the known spreading patterns of the intervening marginal basins, and accordingly must be rejected. The alternative is that the Labang and Kulapis formations, and the Gomantong Limestone, belong to one-and-the-same terrane as the West Crocker and that the 'Kinabalu Suture' did not exist. This was the interpretation of Noad (1998, p. 355). Figure 89 presents a geologically realistic palaeogeographic map for the end of the Oligocene and beginning of Miocene that largely agrees with that of Noad (1998). The Labang and Kulapis Formations and the Gomantong Limestone were deposited in a fore-arc basin that Hutchison (1992a) called the Central Sabah Basin.

The West Crocker, the Kulapis, Labang and Kalumpang formations are all the same age. The first two are wholly turbiditic, the Labang is also turbiditic but ranges southwards to shallower conditions represented by the Gomantong Limestone. The Kalumpang is in transition to shallower water on the apron of the volcanic arc, and it includes volcanic and pyroclastic material. Like Noad (1998), the shallowing is taken to be towards a volcanic arc (Figure 89). Based on trace element mineral

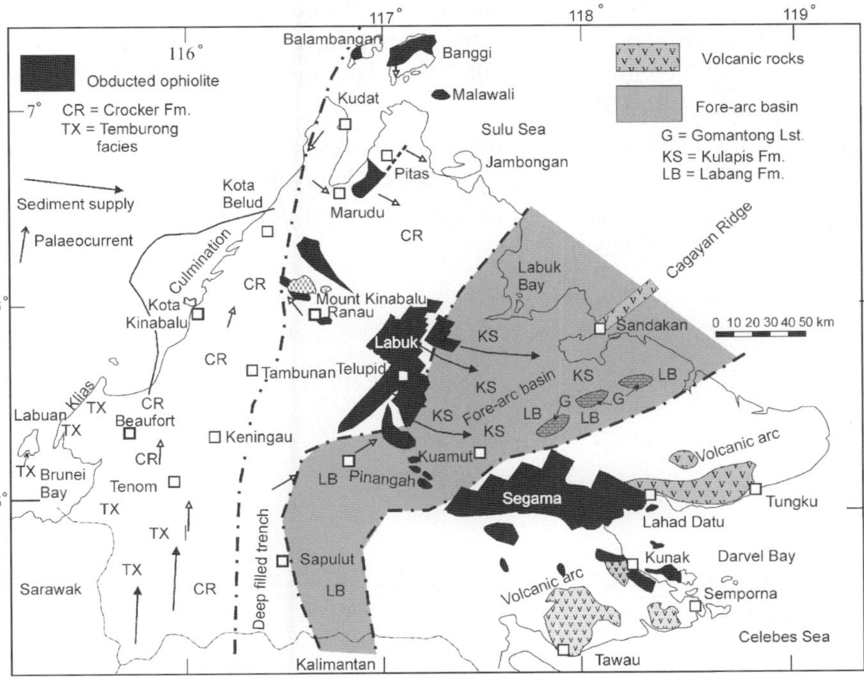

Figure 89. Suggested palaeogeography for the beginning of the Lower Miocene. The Labang, Kulapis and Gomantong Limestone were deposited in a fore-arc basin. The Crocker Formation completely filled the trench and its basin expanded westwards to form a foredeep over the Dangerous Grounds.

chemistry, Omang (1995) has shown that the Darvel Bay (Segama) ophiolite ultramafites were formed in a fore-arc region and fit into an Island-Arc Geotherm. By contrast, Imai and Ozawa (1991) showed that the ultramafites around Mount Kinabalu are more akin to an Ocean Geotherm. The salient features of the palaeogeography are:

- The trench is located deeply buried along a zone trending SW between Ranau and Telupid, hence trending southwards towards the Adio Suture of Kalimantan (see Figure 5.18 of Hutchison, 1989). Subduction was active only into the Early Miocene, and the Trusmadi Formation may represent a part of the original accretionary prism.
- The trench is located by the high-pressure glaucophane–piedmontite meta-sandstones and meta-basalts now exumed near Telupid (Hutchison et al., 2000).
- The high influx of West Crocker Formation turbidites from the south completely infilled the trench, spreading out north-westwards over the attenuated continental crust of the Dangerous Grounds. The same turbidites also sedimented much of the fore-arc basin. Shallowing of the basin towards the volcanic arc is indicated by the Labang Formation and Gomantong Limestone.
- It has been suggested by Hutchison (1992a), later agreed by Clennell (1996) and Noad (1998), that the red and pink colouration of the Kulapis Formation resulted because of their provenance in uplifted Labuk ophiolites in the NW. Uplifted ophiolites, overlain by serpentinite conglomerates, have been described by Newton-Smith (1967) and Hutchison and Tungah (1991) in the eastern Bidu-Bidu Hills.
- Much of the Dent and Semporna volcanic rocks are known to be of Miocene age, and Oligocene andesites formed the Cagayan Ridge, also found beneath the mosque in Sandakan, which became rifted during the opening of the SE Sulu Sea marginal basin.

Chapter XVIII

Crocker Formation (including the Temburong Fm.)

It has been shown above (Figure 87) that the lower part of the Crocker Formation is of Palaeocene to Middle Eocene age. However most Crocker Formation outcrops, being predominantly of sandstones, have yielded only arenaceous Foraminifera of little age significance. The western and north-western outcrops have been more studied, being more readily accessible. This region has commonly been referred to as the 'West Crocker Formation', which has been accurately dated Oligocene to Lower Miocene only in the Sipitang-Beaufort-Tenom Gorge district. The detailed map of Wilson (1964) forms a basis for introduction to the main features of the Crocker Formation (Figure 90). I have replotted the structural data of the area of Figure 90 as lower-hemisphere equal-area stereograms to show the structural style more obviously (Figure 94). The Crocker Formation exhibits all the features of a major turbidite fan system, but it is difficult to accept that the formation may have extended from the Palaeocene to the Lower Miocene as a single system, and unconformities are suspected, though not proven. The most significant paper on the Crocker Formation is that of Stauffer (1967). Unfortunately his outstanding work has not been added to. Many authors have pronounced on the nature of the Formation, but few have added significant analysis, notably William et al. (2003).

XVIII.1. LITHOLOGY

Stauffer (1967) highlighted the various sedimentary types within the Formation and they are summarized here.

XVIII.1.1. Flysch sequences

Rhythmic sequences of interbedded sandstone and mudstone are the commonest sedimentary type, universally recognized as flysch in deformed mountain belts such as the Alps, and known as 'Bouma sequences'. The sandstone beds are generally 10–80 cm thick, but they vary from a few centimetres to several metres. The sandstones have a sharp base and are internally graded. The beds have typical and abundant sole marks. The sandstones are grey, weathering to buff, are poorly sorted, composed of rather angular quartz, lithic fragments and feldspar. The lithic grains are usually volcanic and chert, with some metamorphic rock grains. Grain size is 0.5–1 mm. The sandstones have an argillaceous matrix. The interbedded mudstones and shales are dark grey and mostly 5–50 cm thick, but vary from 1 cm to 2 m in thickness.

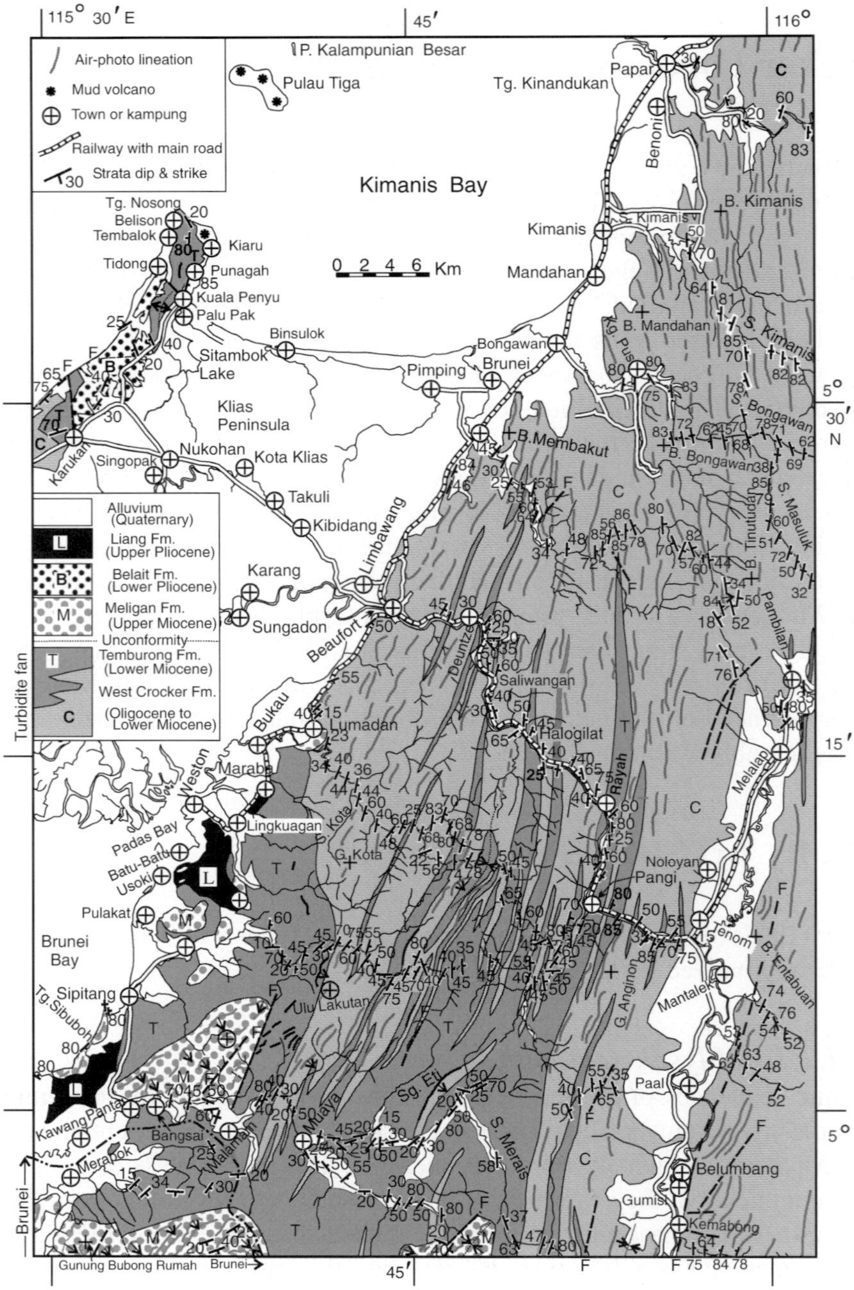

Figure 90. Geology of the West Crocker and Temburong formations (after Wilson, 1964). With permission from Minerals and Geoscience Department, Malaysia.

Penecontemporaneous slumping is extensively developed, commonly confined to one or two beds. The slumping occurs in isolated patches and is concentrated in zones of intense deformation. Later tectonic deformation has been superimposed upon the slumping, obviously tectonic complexity has been attracted to weak slumped zones.

Graded bedding is extremely common in flysch sequences. They represent typical turbidity current deposition. Sole marks are abundant. They include flute moulds, groove marks, lineated lode structures and slide marks. Some beds have ripple marks formed by strong aqueous currents. However they have been seen only towards Papar, where other evidence suggests shallower water conditions of deposition (Stauffer, 1967). Small-scale cross-bedding in the finer-grained flysch and also in the laminite sequences are a poor indicator of current direction because of their variability.

XVIII.1.2. Laminite sequences

Laminated beds of fine sandstone and siltstone are interbedded with mudstone, itself laminated and commonly red or green. Bed boundaries are commonly subdued and gradational. Laminites lack typical turbidite features. Graded bedding is rare to absent. Slump beds are uncommon. Laminites are interpreted to have been deposited in more stable quiet conditions by a less episodic environment compared with flysch. Sometimes, not frequently, flysch and laminite coexist. The environment of laminite sequences is unclear, but Stauffer (1967) suggests they result from weak turbidity or from normal bottom currents.

The frequency of laminite occurrence increases eastwards away from the Sabah coastline, especially above the mapped unconformity (Stauffer, 1967). However, the Temburong Formation of SW Sabah and adjacent Brunei represents laminite facies of the Crocker Formation.

XVIII.1.3. Red and green mudstone

Mudstone beds in the laminite (less frequently in the flysch) may be red, or green. Usually the coloured beds are thicker than 20 cm. The coloured sections are sand-free. Frequently the bed top is green, the lower part red, with irregular boundaries between the colours, suggesting a secondary or diagenetic origin, not yet understood. Stauffer (1967) hoped that the coloured beds could be used as ‘marker horizons’ in this monotonous formation, but unfortunately there are too many to correlate.

XVIII.1.4. Mass-flow sandstone

Thicker beds, especially in the flysch sequence, indicate mass-flow mechanisms resulting from submarine slides. The beds range from 1 to 60 m thickness. They lack turbidite features and are devoid of internal structures. Some are so homogeneous that they have developed large perfectly spherical concretions, some as large as “cannon balls”. Usually the sandstones are medium-grained, but may be coarse and pebbly. Such beds are common in the Kota Kinabalu region, and are quarried along the Kota Kinabalu to Tuaran road (Figure 91). William et al. (2003) concluded that the

Figure 91. Steeply dipping (strike 25°, dip 63° E) massive homoclinal mass-flow sandstone, with sparse thin mudstones of Crocker Formation. Quarry 4 km north of Kota Kinabalu. 2 m tall geologist for scale.

immature composition and texture and poor sorting necessitates a short transport distance for these high-density turbidites, followed by deposition as sand-rich lobes on a slope apron.

XVIII.1.5. Slumped zones

Slumping is sometimes intense and involves several beds, resulting from submarine sliding. There are scattered sandstone blocks, many bent or twisted, even rolled croissant-style. The original bedding is destroyed beyond recognition. Such slumped zones are usually associated with flysch and with mass-flow sandstone bodies.

XVIII.2. TEMBURONG FORMATION

The Temburong Formation, which occurs in eastern Brunei and south of Beaufort, is an important facies of the upper part of the Crocker Formation (Wilson, 1964). It occurs as conformable intercalations within the more sandy Crocker Formation flysch. The formation is dominantly argillaceous, composed of a laminite sequence of siltstone and shale and is in fact the laminite sequence of Stauffer (Section 18.1.2 above).

The Formation is remarkably uniform in lithology, being mainly a flysch deposit, with common intercalations of slightly calcareous pelagic shale. The siltstone is quartz-rich and poorly sorted. A typical sequence is of regular bedding of layers 2 cm to 0.3 m thick. This type of sequence is well exposed in the Tenom Gorge. Massive cleaved dark grey shale occurs also in the Tenom Gorge and elsewhere. Usually it contains pyrite concretions. Small bioclastic Foraminiferal limestones occur in a few localities.

Wilson (1964) has described turbidite sedimentary structures for the siltstones, such as graded bedding and flute and groove casts. Some of these are seen in the Tenom Gorge near Pangi and Halogilat.

XVIII.3. PALAEO-CURRENT DIRECTIONS

A large number of sole mark palaeo-current measurements have been made by Stauffer (1967) and by his students that he supervised in the field, Mahendran (1980), Chua (1980) and Kamaluddin (1980). The results are shown on Figure 92. Each arrow from Stauffer (1967) represent the direction averaged from an average of 21 individual readings. The pattern is simple, indicating that the sandstones of the Crocker Formation were deposited as an integral part of a major turbidite fan-shaped trough directed towards the north. The provenance of the Crocker Formation sands lay to the south. Stauffer (1967) concluded, therefore, that the Crocker Formation deposits should be systematically younger northwards. This important conclusion has not been given due attention by subsequent workers.

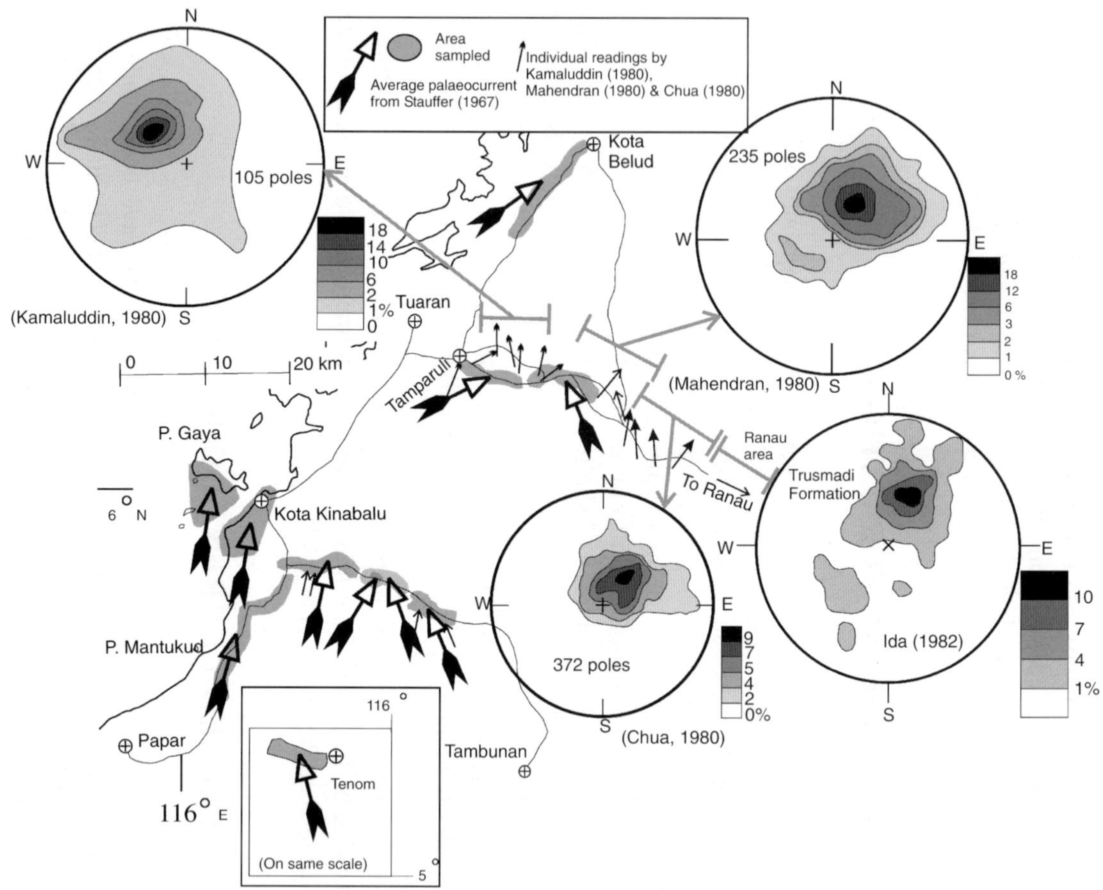

Figure 92. Palaeocurrent determinations (arrow points away from provenance) for Crocker Formation. Averages of many readings from Stauffer (1967) and individual measurements along. Ranau Road from Kamaluddin (1980), Mahendran (1980), Chua (1980) and Ida (1982). Lower-hemisphere equal-area stereograms of poles to bedding along the Tamparuli to Ranau road from the same authors. See also Figure 70 for additional palaeocurrent data.

XVIII.4. AGE OF THE CROCKER–TEMBURONG FORMATION

The Temburong Formation, being a more shaly facies of the Crocker Formation, has been the most productive for age-diagnostic fossils (Wilson, 1964). The age is concluded to be Tcd to Te, Oligocene to Lower Miocene. The following Foraminifera have been identified from Temburong Formation shales: *Bulimina* sp., *Cristellarea* sp., *Gaudryina* sp., *Globigerina* spp., *Globigerina binaensis* Koch. var., *Globigerina* cf. *ciperoensis* Bolli, *Globigerina dissimilis* Cushman and Bermudez var., *Globigerina cf. increbescens* Bandy, *Globigerinoides* spp., *Globigerinoides index* Finley, *Globigerinoides "rubra* group", *Globigerinoides semi-involuta* Keijzer, *Globigerinoides "triloba* group", *Globoquadrina* sp., *Globoquadrina venezuelana* (Hedberg), *Globorotalia centralis* Cushman and Bermudez, *Globorotalia meyeri* Cushman and Ellisor, *Hantkenina alabamensis* Cushman, *Hastigerina micra* (Cole), *Lepidocyclina* sp., *Miogypsinoides* sp., *Operculina* spp., *Quinqueloclina* sp., *Uvigerina* sp. and *Virgulina* sp. The Foraminiferal assemblage indicates an age range from late Tcd to Te_5: Lower Oligocene (Rupelian and Lattorfian) to Lower Miocene (Aquitanian and Burdigalian).

Limestones within the Temburong Formation have yielded Lower to Upper Oligocene Foraminifera (Td to Te_{1-4}). The fauna is *Amphistegina* sp. *Lepidocyclina* (*Eulepidina*) cf. *ephippioides* (Jones and Chapman) A and B forms, *Gypsina* sp. *Heterostegina borneensis* van der Vlerk and *Operculina* sp.

Arenaceous benthonic flysch-type Foraminifera occur both in the Temburong facies and in the Crocker Formation. The fauna is not age-diagnostic, and may range from Eocene to Miocene: *Ammobaculites* spp., *Ammodiscus* sp., *Bathysiphon* sp., *Biloculina* sp., *Bulimina* sp., *Cristellaria* sp., *Cyclammina* sp., *Gaudryina* sp., *Haplophragmoides* sp., *Trochammina* sp., *Valvulina* spp and *Verneuilina* sp.

XVIII.5. STRUCTURE

The Crocker Formation structure has important regional variations. There is, however, a remarkable consistency over quite large districts.

The first order large-scale folds have never been seen because there are no mountain-scale outcrops. Nevertheless, the consistency of air photo and satellite lineaments, and of strike directions measured in the field, show that there are regular fold patterns, which could be understood, but until now remain incompletely resolved. Wilson (1964) shows a photograph of an outcrop-scale anticline in the Tenom Gorge (Figure 93). It is a tight upright fold in Timburong facies. The compression has been so tight that the axial zone is imbricated, with a slight thrust displacement. This is a parasitic fold, but it gives a clue to nature of the large-scale fold style. The fold axis appears to be approximately horizontal.

Figure 93. Tight upright anticline in the Temburong facies of the Crocker Formation. The axial area is faulted and imbricarted. Half kilometre east of Pangi (from Wilson, 1964). Scale given by knife, to right of added white bar. With permission from Minerals and Geoscience Department, Malaysia.

About 3 km east of Pangi, an asymmetrical fold in siltstone and shale is overturned towards the west. Some asymmetrical folds have very steep easterly dips. The thicker beds of the Crocker Formation show regular dips while the interbedded Temburong more shaly thin-bedded facies shows complexity of small folds such as shown in Figure 93. Wilson (1964) therefore concludes that the Temburong facies sequences have provided décollement and compression zones within less deformed thicker Crocker flysch facies. Of course, some of the deformation in the mudstone beds is of penecontemporaneous slumping, but it has been overprinted by tectonic folding.

The conclusion of Wilson (1964) is worthy of repetition here: "the basic problem concerning the regional easterly dip of the Crocker Formation remains unsolved. If this regional dip is that of a tilted but otherwise complete succession, then the Crocker and

Temburong Formations must together exceed 60,000 feet [18,300m] in thickness". This is grossly excessive and the succession must have been repeated by isoclinal folding or strike faulting, or both. The outcrops suggest otherwise, and the succession is overwhelmingly right-way-up. Major strike faulting has never been demonstrated. Wilson (1964) expressed concern also about the palaeontological age—Oligocene to Lower Miocene in the Beaufort–Tenom area and Palaeocene to Eocene in the Sapulut district. The succession, therefore, appears to 'young westwards' but the predominant eastwards dip suggests the reverse. Wilson (1964) therefore made a careful study of 'way-up' criteria such as graded bedding and sole marks in the Tenom Gorge. The only clear evidence of overturning is in localized contorted strata of the Temburong Formation, never in the Crocker Formation flysch. He concluded therefore that the eastwards-dipping contorted zones of Temburong Formation represent décollement zones, parallel to the regional structure, along which thrust movements took place upwards towards the west. This speculation is, of course, not substantiated by actual field observation, for the faulting (Figure 93) is of minor character.

The lower-hemisphere equal-area stereograms constructed from the field dips and strikes of Wilson (1964) show at a glance that the regional strike is consistently NNW–SSE, swinging slightly to due N–S towards Papar. In the south (SE of Sipitang) the succession is dominated by ESE directed dips. There are some upright folds and the stereogram strongly suggests that the folds tend to be recumbent with axial planes dipping towards the ESE (Figure 94). Along and around Tenom Gorge, the strike continues NNE–SSW, the folding more upright, and more westwards-dipping strata can be seen. To the north, SE of Papar, the strike becomes due N–S and the folds are upright. However more steeply eastwards-dipping strata are recorded, but steeply dipping westwards strata are also quite common (Figure 94). From Kota Kinabalu to Tuaran and Tamparuli, the regional strike has swung north-easterly. From Tuaran to Kota Belud, the swing has continued, but here the sequence is less monoclinal and there are almost as many dips towards the NW as there is towards the SE. In the Sugut River and Bongaya region of northern Sabah, the strike has swung towards ESE. Hence, from Sipitang to Sungai Sugut, the Crocker Formation strike has swung gradually through slightly more than a right angle.

The Crocker Formation has been arbitrarily divided into North, East, West and South (Figure 69). These subdivisions have no geological basis, but are convenient for the purposes of discussion and communication.

The pronounced swing from northerly directed around Sipitang in the West Crocker and Temburong facies, to ESE in the North Crocker has suggested to Tongkul (1997) that the whole Crocker Formation has been folded into a gigantic syncline, whose axis lies between Ranau and Tambunan, and plunges towards the SE. Although the structure is broadly synformal, it cannot be a simple syncline, because the limbs of the great fold are folded into anticlines and synclines (Figure 93). The old term synclinorium would be more appropriate. A number of major thrusts have been proposed (Tongkul, 1997). Unfortunately major thrusts do not outcrop and must be inferred. The thrusts themselves would need to have been folded.

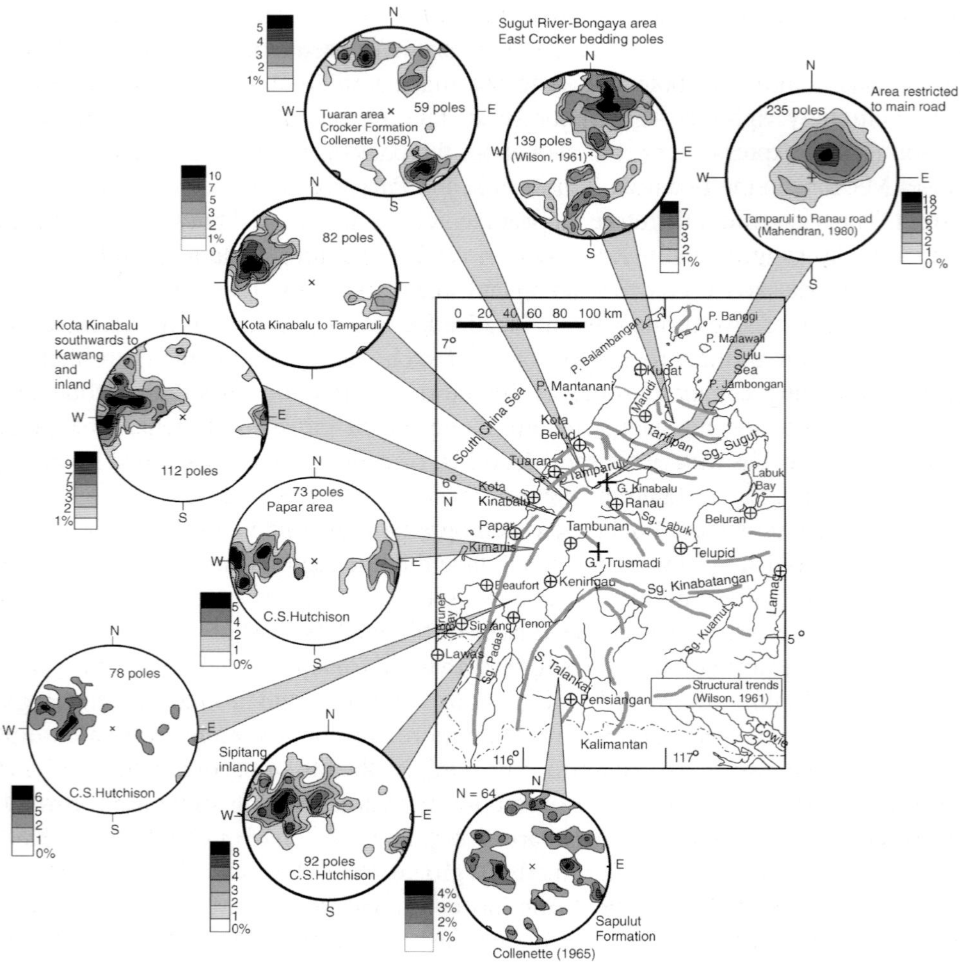

Figure 94. Equal-area lower-hemisphere plots of poles to bedding planes of the Crocker Formation, plotted from various publications. General locations identified on a map of structural trends by Wilson (1961).

It seems very appropriate to suggest that the North Crocker Formation has been thrust SSW over the East Crocker Formation. The proposed locality of the thrust (Tongkul, 1997) is remote country and the theory is unlikely ever to be tested by outcrop study.

The strike swing from N–S near Beaufort, through NE–SW around Tuaran, to ESE–WNW in the North Crocker seems to suggest a large oroclinal bend of what formerly had been a more linear fold-belt. The palaeocurrents of the North Crocker are ESE-directed (Tongkul, 1994). This is in contrast to the northerly directed palaeocurrents of the West Crocker (Figure 92). The suggestion is that these two terrains were once aligned in the same orientation, subsequently folded and oroclinally bent.

XVIII.5.1. Penampang road structure

In an attempt to analyse the structural style of the Crocker Formation, Stauffer (1967) made detailed observations of the road from Kota Kinabalu to Kampung Moyog, at the time of construction (Figure 95). The road continues to Tambunan, and its location is shown in Figure 92. The section of the West Crocker along the Penampang Road is dominantly monoclinal with eastward dip and younging, in general. In fresh outcrops, the way-up criteria are clear, and although much of the section is monoclinal and correct way-up, the folded sections show recumbent folds with localized younging-westwards directions. There are zones of tight isoclinal folds and strike faults. It was hoped that the red mudstones would serve as marker horizons, but they are multiple and cannot be used. Laminites are prominent at Kampung Babagon, where the strike changes abruptly from 190° on the west to about 150° on the east side. An unconformity has been tentatively postulated (Figure 95). Less than 1 km west of Kampung Moyog, there are closely spaced shear zones. To the east is a complex of faults and folds, some with vertical axial planes. The Kampung Mayog locality may well represent a major fault zone.

XVIII.5.2. Border area with Sarawak, near Lawas

The lower cross-section of Figure 96 is taken from Wilson (1964), where much of the Crocker Formation is shaly and thin-bedded and known as the Temburong Formation. The folded Crocker Formation is unconformably overlain by the Meligan Formation. The structure is essentially of an anticline in which the eastern limb is of sandy flysch and the western limb more shaly and thin-bedded, known as the Temburong Formation. The difference is because of a facies change. The sandy flysch is either monoclinal or broadly folded into an anticline. By contrast, the Temburong thin-bedded facies has a complex structure, with asymmetric folds overturned towards the west. Wilson (1964) concluded that "the Temburong Formation has provided zones of movement and compression within the regularly dipping Crocker Formation". Obvious signs of major faults are uncommon. Major faults are suspected, parallel to the regional strike, because of zones of brecciation and shearing and local distortion of the regional strike. Minor faults are commonly seen and generally dip westwards.

XVIII.5.3. Main road from Tamparuli to Ranau

During 1980, while the road was being widened and realigned, the whole section was carefully mapped by Kamaluddin (1980), Mahendran (1980) and Chua (1980) under the close supervision of P.H. Stauffer and taking advantage of abundant way-up criteria that have since decayed. The road section is predominantly monoclinal with a "Sulu Trend" strike and predominant moderate to low dip towards the SSW. However, the Crocker Formation is structured in a more complicated fashion (Figure 95).

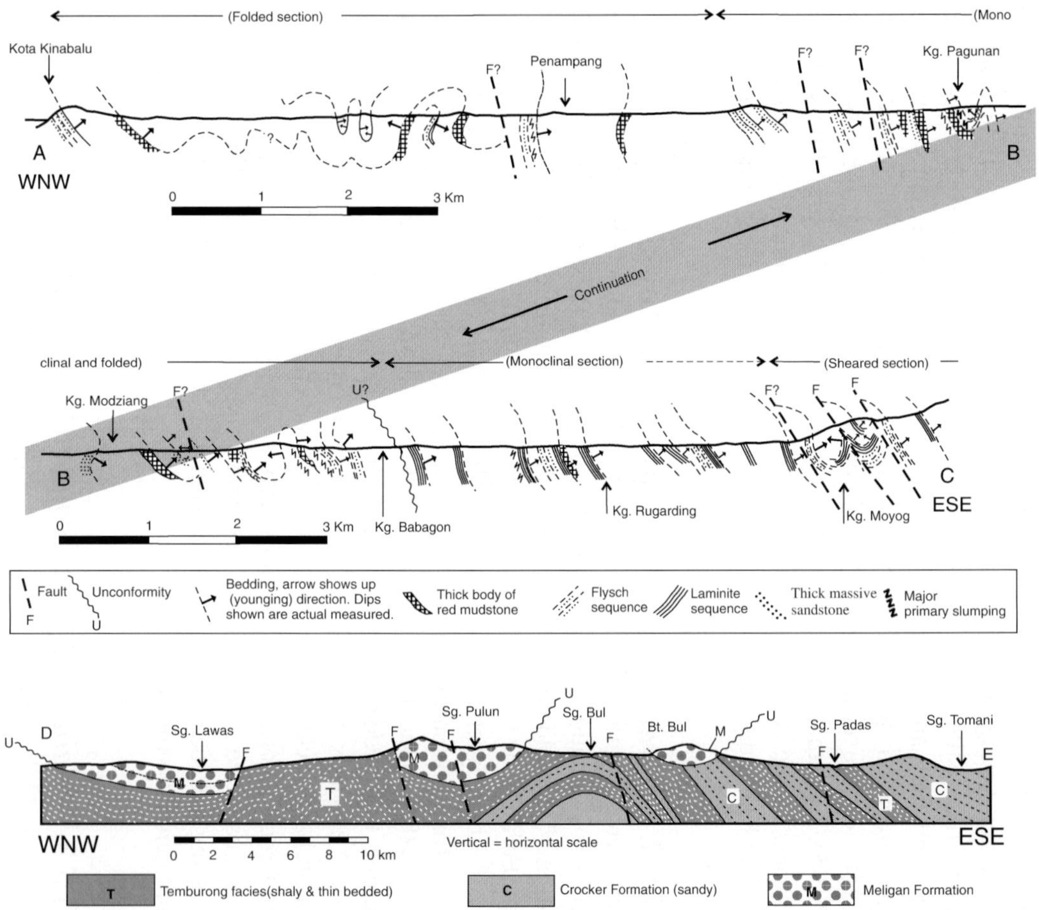

Figure 95. Cross-sections showing structures in the Crocker Formation. (A–C) measured section from Kota Kinabalu to Kampung Moyog, along the road towards Tambunan (from Stauffer, 1967). (D–E) Section across Temburong facies, Crocker Formation and Meligan Formation east to Sipitang (from Wilson, 1964). With permission from Minerals and Geoscience Department, Malaysia.

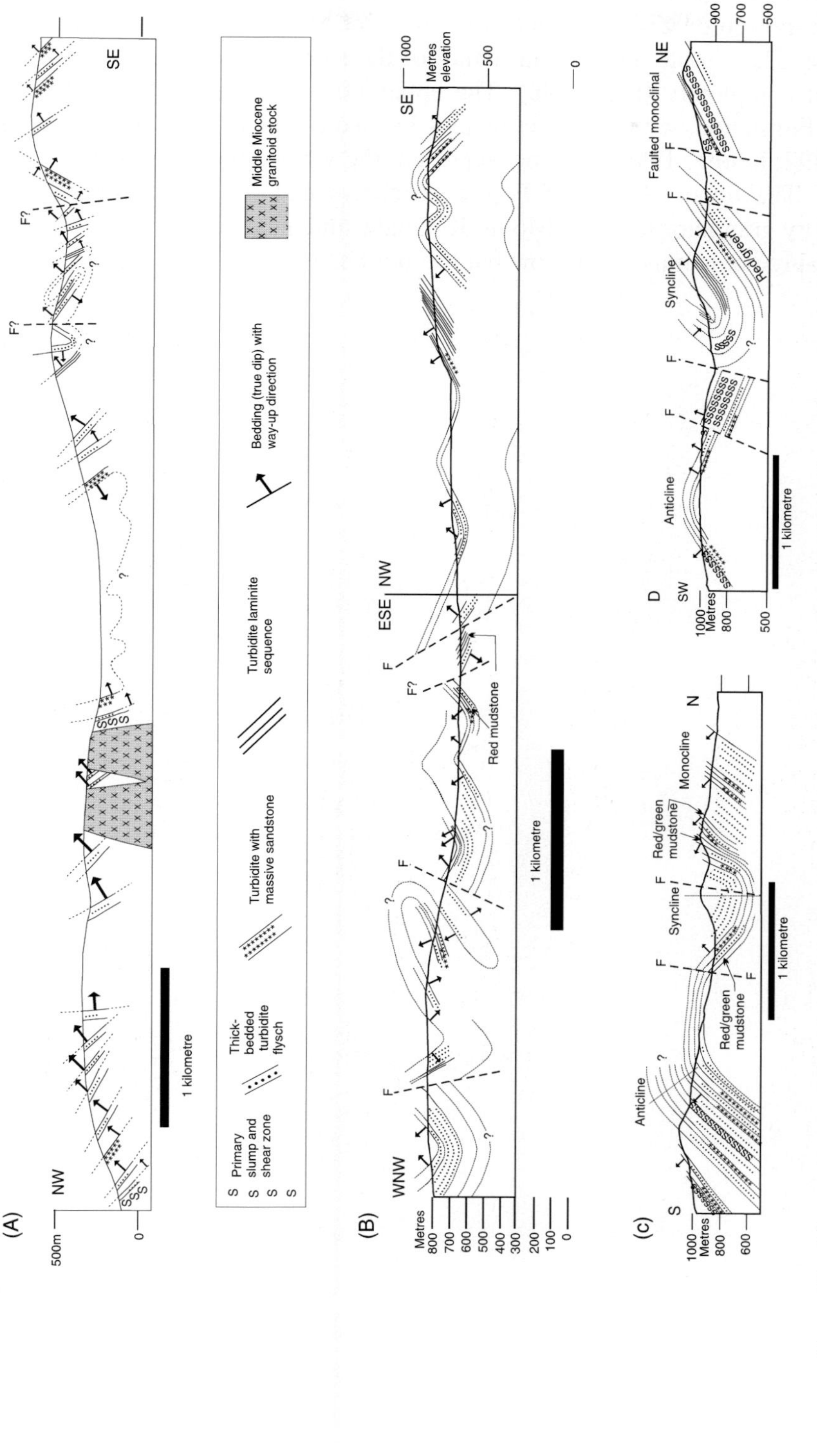

Figure 96. Measured cross-sections through the Crocker Formation along the road from Tamparuli towards Ranau. (A) after Kamaluddin (1980); (B) after Mahendran (1980); (C and D) after Chua (1980). These sections were measured when the sections were fresh during road construction. They have since decayed. With permission from Minerals and Geoscience Department, Malaysia.

There are large anticlines and synclines. Some folds are recumbent as shown by a few spectacular outcrops of inverted beds with prominent sole marks on the upper (inverted) side of a sandstone bed. The locations of the cross-sections of Figure 94 are given in Figure 92 along with lower-hemisphere Schmidt-net plots of the structural data. The simple monoclinal nature of the stereograms (Figure 94) belie the structural complexity (Figure 96). The apparently synclinal nature of the whole Crocker Formation, with the syncline axis located to the south of Ranau (Tongkul, 1994, 1997; Leong, 1999) does not represent the whole story. This is not a simple syncline. The upper section of Figure 96 shows that the Crocker Formation is intruded by granodiorite of the Mount Kinabalu suite near Koporingan. If there is a thermal aureole, is must be narrow but has not been seen.

Chapter XIX

Uplift of the Crocker and Trusmadi Formations

A 10-day field collection expedition was made across Sabah in 1994. One of the objects was to carry out fission-track analyses on both apatite and zircon crystals separated from the rocks to determine the cooling and uplift history (Swauger et al., 1995). The localities over the Western Cordillera and Labuk Highlands are given in Figure 97.

The fission-track ages of the separated apatite and zircon crystals were determined following standard procedures (Swauger et al., 1995). The optimum number of crystals from each rock sample was 20. The results are summarized in Figure 98 together with the standard deviation of each specimen. It has been established that apatite has a partial annealing temperature range of *c.* 60–110°C and zircon has a range of *c.* 200–350°C (Gallagher et al., 1998).

XIX.1. APATITE FISSION-TRACK AGES

The fission-track data (Figure 98) indicate strong uplift of the rocks of the Western Cordillera (Crocker and Trusmadi Formations). The apatite ages are all Middle to Upper Miocene and have been totally reset—the apatite crystals are detrital and have been sedimented within the rocks which are NOT younger than Lower Miocene. If there had been no resetting, the apatites would have preserved their ages from their time of release and transport from their original provenance. This erosional and transport history has been wiped out. This feature suggests that the rocks have been buried beneath 4–8 km of overburden causing heating to >120°C after deposition. This conclusion is supported by vitrinite reflectance from locality 4D (Figure 97), where kerogen yielded a reflectance R_0 of $1.42 \pm 0.06\%$ ($n = 34$) (Hutchison et al., 2000). The data of Figure 98 suggest that after burial, the strata of the Western Cordillera were exhumed and cooled in the Upper Miocene. Exhumation was probably facilitated along one or more major faults, and several NW–SE faults have been identified on SAR radar imagery. Mean fission-track lengths in the range 13–15 μm characterize rocks that have been cooled rapidly, at >10°C Ma^{-1} through a temperature range of 120–60°C. The Crocker Formation apatite fission-track lengths range from 14.5 to 15 μm (Hutchison et al., 2000). The apatite fission-track data (Figure 98) suggest extremely rapid exhumation rates for the rocks of the Western Cordillera of *c.* 600 ± 100 m Ma^{-1} (0.5–0.7 mm a^{-1}). Such values are comparable with other well-known collisional mountains belts.

XIX.2. ZIRCON FISSION-TRACK AGES

The zircon ages (Figure 98) all pre-date the depositional age of the strata. They generally represent a Cretaceous and/or Jurassic provenance from which the zircon grains were eroded and transported. The Trusmadi Formation sample (16A) even contains much older zircon crystals. Burial of the Crocker and Trusmadi Formations did not anneal these provenance ages.

It is likely that the Crocker and Trusmadi formations had their provenance in the uplifted Rajang Group and Embaluh Group of Sarawak and Kalimantan, which in turn had their provenance probably in Indo-China. It is very likely that the proto Mekong River was instrumental in initially bringing these turbiditic sediments into the Borneo region. The very sandy West Crocker Formation probably was provenanced from uplifted areas in the Kelabit Highlands of Sarawak. The very sandy Nyalau Formation, which has the same age as the West Crocker (Oligocene to Lower Miocene) represented a time when the palaeocoastline of Sarawak was oriented N–S, with the Sundaland landmass extending from SE Asia into western Sarawak. The sea deepened eastwards and it is possible that the West Crocker turbidites represent a fan system at the furthest reach of the Nyalau Formation transport and deposition system. The depocentre for the Trusmadi and Crocker Formations is thought to underlie the region around Telupid, where glaucophane and piedmontite in meta-sandstones indicate that static metamorphism took place under 7–8 kbar at a low geothermal gradient, then dramatically inverted and exhumed from a depth of *c*. 20 km. There has been spectacular erosional removal of overburden (Trusmadi and Crocker Formations) from this district.

Chapter XX

East Sabah Mélanges

A remarkable feature of eastern Sabah is the great extent of mud-matrix mélange units and associated chaotic rocks. They outcrop over an area of 12,000 km^2 in eastern Sabah from Telupid to Sandakan and southwards towards Lahad Datu. They have been studied and described by Clennell (1991, 1992). Although all one, they have been given district formation names—Garinono (Collenette, 1966), Ayer, Kuamut, but these units cannot be defined as formations. Each of them represent an event that resulted in broken formations associated with and grading into mudstone-matrix mélanges. Clennell (1991, 1992) concluded that the mélanges were mostly produced by submarine slope failures triggered by tectonic rearrangement of the Central Sabah Basin at the end of the Lower Miocene. Parts of the mélanges are of olistostrome origin, associated with slump structures. Balaguru and Nichols (2004) have proposed that the mélange formations are genetically related to rifting in the fore-arc basin in which the Tanjong Formation was deposited.

XX.1. GARINONO MÉLANGE

This is the best known mélange, outcropping over 3800 km^2 mostly along the Labuk Road. Its mudstone matrix unfortunately invariably results in rapid outcrop erosion. The best exposures were on steep road cuts that have eroded partially or completely or been covered in grass. Generally the mélange has faulted contacts with broken Kulapis Formation, or may grade into Kulapis Formation through a zone of broken formation. Near the Kolapis River and north of the Segaliud River, the Garinono Mélange structurally overlies strongly folded Kulapis Formation with a marked unconformity (Clennell, 1991). In the Segaliud area, bedded Tanjong Formation mudstones unconformably overlie the mélange. Likewise, the Sandakan Formation unconformably overlies the mélange (Lee, 1970). A few kilometres east of Telupid, the ophiolite overthrusts the mélange (Clennell, 1992).

The mélanges are formed by local faulting of the strata through an intermediate stage of broken formation. Where the clasts are entirely of the redbed Kulapis Formation, the matrix of the mélange is red. Where the clasts are of Labang Formation, the matrix is grey. Along the Telupid Road, the grey matrix mélange generally overlies the red.

Clennell (1991) concluded that the Garinono Mélange, and the other named mélanges, were formed by submarine slumps and slides into a deep marine rift basin. This was the earlier interpretation of Hutchison (1992a), who had independently visited the newly created Labuk road mélange slopes, and concluded that the deep marine rift basin was the extrapolation into Sabah of the SE Sulu Sea marginal

Basin (Figure 99). The model is firmly based on that of Karig (1972) to explain a basin between a remnant arc (extinct) and an active volcanic arc. The trench migrates forwards and away from the rifting volcanic arc. In this case the rifting arc is clearly identified as the Cagayan Ridge, which extends towards Sandakan. The mélanges closer to the active arc contain abundant volcanic material (Ayer Mélange).

XX.1.1. Matrix

XX.1.1.1. Lithology

The matrix to the contained scattered clasts forms 70–80% of the rock. It is a bluish-grey plastic clay that is generally scaly. It breaks into lensoid fragments with slickensided surfaces. This scaly foliation wraps around the clasts. However, cleavage is totally absent and bedding is obscure, but there may be slump zones. They are not tectonic and show no regular pattern.

The clay mineralogy is illite, micas, chlorite and kaolinite. Portions of the mélange matrix are red and green in colour and are smectite rich. They are thought to have resulted from bands of volcanic ash (Clennell, 1992). Similar layers also characterize the Crocker Formation.

Vitrinite studies of the matrix show a range of $R_0 = 0.55–0.96\%$ indicating that the melange remained below 60°C and was never hot (Clennell, 1991).

XX.1.1.2. Palaeontology and age

Clennell (1992) listed the following palynomorphs from the matrix of the Garinono mélange, collected from the Labuk and Telupid roads and also from Segaliud:

Acrostichum, *Alangium*, *Anacolosa*, *Antidesma*, *Barringtonia*, *Brownlowia* sp., *Canthium*, *Casuarina*, *Cephalomappa* sp., *Crudia* sp., *Dactylocladus* sp., *Ephedra*, *Eugeissona insignis*, *Ficus*, *Florschuetzia levipolii*, *Florschuetzia semilobata*, *Florschuetzia trilobata*, *Graminae*, *Laevigatosporites*, *Longetia*, *Magnastriates howardii*, *Meyeripollis*, *Myrtaceae*, *Polypodisporites*, *Polypodisporites usmensis*, *Pometia*, *Pteris* type, *Rhizophora*, *Spinozonocolpites echinatus*, *Stemonurus* and *Timonius* type. This flora best indicates of mid-Lower to mid-Middle Miocene age. Some of the species are Oligocene, interpreted to be reworked. Both the grey matrix and the red matrix mélanges were sampled; they are the same age.

Lee (1970) reported the following Middle Miocene planktonic Foraminifera from bluish-grey mudstone matrix at mile 31 on the Labuk Road: *Globorotalia lohsi barisanensis* LeRoy, *Globigerinoides sacullifere* (Brady), *Globigerinoides* spp. and reworked *Globotruncana* spp. Localities in Sungai Manila yielded many arenaceous Foraminifera, such as *Bathysiphon* sp., *Haplophragmoides* sp. and *Reophax* sp. probably of Oligocene age but reworked.

Lee (1970) recorded several localities of bedded mudstone and calcarenite closely associated with the Garinono Formation around Sandakan. They may be interpreted as an integral part of the Garinono Mélange, although the localities do

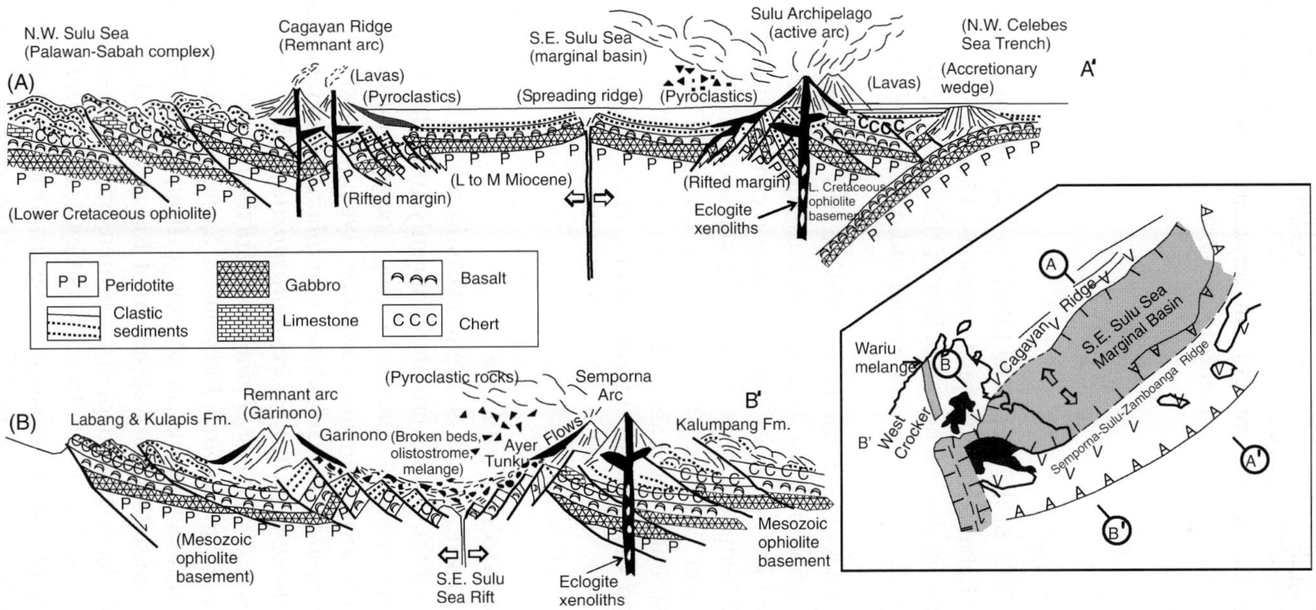

Figure 99. Schematic cross-sections to illustrate the Lower to Middle Miocene active rifting stage of the SE Sulu Sea (A–A′) and its onlandextension into Sabah (B–B′). No inferred scale. The melanges formed in this rift basin. This cartoon fits the model of Karig (1972) for marginal basins.

not contain clasts. The following Foraminifera were recorded: *Globigerina bulloides* d'Orbigny, *Globigerina* spp., *Globigerinoides trilobus* group, *Globigerinoides sacculifera* (Brady), *Globigerinoides ruber* (d'Orbigny), *Globoquadrina attispira* (Cushman and Jarvis), *Globorotalia mayeri* (Cushman and Ellisor), *Globorotalia* cf. *scitula* (Brady), *Globorotalia* cf. *lohsi barisensis* LeRoy, *Globorotalia lohsi lohsi* (Cushman and Ellisor), *Orbulina suturalis*, *Orbulina bilobata*, and *Sphaeroidinella* spp.

XX.1.2. Clasts

XX.1.2.1. Lithology

The clasts form 20–30% of the rock. The lithologies are predominantly sandstones and siltstones of the Labang and Kulapis formations. There are also scattered clasts of pillow basalt, serpentinite, gabbro and chert of the ophiolite basement. There is a total absence of rock lithologies that would indicate an underlying continental basement.

The sedimentary rocks that form the clasts were still soft and relatively unlithified before incorporation into the mélange. Deformation of the clasts is common and always pre-dated the mélange. The deformation can therefore be classified as hydroplastic, or soft-sediment deformation. The sandstone clasts have been brecciated and boudinaged. Vitrinite analysis of the Kulapis Formation indicates a range R_0=0.58–0.88% and the Labang Formation 0.53–0.88% (Clennell, 1992). The rocks that form the majority of the clastrs, accordingly, were not deeply buried.

XX.1.2.2. Palaeontology and age

Several sandstone, mudstone and dolomite clasts in the Garinono Mélange were analysed (Clennell, 1992). They were predominantly from the Telupid and Labuk road outcrops, and yielded the following palynomorphs: *Acrostichum*, *Alnipollenites venus*, *Brownlowia*, *Canthium* type, *Casuarina*, *Cyathidites*, *Dacrydium*, *Dicolpopollis*, *Florschuetzia semilobata*, *Florschuetzia trilobata*, *Graminae*, *Inaperturopollenites*, *Laevigatosporites*, *Lycopodium cernum*, *Lygodium phlegmaria*, *Myrtaceae*, *Palaquium*, *Polypodiisporites*, *Polydiisporites usmensis*, *Pteris* type, *Retitricolpites*, *Rhizophora*, *Spinozonocolpites echinatus* and *Zonocostites.* This flora represents an age range from Upper Eocene to Middle Miocene. Therefore the sedimentary clasts are of Kulapis or Labang Formation.

Many rounded blocks of gabbro, of size 1–5 m diameter, litter the hillside in the Segaliud oil palm Estate (Figure 76, locality 38A). Clennell (1992) concluded that the gabbro was intrusive into the mélange. However, Swauger et al. (1995) obtained an apatite fission-track age of 76.3 ± 22.9 Ma, showing that the gabbro blocks predate the mélange and represent clasts incorporated from the Neocomian ophiolite basement.

XX.2. AYER MÉLANGE

The Ayer Mélange is not well exposed. Clennell (1992) made use of road quarries near Lahad Datu. The mélange consists of broken-up beds of sedimentary and tuffaceous cover of the ophiolite basement. The sedimentary rocks are mostly referred to the Kalumpang Formation, a distal turbiditic part of the Labang Formation. The mélange contains broken-up volcanic arc material both andesitic and dacitic. This makes it distinctly different from the Garinono Mélange. Clearly the Ayer Mélange formed adjacent to the Dent volcanic arc which was active at the time of rifting.

XX.2.1. Lithology

The matrix is composed of illite, smectite and chlorite. The clasts and broken beds are of all lithologies of the basement ophiolite and its overlying sedimentary strata, as well as of an active volcanic arc. Bedded tuff beds have been broken and incorporated. The igneous blocks are of gabbro, pillow basalt, serpentinite, basaltic agglomerate and dacitic agglomerate. Radiolarian chert is common and marble clasts also occur.

XX.2.2. Palaeontology and age

Haile and Wong (1965) reported a good Foraminifera fauna from a variety of matrix rocks collected in the southern Dent Peninsula. The lithologies are calcareous mudstone, calcareous tuff, calcareous shale, crystal tuff (both andesitic and dacitic), mudstone, calcarenite, tuffaceous limestone and pebbly mudstone: *Amphistegina* sp., *Austrotrillina howchini* (Schlumberger), *Cycloclypeus* sp., *Globigerina bulloides* d'Orbigny, *G. subcretacea* Lomnicki, *G. dissimilis* Cushman and Bermudez var., *G. binaiensis* Koch, *G. subcretacea* Lomnicki, *G. tripartita* (Koch), *Globigerina.* spp., *Globigerinoides bispherica-glomerosa*, *G. diminuta* Bolli *G. subquadrata* Bronnimann, *G. glomerosa glomeroso* Blow, *G. triloba* group, *G. rubra* (*?*) group, *G. sacculifera* (Brady), *Globigerinoides.* spp., *G. altispira* (Cushman and Jarvis), *Globoquadrina altispira* (Cushman and Jarvis), *G. tripartita* Koch, *Globorotalia opima nana* Bolli, *G. fohsi fohsi* Cushman and Ellisor, *G. fohsi barisanensis* LeRoy, *G. mayeri* Cushman and Ellisor, *Globorotaloides suteri* Bolli, Heterostegina sp., *Lepidocyclina* spp., *Lepidocyelina.* (*Eulepidina*) sp., *Lepidocyelina.* (*Nephrolepidina*) sp., *Miogypsina* sp., *Nummulites* sp., *Operculina* spp., *Orbulina suturalis* Bronnimann, *O. bilobata* (d'Orbigny), *Rotalia* sp., *Sphaeroidinella multiloba* LeRoy, *S.* sp. and *?Spiroclypeus* sp. This fauna is well defined as extending from Lower Miocene to Middle Miocene.

The Ayer Mélange is stratigraphically overlain by well-bedded sandy to conglomeratic Libong Tuff of Upper Miocene age (Haile and Wong, 1965).

XX.3. KUAMUT MÉLANGE

The Segama ophiolite underlies the Kuamut mélange, has been imbricated and blocks of it incorporated into the Lower to Middle Miocene mudstone matrix. Accordingly there is a range from broken beds to mudstone-matrix mélange. Leong (1974) experienced difficulty in mapping the area and stated that it was difficult to ascertain whether a large block, especially of chert, was an inlier of the underlying ophiolite and chert cover, or a block incorporated into the mélange mudstone matrix. Accordingly it must be stressed that the Kuamut Mélange also includes broken formations and therefore is not to be regarded as a formation. Collenette (1965) referred to the mélange as slump breccias. The Kuamut Mélange is overlain by the Tanjong Formation.

XX.3.1. Lithology

The matrix is chlorite-and illite-rich. It contains clasts of a range of size of sandstone, siltstone, mudstone, tuffaceous sandstone and mudstone, chert, limestone breccia, micritic limestone, serpentinite, pillow basalt, basaltic agglomerate and ophicalcite (Clennell, 1992). Closely associated with the mélange is a bedded sequence of shale and siltstone, commonly tuffaceous, that has yielded the same age Foraminifera as the matrix of the mélange. This is accordingly interpreted as the Lower to Middle Miocene sequence that overlies the ophiolite basement and has been broken and incorporated into the mélange matrix.

Many of the limestone clasts have yielded Foraminifera of Upper Cretaceous to Eocene age. The cherty clasts yielded Radiolaria that are non-diagnostic (Leong, 1974).

XX.3.2. Age and palaeontology

Leong (1974), Kirk (1962) and Collenette (1965) recorded the following very rich age-diagnostic Foraminifera fauna from the mudstone matrix of the mélange.

The following were found in both mélange matrix and also in associated bedded rocks: *Ammobaculites* spp., *Ammodiscus* spp., *Amphistegina* sp., *Bathysiphon* spp., *Bolivina* sp., *Cassidulina* sp., *Cibicides* spp., *Cyclammina amplectens*(?) (Gryzbowski), *C. cancellata* Brady, *C. minima* LeRoy, *Epinoides* spp., *Gaudryina* spp., *Globigerina binaiensis* Koch, *G.* cf. *ciperoensis* Bolli, *G. dissimilis* Cushman and Bermudez, *Globigerinoides bisperia* Todd, *G. rubra* (d'Orbigny), *G. subquadrata* Bronniman, *G. trilobata* (Reuss), *Globorotalia mayeri* Cushman and Ellisor, *Globoquadrina altispira* Cushman and Jarvis, *G. venezuela* Hedberg, *Glomospira* spp., *Gyroidina soldanii* (d'Orbigny), *Haplophragmoides* sp., *Lagena* spp., *Nodosaria* spp., *Psammosphaera placenta* (Grzybowsky), *Pullenia bulloides* (d'Orbigny), *Rhizammina* sp., *Sphaeroidina bulloides* d'Orbigny, *Trochammina renzi* nom. nov., *Trochamminoides* spp., *Uvigerina* spp., and *Valvulina* sp.

The following have been found so far only in the mélange matrix: *Angulogerina* sp., *Anomalina* sp., *Bigenerina* spp., *Bulimina* sp., *Chlostomelloides* sp., *Cibicides praecinatus* (Karrer), *Clarulina* sp., *Cristellaria* sp., *Cyclammina* spp., *Elphidium koeboense* LeRoy, *Epinoides umbonatus* (Reuss), *Epistominella pulchella* Husezima and Maruhasi, *Euuvigerina notohispida* (Finlay), *Glandulina* sp., *Gaudryina* spp., *Globigerina obesa* Bolli, *G. perai*(?) Todd, *G. opima nana* Bolli, *G. sellii* (Borsetti), *G.* cf. *tripartite* Koch, *G. unicava primitive/pera*(?), *Globigerinoides diminuta* Bolli, G. *sacculifera*(?) (Brady), *G. triloba* group, *Globorotalia kugleri* Bolli, *G. scitula* (Brady), *Haplophragmoides carinatum* Cushman and Renz, *H. deformis* (Andreas), *H. narivaensis* Bronnimann, *H. walteri* (Grzybowsky), *Harmosina* sp., *Itanzawaia sumitomoi* Asano and Murata, *Lagenammina* sp. *Liticarinata* sp., *Nonion pompilioides* (Fichtel and Moll), *Psammosphaera fusca* (Schulze), *Pleurostommela* sp., *Quinqueloculina* sp., *Rectoglandulina laevigata* (d'Orbigny), *Reophax* sp., *Rotalia* sp., *Sigmoilina* sp., *Siphogenerinoides* sp., *Sphaeroidinella multiloba* LeRoy, *Spiroloculina* sp., *Spiroplectammina* sp., *Textularia* sp., *Triloculina* sp., *Tubulogerina* sp., *Uvigerina hantkeni* Cushman and Edwards, *Vaginulin* sp., *Verneuilina* sp., and *Virgulina* sp.

The following so far have been found only in the associated bedded rocks: *Ammodiscus grzybowskii* Emiliani, *Catapsydraz dissimilis* (Cushman and Bermudez), *Clavulina* sp., *Discocyclina* sp., *Eggerella bradyi* (Cushman), *Frondicularia* sp., *Globigerina* cf. *angulisuturalis* Bolli, G. cf. *ciperoensis* Bolli, *G.* diminuta Bolli, G. parva Bolli, G. subcretacea Chapman, G. venezuelana Hedberg, *Globigerinatella insueta* Cushman and Stainforth, *Globigerinoides glomerosa* Blow, *G. triloba* (Reuss), *Globorotalia centralis* Cushman and Bermudez[reworked], *G. cerroazulensis Cole* [reworked], *G.* cf. *Fohsi barisanensis* LeRoy, G. cf. *opima nana* Bolli, *G. scitula* Brady var., *G. spinulosa* Cushman [reworked], *G. wilcoxensis* Cushman and Ponton [reworked], *Globoquadrina dehiscena* Cushman Parra and Collins, *Haplophragmoides walteri* (Gryzbowsky), *Heterostegina*(?) sp., *Kalamopsis grzybowskii* (Dylazanka), *Lepidocyclina* sp., L. (*Eulepidina*) spp., *Miogypsinoides* sp., *Operculina* spp., *Osangularia walteri* (Parker and Jones), *Porticulasphaera transitora* (Blow), *Psammosiphonella*(?) *latissima* (Grybowsky), *Recurvoides deformis* (Andrea), *Trifarina* sp. and *Trochammina* sp.

The great similarity both of fauna and age between the mélange matrix and the associated bedded rocks shows that the bedded rocks were those disrupted to form the mélange. This fauna indicates earliest Lower Miocene to Middle Miocene. However, there are samples which contain reworked Middle Eocene Foraminifera such as *Gaudryina* sp., *Bigenerina* sp., *Trochamminoides subcoronata* (Rzehak and Grzybowsky), *Globigerina boweri* and *Globorotalia crassa.*

Balaguru (2001) lists the following nannofossils from the Kuamut Mélange, which consistently indicate an Upper Oligocene age: *Braarudosphaera bigelowii*, *Coccolithus* sp., *Coccolithus pelagicus*, *C. eopelagicus*, *C. miopelagicus*, *C. pelagicus*, *Cyclicargolithus abisectus*, *C. floridanus*, *Dictyococcites bisecta*, *Discoaster deflandrei*, *D. saipanensis*, *D. tani*, *Ericsonia farmosa*, *Helicosphaera intermedia*, *Helicosphaera* sp., *H. recta*, *Pontosphaera plane*, *Sphenolithus ciperoensis*, *S. distentus*, *S. microformis*, *S. moriformis*, *S. predistentus*, *S. pseudoradians*, *S. radians*, *S. reticulofenestra umbilica* and *Thorasosphaera* sp.

Chapter XXI

Tanjong Group 'Circular Basins'

A number of 'circular basins' occur on a SW line from Sandakan to the Meliau Basin, thence through the Malibau Basin into the Tarakan Basin of Kalimantan. They have been sedimented, predominantly in the Middle Miocene, by quartz-rich material traditionally thought to have been derived from erosion of the uplifting Western Cordillera, as well as volcaniclastic material from the Semporna Volcanic Arc. The basins occupy the landward extension of the SE Sulu Sea Rift (Hutchison, 1992a). Generally the strata are ascribed to the Tanjong Formation, but local names are applied. The Sandakan Formation basin is younger, predominantly Upper Miocene, but it is convenient to include it in the Tanjong Formation, with which it shares many characteristics, including sedimentation pattern. Although now separated into individual 'circular basins', the Sandakan Formation and Tanjong Formation must have originally been continuous throughout the Central Sabah Basin. Their subsequent separation into individual basins of circular nature has not satisfactorily been explained. The basin margins have steep dips, while the basin centres are flat-lying. Strike patterns are circular or elliptical. The most readily accessible and therefore the best known is the Sandakan Basin, whose magnificent outcrops dominate the topography of Sandakan town and the surrounding area. Unfortunately it is not the best example of the circular basin phenomenon.

XXI.1. SANDAKAN FORMATION

This Upper Miocene formation dominates the eastern Sandakan Peninsula of Sabah. There are two main lithologies: sandstone and mudstone. The structure is simple, with a predominant N–S strike and westwards dip (Figure 100). The sandy facies forms impressive cuestas that dominate the immediate surroundings of the town. The lithologies allow an interpretation of shallow marine to deltaic. The main descriptions are by Lee (1970) and Noad (1998). The formation sits unconformably upon Garinono Mélange (Lee, 1970). At Tanjong Papat, on the coast beneath the Mosque, Oligocene(?) andesitic tuff is exposed. It is overlain by the Sandakan Formation, and has been correlated with the volcanics of the Cagayan Ridge of the Sulu Sea.

Although the Sandakan Basin (*sensu stricto*) is monoclinal, its outliers to the west form isolated 'circular basins', notably the Bidu-Bidu and Manjang basins, which Noad (1998) recorded as outliers of the main Sandakan Formation. They are both less than 10 km in diameter.

XXI.1.1. Lithologies

A comprehensive study allowed Noad (1998) to describe the Sandakan Formation as comprising several lithofacies:

Mudstone facies. It is made of thick cohesive dark-grey mudstone with abundant fossil content. The beds are highly carbonaceous and occasionally contain coal beds up to 5 cm thickness. Scattered sideritic nodules are common. The mudstones contain abundant logs up to 2.5 m length, rooted trees and carbonaceous detritus and a brackish to marine microfauna. Scattered rounded amber (damar) clasts are common. They have yielded spiders, ants and other insects. The mangrove lobster *Thalassina* sp., up to 10 cm length, has been described (Noad, 1998). A brackish water mangrove environment is interpreted.

Channelized trough cross-bedded sandstones. The sandstone is fine-grained to very fine grained, forming beds ranging from 1 to 14 m thickness, interbedded with the Mudstone facies. The sandstone bases are channelized, incising into the underlying grey mudstone with a relief up to 2 m. The channels pinch out laterally and vary in width from a few metres to 150 m. The channels are not stacked and occur singly. They have been measured to trend N–S to NW–SE. The sandstone is usually trough cross-bedded. The main fossils are trace fossils, including a bird footprint The channelized sandstones are interpreted as fluvial.

Thinly interbedded sandstone and bioturbated mudstone. The interbeds are only 0.5–5 cm thick. The sandstone beds are silty and contain abundant mud drapes. The mudstone beds are well bioturbated. The environment is interpreted as a mixed to muddy tidal flat.

Thick stacked sandstone sequences. These dominate the southern and eastern Sandakan Peninsula (Figure 100). They resist erosion and form large scarps reaching more than 100 m in height. The sandstones are both trough cross-bedded and planar cross-bedded. Trough cross-bedded sands are individually 30–50 cm thickness. Planar cross-bedding is less common. The planar sets are on a metre scale and dip at up to 30°.

The sandstones grade up from medium fine grained or very fine-grained. Rippling is common. Paaeocurrent directions trend dominantly to the WNW. Fossils are generally absent; trace fossils abundant. There is a dominance of *Ophiomorpha* burrows. The thick sandstone deposits are interpreted as sub-tidal middle to upper shoreface. A high rate of sediment supply is implied.

Thin to medium bedded sandstone with abundant Skolithos. This facies is of stacked sandstone beds; interbedded mudstones are usually absent. The sands are thinly bedded. The most common trace fossils are the small vertical tubes of *Skolithos*. The environment is thought to have been on the foreshore or a shoaling part of the shelf (Noad, 1998).

Sandstone and mudstone interbeds. This is a very heterolithic facies. It is a set of interbedded very fine-grained sandstone, siltstone and mudstone. The sandstone beds are less than 150 cm and are packed with carbonaceous material. Rippling is

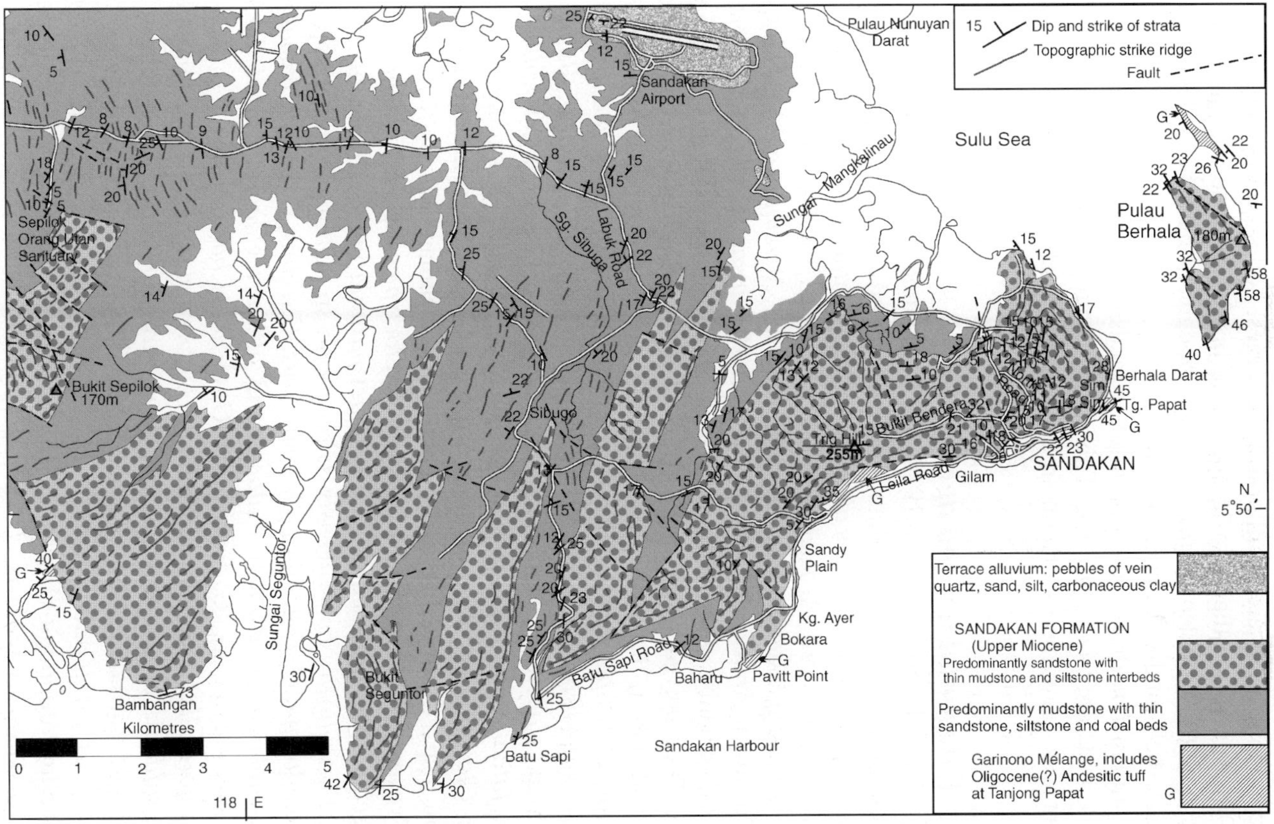

Figure 100. Geological map of the type area of the Upper Miocene Sandakan Formation (after Lee, 1970). The sand-dominant facies forms the higher topography and cuestas and strike ridges. The strata dip predominantly westwards forming a monocline. It is possible that normal faulting causes some stratal repetition. With permission from Minerals and Geoscience Department, Malaysia.

common. Noad (1998) regards these deposits as tempestite, representing abrupt changes in energy level.

Mudstone rich in crabs. Thick grey mudstone with occasional thin siltstones. The most common fossils are small crabs, the largest collected being 7 cm across (Noad, 1998). In addition there are 15 species of bivalves. These deposits resulted from quiet conditions of deposition. They are thought to have been deposited in shallow inner-shelf mudstone environments.

Finely laminated mudstone. Grey featureless mudstone with some thin impersistent siltstone beds only 1–5 cm thick. This facies occurs mainly in the north. The environment was a low energy one on an open marine shelf.

XXI.1.2. Palaeontology and age

Lee (1970), Noad (1998) and Ujiié (1970, 1977) obtained the following rich fauna of Foraminifera from the Sandakan Formation mudstones: *Adercotruma* sp., *Alveovalvulina pozonensis*, *Ammocabulites* sp., *A. strathernsis*, *Ammomargulina* sp., *Ammonia* cf. *beccarii*, *A. koeboeensis* (LeRoy), *A. sandakanensis* n. sp., *A. shutoi* nov. sp., *Amphistegina* sp., *Angulogerina* sp., *Anomalia ammonoides*, *A. glabrata* Cushman, *Arenobulimina* sp., *Articula* aff. *lineate* Brady, *Asterorotalia trispinosa* (Thalmann), *Bathysiphon* sp., *B. arenacea*, *B. filiformis* M. Sars, *Bolivina* sp., *B.* cf. *amygdalaeforme* Brady, *B.* cf. *canullata*, *B. elliptica* nov. sp., *B. formosana* Nakamura, *B. longicostata* LeRoy, *B. plicata*, *B. robusta* Brady, *B.* cf. *schwagerrana*, *B.* cf. *spathula* (Williamson), *B. subaenariensis mexicana*, *B. tenuistriata*, *B.* cf. *tikutoensis* Nakamura, *B. victoriana* Cushman, *Bulimina* sp., *Bulimina* sp., *Cancris* sp., *Cancris panamaensis*, *Cassidulina* sp., *C. crassa*,*Cellanthus* sp., *Cibicoides falconensis*, *C. praeciactus* (Karrer), *C.* aff. *Pseudoungerianus* (Cushman), *Clavulina* sp., *Cribrononion hispidum* Ujiié, *C. advenum* (Cushman), *C. hispidum* n.sp., *C.* aff. *Hughesi foraminosum* (Cushman), *C. multicameratum* nov. sp., *Cyclammina cancellata*, *C. planus*, *Cycloclypeus* sp., *Elphidium* sp., *E. crat.*, *Entoselenia* sp., *Epistomina* sp., *Epinoides paratilarum*, *E. praecintus*, *E.* cf. *praecintus*, *Flosculinella bontangensis*, *Globigerina* sp., *Globigerinoides* spp., *G. ballii*, *G. conglobata*, *G. conglobus*, *G. immaturus*, *G. rubra*, *G. triloba immature*, *G. trilobus*, *Globoquadrina altispira altispira*, *G. altispira globosa*, *Globobulimina perversa*, *Globorotalia* sp., *G. fohsi fohsi*, *G.* cf. *fohsi barisanensis*, *G. mayeri*, *Glomospira gordialis*,*Gyrinoides planulatus*, *G. venezuelana*,*Gyroidina nipponica* Ishizaki, *G. soldanii*, *G. suturalis* nov. sp., *Haplophragmoides* sp., *H. carinatus*, *H.* cf. *carinatum*, *H. carinatum* Cushman and Renz, *H. coronatum*, *H. emaciatum*, *H. naraivaensis*, *H. obliquicameratus*, *H. sphaeriloculous*, *H.* cf. *wilsoni*, *Hippocrepinella carapitana* (Hedberg), *Hormosina* sp., *Karrierella microgranulosa*, *Lagenonodosaria scalaris*, *Lenticula* sp., *Lenticulina americanus* sp., *L. americanus* var. *grandis*, *L. asanoi* nov. sp., *L. gemmeta*, *Lepidocyclina* (*Nephrolepidina*) sp., *Loxostomum* sp., *L. amygdalaeformae*, *L. karrerianum*, *L.* aff. *limbatum*, *Margulina* cf. *superba*, *Martinotiella* sp., *Millammina* sp., *Miogypsina* spp., *Neoeponides berthelotianus* (d'Orbigny), *Nodosaria insecta, N. longiforma, N. vetebralis, Nonion incisum, Operculina* sp., *Orbitulina* sp., *O. suturalis, O. universa, Ostrachoda* sp., *Planulina* sp., *Planularia gemmata* (Brady), *Plectofrondicularia californica, Pseudorotalia borneensis* nov. sp., *Rectobolivina bilfrons, R. multicosta* LeRoy, *R. viform* var., *Recurvoides contortus, Reussella striatula, Rhizammina* sp., *Robulus calcar, R. inornatus, R. orbicularis, , Robulus* sp., *Rotalia ceocardii, R.* cf. *nipponica, R.* cf. *papillosa, Sacammina* sp., *S. sphaerica, Spiropsammina* sp., *S. uhligi, Sponides berthelotiana, S. praecintus, Trochammina* sp., *T. globigeri, T. nanaformis, T. pacifica, T.* cf. *squamata, T.* cf. *variola, Uvigerina laevis, U. proboscidae, U. hispida, U.* cf. *schwagerina, U. soendaensis*, and *Virgulina exilis.*

The fauna ranges from Middle Miocene, but is overwhelmingly Upper Miocene. However, Ujiié (1977) stated that the fauna he described is of Middle Miocene age. He further noted that the Foraminifera collected from rocks along the south coast are brackish water forms. Those collected along the Labuk Road latitude are inner neritic, while those from the Sungai Manila region are outer neritic.

The Sandakan Formation also yielded an impressive range of macrofossils, indicating an environment ranging from shallow to open marine (Noad, 1998):

Gastropods: *Bursid, Batillaria,* ribbed *Cerithid, Muricids, Thaidae, Nerita, Naticidae, Cerithiacea, Atys, Turridae, Cassidae, Phalum, Conus, Cyprea* (Cowrie), *Terebridae, Olividae, Pyrazius aucernor, Terebralia, Vicarya* and *Vicarya verneuli,*

Bivalves: *Corbiculidae, Oyster, Hiatula, Venerid, Pitar, Teredinae, Myclid, Pectinid, Chlamys, Plicatula, Tellinidae, Teredinae, Mactrodae, Nucula, Glycimeris, Dosinia, Brechites, Cardiola, Modiolini, Tapatini, Psamm, Mytilidae, Veneracea, Dendrostrea, Fimbria, Lucinidae, Nucula, Arcadae, Anomalocardia, Cardidae* and *Anadra.*

Argonauta: *Izumonauta* sp. nov. (indicates deep-marine conditions).

Annelida: Keeled worm tubes and *Serpula.*

Crustacea: Ranina, Thalassina, Raninoides, Portunus woodwardi, Portunus obvalatus, Dorippe, Charybdis, Portunus sp., Xanthid sp., Parthenope, Calappa, Iphiculus sexspinosus, Pariphiculus, Pagurid, Costacapluma, Retroplumid, Pinnixia, Amplieura, Oxyrynch, Opthalmaplax, spider crab species, Myra, Leucosia, Nucia, and Typilobus.

Insecta: Platygastroidea, Formicidae, Cecidomyiidae, Thysanopteridae and Arachneida.

Vertebrata: *Trionyx, Coprolite, Plover* (?) footprints and Fish vertebra.

Palynology: The following palynomorphs have been recorded by Clennell (1992) from the Labuk Road outcrops, and also from two outliers at km 3 on the Sungai Manjang road and on the Api road: *Acrostichum, Alangium* sp., *Alnus, Anacolosa, Arenga, Avicennia, Barringtonia, Brownlowia* sp., *Canthium, Casuarina, Cephalomappa* sp., *Cicatricosisporites* sp., *Crudia* sp., *Dacrydium, Dactylocladus* sp., *Dipterocarpus, Durio, Ephedra, Eugeissona minor* sp., *Ficus, Florschuetzia trilobata, Florschuetzia semilobata, Graminae, Hibiscus* sp.,

Inocarpus sp., *Laevigatosporites*, *Longetia*, *Lycopodium phlegmaria*, *Myrtaceae*, *Pandanus*, *Picea*, *Pinus* sp., *Podocarpus polystachyus*, *Polypodiisporites*, *Pometia*, *Rhizophora*, *Shorea* sp., *Sonneratia*, *Stemonunus*, and *Stenoclaena lauritolia*.

This flora indicates a Middle to Upper Miocene age.

Nannofossils. Clennell (1992) recorded the following Middle Miocene fauna from Sungai Manila: *Discoaster exilis* and other zonal correlatives.

XXI.1.3. Fission track and reflectivity studies

Two specimens, 32B and 36B, were collected from the Sandakan Formation sandstones by Swauger et al. (1995) and their apatite and zircon crystals analysed for fission-track dating. The localities were given in Figure 76 and the results are shown in Figure 101.

The Sandakan Formation has not been deeply buried and not therefore ever heated by burial, so that the apatite fission tracks were not annealed in the way that rocks of the Western Cordillera had been. The ages obtained for the apatites are similar to those of the zircons. Therefore these crystals preserve their provenance ages, which are predominantly Cretaceous with a suggestion of some Jurassic ages. The view of Hutchison (1992a) that the provenance of the Sandakan Formation sands must have been the uplifting and eroding Crocker and Trusmadi formations of the Western Cordillera can no longer be maintained in view of the fission-track data. It is much more likely that the fluvial system originated in the Schwaner Mountains of western Kalimantan that are dominated by Cretaceous granitoids and volcanic rocks (Hutchison, 1996a).

Three samples were analysed for kerogen microscopy by Wallace G. Dow (Swauger et al., 1995). The following results were obtained from low-rank coaly material in all three:

Sample	Number of readings	Interpreted maturity	Standard deviation
32A	35	Ro = 0.45	± 0.05
32D	30	Ro = 0.48	± 0.02
32E	30	Ro = 0.48	± 0.01

XXI.1.4. Palaeogeography

A large amount of palaeocurrent data were obtained in the field by Noad (1998). A more limited but complementary study was made by Stauffer and Lee (1972). Noad's interpretation is given in Figure 102, together with the measurements of Stauffer and Lee (1972). The palaeocurrent data fit the distribution of environments interpreted from the different lithofacies. The overall environment is a fluvial system entering the sea and being redistributed by longshore currents. Support also comes from the environmental indications of the benthonic Foraminifera (Ujiié, 1977). He showed that the fauna from south coast outcrops indicate coastal zone brackish water, fauna from the latitude of the Labuk Road indicate inner neritic conditions,

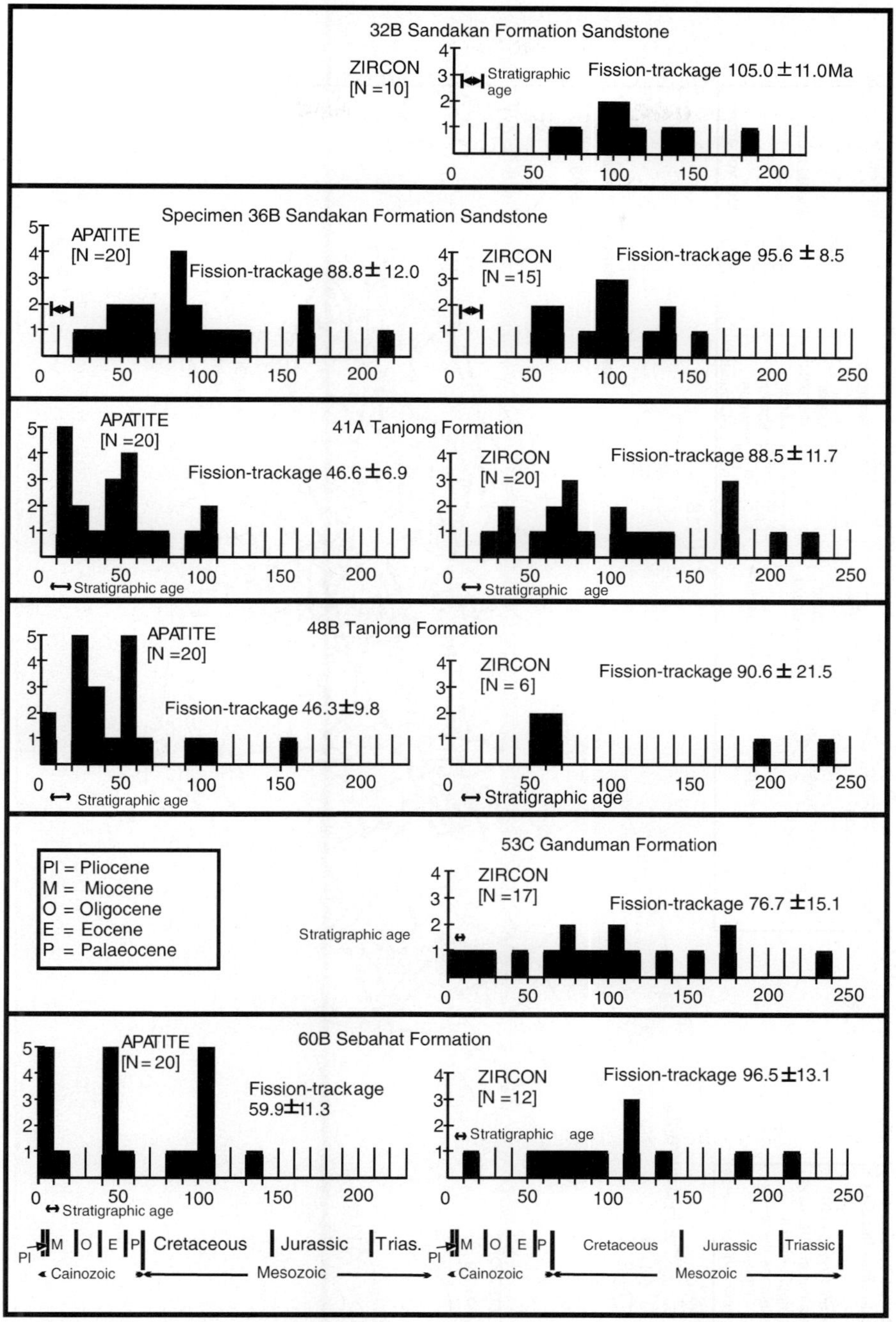

Figure 101. Fission-track data on the Sandakan, Tanjong and Dent Group formations. Redrawn from Swauger et al. (1995).

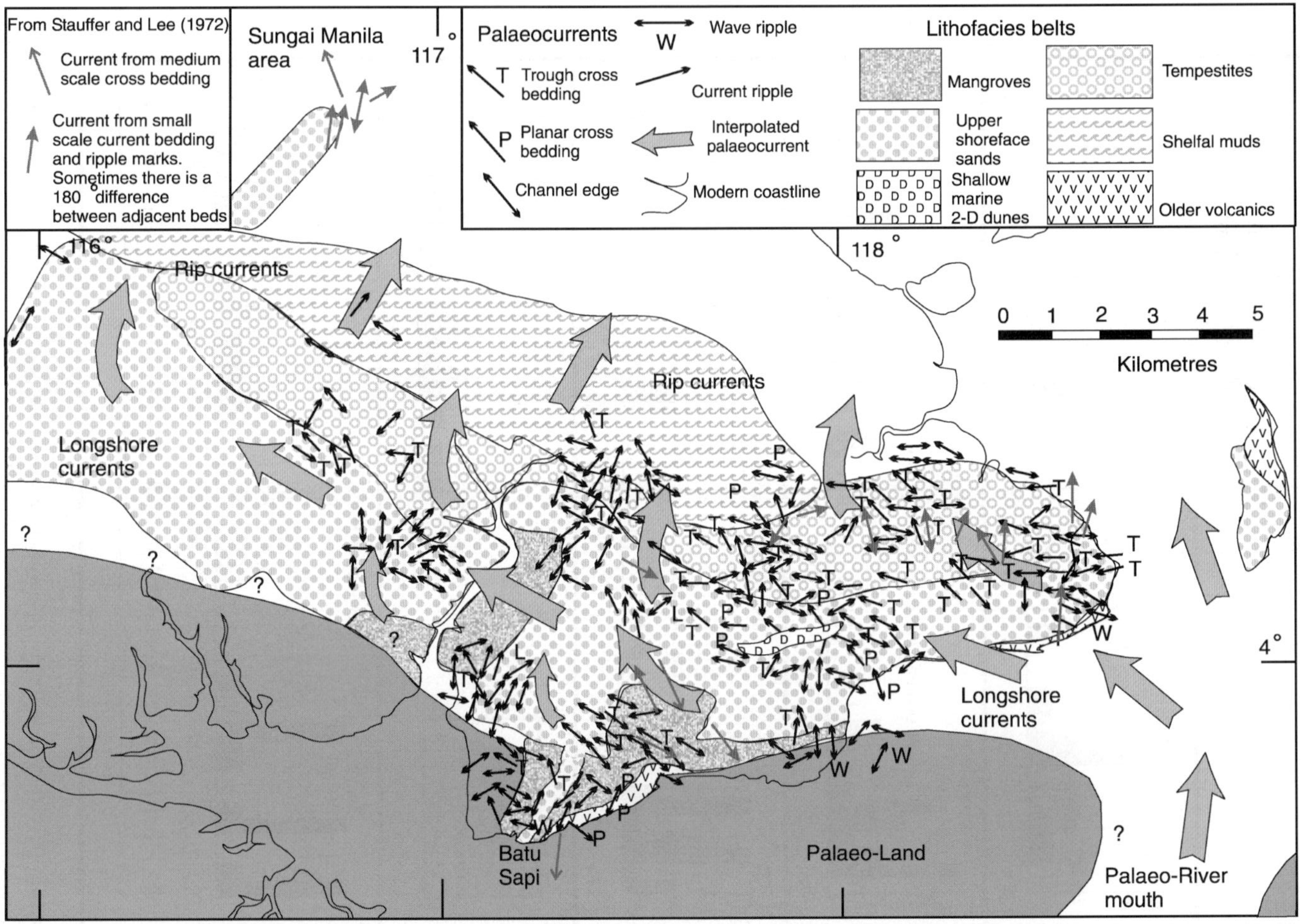

Figure 102. Interpreted palaeogeographical reconstruction of the Sandakan Formation based on measured palaeocurrent directions made by Noad (1998). Additional measurements by Stauffer and Lee (1972) are shown by different symbol.

and fauna from the Sungai Manila area to the north indicate outer neritic conditions. Sands also die out northwards away from the palaeo-coast. The implications are that the sands were transported along the Central Sabah Basin by a fluvial system originating far to the west. The fission track data suggest that the provenance for the sands may have been the Cretaceous outcrops of the Schwaner Mountains and not the uplifting Western Cordillera. The generally fine-to very fine-grained nature of the Sandakan Formation sandstones suggests such a long transport or sandstone provenance that itself is medium-to fine-grained, such as turbidites of the West Crocker outcropping in the Western Cordillera. Stauffer and Lee (1972) concluded that the fine-grained nature of the sandstones, with a total lack of coarse material, indicates no rugged relief and a low lying river entering the sea.

A good analogue for the Sandakan depositional environment is the rivers Klang and Langat entering the Strait of Malacca. A delta did not build out, and the sediments were distributed parallel to the coast by tidal currents, as described by Coleman et al. (1970).

XXI.2. BONGAYA FORMATION

The Bongaya Formation Basin is transacted by the Bongaya River that flows into Labuk Bay. The mainland occurrence is in the Bengkoka Peninsula. The same formation also forms the whole of Jambongan Island and part of Banggi Island. The formation has been mapped by Wilson (1961). The South Banggi Formation is included, so that the age range is Lower to Middle Miocene. The Kudat Formation forms part of the same depositional system.

Wilson (1961) concluded that the Bongaya Formation rests unconformably upon the Crocker Formation as well as upon the ophiolite. The structure changes from north (Banggi) to south (mainland). In the north the formation unconformably overlies the ophiolite. The strike is northerly and it is cut by N-trending fractures with down-faulted segments into the underlying ophiolite. Dips are westerly at 45°, between outcrops of brecciated and mylonitized basement ophiolite.

There is a wide basinal structure on the Benkoka Peninsula. Dips of 45–70° are directed inwards. The basin margins are displaced by transcurrent faults in the west. The formation is folded, with easterly trending axes becoming ESE southwards, where the dips seldom exceed 40°. On Jambongan Island two symmetrical synclines pitch ESE, with beds dipping 5–15°.

XXI.2.1. Lithology

XXI.2.1.1. South Banggi Formation

This formation has been mapped separately by Wilson (1961) on the basis of its distinct lithology at outcrops in Banggi Island as well as on the Bengkoka Peninsula. It is composed of grit and flagstone. The grit has grains up to 0.35 cm across. The grains are predominantly angular quartz, limestone and chloritized gabbro or basalt. There are also grains of serpentinite, chert and plagioclase. The strata lie unconformably upon a basement of ophiolite, which by Lower Miocene time was uplifted and contributing erosional material to the South Banggi Formation. This is a similar situation to the serpentinite conglomerates and sandstones of the eastern Bidu-Bidu Hills to the south (Newton-Smith, 1967; Hutchison and Tungah, 1991). The Formation also includes a limestone on the west coast of South Banggi (Wilson, 1961). At Turong Puteh on the mainland at Bengkoka Peninsula, the formation includes limestone, calcareous grit, sandstone and siltstone. The limestone contains larger Foraminifera, indicating a range from late Te to lower Tf.

XXI.2.1.2. Bongaya Formation (sensu stricto)

This is a sandstone–shale sequence. A broad synclinal structure on Bengkoka Peninsula trends SW. Thickly bedded or massive fine-to medium-grained sandstone gives way upwards to 0.7 m thick beds interbedded with shale. Cliffs of current-bedded buff quartz sandstone are well sorted and contain >90% quartz. Along the Bongaya River, 60–100 m of fine-grained sandstone is interbedded with up to 3 m thick grey shale beds. Thick sandstones, 150 m thick, are interbedded with siltstone containing coaly laminations. Pebbly grit contains bivalves and fragmented gastropods. Clasts of amber occur. Good exposures on Jambongan Island of clay, 10–15 cm thick, are interbedded with sandstones composed of >90% quartz. There are carbonaceous layers.

XXI.2.1.3. Balambangan Limestone Member

This pure Middle Miocene limestone is cliff forming.

XXI.2.2. Palaeontology and age

The outcrops mapped as the *South Banggi Formation* have yielded the following fauna (Wilson, 1961): *Austrotrillina howchini* (Schlumberger), *Biplanispira*, *Cycloclypeus* sp., *Gypsina* spp., *Lepidocyclina* sp., *L.* (*Eulepidina*) spp., *L.* (*Nephrolepidina*) spp., *Miogypsina* sp., *M. primitive* Tan Sin Hok, *Miogypsinoides abuensis* Tobler, *Neoalveolina* sp., *Operculina* sp., *Operculinella* sp., *Orbitulina* sp., *Rupertilidae* sp. and *Spiroclypeus sp.* This fauna is Lower Miocene but may range into the Middle Miocene (late Te but ranging to Tf).

The following are reworked Eocene Foraminifera: *Biplanispira*, *Discocyclina*, *?Pellatispira* and *Nummulites.*

The outcrops mapped as Bongaya Formation on Banggi and Balambangan islands have yielded the following Middle Miocene fossils (Wilson, 1961) (Tf): *Globigerina subcretacea* Chapman, *Globigerinoides subquadratus* Bronniman, *G. rubrus* (d'Orbigny), *Globoquadrina altispira* Cushman and Jarvis, *G. dehiscens* Chapman Parr and Collins, *Globorotalia mayeri* Cushman and Ellisor, *G. praemenardii* Cushman and Stainforth, *Orbulina bilobata d'*Orbigny, *O. suturalis*

Bronniman, *O. universa* d'Orbigny, *Sphaeroidinella multiloba* LeRoy, and *S. seminulina* (Schwager).

The outcrops mapped as the *Balambangan Limestone Member* yielded the following Middle to Upper Miocene (T_{1-3}) fauna: *Cycloclypeus* sp., *Gypsina* sp., *Lepidocyclina* sp., *Miogypsina* sp., *M. tuberosa* Tobler, *M. tuberosa* Tobler cf. *musperi* Tan Sin Hok, *Miogypsinoides* sp., *Operculina* sp. and *Orbulina* sp.

XXI.3. KUDAT FORMATION

The Kudat Formation occupies most of the Kudat Peninsula. It shows characteristics both of deep-and shallow-water deposition (Tongkul, 1994). It is composed of interbedded sandstones and mudstones with small Foraminiferal limestone and breccia lenses (Stephens, 1956). Small occurrences of andesite and agglomerate have been mapped at Tanjung and Pulau Sirar. The structure is not completely understood but Tongkul (in press) interprets it to be infolded with the Eocene Crocker Formation. The folds trend NW–SE across the Peninsula, parallel to the infolded Eocene Crocker Formation (Tongkul, in press).

XXI.3.1. Palaeontology and age

Stephens (1956) listed a fauna of Foraminifera that indicates an age extending from Lower to Upper Miocene: *Actinocyclina* sp., *Alveolina* sp., *Amphistegina lessonii* d'Orbigny, *Biplonispira* sp., *Camerina* sp., *Carpenteria* sp., *Cycloclypeus* sp., *Discocyclina* sp., *D.* (*Aktinocyclina*) sp., *Globigerina* sp., *Gypsina* sp., *G. globus* Reuss, *Heterostegina* sp., *H. borneensis* V.D.Vlerk, *Lepidocyclina* sp., *L. angulosa* (Provale), *L.* (*Eulepidina*) sp., *L.* (*Eulepidina*) *gluberculata* V.D.Vlerk, *L.* (*Eulepidina*) *dilatata* Mich., *L.* (*Eulepidocyclina*) sp., *L. flexuosa* Rutten, *L.* (*Nephrolepidina*) sp., *L.* (*Nephrolepidina*) *borneensis* Provale., *L.* (*Nephrolepidina*) *isolepidinoides* V.D. Vlerk, *L.* (*Nephrolepidina*) *sumatrensis* Brady, *L.* (*Trybliolepidina*) *transens* Umber., *L.* (*Nephrolepidina*) *verrucosa* Schlumberger, *Lockhartia* sp., *Miogypsina* sp., *M.* (*Miogypsinoides*) sp., *M.* (*Miogypsinoides*) *complanata* Schlumberger var. *bantamensis* Tan, *Miogypsinoides* sp., *M. ubagsi* Tan Sin Hok, *Neoalveolina* sp., *N. pygmaea* (Verbeek), *Nummulina* sp., *Nummulites* sp., *Operculina* sp., *Operculinella* sp., *O. venosa* F. & M., *Pellatispira* sp., *P. inflate* Umbgr., *Spiroclypeus* sp., *S. leupoldi* V.D. Vlerk, *S. pleurocentralis* (Carter), *S. tideonganensis* V.D. Vlerk, and *Trillina howchini* Schlumberger.

XXI.4. BUKIT GARAM BASIN TANJONG FORMATION

The Tanjong Formation was first described as a thick sequence of sandstones, mudstones and siltstones, with lenses of limestones and conglomerates (Collenette,

1965) but he was not describing the Bukit Garam Basin. Tham (1984) made one of the few early studies. A more recent and more comprehensive study of the Bukit Garam Basin was made by Noad (1998). This is a typical circular basin, and the main highway from Kota Kinabatangan to Lahad Datu traverses part of the basin. The Kinabatangan River traverses it and the river town of Bukit Garam lies within the basin. However, outcrops are good only along its tilted margins.

XXI.4.1. Lithology

Noad (1998) described nine different lithofacies which occur in the Tanjong Formation:

Thick swaley-bedded sandstones. Individual packets, generally 30–100 cm thick, occur towards the base of the formation, well seen in Sentosa and Malibau Estates, to the east of the main road between Kota Kinabatangan and the river. They form prominent sandstone ridges. They appear massive, but have some hummocky or swaley cross-bedding. Trace fossils such as *Ophiomorpha nodosa* are common. Amber clasts up to 10 cm also occur. The sandstones are considered to be storm deposits in water shallower than 50 m on the shoreface.

Laminated mudstone. Featureless mudstone beds may contain millimetre-scale siltstone beds. Pieces of wood up to 20 cm length occur. Amber clasts are common. These deposits are interpreted as shelf in a low-energy environment.

Interbedded sandstone–mudstone sequences with swaley bedding. The thick mudstone beds pass upwards into interbedded sandy siltstone and mudstone. The siltstone beds thicken and coarsen upwards to fine sandstone to a thickness of 50 cm. This sequence is followed by thick featureless mudstone. Amber clasts are common. They have yielded millipedes, ants and spiders. The environment is one of fluctuation. An inner shelf with strong storm currents is suggested (Noad, 1998).

Laterally extensive silty sandstone beds with grooved bases. These very fine-grained sandstone beds, 30–100 cm thick, are tabular with flat bases and tops, interbedded with mudstone. In the Sentosa Estate, the basal grooves indicate a palaeocurrent towards the NNE. An outer shelf environment, possibly with turbidity current origin, is suggested (Noad, 1998).

Channelized sandstone bodies with mud clasts. An example is well seen at Sentosa Estate and along the Kinabatangan River. The deeply incised Sentosa channel is around 15 m thickness, thinning to 4 m over a distance of 50 m. The channel fill is of fine-to medium-grained sandstone containing mudclasts. There are also intraformational mudflake conglomerates.

Contorted mudstones. These slumped mudstones overlie thick sandstones. At Sentosa they contain molluscs and crabs. They are slumped throughout their 12 m thickness. The overlying sandstone is undeformed. The muds were therefore deformed soon after deposition representing a debris flow mechanism.

Deeply incised channels. These channels cut down 900 cm over a distance of 1650 cm giving a channel edge slope of 30°. The fill is composed of large mudstone intraformational clasts up to 60 cm length, as well as well-rounded extraformational clasts up to 4 cm diameter of chert and vein quartz indicating ophiolite basement erosion. Limestone clasts also indicate erosion of Gomantong Limestone. The channels have lag deposits at the base.

Massive to thinly bedded siltstone. The siltstone occurs massive, ranging in thickness 20 cm, passing laterally into 2–5 cm thick beds of siltstone interbedded with grey mudstone. The thinner beds have rippled tops. The environment was low energy. Preserved leaves indicate an inner shelfal environment.

Thin interbedded siltstone and mudstone. The interbedded sequence is carbonaceous-rich. The trace fossils *Paleodictyon* and *Cosmoraphe* might suggest deeper water but the carbonaceous content may suggest proximal conditions. Noad (1998) generally concludes that the Tanjong Formation was deposited in a storm-dominated shelfal environment. The base of the formation is structurally conformable upon turbiditic Labang Formation at Sentosa Estate.

XXI.4.2. Palaeontology and age

The Bukit Garam Basin strata have been dated Lower Miocene to top Middle Miocene and is therefore the correlative of the Sandakan Formation, with which it shares similar lithologies and a similar environment of deposition. Tham (1984) listed the following fossils from the Bukit Garam Basin:

Foraminifera. *Adercotryma glomeratum, Amodiscus incertus, Amphistegina* sp., *Arenobulimina* sp., *Asterigerina, Bathysiphon* sp., *Bulimina alazanensis, Cornusphiroides primitives, Cyclammina cancellata, C. trullisata, Cycloclypeus postindopacificus* (Tan Sin Hok), *Dorothia* sp., *Eggerella* sp.,*Eponides* sp., *Globigerinoides quadrilobatus sacculifer, Globorotalia* sp., *Glomospira gordialis, Haplophragmoides coronatum, H. rotulatum, H. subglobosum, Heterostegina, Lepidocyclina, Miogypsina, Miolepidocyclina, Miogypsinoides, Nephrolepidina galliennei* (Lemoine and Douville), *Operculina ammonoides, Plectina* sp., *Plunuline wuellerstorfi, Pullenia bulloides, Pyrgo, Sigmoilina tenuis, Valvulina* sp. and *V. flexilus.* The Foraminifera fauna indicate an age range from uppermost Oligocene (Te_{1-4}) to the top of the Middle Miocene (Tf_{2-3}). It is doubtful if the formation extends downwards into the Oligocene.

Nannofossils. *Coccolithus* cf. *miopelagicus, Corococyclus nitescens, Coronocylus nitescens, Cyclicargolithus floridanus, Discoaster deflandrei, D. variabilis* sp., *Reticulofenestra pseudombilica, Sphenolithus* sp., *S.* cf. *belemnos, S. ciperoensis, S.* cf. *heteromorphus, S. moriformis* and *Umbilicoasphera* sp. The nannofossils indicate an age of early Middle Miocene, including some reworked Oligocene species.

Clennell (1992) presented palaeontology suggesting a Lower to Middle Miocene age. His localities are from Kota Kinabatangan and along the roads to Bukit Garam, Lahad Datu and a Sukau outlier:

Foraminifera. *Barringtonia, Bathysiphon* sp., *Cyclammina* sp., *Haplophragmoides narivaensis* (Bronnimann), *Psammosiphonella carapitana* Hedberg, and *Trochammina* spp.

Palynomorphs. *Acrostichum*, *Alangium*, *Anacolosa*, *Arenga*, *Brownlowia* sp., *Casuarina*, *Cephalomappa* sp., *Crudia* sp., *Cyclophorus* sp., *Dactylocladus* sp., *Dipterocarpus*, *Durio*, *Florschuetzia trilobata*, *Graminae*, *Laevigatosporites*, *Lycopodium cemuum*, *L. phlegmania*, *Pinus* sp., *Podocarpus polystachyus*, *Polypodiisporites*, *Pometia*, *Rhizophora* sp. and *Spinozonocolpites echinatus*. An age of Lower to Middle Miocene is the most likely.

XXI.4.3. Fission track and maturity data

Two samples from the Bukit Garam Basin (Figure 76) were collected for fission track analysis by Swauger et al. (1995). The results are shown in Figure 101. The zircon ages (88.5±11.7 and 90.6±21.5 Ma) indicate that the crystals have maintained their provenance ages. The crystals grouping <60 Ma might indicate derivation from outcrops of Rajang Group. The Trusmadi Formation could well have supplied the clasts of vein quartz in Tanjong Formation channel conglomerates. Crystals as old as Triassic could well have been derived from the western or Sundaland part of Kalimantan, either directly or via cannibalization of intermediate age sandstones.

The apatite ages (46.6±6.9 and 46.3±9.8 Ma) suggest partial annealing caused by enough burial to cause the ambient temperature to have reached around ~80°C. Only specimen 42B was analysed for kerogen maturity (Swauger et al., 1995): $n = 29$ readings, maturity $R_0 = 0.54\% \pm 0.05$ standard deviation. The value appears to be valid but a small percentage of the vitrinite may by lipid-rich. These values are marginally higher than those obtained for the Sandakan Formation, indicating fractionally deeper burial.

XXI.4.4. Paleocurrent data

Noad (1998) presented only a limited palaeocurrent analysis of the Bukit Garam Basin.

Bidirectional features			Unidirectional features	
Grooves (n = 20)	Channel edges (n = 8)	Wave ripples (n = 10)	Current ripples (n = 4)	Flutes (n = 4)
30–210°	85–265°	25–205°	40°, 60°, 245°	155°

From these measurements, and the facies distribution, he presented a model of land and nearshore deposits lying on the west, with channels directed from it in an easterly direction into deeper water.

XXI.5. ORIGIN OF CIRCULAR BASINS

The varieties and possible origin of circular geological structures has been reviewed by Stewart (1999). None of his categories exactly fits the remarkable circular basins of Sabah, of which he was apparently unaware. However there are aspects of his categories and mechanisms that are attractive and likely to be applicable to Sabah. Pull-apart basins in an extensional regime are characteristically associated with strike-slip faults that show as flower structures on seismic records. Shale/mudstone diapirism is also an attractive mechanism. A combination of both seems most appropriate.

A seismic section across the Pad Basin, some 45 km due east of Sandakan, throws some light on the possible origin of circular basins (Leong and Azlina, 1999). The section (Figure 103) shows a pronounced unconformity between the flat-lying Pliocene Togopi Formation and the underlying Middle to Upper Miocene Tanjong Formation. Considerable sandstone erosion took place during this unconformity.

The Tanjong Formation was deformed into a typical circular basin morphology, then uplifted, before deposition of the Dent Group Togopi Formation. There are NE–SW trending wrench related flower structures (Figure 103). The anticlinal

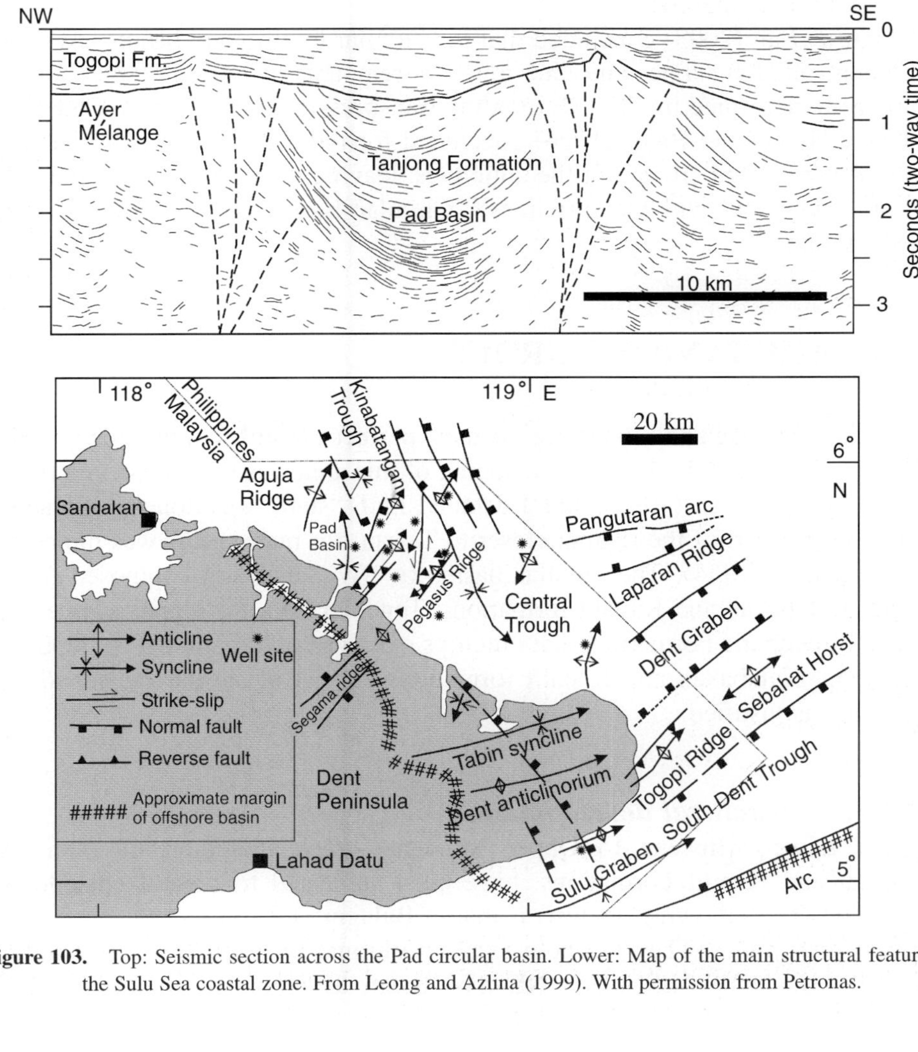

Figure 103. Top: Seismic section across the Pad circular basin. Lower: Map of the main structural features of the Sulu Sea coastal zone. From Leong and Azlina (1999). With permission from Petronas.

flower-structured areas were modified by cores of mobile diapiric mudstone from the underlying Ayer Mélange. These features are interpreted as compressive wrench anticlines (Leong and Azlina, 1999). These compressive wrench and diapiric events resulted in uplift and erosion of the Upper Miocene Ganduman Formation and upper section of the Sebahat Formation in the region NW of the Pegasus Ridge.

The resulting Pad circular basin (Figure 103) has steep dips along its margins that are bounded by the anticlinal wrench-fault and diapiric areas. Underlying mudstone is withdrawn from beneath the sand-dominated Tanjong Basin into the anticlinal areas, and regional uplift helps in the foundering of the sand strata downwards. The Pad Basin is similar to onland Tanjong Formation basins, except that the good seismic (Figure 103) shows the whole morphology. The strata are flat-lying only in a restricted central area, and dips steepen towards the bounding wrench–anticlinal–diapiric areas.

Balaguru et al. (2003) concluded that the circular to elliptical basins are structurally controlled synclines and interpreted as remnants of a single large Tanjong Group basin, deformed in NW–SE trending transpressional fault zones. Chambers and Daley (1997) and Bates (1996) concluded for the Kutai Basin that inversion and uplift has caused erosion of the anticlines, leaving the original synclinal remnants as "synclines without anticlines". Circular basins are synclines without related anticlines.

XXI.6. THE TANJONG GROUP

The Meliau, Malibau and Sibuda are the most perfect examples of circular or elliptical basins (Figure 104). The early mapping of this area was by Collenette (1965), followed by Tjia et al. (1990), and a detailed Ph.D. study was done by Balaguru (2001). The SE part of the region, closer to Tawau, for many years was mined for coal (Collenette, 1954). The circular basins of SW Sabah are composed of the Tanjong, Kalabakan and Kapalit formations. Together with the Upper Miocene to Pliocene Simengaris Formation, that outcrops over a small area in SE Silimpopon, the Tanjong, Kalabakan and Kapalit formations consist of similar facies, so that group status is justified.

XXI.6.1. Structural development

The late Lower Miocene to Upper Miocene strata, including the Tanjong, Kalabakan and Kapilit Formations, have been subjected to several episodes of deformation. The following analysis is that of Balaguru (2001).

Extensional faults. They are both syn-depositional and post-depositional, and trend NW–SE, WNW–ESE, and NE–SW. Basin development was effected by

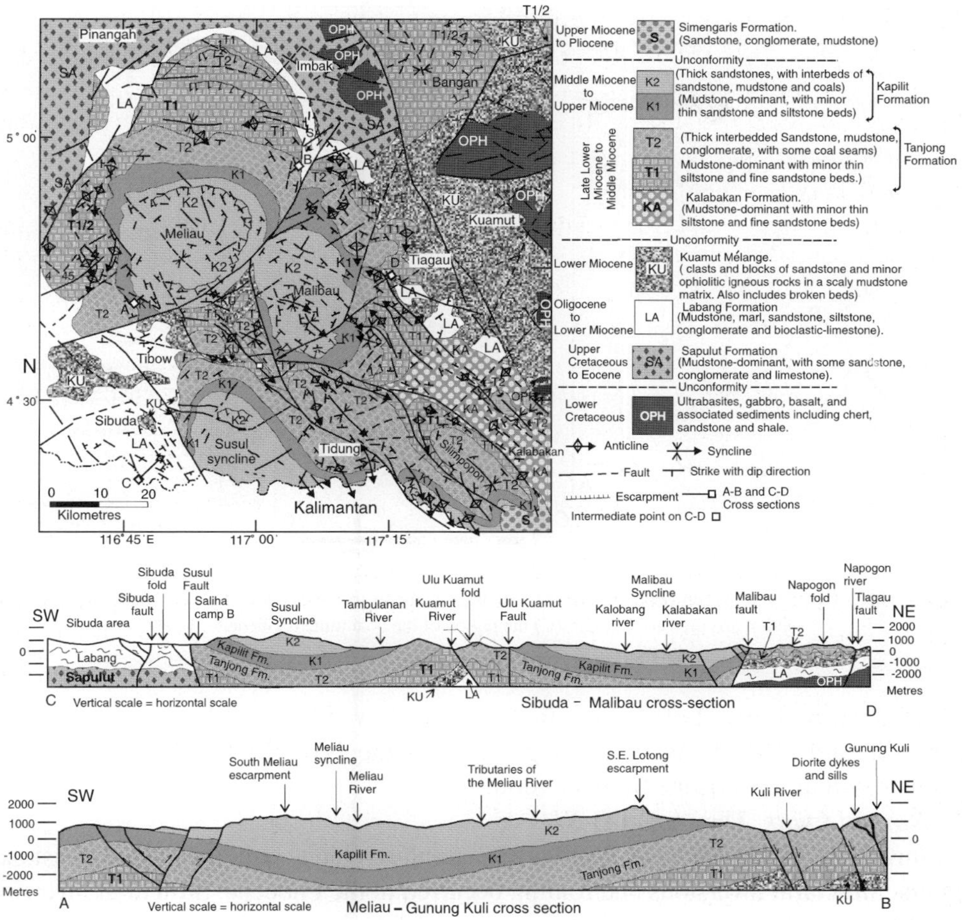

Figure 104. Geological map of the Kapilit and Tanjong formations of the Meliau and Malibau circular basins of SW Sabah. Redrawn after Collenette (1965); revised by Balaguru (2001).

syn-depositional extension up to the late Miocene. These faults occur locally but also are important on a regional scale (Figures 104 and 105).

Fold-fault association. Major folds, such as the Silimpopon Syncline trend NW–SE, sub-parallel to the major faults. Minor folds trend N–S to NNW–SSE and are formed oblique to the main fault. Narrow bands of intense deformation and tight anticlines occur between the broad synclines. Local transpression and transextension structures have been observed.

It appears that left-lateral strike-slip fault movement and transpression produced positive flower structures during the Pliocene. The Sun Oil seismic section (Figure 105) shows a good flower-structured fault, indicative of strike-slip faulting. The strike-slip fault system was clearly compressive, causing the faults to be reverse faults, and to link up at depth. The folds are breached by the faults.

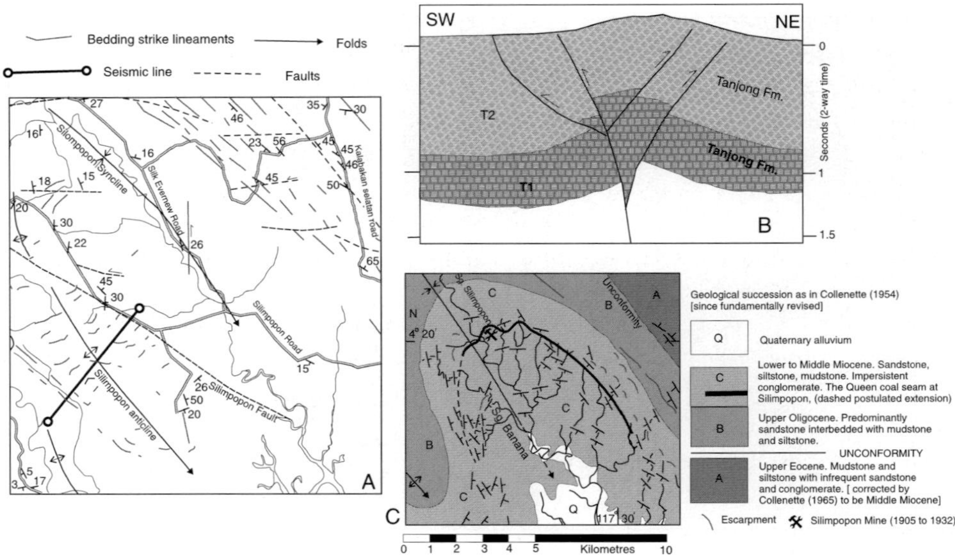

Figure 105. Geology of the Silimpopon area. (A) Structural elements from Balaguru (2001); (B) Interpreted seismic section acquired by Sun Oil; (C) the region of the coal mine (Collenette, 1954).

Strike-slip faults. Left-lateral faults trend NW–SE, right-lateral faults trend NE–SW and NNE–SSW. It is suggested that the NW–SE trending right-lateral faults resulted in Pliocene transpression.

Normal faults. They trend E–W.

Inversions. NW–SE and WNW–ESE trending faults are inverted extensional faults. Structural inversions and folding occurred in the Pliocene. Structural inversion must be considered important in the process of forming circular basins. The escarpments of both the Meliau and Malibau basins commonly exceed 1000 m elevation.

Post-folding extensional faults. These late stage extensional faults trend WNW–ESE, NW–SE and NE–SW (Balaguru, 2001). The renewed extension occurred in the Pliocene.

XXI.6.2. Tanjong Group stratigraphy

The Group begins with the *Tanjong Formation*, which is a coarsening-upwards megasequence. The Tanjong Formation rests unconformably upon Labang Formation and Kuamut Mélange, and is conformably overlain by the Kapilit Formation. Balaguru and Nichols (2004) emphasize the Lower Miocene unconformity between the Kuamut Formation and its mélange with the overlying Tanjong Formation. Uplift occurred between 22 and 16 Ma (Lower to Middle Miocene), correlated with the deep regional unconformity (DRU). Before the unconformity, Sabah was dominated by deep water, predominantly turbiditic sedimentation, thereafter by shallow water.

It is subdivided into a Lower Member and an Upper Member (Balaguru, 2001). The Lower Unit is estimated to be 1200 m and the Upper Unit 1600 m thick. Vitrinite reflectance of the coal lithologies is $R_0 = 0.85 - 0.93\%$, indicating that former burial before erosion was of the order of around 3–4 km. This is considerably higher reflectance values than the Tanjong Group at Sandakan or Bukit Garam basins, where burial was considerably less.

The Lower Unit is dominated by mudstone with siltstones, the Upper Unit by sandstones, forming arcuate escarpments, and mudstone. It also contains conglomerate and coal beds up to 5 m thick.

The *Kalabakan Formation* is of bedded mudstone with rare siltstone beds. It is similar to the mudstone of Tanjong Unit 1. This formation presents poor outcrops and is tightly folded. Thickness was estimated by Collenette (1965) at 1500 m. It overlies the Kuamut Mélange and interfingers with the Tanjong Formation.

The *Kapilit Formation* is a coarsening-upwards fluvio-deltaic formation, similar to the underlying Tanjong. It is divided into two divisions. Lower Unit 1 is dominated by mudstone with siltstone beds. Unit 2 is sandstone-dominant with mudstone and coals up to 1 m thickness. The sandstones form impressive arcuate escarpments. Vitrinite analysis indicates R_0 values of $0.58 - 0.67\%$, indicating pre-erosion burial the order of 3 km. Preserved thickness is 1400 m for Unit 1 and 1800 m for Unit 2.

XXI.6.3. Tanjong Group lithologies

The following lithofacies occur in all three of the Tanjong Group formations, but in different proportions, abbreviation code is as used in Figure 106:

Clast supported conglomerates [*Gcl*]. The facies is of pebble to cobble conglomerate and medium to coarse pebbly sandstone. Beds are thick to massive, 1–15 m thick, with a sharp erosive contact. Clasts are of sandstone and chert with minor quartzite and mudstone. The lithic arenite matrix is composed of coarse sand, chert and muddy sand. The conglomerates are associated with cross-bedded sandstone beds and interpreted as braided river deposits (Balaguru, 2001). However *Ophiomorpha* trace fossils suggest a delta lobe setting.

Channelized trough cross-bedded sandstone interbedded with carbonaceous mudstone [*Sct*]. The channels, of multiple stacked fining-upwards sandstones interbedded with pebbly sandstones, incise deeply into the underlying carbonaceous muds and coals. Channel edges have been measured to dip 35° and the channels trend NE–SW. Maximum thickness is 10 m, channel width up to 50 m. They are trough and planar cross-bedded. Basal lag deposits are of chert pebbles, with mud and coal clasts and mud drapes. There are common *Ophiomorpha* and *Skolithos* indicating they represent distributory channels of a delta.

Thick to massive fine-grained trough cross-bedded sandstone [*St*]. These erosion-resistant rocks form high escarpments reaching 150 m height. The trough cross-bedded sandstones are interbedded with thin grey mudstones. *Ophiomorpha*,

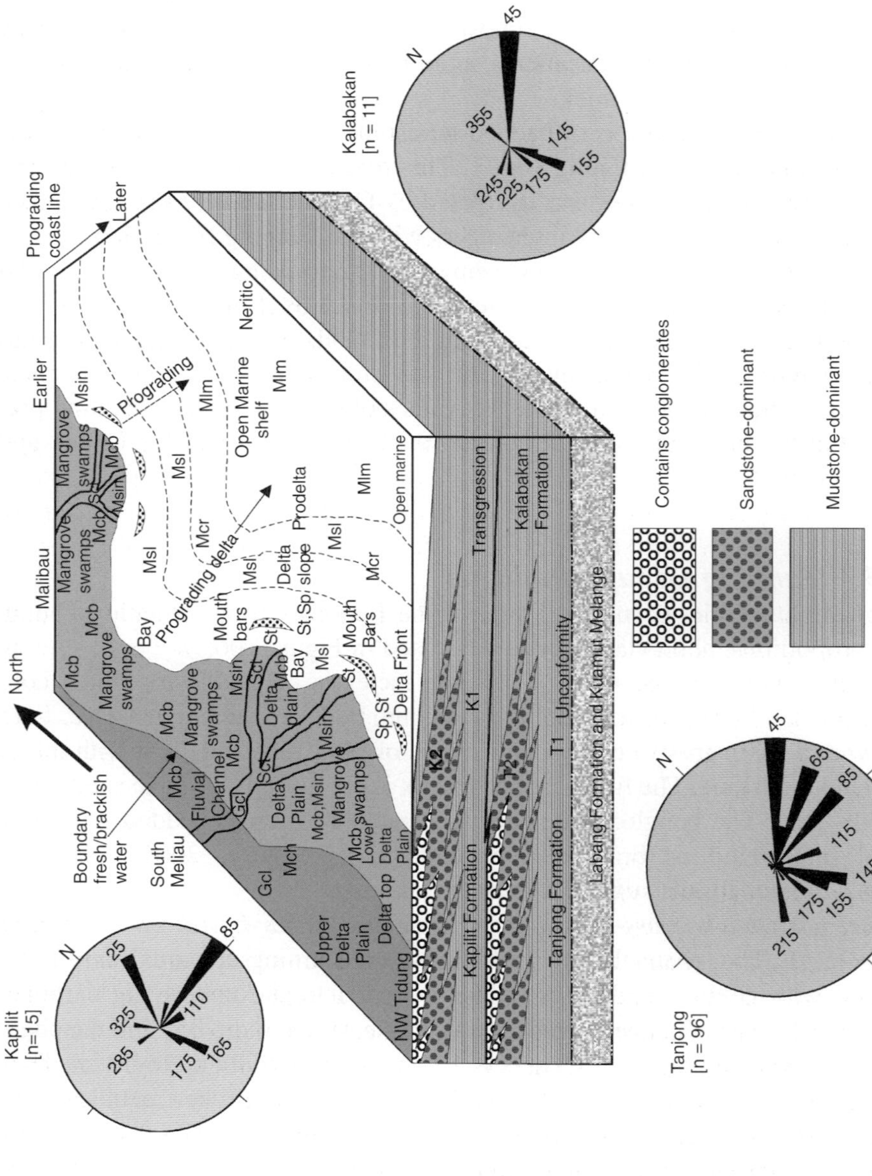

Figure 106. Palaeoenvironment reconstruction for the Tanjong Group of SW Sabah. The rose diagram plots are of palaeocurrent directions measured by Balaguru (2001). Facies as described in text. Note that NORTH is rotated 45° clockwise.

Planolites and bioturbation are common. The environment is interpreted (Balaguru, 2001) as a sandy shoreface.

Planar cross-bedded sandstone [*Sp*]. These medium-grained coastal marine sandstones are common in Silimpopon, typically 0.3–3 m thick. Sometimes they are interbedded with grey rippled mudstone. *Ophiomorpha* occurs.

Horizontally bedded and laminated sandstone [*Sh*]. These massive fine-grained sandstones are 1–9 m thick and form high cliffs and escarpments. The trace fossil *Helminithoida* is common. An offshore/shelf transition with storm influence is favoured.

Bedded carbonaceous mudstone with coal beds [*Mcb*]. The dark grey highly carbonaceous mudstone contains rooted trees and a brackish water microfauna. The coal is sub-bituminous and up to 5 m thick. There are silty interbeds. Petrified wood, fossil leaves, amber clasts and molluscs are common. The envirooment is paralic shallow intertidal mangrove swamps.

Heterolithic mixed sandstone, siltstone, mudstone sequences [*MSin*]. The bedding is thin, up to 20 cm. Ripples and flaser bedding are common. Shallow intertidal conditions are suggested.

Mudstone with crabs [*Mcr*]. Thick light-grey mudstones containing crabs suggest a quiet inner-shelf environment.

Thick mudstone with thin siltstone intercalations [*Msl*]. The siltstones have parallel to cross lamination and wavy ripples. Trace fossils are common: *Nereites, Scolicia, Helminthoida, Chondrites, Palydictyon, Planoeites* and *Thalassinoides*.

Thick to massive laminated grey laminated mudstone [*Mlm*]. Dark grey mudstone with waxy conchoidal fractures. Bedding is obscure. Deposited in quiet conditions, shelf to open marine.

Limestone thin beds and lenses [*L*]. These are rare, found only in Malibau. They are dense coralline algal packstone and wackstone. *Lepidocyclina* occurs in a muddy cemented matrix. Environment is shallow-water shelf or lagoon.

XXI.6.4. Palaeocurrents and palaeoenvironment

Palaeocurrent directions were measured in the field mainly on planar and trough cross-bedded sandstones (Balaguru, 2001). They are shown in Figure 106 together with a suggested palaeoenvironment scheme for the Kapilit Formation. The underlying Tanjong Formation is very similar and Figure 106 may be taken to represent the whole Tanjong Group. Most palaeocurrents are directed north-easterly to easterly, as previously demonstrated by Tjia et al. (1990). The sediment source and land lay on the west. The river system flowed easterly, building out a sandy delta system into a muddy marine environment, represented by the Kalabakan Formation. Provenance of the sediments must have been the Trusmadi and other Rajang Group formations.

XXI.6.5. Palaeontology and age of the Tanjong Formation

Collenette (1965) gave the age of the Tanjong Formation as Te_5 to Tf , Lower to Middle Miocene. The majority of the Foraminifera were collected from the Pinangah River and from the northern side of the Malibau Basin. The larger benthonic Foraminifera include: *Amphistegina lessoni* d'Orb, *Amphistegina* sp., *Austrotrillina howchina* (Schlumberger), *Carpentaria* sp., *Cycloclypeus*, *Gypsina globulus* (Reuss), *Heterostegina borneenesis* van der Vlerk, *Heterostegina* sp., *Lepidocyclina* (*Eulepidina*) sp., *L. flexuosa* Rutten, *L.* (*Nephrolepidina*) *douvillei* Yabe, *L.* (*Nephrolepidina*) *sumatrensis* Brady, *L.* (*Nephrolepidina*) spp., *Lepidocyclina* spp., *Miliolidae*, *Miliolina* sp., *Miogypsina primitiva* Tan Sin Hok, *Miogypsina* sp., *Miogypsinoides* (*Conomiogypsinoides*) *abuensis* Tobler, *M.* cf. *dehaarti* van der Vlerk, *Miogysinoider* spp., *Operculina venosa*, *Operculina* sp., *Peneroplidae* sp. and *Rupertiidae.*

The mudstone samples contain the pelagic Foraminifera (Collenette, 1965): *Globigerina subcretacea* Chapman, *Globigerinoides rubrus* group spp., *G. subquadranus* Bronnimann, *G.subquadranis* cf. *subquadratus* Bronnimann, *G. trilobus* (Reuss) group spp., *Globoquadrina altispira* Cushman and Jarvis, *G. altispira* cf. Cushman and Jarvis, and *Globorotalia mayeri* Cushman and Ellisor.

The following molluscs were obtained from the Tanjong Formation in the Pinangah River section: *Turritella* sp., *Terebellum* (?) sp., *Strombus* (*Doloena*) cf. *sedanensis* Martin, *Strombus* sp., *Globularia carlei* (Finlay), *G.* cf. *fluctuate* (Sowerby), (?) *Murex* sp., *Vexillum* (*?pusia*) sp., *Cardium* sp. and *"Venus"* sp. A Lower to early Middle Miocene age is confirmed.

Balaguru (2001) obtained the following nannofossils from the Tanjong Formation, confirming the Lower to Middle Miocene age: *Calcidiscus macintyrei*, *Chiasmolithus* sp. *[reworked]*, *Coccolithus* sp., *C. pelagicus*, *Cyclicargolithus floridanus*, *Discoaster* spp., *D. bollii*, *D.r deflandrei*, *D. pentaradiatus*, *D. variabilis*, *Heliscospheara* sp., *H.* (?) *intermedia*, *H. kamptneri*, *Sphenolithus* sp. *cf.*, *S. belemos*, *S. heteromorphus* and *S. moriformis.*

XXI.6.6. Palaeontology and age of the Kalabakan Formation

The mudstones have yielded the following pelagic fauna (Collenette, 1965), indicating a Lower to Middle Miocene age: *Globigerina binaiensis* Koch, *G. dissimilis* Cushman and Bermudez var., *G. subcretacea* Chapman, *Globorotalia mayeri* Cushman and Ellisor and *Globoquadrina altispira* Cushman and Jarvis. The arenaceous fauna is more common, but less precise and is composed of the following (Collenette, 1965): *Ammodiscus* sp., *Bathysiphon* sp., *Cibicides* sp., *Cyclammina* sp., *Discocyclina* sp., *Gaudryina* sp., *Glomospira* sp., *Gyroidina* sp., *Haplophragmoides* sp., *Nodosaria* sp., *Rhenophax* sp., *Sigmoilina* sp., *Spiroplectoides* cf. *attenuata* Cushman, *Trochammina* sp. *and Trochamminoides* sp.

Balaguru (2001) listed the following Lower Miocene nannofossils: *Calcidiscus macintyrie, Cocolithus pelagicus, Coronocyclus nitescens, Cyclicargolithus floridanus, Discoaster spp., D. deflandrei, D. variabilis, Helicospheara intermedia, H. kamptneri, Spheenolithus heteromorphus* and *Spheenolithus moriformis.*

Tjia et al. (1990) listed the following Lower Miocene palynmorphs: *Alangium sp., Barringtonia sp., Brownlowia sp., Casuarina sp., Cephalomappa sp., Dacrydium sp., Durio sp., Florschuentzia triloba, Gonystylus sp., Lycopodium phlegmaria, Hycopodium cernuum, Picea* type, *Pinus* type, *Rhizophora sp.*, and *Stenochlaena areolaris.*

XXI.6.7. Palaeontology and age of the Kapilit Formation

Collenette (1965) recorded the following Lower to Middle Miocene (Te_5–Tf) Foraminifera from the Kapilit Formation—Pelagic Foraminifera: *Globigerina binaeinsis* Koch, *Globigerinoides glomerosa* Blow, *G. rubus* group, *G. subquadratus* Bronnimann, and *G. trilobus* group. Arenaceous Foraminifera: *Ammobaculites, Cyclammina, Gaudryina, Glomospira, Haplophragmium, Halpophragmoides, Textularia* and *Trochammin.* Rare calcareous benthonic species: *Bolivina ligularia* Schwager, *Cibicides* sp., *Elphidium* spp., *Glandulina laevigata* (d'Orbigny). *Hanzawaia tapanoeliensis* (LeRoy), *Nonion japonicum* Asano, *Nonion* sp., *Operculina* sp., *Reusella simplex* (Cushman), *Robulus* sp., *Rotalia* spp., *Quinqueloculina* sp., and *Spiroloculina* sp.

Balaguru (2001) listed the following Foraminifera from the Lower Unit (Kp1) of the Kapilit Formation: *Globigerinoides* sp., *G. immaturus, G. sicanus, Globigerinoides steanus, G. subquadratus, Globoquadrina* sp., *G. altispira* and *Globorotalia* sp.

The following Lower to Middle Miocene Nannofossils are listed by Balaguru (2001) from the Lower Unit of the Kapilit Formation: *Coccolithus miopelagicus, C. pelagicus, Coronocyclus nitescens, Cyclicargolithus abisectus* [reworked], *C. floridanus, Discoaster* sp., *D. diflandrei, D. exilis* (?), *D. perplexus, D. variabilis, Discolithus multipora, Helicosphera kamptneri, H. intermedia, Reticulofenestra* sp., *Sphenolithus heteromorphus, S. moriformis* and *Triquetrorhadbulus carinatus.*

Kapilit Formation Upper Unit yielded the palynomorphs (Balaguru, 2001): *Stenochlaena palustris* and *S. areolaris,* indicating a Late Middle Miocene age.

XXI.6.8. Chronostratigraphy

A careful compilation of the palaeontological ages of the Tanjong Group and related formations shows that false conclusions have been made (Figure 107). Thus, Tate (2001) shows the Sandakan and Bongaya formations as Pliocene. They are indeed Miocene and no case can be made for suggesting that the Tanjong Group youngs towards the Sulu Sea.

The following are the important conclusions to be drawn from Figure 107:

- The creation of broken beds and mélanges significantly overlaps in time with deposition of the Kulapis, Kalumpang and upper carbonate-bearing part of the Labang Formation (Gomantong Formation). This was already stressed by

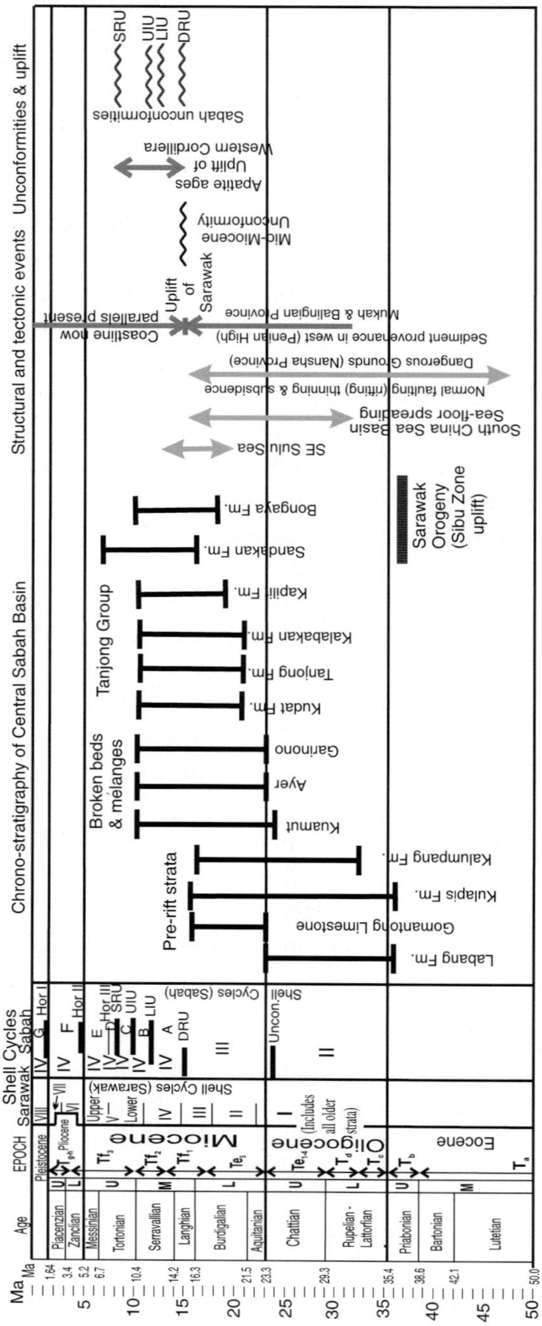

Figure 107. Chrono-stratigraphy of the formations in the Central Sabah Basin and related tectonic events.

Clennell (1992), who reminded us that these are not stratigraphic 'formations' but the results of tectonic events.

- Sea-floor spreading occurred in the SE Sulu Sea at the same time and therefore the broken beds and mélanges resulted from an important rift system that extended south-westwards into Sabah as far as the Malibau and Meliau basins (Hutchison, 1992a; Balaguru, 2001).
- The Tanjong Group (Tanjong, Kalabakan, Kapilit, Sandakan and Bongaya formations) were deposited in this rift basin, and rifting was still very active at least during the early stages of Tanjong Group deposition.
- The Tanjong Group does not young towards the NE. It was deposited throughout during the Lower, Middle and Upper Miocene. The lithofacies are remarkably similar throughout. An Upper Miocene section is preserved only in the Sandakan Formation, but it has been eroded from the Malibau–Meliau areas, where uplift and erosion have been more pronounced as shown by vitrinite reflectance studies.
- Uplift of the Rajang Group (Sibu Zone) of Sarawak at the Mid-Miocene Unconformity (DRU of Sabah) and the subsequent dramatic uplift of the West Crocker Formation to form the Western Cordillera created an erosional provenance for the Tanjong Group strata (Figure 107).

XXI.6.9. Simengaris Formation

This very minor formation occurs near the coast in the Silimpopon area. It is said to be slightly unconformable upon the Kapilit Formation (Collenette, 1965). The basal beds are of 60 m of soft blue highly fossiliferous plastic mudstone. Upwards the formation includes fine-grained sandstone, tuff, conglomerate and coal. This formation represents the closing stages of the Miocene Basin sedimentation in SW Sabah, and is accordingly included within the Tanjong Group.

XXI.6.9.1. Palaeontology and age

A Middle to Upper Miocene age (Te_5 to Tf) is indicated by the Foraminifera, listed by Collenette (1965): *Ammobaculites* sp., *Angulogerina* sp., *Anomalina* sp., *Bigenerina* spp., *Bolivina* sp., *Cibicides* spp., *Cristellaria* sp., *Cyclammina* sp., *Elphidium* spp., *Globorotalia* sp., *Nonion* sp., *Operculina* spp., *Reusella* sp., *Rotalia* spp. and *Siphogenerina* sp.

The following macrofossils have been identified: *Arca* sp., *Cardium* spp., *Cerithium* sp., *Clementia* sp., *Lucina* sp., *Macrocallista* sp., *Martinocarcinus* sp., *Natica* sp., *Nucia* sp., *Natica* cf., *N. fennemai* (Böhm), *Nuculana* sp., *Rimella* sp., *Tellina* sp. and *Turritella* sp.

XXI.6.10. Miocene palaeogeography

A very significant analysis of the palaeogeography was written by Balaguru (2001) and Figure 108 is largely based on his work. None of these palaeogeographic maps

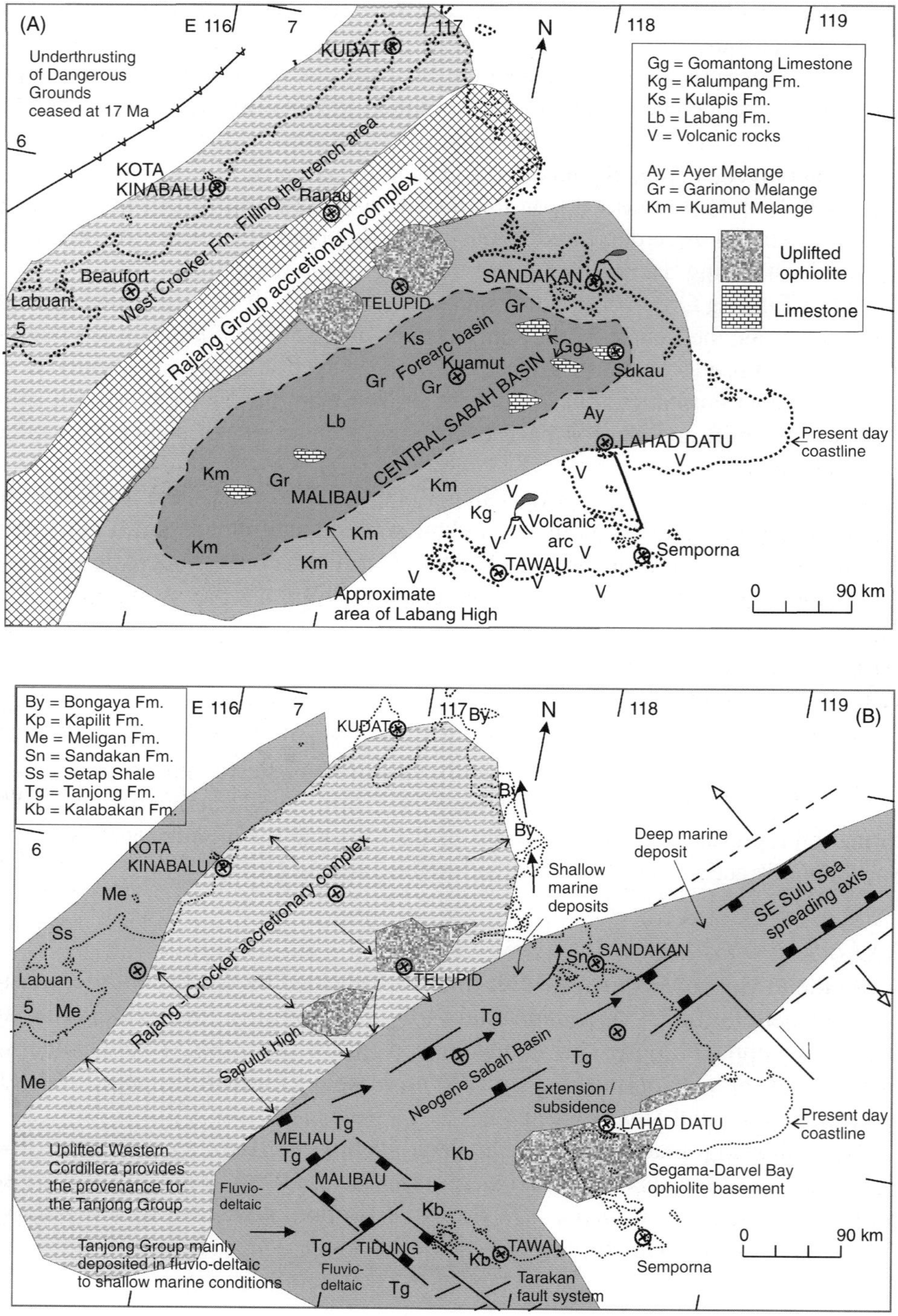

Figure 108. Approximate palaeogeographic reconsctuction of the Central Sabah Basin. (A) Early Miocene broken and mélange formation and establishment of a forearc basin. (B) Middle to Late Miocene deposition of the Tanjong Group. Arrows show sediment transport direction. Modified after Balaguru (2001).

support the concept of Tjia (1988, 1999) that a 'suture' extends approximately N–S linking Kudat, through Telupid, to Darvel Bay. On the contrary, the structural and sedimentary elements of Sabah extend SW–NE throughout the Oligocene and Miocene. It had earlier been shown by Hutchison (1975) that the uplifted ophiolite masses of Sabah did not represent a suture.

During Oligocene time Sabah did not exist as a landmass. It is, however, possible that the Rajang Group, which was uplifted throughout Sarawak at the end of the Eocene, may have been similarly uplifted to form the NE-trending ridge shown in Figure 108(A) as the "Rajang Group accretionary complex". It would predominantly have been constructed of the Trusmadi and Sapulut formations. The early part of a Dent–Semporna volcanic arc was established. A natural basin was bounded by the volcanic arc along the SE side and by the Rajang accretionary complex along the NW side. Such a basin may be classified as a fore-arc basin. It was sedimented by turbidites, but the seas shallowed as it filled and by the beginning of the Middle Miocene the Gomantong Limestones were being deposited (Figure 108).

Active rifting began in the Lower and extended to the Middle Miocene. The strata of the fore-arc basin were broken and disrupted to form extensive mudstone-matrix mélange deposits. Sea-floor spreading in the SE Sulu Sea is the most readily recognizable expression of the rift system that extended south-westwards through Sabah (Hutchison, 1992a). Rivers, such as the proto-Kinabatangan, eroded abundant material from the actively uplifting Western Cordillera, caused by a change from subduction of the proto-South China Sea oceanic crust to a collision situation as the stretched continental crust of the Dangerous Grounds was underthrust south-eastwards.

The predominantly Middle to Upper Miocene Tanjong Group was deposited mainly by fluvio-deltaic and shallow marine conditions into this actively rifting Central Sabah Basin. Palaeocurrents indicate the transport was predominantly from the Malibau–Meliau areas towards Sandakan, thence northwards towards Bongaya. Where there were deltas, the sediments were sandy but of fine-to medium-grain size because the provenance was of fine to medium grained turbidites forming the Western Cordillera. Away from the deltas, the Tanjong Group is predominantly of mudstone. Sedimentary transport was also southwards into the Tarakan Basin where the delta deposits are oil-productive. This Neogene Sabah Basin was tectonically unstable well into the Middle Miocene, resulting in the structural development of "elliptical" to "circular basins". Uplift and erosion of the Tanjong Group was more active in the Malibau-area, becoming less towards Sandakan. Hence vitrinite in the former is over-mature, and within the oil-window in Sandakan. Tanjong Group sediments were carried into the deep-water SE Sulu Sea as turbidite deposits.

Chapter XXII
Dent Peninsula Volcanics and Pyroclastics

Volcanic and pyroclastic rocks occur in the southern Dent Peninsula and the whole of the Semporna Peninsula. The furthest west occurrences are of andesite porphyry and dacite intrusives into the Middle Miocene Simengaris Formation in the Silimpopon Syncline (Collenette, 1965). The volcanic arc does not extend any further westwards. The volcanic materials have been disrupted into pyroclastic formations, included along with plant-bearing strata within the Tungku Formation (Figure 109). Volcanic clasts also occur within the Ayer Mélange.

XXII.1. RADIOMETRIC DATING

From a tabulation of K:Ar dating of the volcanic rocks (Table 24), the rocks of the Dent Peninsula cannot be separated from those on the Semporna Peninsula. The following is a summary:

	Number of analyses	Average (Ma)	Standard deviation (Ma)
Kunak basalts	2	2.95 (Pliocene)	± 0.23
Dent Volcanics	8	13.95 (Middle Miocene)	± 3.25
Tawau-Wullersdorf	5[a]	12.67 (Middle Miocene)	± 2.50

[a]Excluding two other values of 6.4 and 1.62 Ma, assumed to be of younger lava flows (Lim, 1988).

XXII.2. LIBONG TUFFITE FORMATION

This is a well-bedded sequence forming well-defined synclinal basins. The strata are moderately dipping 20–30°, locally as steep as 70–80°. It differs from the underlying Ayer Mélange in being well bedded. A geographical association with active volcanism is clear because the strata are all tuffaceous, generally crystal-lithic. The varieties are tuffaceous conglomerate and sandstone and well-bedded shale. The conglomerate and sandstone contain clasts and grains derived from the ophiolite basement. Haile and Wong (1965) consider that the Libong Tuffite Formation underlies the Tungku Formation in the Tabanak Syncline (Figure 109).

XXII.2.1. Lithologies

The tuffaceous sandstones are dark in colour because of their content of hornblende crystals. Common grains are hornblende (up to 25%), plagioclase (15–40%), antigorite (up to 25%) and augite (up to 15%). They also contain fragments of chert, limestone and andesite porphyry.

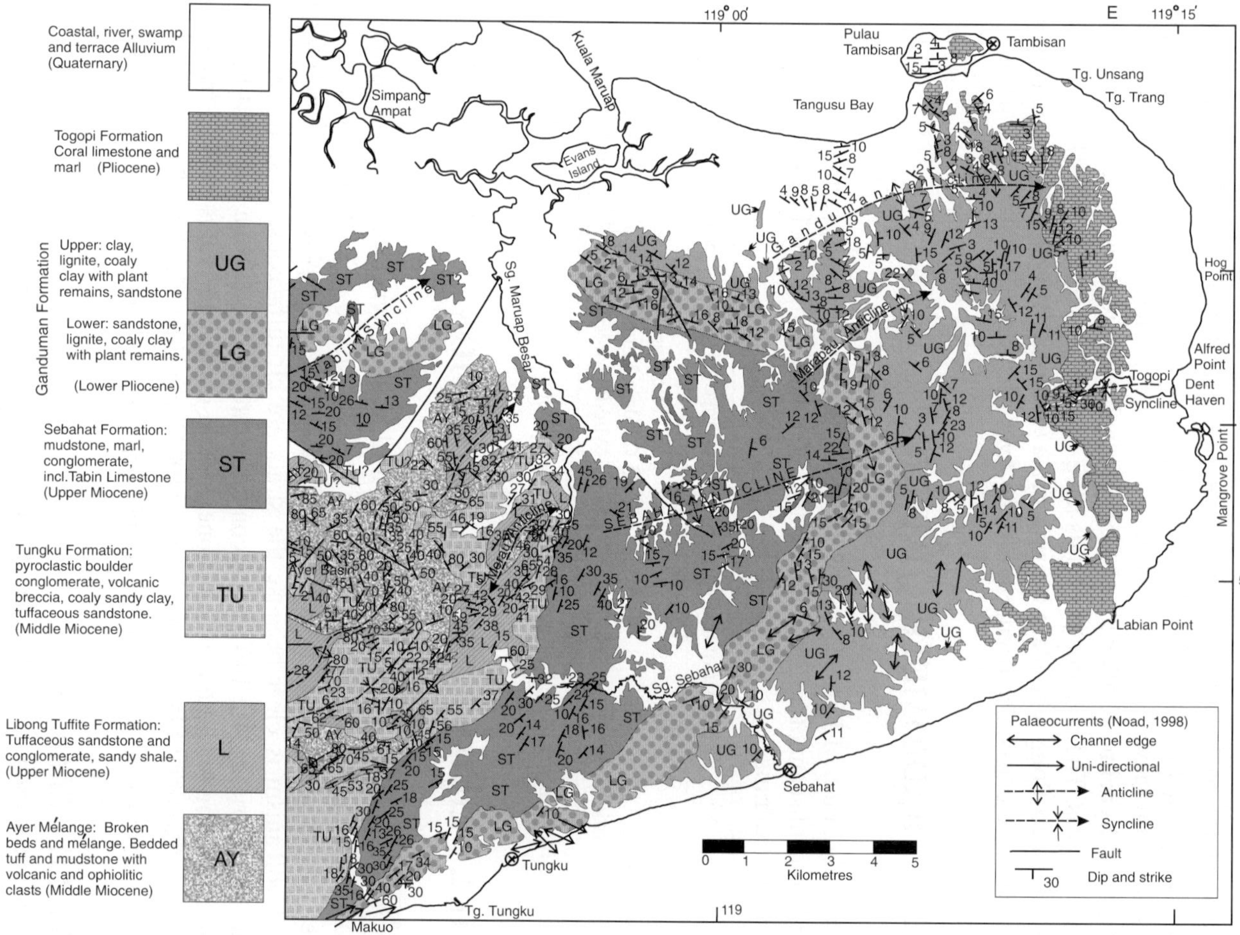

Figure 109. Geological map of the Dent Peninsula (redrawn from Haile and Wong, 1965). The stratal sequence youngs towards the ENE in a gently plunging anticline. With permission from Minerals and Geoscience Department, Malaysia.

Table 24. K: Ar radiometric dates of volcanic rocks of east and SE Sabah

Specimen	Source	Rock-type	Locality	Mineral	K (wt%)	Age with error (Ma)	Geological age
94SB31A	2	Andesite tuff	Coast under Sandakan Mosque	Plagioclase	0.10	76.6±4.1	Campanian U. U. Cretaceous
S87-34	4	Diorite block in Ayer Melange	8 km north of Segama River on Lahad Datu road**	Whole rock*	0.48	11.47±0.57	Serravallian Middle Miocene
94SB62B	2	Dacite tuff breccia	Lahad Datu-Tungku road	Plagioclase	0.31	17.9±1.2	Burdigalian L. Miocene
				K-feldspar	2.26	18.8±0.6	Burdigalian L. Miocene
S267	4	Massive andesite flow	Membatu Bridge, Tungku	Whole rock*	1.84	11.07±0.55	Serravallian, M. Miocene
S268	4	Moderately	Membatu Bridge altered massive andesite	Whole rock* M. Miocene	2.32	11.69±0.58	Serravallian,
S266	4	Aphanitic volcanic rock in breccia	Membatu Bridge, Dent Peninsula	Whole rock*	2.11	16.56±0.83	Burdigalian Lower Miocene
S138-1	4	Porphyritic andesite in shale sequence	South of Kunak	Whole rock*	1.66	12.92±0.65	Seravallian Middle Miocene
S285-2	4	Porphyritic andesite in shale sequence	South of Kunak	Whole rock*	1.76	11.80±0.59	Seravallian Middle Miocene
264-1	4	Andesite-dacite blocks in conglomerate	Cape Membatu, Tunku, Dent Peninsula	Whole rock*	1.49	11.53±0.58	Serravallian Middle Miocene
S264-3	4	Hornblende tuff block in conglomerate	Cape Membatu, Tunku, Dent Peninsula	Whole rock*	2.56	12.58±0.63	Serravallian Middle Miocene
94SB82A	2	Andesite porphyry dome	Kukusan quarry, Tawau	plagioclase	0.43	14.4±0.9	Langhian M. Miocene
S87-90	4	Massive andesite	Quarry, Tawau	Whole rock*	2.53	11.61±0.58	Seravallian M. Miocene
S87-91	4	Andesite plug	Quarry, Tawau	Whole rock*	2.31	9.01±0.45	Tortonian U Miocene
94SB83A	2	Vitric andesite lava	Batu Payong, Tawau coast	plagioclase	0.36	16.3±1.1	Langhian M. Miocene
	5	andesite	Gunung Andrassy	Whole rock		6.40 ± 0.20	Messinian (U. Miocene)
	5	dacite	Gunung Maria	Biotite		1.62 ± 0.70	Pleistocene–Pliocene
94SB85B	2	Andesite flow banded lava	Mile 51 quarry Mount Pock	Plagioclase	0.37	18.2±1.2	Burdigalian L. Miocene

Continued

Table 24 (*Continued*)

Specimen	Source	Rock-type	Locality	Mineral	K (wt%)	Age with error (Ma)	Geological age
S129	4	Basalts extruded from 130° -trending faults	Mostyn Estate, Kunak	Whole rock*	0.30	2.79±0.42	Piacenzian U. Pliocene
S87-92	4	Tholeiitic vesicular basalt	Mostyn Estate	Whole rock		3.11±0.93	Piacenzian U. Pliocene

Data Source: 1, Kirk (1968); 2, Swauger et al. (1995), Bergman et al. (2000); 4, Rangin et al. (1990), Bellon and Rangin (1991) and 5, Lim (1988).

** The authors state the locality to be "8 km north of kinabatangan River along the Lahad Datu road". Such a locality is impossible for the outcrops there are all of the Tanjong Formation. It is more than likely that the confused authors meant the Segama River, for at that locality there are extensive outcrops of Ayer Formation Mélange, containing a variety of blocks of ophiolitic as well as volcanic rocks, as well as diorite. Clennell (1992) agrees with this interpretation. * whole rock wt. % quoted in K_2O.

Tuffaceous dark conglomerate contains rounded pebbles of meta-gabbro, andesite, serpentinite, sandstone, shale, limestone and vein quartz, in a matrix of tuffaceous sandstone. Well-bedded calcareous shale contains lenticular bands of dark green hornblende-rich sand. Thin lenticular limestones occur within sandstone and shale sequences.

XXII.2.2. Palaeontology and age

There is a considerable overlap in fossil assemblage between the Libong Tuffite and Tungku formations (Table 25). Based on the assemblage, an allocation to the *Globorotalia praemenardi* zone with a range from late Tf_1 to early Tf_{2-3} (late Middle Miocene to early Upper Miocene).

Haile and Wong (1965) found many similarities between the fossil contents of the Libong Tuffite and the Tungku formations, as shown in Table 25.

Haile and Wong (1965) also reported the following fossils — Gastropods: *Delphinula* cf. *fossilis* Martin, *Trochus* cf. *sondeianus* Martin, *Trochus* sp., *Cerithium* sp. and *Cyprea* sp. Bivalves: *Thracia* sp. and *Tellina* sp. Corals: *Acanthastraea polygonalis* Martin, *Cyphastraea monticulifera* Felix, *Maeandrina* sp., *Stylophora pistillata* E.H., *Fungia* (*Cycloseris*) *martini* Felix, *F.* (*Cycloseris*) *stammi* Felix, *F.* (*Cycloseris*) *subcyclolites* Felix, *F.* (*Cycloseris*) *wanneri* Felix, *Porites* sp., *Alveopora* sp. and *Goniaraea anomala* Reuss.

XXII.3. TUNGKU FORMATION

The Tungku Formation is dominated by andesitic pyroclastic rocks. It outcrops along the main road from Cape Membatu to Tungku, with the main outcrops inland along the Tungku River (Figure 109). The rocks of this formation are variable from

Table 25. Identified Foraminifera in both the Libong Tuffite and Tungku formations

Foraminifera (pelagic)	Libong	Tungku
Globigerina bulloides (d'Orbigny)	X	X
Globigerina nepenthes (Todd)	X	X
Globigerina sp.	X	X
Globigerina subcretacea (Lomnicki)	X	X
Globigerinoides rubra (?) *group*	X	X
Globigerinoides sacculifera (Brady)	X	X
Globigerinoides spp.	X	X
Globigerinoides subquadrata (Bronnimann)	X	X
Globigerinoides triloba group	X	X
Globoquadrina altispira (Cushman & Jarvis)	X	X
Globorotalia fohsi barisanensis (LeRoy)	X	X
Globorotalia fohsi fohsi (Cushman & Ellisor)		X
Globorotalia fohsi lobata (Bermudez)		X
Globorotalia mayeri (Cushman & Ellisor)	X	X
Globorotalia sp.	X	X
Hastigerina aequilateralis (Brady)		X
Orbulina bilobata (d'Orbigny)	X	X
Orbulina suturalis (Bronnimann)	X	X
Orbulina universa (d'Orbigny)		X
Sphaeroidinella multiloba (LeRoy)	X	X
Foraminifera (benthonic)		
Alveolinella bontangensis (Rutten)		X
Alveolinella sp.		X
Ammdiscus sp.		X
Bolivina spp.	X	X
Bolivinita quadrilatera (Schwager)		X
Bolivina spp.	X	X
Bolivinita quadrilatera (Schwager)		X
Bulimina spp.	X	X
Cassidulina spp.		X
Cibicides spp.	X	X
Cycloclypeus sp.		X
Elphidium sp.	X	X
Eponides sp.	X	X
Gyroidina sp.	X	X
Katacycloclypeus sp.		X
Miogypsina cushmani vaughan var. indonesiensis (Tan)		X
Miogypsina sp.		X
Nodosaria sp.	X	X
Nonion sp.	X	X
Operculina sp.	X	X
Pullenia spp.	X	X
Robulus sp.	X	
Robulus spp.	X	X
Rotalia spp. (Streblus)	X	X
Rotaliatina sp.	X	
Sphaeroidina sp.	X	X
Trochammina sp.		X
Uvigerina spp.	X	X

X = Foraminifera present in samples of that formation.

coarse volcanic conglomerate to tuffaceous strata containing plant remains (Haile and Wong, 1965).

XXII.3.1. Tuffaceous strata

Well-bedded tuffaceous and carbonaceous sandstone, conglomerate, and clay with plant remains are the main components. The rocks share a similarity with the Libong Tuffite Formation.

Shale is well-bedded, bluish-grey, containing fossil leaves and abundant Foraminifera. Thin layers of fine-to medium-grained sandstones occur within the shale.

Tuffaceous sandstone. Well-bedded, sometimes massive or cross-bedded. There is a strong similarity to the Libong Tuffite Formation. The grains are of basaltic and andesitic clasts, angular fragments of labradorite, pyroxene, hornblende, biotite and iron oxide. The matrix of the tuff may be of devitrified volcanic glass.

Conglomerate. The pebbles and boulders are well rounded, mostly 10–30 cm across. The main clasts are amphibolite (from the ophiolite basement), limestone, pelagic marl, red claystone, chert, sandstone, quartz and andesite. The matrix is generally sandy clay with plant remains, or tuffaceous sandstone.

XXII.3.2. Eclogite

Eclogite boulders occur in the Tungku River and along its tributary the Pungulupi. These unusual rocks have never been seen *in situ*. The descriptions are by Reinhard and Wenk (1951) and later by Haile and Wong (1965). Their origin is of subducted basalt, metamorphosed to eclogite as a result of increased pressure, then subsequently brought to the surface through supra-subduction zone volcanism. The high density of eclogite explains its rarity in crustal exposures.

These dark-green rocks contain porphyroblasts of dark-red garnet of 1–3 cm diameter. The garnet is accompanied by approximately equal amounts of omphacite and carinthine hornblende. Plagioclase of ~15% mode is of An_{55} to An_{80} composition. Accessory minerals are ilmenite, epidote, zoisite, apatite, rutile and sphene. The Omphacite has $\gamma \wedge c$ = 41–49° and 2V = 60–65°. The hornblende has $\gamma \wedge c$ = 13–17° and 2V ~80–85°. It is pleochroic from light-yellow to dark olive green.

A single chemical analysis (NB11337A, from the Pungulupi tributary of Sungei Tungku) is given by Haile and Wong (1965): SiO_2 = 45.0%, TiO_2 = 1.65%, Al_2O_3 = 14.3%, Fe_2O_3 = 3.37%, FeO = 9.80%, MnO = 0.27%, MgO = 8.70%, CaO = 13.45%, Na_2O = 2.20%, K_2O = 0.27%, H_2O^+ = 0.82%, H_2O^- = nil, CO_2 = 0.17%, P_2O_5 = 0.12% and Total = 100.12%.

XXII.3.3. Palaeontology and age

The identified Foraminifera are listed in Table 25. The age of the Tungku Formation is considered to be late Middle Miocene to Pliocene (Tf_{2-3} to Tgh). The fauna is characteristic of the *Globigerinoides subquadratus* association. The Tungku

Formation does not contain *Globigerinoides oblique* Bolli, which is ubiquitous in the Sebahat Formation.

XXII.3.4. Volcanic boulder conglomerate (Bagahak Pyroclastic Member)

The Bagahak Pyroclastic Member (Haile and Wong, 1965) is made up of massive to bedded poorly sorted volcanic boulder conglomerate. The water worn clasts are mainly of andesite, ophiolitic rocks, limestone claystone, chert, sandstone and vein quartz. Clasts are as much as 90 cm across. Outcrops extend inland from Tanjung Membatu to the Tungku River.

The volcanic rocks are best exposed along the Bagahak Range inland from Tanjung Membatu, where the succession is of:

- Interbedded conglomerate, tuffaceous sandstone and shale (760 m)
- Boulder conglomerate with some interbedded shale or tuff (300 m)
- Andesitic breccia (the Silabukan Volcanic Breccia) (120 m)

XXII.3.4.1. Volcanic rocks

No lava flows have been found in the Tungku Formation. The volcanic rocks occur mostly as rounded boulders of basaltic andesite and andesite in the Bagahak Pyroclastic Member. In addition the basal Silabukan Volcanic Breccia is rich in andesite blocks. The blocks and clasts are all porphyritic with a glassy groundmass. The andesites of Tungku Formation have been described by Reinhard and Wenk (1951) as fine-grained. The common phenocrysts are euhedral hornblende, biotite, plagioclase (labradorite –bytownite), augite, and iron oxides in a glassy matrix rich in plagioclase microlites. The glassy matrix is commonly devitrified. The universal presence of small phenocrysts of biotite suggests the andesites belong to the high-K calc-alkaline series. Potassium would also be contained in the hornblende.

A selection of whole-rock chemical analyses of the andesitic clasts within the volcanic conglomerate is given in Table 26. The most complete study is that of Chiang (2002). All her analyses, together with those of Haile and Wong (1965) and of Swauger et al. (1995) have been plotted on a K_2O versus SiO_2 diagram (Figure 110).

The volcanic clasts are of basaltic-andesite and andesite, and predominantly of the high-K calc-alkaline series (Figure 110). Those few samples that appear in the dacite field are more pyroclastic. The high-K character is consistent with the presence of biotite and hornblende phenocrysts. These are volcanic-arc andesites, formed over the deeper parts of a Benioff Zone, farther from the trench than the calc-alkaline series. The whole-rock K:Ar average gives a Middle Miocene age of 13.4 ± 2.9 Ma, totally supportive of the late Middle Miocene to Pliocene age of the fossiliferous strata of the Tungku Formation.

Table 26. Chemical analyses of Miocene volcanic rocks of the Tungku Formation, Dent Peninsula

Specimen	a	b	c	d	e	f	g	h
Source	3	1	2	2	3	3	1	2
SiO_2	51.66	54.4	54.4	55.0	58.37	61.12	62.82	64.9
TiO_2	0.94	0.72	0.69	0.63	0.75	0.61	0.75	0.61
Al_2O_3	18.13	18.0	17.1	17.10	18.15	17.94	16.16	15.4
Fe_2O_3	9.23	4.65	5.30	4.68	6.73	5.41	5.89	2.68
FeO	*	3.15	1.40	1.60	*	*	0.19	1.10
MnO	0.15	0.18	0.11	0.13	0.13	0.12	0.07	0.07
MgO	4.37	3.50	3.12	3.58	3.62	2.97	1.82	1.44
CaO	10.04	8.30	6.33	6.70	7.28	6.39	5.75	6.04
Na_2O	2.56	2.80	2.31	2.58	2.69	2.64	3.12	2.86
K_2O	2.11	1.10	1.54	1.09	1.94	2.28	2.06	2.10
H_2O+	+	1.88	2.70	2.90	+	+	0.27	1.00
H_2O-	+	0.61	3.90	3.70	+	+	0.49	1.20
CO_2		0.02	<0.01	<0.01			—	<0.01
P_2O_5	0.28	0.25	0.21	0.21	0.25	0.21	0.34	0.16
total	99.47	99.60	99.10	99.60	99.90	99.68	99.73	100.0

Data Source: 1, Haile and Wong (1965); 2, Swauger et al. (1995) and 3, Chiang (2002). Specimens: a = SBK35, andesite clast in Tungku Formation volcanic conglomerate [loss on ignition = 0.43], b = NB8795 andesite, Tungku at mouth of Pungulupi tributary. c = 94SB-63A, Membatu Road, andesite tuff breccia. d = 94SB-62B, andesite tuff breccia, Lahad Datu-Tungku Road. e = SBK56, andesite clast in Bagahak Pyroclastics [loss on ignition = 0.52], f = SBK41, andesite clast in Tungku Formation volcanic agglomerate [loss on ignition = 2.33]. g = WK33, dacite, Sungai Makuo. h = 94SB-62C, dacite clast in agglomerate, Lahad Datu-Tungku Road.
* Total iron measured as Fe_2O_3, + Water not measured; loss on ignition measured.

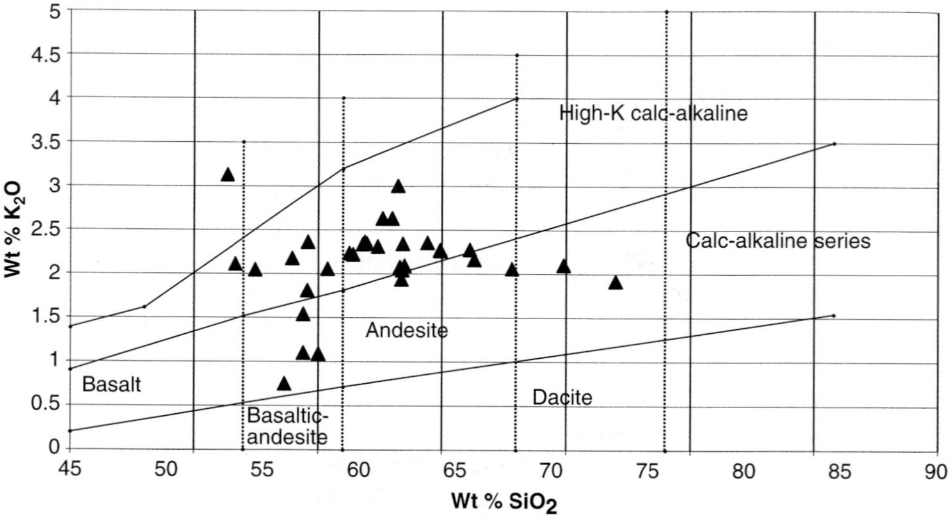

Figure 110. Dent Volcanics K_2O versus SiO_2 plot.

The sloping pattern (Figure 111), resulting from light REE enrichment, is typical of calc-alkaline volcanic arc andesites (Swauger et al., 1995). The incompatible element contents were normalised against primitive mantle for the more acid samples

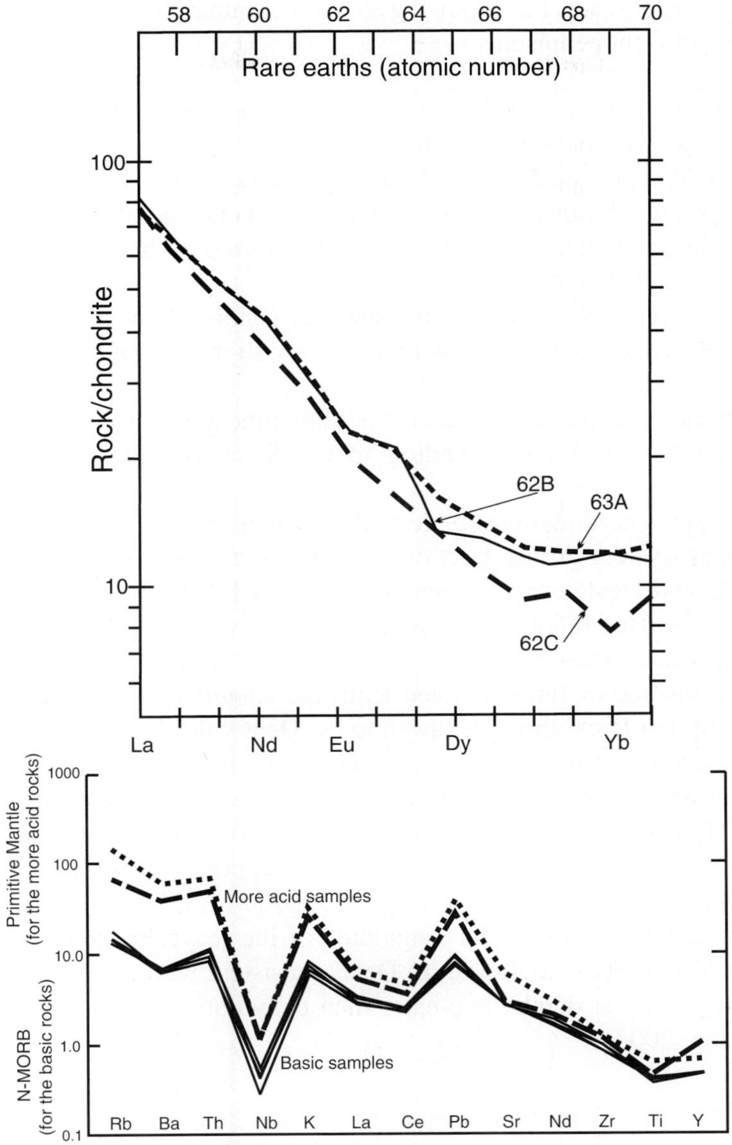

Figure 111. Trace element geochemistry of the Dent Volcanics. Top: Rare earths from Swauger et al. (1995). Bottom: Incompatible elements from Chiang (2002).

(SiO_2 (wt%) 54.5–59.9). The order of the elements is in increasing compatibility from left to right (Chiang, 2002). The more basic samples (higher MgO) were normalized against N-MORB (Figure 111). The Dent volcanic rocks show very little variation and all show a distinct negative Nb anomaly, a slight negative Ba anomaly and an enrichment in K and Pb. The Dent samples all have a $^{87}Sr/^{86}Sr$ isotope ratio of 0.7045 irrespective of the SiO_2 content.

Chiang (2002) made the following general summary of the volcanic rocks of Semporna and Dent peninsulas:

- The igneous rocks of SE Sabah range compositionally from basalt to dacite, and are predominantly andesite.
- The field appearance is that of Island-Arc character.
- Plagioclase, clinopyroxene, orthopyroxene, magnetite, hornblende and biotite were the main phenocryst phases that crystallized at shallow crustal levels due to fractional crystallization.
- The Miocene–Pliocene magmatism was generated by partial melting of a mantle wedge that was very similar, or very slightly depleted, compared to the MORB source.
- Fluids derived from the subducted oceanic lithosphere triggered melting in the Middle Miocene, but extending in the Semporna Peninsula to the early Pliocene.
- Trace element modelling suggests that a small degree of auto-assimilation in an open system during fractional crystallization affected the magmas from Tawau and Dent centres, whereas Mount Pock lavas interacted with a wider range of crustal lithologies, leading to greater variability in $^{87}Sr/^{86}Sr$ and trace element concentrations.
- The influence of fluids derived from the subducted lithosphere diminished from Mount Pock, through Tawau to the Dent volcanic centre (also shown by the common occurrence of biotite phenocrysts in Dent Volcanics). This is consistent with Dent lying furthest away from the volcanic front and suggests that the Miocene–Pliocene arc resulted from NW-directed subduction of the Celebes Sea beneath SE Sabah. This supports the early suggestion of Hutchison (1992a).
- There is an increase in the magnitude of the recycled sediment component from Mount Pock, to Tawau, to Dent, consistent with models in which sediment melting at depths is greater than dehydration of altered oceanic crust (Chiang, 2002).

Chapter XXIII
Dent Group

The Dent Group forms arcuate outcrops on the eastern Dent Peninsula, divided into three formations: Sebahat (oldest), Ganduman, and Togopi (youngest) (Haile and Wong, 1965). Additional studies have been published by Wong (1993), Noad (1998), and Ismail (1994) who consistently misspelt Sabahat.

XXIII.1. SEBAHAT FORMATION

This is the basal formation of the Dent Group. Dips are generally within the range 20–30° eastwards. It was deposited unconformably upon an irregular surface formed by volcanic and pyroclastic rocks of the Tungku Formation and locally upon broken and disrupted rocks of the Ayer Mélange. Accordingly the thickness is highly variable. The depth to the magnetic basement has been calculated from two aeromagnetic surveys (Figure 112). Several topographic highs trend SW–NE, upon which the Dent Group is <1000 m thickness. The offshore extrapolation is more complex and there are depocentres >5 km depth in which the Dent Group is thickest.

The variable thickness of the Sebahat Formation is also shown on the NW–SE seismic section parallel to the Sulu Sea coast, through the Central Trough (Figure 113). The Sebahat Formation is strongly transgressive upon a surface formed by volcanic and pyroclastic rocks of the Tungku Formation (Figure 113).

XXIII.1.1. Lithology

The Sebahat Formation is of dark grey to black mudstone with subordinate marl, argillaceous limestone, sandstone and conglomerate (Haile and Wong, 1965). The basal section is exposed in the Tungku River. Twelve metres of lower conglomeratic mudstone contains clasts of chert, andesite, basalt and tuff of the underlying formations. The conglomeratic mudstone is overlain by calcareous tuffaceous sandstone.

Thick grey mudstone beds, with thin well cemented and highly fossiliferous calcareous horizons and rare siltstone beds, form the main lithology of the Sebahat Formation (Noad, 1998). Amber clasts may reach a 10 cm diameter. The highly fossiliferous mudstones suggest a well-oxygenated shelf environment of deposition.

XXIII.1.2. Fission-track data

Swauger et al. (1995) reported the following fission-track data from brown organic-rich sandstone 94SB60B collected from the W. Sahabat Palm Estate (Figure 76): Apatite [n = 20] 59.9 ± 11.3 Ma; Zircon [n = 12] 96.5 ± 13.1 Ma. These rocks have

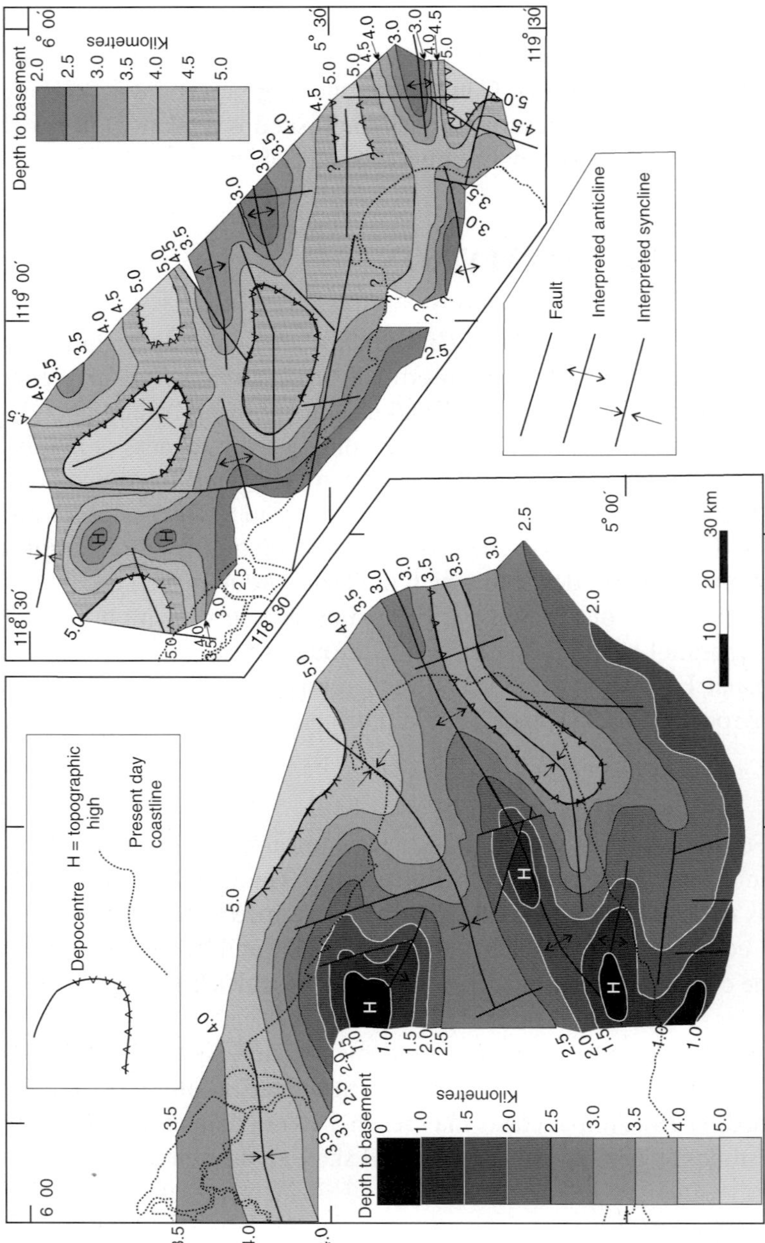

Figure 112. Interpreted depth to magnetic basement from two independent aeromagnetic surveys (left, onshore; right, offshore), showing the irregular basement on which the Dent Group was deposited. From Leong and Azlina (1999). With permission from Petronas.

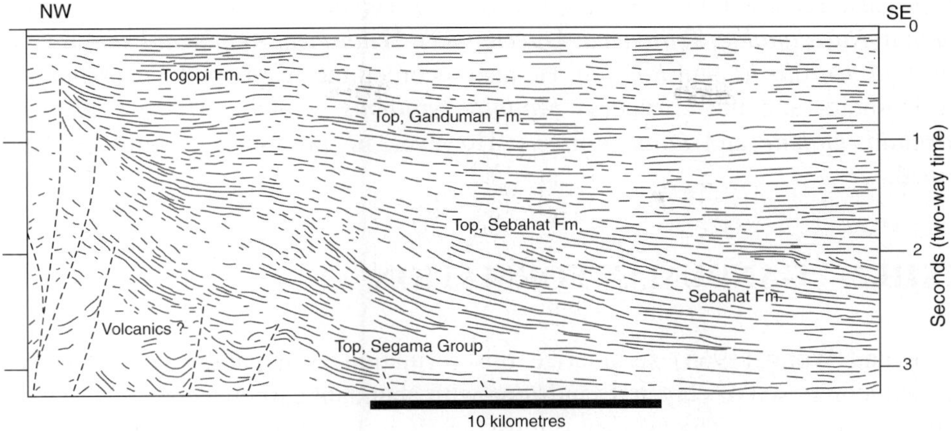

Figure 113. NW to SE seismic cross-section parallel to the Sulu Sea coast, through the Central Trough, showing the Sebahat Formation prograding over a basement of volcanic and pyroclastic rocks of the Tungku Formation. The Togopi Formation at the top of the Dent Group is flat-lying. From Leong and Azlina (1999). With permission from Petronas.

not been heated by burial and hence the apatite crystals preserve their Cretaceous provenance age, as confirmed by the zircon fission-track data (Figure 101).

XXIII.1.3. Palaeontology and age

Haile and Wong (1965) listed the following Foraminifera that indicate a late Tf to early Tg age (Upper Miocene to early Pliocene): *Bolivinita quadrilateral* (Schwager), *Globigerina subcretacea* Lomnicki, *Globigerinoides* spp., *G. obliqua* Bolli, *G. sacculifera* (Brady), *G.* cf. *subquadratus* Bronnimann, *G. triloba* group, *Globoquadrina altispira* (Cushman and Jarvis), *G. dehiscens* (Cushman, Parr and Collins), *Globorotalia fohsi robusta* Bolli, *G. mayeri* Cushman and Ellisor, *G. praemenardii* Cushman and Stainforth, *G. scitula* (Brady), *Orbulina suturalis* Bronnimann, *O. universa* d'Orbigny, *Rotalia* cf. *hamiltonensis* Parr, *R.* cf. *tuvuthaensis* Kleinpell, *Sphaeroidinella* sp., *S.* cf. *dehiscens* (Parker and Jones) and *S. multiloba* LeRoy.

The Tabin *Limestone Member* occurs only in the Tabin Syncline. It contains the following Foraminifera (Haile and Wong, 1965): *Amphistegina*, *Cycloclypeus*, *Heterostegina*, *Katacycloclypeus*, *Lepidocyclina*, *Operculina*, *Miogypsina* and Spiroclypeus. This fauna indicates neritic conditions, and the presence of *Katacyclocypeus* sp. indicates Pliocene (Tf) age.

Noad (1998) identified the following gastropods: *Callistoma*, *Volutid*, *Xenophora*, *Tibia*, *Terebridae*, *Cancellaria*, *Seraphus*, *Epitonidae*, *Voluta*, *Muricidae*, *Natica*, *Militoidae*, *Ringicula*, *Olivella*, *Turritellidae*, *Hemijusus*, *Murex*, *Nassarius*, *Fusinus*, *Terebellum*, *Conidae*, *Cassidae*, *Marginellidae*, *Turridae*, *Gemmula*, *Cerithidae*, *Cancellariidae*, *Bursidae*, *Polinices*, *Nassaria*, *Oliva*, *Impages*, *Turris*, *Turridrupa*, *Hyalina*, *Galeodea*, *Atys* and *Volema*. Noad

(1998) also recorded the following bivalves: *Spondylus*, *Chama*, *Gari*, *Callanaitis*, *Nucula*, *Nuculanidae*, *Tellinacea*, *Laternula*(?), *Veneracea*, *Anadara*, *Psamobiidae*, *Cardilia*, *Cardiacia*, and *Bassina*. He also recorded the scaphopod *Fissidentalium*, an echinoid, and the crustacea: *Ranina*, *Typilobus*, *Cancer*, *Myra*, *Callianasa*, *Portunid*, *Galene* and *Palaeograpsus*. However, no precise age could be determined.

XXIII.2. GANDUMAN FORMATION

Haile and Wong (1965) summarize this formation as a paralic coastal swamp deposit, composed of clay and sandstone with abundant carbonaceous material and some lignite seams. The average dip eastwards is 15°. It is conformable upon the Sebahat Formation but also onlaps onto the Tungku Formation basement (Figure 109). It is unconformably overlain by the Togopi Formation.

XXIII.2.1. Lithology

Noad made a detailed study of the sedimentology of the Ganduman Formation along the Sahabat loop road which leads to the Sahabat Beach Resort on the Sulu Sea. The following lithofacies have been detailed:

Trough cross-bedded sandstone. This facies is of stacked trough cross-bedded green–grey very coarse to medium-grained sandstone. The stacked set can reach 10 m or more in thickness. Clasts of extraformational chert and vein quartz suggest a provenance from the ophiolite basement and the Western Cordillera. The clasts, usually <10 cm diameter, form lags at the trough bases. There are also subrounded mud clasts. The cross-beds dip 20–34° towards the east-north-east (ENE). The bounding surfaces between the cross-beds are undulating and sub-parallel. Deposition is interpreted to have been in fluvial channels. The total absence of bioturbation argues against a marine environment

Sandstone fining upwards to mudstone. Clean, very coarse-grained thick sandstone, containing mudstone and small chert clasts, shows channelized bases and conglomeratic lags. *Ophiomorpha* trace fossils are common. The sandstones show trough cross-bedding and pass up into rippled muddy siltstone and then into grey mudstone containing vertical burrows. The environment is interpreted as fluvio-estuarine.

Grey-green mudstone beds. Rooted trees and peat horizons are seen. The setting was probably terrestrial.

Thick highly erosive channelized sand bodies. They may be of steeply inclined (20–35°) thinly bedded fine-grained sandstone and mudstone stratification filling wide (as much as 400 m) flat-based channels. *Ophiomorpha* are common. These deposits have been interpreted as point bars and an estuarine environment. There may also be thin sets of stacked trough cross-beds, with abundant *Ophiomorpha*, passing up into rippled siltstone. These sediments are interpreted as the fill of large fluvio-estuarine channels.

Channel fills of sandstone–mudstone interbeds. They form channels only around 5–30 m across. The fill is of finely parallel laminated rhythmites. These channels are interpreted as tidal inlet or creek deposits under low-energy conditions (Noad, 1998).

Thin planar cross-bedded sandstone. The beds are very fine-grained and rich in *Ophiomorpha.* Deposition was as dunes in a tidal-dominated upper shoreface coastal environment.

Carbonaceous mudstone. The dark-grey channel-fill beds contain abundant woody debris and occasional macrofaunal casts. There are interbedded coals. Plant debris includes whole leaves and coal seams up to 28 cm thickness. Palaeocurrents are towards the east. Some sections are bioturbated. They are interpreted as the remnants of ox-bow lakes and meandering tidal creeks.

Very fine-grained silty sandstone. The beds are laterally continuous. Very fine-grained silty calcareous sandstone predominates, interbedded with muddy siltstone beds. Bioturbation and accumulations of mollusc shells are common. A shallow marine environment is interpreted with shell accumulations washed in by wave action. Close to the mouth of an estuary is possible.

XXIII.2.2. Palaeontology and age

Haile and Wong (1965) reported that the Ganduman Formation is impoverished in Foraminifera. However, *Globigerinoides obliqua* Bolli has been found and it indicates a Tgh (Pliocene) age.

Noad (1998) recorded the palynomorphs *F. claricolpata*, *F. levipoli*, *M. diversus* and *M. annulatus* from which an age of Lower Miocene or younger may be deduced. Noad (1998) also identified the following gastropods: *Tibia*, *Turritella*, *Conus*, *Hemifusus*, *Bufonaria*, *Murex*, *Strombus*, *Natica*, *Nassaridea* and *Opisthobranch.* He also identified the bivalves: *Anadara*, *Corbula*, *Placuna* and *Venerid.* No age conclusion could be made.

XXIII.2.3. Fission track and thermal maturity

The Ganduman Formation contains only poor to fair organic carbon contents with values ranging from 0.1% to 1.3%. Coal and shale samples have yielded vitrinite reflectivity values R_0% ranging from 0.23 to 0.44 (Khalid et al., 2003). The sediments are thermally immature. Biomarker parameters such as carbon preference index, pristine/phytane ratios, trisnorneohopane/trisnorhopane ratios, oleanane/C_{30} hopane as well as the hopane and sterane isomerization ratios, support the results of the vitrinite reflectivity (Khalid et al., 2003). Swauger et al. (1995) reported an R_0 value of 0.23±0.03% for a lignite (54B) of the Ganduman Formation.

Only one Ganduman Formation sample (53C) was analysed for fission track. Seventeen zircon crystals gave a value of 76.7±15.1 Ma (Swauger et al., 1995). The crystals preserve their provenance age, which is predominantly Cretaceous (Figure 101) with several older crystals. The pattern is similar to the Sebahat Formation.

XXIII.3. TOGOPI FORMATION

The Togopi Formation is of loosely cemented rubbly reef limestone, calcareous sandstone, clay and marl, with a general dip of 3–4° towards the NE (Haile and Wong, 1965). However, the formation thickens towards the Sulu Sea and becomes flat-lying (Figure 110). It is usually said to be unconformable upon the Ganduman Formation, but Figure 110 shows it to be conformable.

XXIII.3.1. Lithology

Noad (1998) made a study of outcrops along the Sahabat Road and described the following facies:

Muddy fossiliferous calcarenite limestone. It is mainly a bioclastic calcarenite framestone with a muddy component. It contains abundant *in situ* corals of 15 or more species. Bivalves and gastropods are abundant. The rocks were deposited in a marine reef-flat environment protected from the open sea.

Molluscan packstone limestone. Bivalve fragments dominate the packstone. Other components include echinoderm fragments and small pieces of gastropods and ostracods. The environment was a shallow subtidal setting where medium energy caused rounding of the fossil fragments.

Marly limestone containing rounded coral clasts. The coral clasts are set in a wackestone matrix containing a few small Foraminifera. The setting was probably a foredeep on the seaward side of the reef complex.

XXIII.3.2. Palaeontology and age

Haile and Wong (1965) listed the following Foraminifera, indicating a Pliocene to Pleistocene age: *Amphistegina* spp., *Calcarina spengleri* (Gmelin), *Cibicides* sp., *Elphidium* cf. *koeboeense* LeRoy, *E. craticulatum* (Fichtell and Moll), *E.* cf. *decipiens – hispidulum* group, *Nonion japonicum* Asano, *Operculina complanata* (Defrance), *Peneroplis* sp., *Quinqueloculina* spp., *Steblus annectens* (Parker and Jones), *S. ketienziensis* Ishizaki, *Steblus* spp. and *Textularia* spp.

The following *Otoliths* (ear bones of fish) were recorded: *Setippinna retusa* Stinton, *Coilia planata* Stinton, *Myctophym circularis* (Frost), *Coryphaenoides bipartitus* Stinton, *Myripristis trigonis* Stinton, *Apogon sictatus* Stinton and *Lutianus geminans* Stinton.

The following Echinoderms were listed: *Prionocidaris* cf. *bispinosa* (Lamark) and *Echinocyamus* cf. *planissimus* Clark.

The above fauna suggests littoral to inner neritic conditions with intermittent reef growth.

Nuttall (1965) identified a large fauna of *Presobranchiata* and *Bivalvia*, which generally confirm a Pliocene to Pleistocene age, although a few species were long-lived, dating back to the Early Miocene.

Prosobranchiata: *Ancilla ampla* (Gmelin), *Apollon bituberculare* (Lamark), *Architectonica modesta* (Philippi) subsp. *rutteni* Cox, *Aulica aulica subsp. martini* Cox, *Cancellaria* aff. *bifasciatum* Deshayes, *Cantharus* aff. *ventriosus* (Martin), *Clava attenuatum* (Philippi) var. *reinhardi* Cox, *Cl. jonkeri* Martin var., *Cl. trailli* (G. B. Sowerby) var., *Clavus unifasciata* (Smith) = *C. tjibaliungensis* (Martin), ?*Clypeomorus verbeeki* (Woodward), *Conus mucronatus* aff. Subsp. *socialis* (Martin) = ?*Conus* sp. *indet. of* Nutall, *C. mucronatus* Reeve = *C. socialis* Cox non Martin, *C. mucronatus* Reeve = *C.* sp. nov. aff. *mucronatus* of Nutall, *Conus* sp. ? *extinct*, *C.* aff. *ngavianus* Martin, *C.* cf. *croceus* Smith, *C. djarianensis* Martin, *C. sulcatus* Hwass subsp. *vlerki* Cox, ?*C sulcatus* aff. subsp. *undulatus* G. B.Sowerby, *C. varius* Linné var. *extinct*, *Conus?* sp. nov. *Cryptospira tricincta* (Hinds), *C. pilzi* Cox, *Dolomena togopiensis* (Cox), *D. labiosus teschi* (Cox), *Eburna spirata* (Linné) = *E. canaliculata* (Schumacher), *Erosaria sabahensis* Cox, *Erroinea (Adusta) cowiei* Cox, *E.* (*A*) *posewitzi* Cox, *Ficus subintermedius* (d'Orbigny), *Fusinus menengtenganus* (Martin), *Gemmula* (*Unedodogemmula*) sp., *Gyrineum spinosum* (Lamark), *Hemifusus ternatanus* (Gmelin), *Jezzleri* Cox, *Labiostrombus denti* (Cox), *L. varinginensis* (Martin), *L. varinginensis* martini(Oostingh), *L. varinginensis overbecki* (Cox), *Littorina scabra* (Linné), *Lampusia pilleare borneana* Cox, *Lyria jugosa* (J. de C Sowerby) var., *Marginella* (*Eratoidea*) cf. *karikalensis* Crossman = *M.* (*E.*) *ringicula* Beets *non* Sowerby, *M.* (*E.*) *bonneti* Crossman, *Mitra* (*Tiara*) *flammea* Quoy, *M.* (*T.*) *flammea interlirata* Reeve, *M.* (*T.*) *indentata* G.B.Sowerby, *Murex bataviana* Martin, *M. ternispina* Lamark, *M. trapa* Röding = *M. martiniamus* Reeve, *M. macgillivrayi* Dohrn = *M. brevispina* Cox *non* Lamark, *Nassaria acuminate* (Reeve) = *N. gendinganensis* (Martin), *Nassarius* cf. *verbeeki* (Martin), *Natica* cf. *marochiensis* Gmelin, *N. vitellus* (Linne) = *N. rufa* Born, *N. vitellus helvacea* Lamark = *N. globosa* Chemnitz, *N.* aff. *Zebra* Lamark, *Neosimnia javana* (Martin), *Neverita* cf. *didyma* (Roding), *Oliva accuminata* Lamark (?), *Oliva hattoni* Cox (?), *O. funebralis* Lamark, *O.* cf. *carneola* Gmelin, *O.* (*Anazola*) *subulata* Lamark, *Peristernia pulchella* (Reeve), *Phalium dalrymplei* Cox, *Phalum bisculata* (Schubert and Wagner) = *P. pila* (Reeve), *Polinices perselephanti* (Link), *Pterynotus pinnatus* (Swainson), *Ranularia* aff. *clavator* (Dillwyn) = *Cymatium gallinago* Cox *non* Reeve, *Rimella crispate* (G.W.Sowerby), *R. cancellata spinifera* Martin, *R. javana* Martin var., *Rissoina* (*Phosinella*) *clathrata* A. Adams = *R.* (*Phosinella*) *clathrata* var. *beetsi* (Cox), *R.* (*P.*) *nitida* A. Adams, *R.* (*Eurissolina ?*) cf. *plicata* A. Adams, *R.* (*Zibinella*) sp., *Siphonalia* (*Pseudoneptunea*) aff. *varicose* (Anton), *Tenogadus* cf. *multistriatus* Nuttall *non* Deshayes, *Terebellum terebellum* (Linné), *Thais* (*Stramonita*) aff. *Carinifera* (Lamark), *Tibia wenki* Cox, *Tonna* cf. *zonata* (Green), *Trigonostoma crenifera* (G. B. Sowerby) = *T. crispatum* Crossmann and Martin *non* Sowerby, *Triphoris micans* Hinds, *non T. micans* Hinds of MacNeil, Okinawa, *Turricula nelliae* (Smith) = *T. byorituensis* Nomura, *Turris* (s.s.) *garnonsii* (Reeve), *T.* (*s.s.*) cf. *crispa* (Lamark), *Turritella* aff. *jenkinsi* Crossmann, *T. terebra kendengensis* Altena, *T. terebra* cf. subsp. *spectrum* Reeve, *T. cingulifera* G. B. Sowerby,

Turritella sp. extinct, *Vermetus javanus* Martin, *Vexillum batavianum* (Martin), *V. schurmanni* Cox, *V.* cf. *taeniatum* Nuttall *non* (Lamark), *Vexillum* sp. (?) extinct, *Vitrinella* (*Lydiphnis*) *cingulifera* A. Adams ?= *V.* (*L.*) *novemcarinata* (Melvill), *V. reeviana* (Hinds) subsp. *henjamense* (Melvill and Standen), *V. sculptillis* (Garrett), *Xenophora solarioides* (Reeve) subsp. and *X. calculifera* (Reeve).

Opisthobranchiata: *Ringicula assularum* Watson, *Ringicula kochiana* Sowerby,

Solidula cf. *pudica* (Adams), *S. solidula* (Linné), and Sulcoretusa concentrica (*Adams*).

Bivalvia: *Amussium hulshofi* (Martin), *A.* (*Pseudentolium*) *borneanum* Cox, *Anadara* cf. *philippiana* (Dunker), *A. tambacana* (Martin), *A. wendti* (Lamy) = *A. oostinghi* (Beets), *A. antiquate* (Linné), *Antigona chemnitzi* (Hanley), *Arcopsis compressa* (Martin), *A. sculptilis* (Reeve) = *A. menkrawitensis* Beets, *A. sculptilis* (Reeve) var.(living),

Apolymetis cf. *contorta* (Deshayes) = ? *A. kerheri* (Oostingh), *Arcopagia* (*Arcopaginula*) *inflata* (Gmelin) = *A.* (*A.*) *hippopoidea* (Lamark), *A.* (*Pinguitellina*) *pudica* (Hanley),

Arcopagia martensi (Lynge), *Barbatia fusca* (Bruguiere), *B. wenki* Cox, *Cadella semen* (Hanley), *Chama* sp., *Cardita ovalis* Reeve, *C. javana* Martin, *Chione tiara* (Dillwyn)= *C. isabellina* (Philippi)= *C. chlorotica* (Philippi)= *C. karikalensis* (Cossmann), *Chlamys senatoria* (Gmelin), *C. singaporina* (Reeve), *C.* aff. *albolineata* (G. B. Sowerby), *Corbula modesta* Hinds, *C. crassa* Hinds = *C. lamellate* Fischer, *Corbula* aff. *scaphoides* (Hinds) = *C.* aff. *erythrodon* Nuttall *non* Lamark, *Corbula* aff. *fortisulcata* Smith, *Dosinia histrio* (Gmelin) = *D. steinmanni* Fischer, *Crassatella radiata* (G.W.Sowerby), *C. foveolata* (G.W.Sowerby), *Circe scripta* (Linné), *Circe ?*sp. nov., *Clementia papyracea* (Gray), *Cuspidaria* (*Cardiomya*) sp., *C.* (Plectodon) sp., *Excellichlamys histrionicus* (Gmelin) = *E. spectabilis* (Reeve), *Gari pallida* (Deshayes), *Gouldia jucunda* (Smith) = *G. pryeri* Cox, *Hemicardium* aff. *guichardi* (Bernardi), *Hippopus hippopus* (Linné),

Lammelliconcha phillippinarum (Hanley) = *L. molengraaffi* (Tesch), *Leda* (*Jupiteria*) aff. *confusa* Hanley, *L.* (*J.*) aff. *fulgida* Adams, *L.* (*Thestyleda*) aff. *ramsayi* Smith, *Lima lima* (Linné) var. *paucicostata* G.B. Sowerby = *L. squamosa* Lamark, *L. lima* var. *multicostata* G.B. Sowerby = *L. squamosa* Lamark, *Limopsis multistriata* (Forskal), = *L. cancellata* (Reeve) = *L. venusta* (Martin), *Lima* (*Limaria*) aff. *orientalkis* Adams and Reeve,

Lucina (*Here*) *gemma* Reeve = *Phacoides tegalense* Oostingh, *L.* (*Pleurolucina*) aff. *pisum* Reeve, *Macoma cuspidate* (Dehayes), *M.* (*Psammacoma*) *galathea* (Lamark), *Macalia bruguleri* (Hanley), *Mactra* sp. aff. *mera* Reeve, *Myodora trigona* Reeve,

Nucula cumingi Hinds, *Paphia undulate* (Born), *Paphia cf. amabilis* (Philippi), *P. textile* (Gmelin), "*Pecten*" aff. *inaequivalvis* (G.B. Sowerby) = *Minnivola cf. pyxidata* Nuttall *non* (Born), *Miocardia vulgaris* (Reeve), *Nemocardium* (*?Lyrocardium*) *bechei* (Reeve),

Ostrea hyotis Linné, "*Pecten*" cf. *asper* (G.B. Sowerby), "*P.*" *leopardus* (Reeve),

Pecten sp. ?extinct, *Pitar belcheri* (G. B. Sowerby), *Placenta ephippium* Philippson = *P. sella aucct, Pliculata plicata* (Linné), *P. imbricate* Menke = *P. essingtonensis* Reeve, = *P. philippinarum* Reeve, *Quadrans belcheriana* (G.W.Sowerby), *Semelangulus tenuillirata* (G.W.Sowerby), *S. diluta* (Smith) = ?*S. prototenuilirata* (Nomura),

Pseudarcopagia elegantissima (Smith), *Sinodia sinuate* (Gmelin) = *S. nana* (Reeve) = *S. excise* (Chemnitz), *Solenocurtus aff. philippinarum* Dunker, *Spisula* sp. aff. *trigonella* (Lamark), *Spondylus imperialis* Chenu, *Tellinides* cf. *ovalis* (G.W. Sowerby), *Timoclea subnodulosa* (Hanley) var., *T. bataviana* (Martin), *Topidoleda* aff. *lata* (Hinds), *Trisidos semitorta* (Lamark), *Vasticardium lacunosum* (Reeve) = *Cardium* sp. nov. aff. *lacunosum* Nuttall, *V. denticostulatum* (Beets) and *Zozia coarctatus* (Gmelin),

Noad (1998) identified the following gastropods: *Lanbius*(*?*), *Opisthobranch*, *Neogastropod*, *Turbo* (*?*), *Cyprea* (*Cowrie*), *Natica*, *Conus* and *Voluta*. He identified the following bivalves: *Anadontia* (*?*), *Pitar*, *Periglynta* (*?*), *Carditidae*, *Spondylus*, *Chama*, *Clementia* and *Cardiiae*.

Chapter XXIV
Semporna Peninsula Volcanism

Neogene volcanism occurs around three principal areas: Tawau – Gunung Wullersdorf area, Gunung Pock – Semporna area, and Mostyn Estate near Kunak. Each has its characteristic differences, and will be described separately.

XXIV.1. TAWAU – GUNUNG WULLERSDORF AREA

The first comprehensive regional description is by Kirk (1962). Subdistricts have been described by Lim (1981, 1988). The geochemistry has been detailed by Chiang (2002). The volcanic geology of the area is shown in Figure 114, (redrawn after Lim, 1981, 1988).

XXIV.1.1. Age of volcanism

Whole-rock K:Ar dating is listed in Table 24. The andesites and diorites have yielded ages indistinguishable from the lavas of the Dent Peninsula, as follows:

Tawau–Wullersdorf	5*	12.67 (Middle Miocene)	± 2.50

*Excluding two other values of 6.4 and 1.62 Ma, assumed to be of younger lava flows (Lim, 1988).

These dates, as with the Dent Peninsula, indicate a Middle Miocene age and the identical geochemical signatures, as shown below, support the conclusion that the older volcanism of Tawau and Mount Wullersdorf was Middle Miocene and belonged to the same province as the southern Dent Peninsula.

However, Lim (1988) reported two distinctly younger ages, whole-rock 6.4 Ma (Upper Miocene) for andesite of Gunung Andrassy. Biotite extracted from dacite at Gunung Maria yielded an age of 1.62 Ma. The younger volcanic rocks of the Tawau–Gunung Wullersdorf area have been shown to be geochemically distintly different from the Miocene rocks. They are predominantly basaltic-andesite, may contain olivine and geochemically resemble the fissure-related Pliocene eruptions of the Mostym Estate of Kunak. Accordingly they are classified together with them, indicating a young age of volcanism. Even on the slopes of Quoin Hill, recognized as a younger volcanic edifice, the volcanic rocks are weathered to a depth of 6 m, so that Kirk (1962) concludes that the volcanism did not continue into the Quaternary, as some writers have suggested. Only lava fragments, in various stages of decomposition, occur within the soils. It is untrue that they contain the so-called 'volcanic bombs'. However in the Burut Estate, 13.7 km NNE of Tawau, a charred tree stump in growth position was found within a 1.2 m thick bed of dacite breccia. The breccia itself was overlain

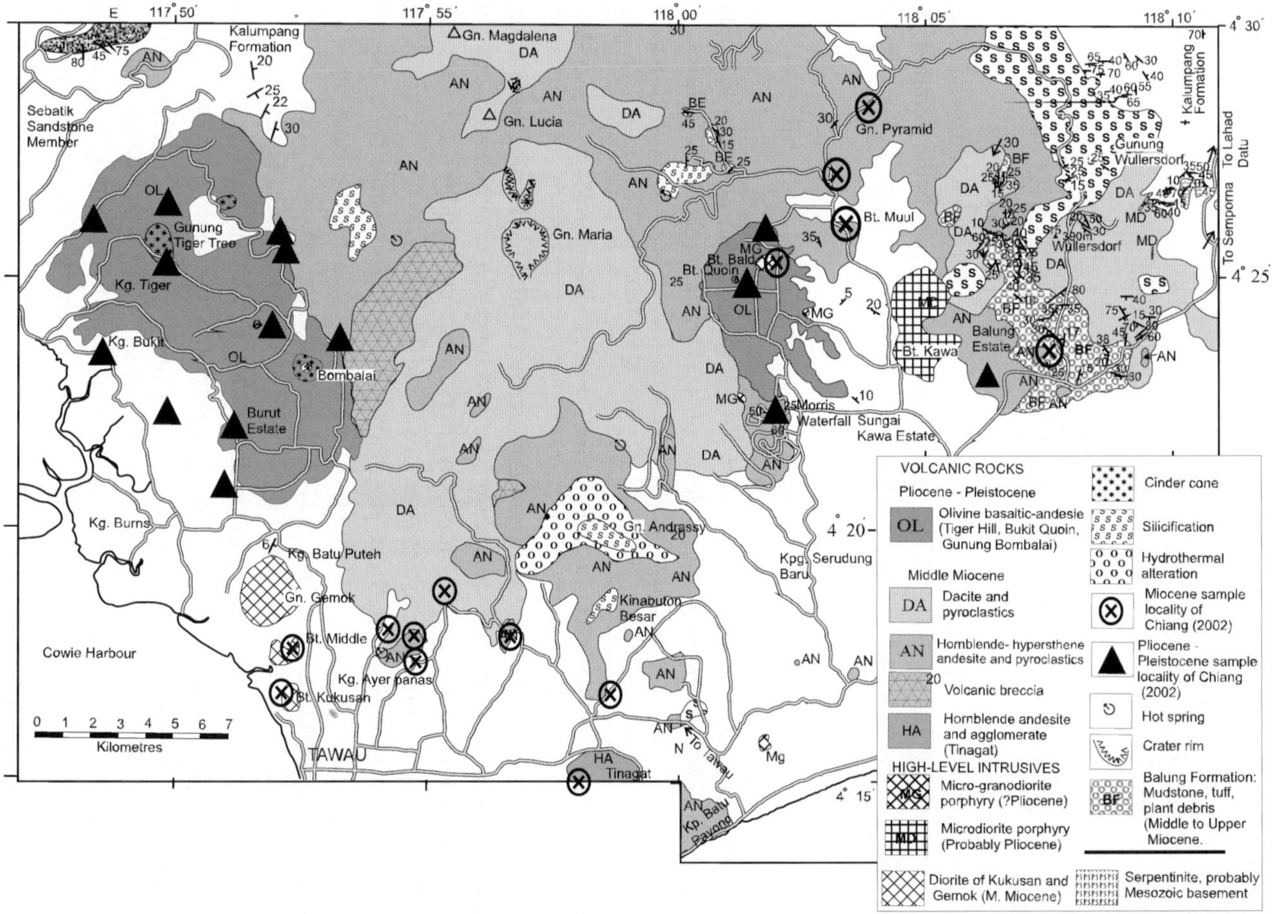

Figure 114. Geological map of the Tawau and Gunung Wullersdorf volcanic area of the Semporna Peninsula (based on Lim, 1981, 1988). Geochemical sample sites from Chiang (2002). The outcrops of Lower to Middle Miocene Kalumpang Formation in the NE and NW are unshaded. Unshaded southern region is underlain by Quaternary Alluvium. With permission from Minerals and Geoscience Department, Malaysia.

by vesicular basalt. The charred wood is of the species *Dialium*, known today locally as Keranji. The charred wood gave a radio-carbon date of 27,000 ± 500 years (Wilford et al., 1967). It appears therefore that volcanism continued locally in the Tawau area into Pleistocene time. There are no fumeroles in the Tawau–Wullersdorf volcanic area. There are relict volcanic craters and a few hot springs, but the system is extinct.

Only two volcanic rocks have been sampled for fission-track dating by Swauger et al. (1995); Lim (1988) reported a data set from the Tingat area. The results are as follows (see Figure 77):

Specimen	Rock type	Locality	Zircon age (Ma)	No.	Apatite age (Ma)	No.
94SB31A	Andesitic tuff	Mosque, Sandakan		0	33.9 ± 7.7 (Lower Oligocene)	20
94SB82A	Andesite dome (hypersthene diorite)	Kukusan quarry, Tawau	12.4 ± 0.9 (Middle Miocene)	17	16.8 ± 3.5 (Middle Miocene)	20
	Hornblende andesite	Tinagat area	17.0 ± 2.9 Ma (Lower to Middle Miocene)	22		

The apatite fission-track results from the large Kukusan quarry at Tawau have not been annealed, and they confirm the zircon fission-track age and fully agree with the Middle Miocene age determined by whole-rock K:Ar dating (Table 24). The hornblende andesite from the Tanagat area (Lim, 1988) gives rather confusing zircon fission-track results for a lava: 5 crystals lie within the range 10–15 Ma, 7 within 15–20 Ma, 3 within 20–25 Ma, 1 within 25–30 Ma, 2 within 30–35 Ma, 1 within 35–40 Ma, 2 within 55–60 Ma and 1 within the range 70–75 Ma. This is an unacceptable range for a cooled magma, and therefore it must be concluded that the rock is a tuff, and that some of the zircons are detrital, maintaining their provenance age. Lim (1988) took the youngest group 13.6–21.4 as representing the magmatic age, with an average 17.0 ± 2.9 Ma (Lower Miocene).

XXIV.1.2. Middle to Late Miocene volcanism

The Island-Arc volcanic edifices were built up over a surface composed of Lower to Middle Miocene Kalumpang Formation and Middle to Upper Miocene Balung Formation, as described by Lim (1981, 1988) and shown in Figure 114. The lavas are calc-alkaline and extend in composition from andesite to dacite. There are also exposures of high-level intrusives ranging from granodiorite to diorite. The larger edifices, such as Gunung Maria, have preserved geomorphological features such as crater rims.

XXIV.1.2.1. Geochemistry

Many whole-rock analyses have been published by Kirk (1962, 1968), Lim (1981) and Chiang (2002). A selection of analyses of volcanic and sub-volcanic rocks is given in Table 27.

Table 27. Analyses of selected Miocene volcanic and sub-volcanic rocks of Tawau–Wullersdorf area

Specimen	a	b	c	d	e	f	g	h	i
Source	1	4	2	2	3	1	2	1	4
SiO_2	56.24	57.36	60.90	61.74	61.80	62.31	63.04	63.13	98.60
TiO_2	0.68	0.54	0.68	0.38	0.57	0.56	0.71	0.53	0.51
Al_2O_3	19.10	17.27	17.97	16.07	16.40	16.72	15.05	16.52	0.20
Fe_2O_3	7.33	4.10	1.18	2.71	2.36	6.14	3.21	5.96	0.03
FeO	*	3.38	4.68	2.73	2.90	*	2.27	*	0.21
MnO		0.13	0.12	0.11	0.13	0.12	0.14	0.11	0.01
MgO	3.31	3.41	0.99	1.93	3.02	3.21	2.95	2.71	0.08
CaO	8.99	7.32	7.10	5.82	6.34	6.22	5.82	5.79	0.09
Na_2O	2.26	2.16	2.42	2.40	2.77	2.41	2.36	2.63	<0.01
K_2O	1.80	2.06	1.91	2.20	1.85	2.04	1.52	2.47	0.01
H_2O+		1.12	1.50	2.70	0.40		1.42		0.17
H_2O-		0.90	0.09	0.63	0.40		0.97		0.07
CO_2			0.04	0.04	<0.01				
P_2O_5	0.19	0.52	0.20	0.18	0.17	0.15	0.61	0.13	0.03
Total	100.05	100.27	99.78	100.31	99.40	99.87	100.07	99.99	100.02

Data Source: 1, Chiang (2002); 2, Lim (1981); 3, Swauger et al. (1995) and 4, Kirk (1968).
a = SBK 87, basaltic-andesite. North side of hill north of Mull Hill, Wullersdorf area (Loss on ignition 0.66). b = andesite, NB 6685, Mount Lucia. c = microdiorite porphyry, J13604, Southern slope of Bukit Kawa. d = microgranodiorite porphyry, J13639, SE of Bukit Bald. e = andesite, 94SB-83A, andesite, Batu Payong, east of Tawau. f = SBK 11, hornblende andesite, Tinagat, east of Tawau (loss on ignition 1.01). g = NB 6690, dacite, near summit of Gunung Maria. h = diorite, SBK1, Kukusan Hill quarry, Tawau (loss on ignition 1.61). i = silicified breccia, NB6720, Mount Wullersdorf.
* Total iron as Fe_2O_3.

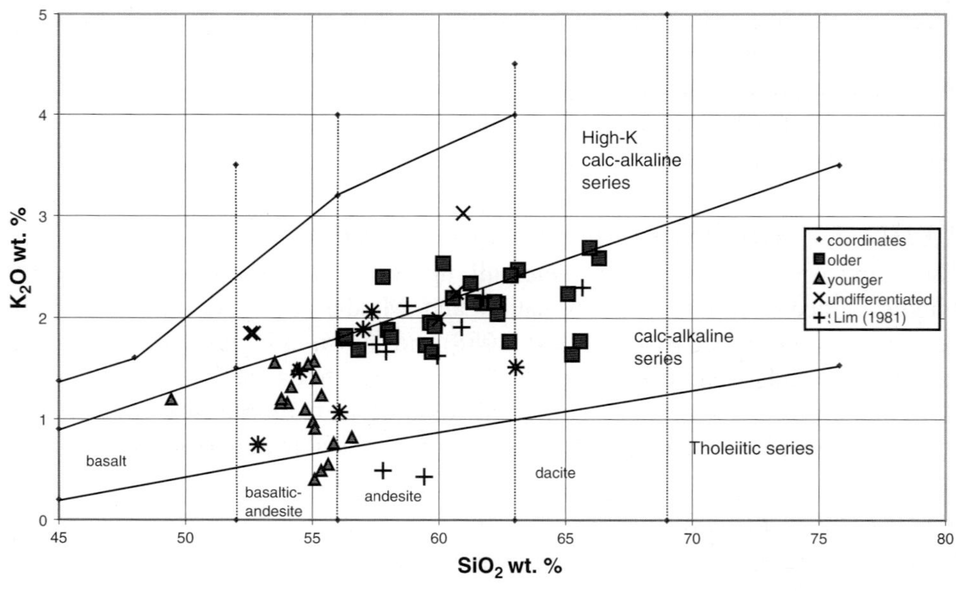

Figure 115. K_2O versus SiO_2 for Tawau district.

All available analyses have been plotted on a K_2O (wt%) versus SiO_2 diagram (Figure 115). The analysed rocks extend over the whole andesite range and half way through the dacite range. They lie within the calc-alkaline field, and into the lower part of the high-K calc-alkaline field (Figure 115).

Rare-earth analyses were published for three samples: 82A, Kukusan Hill quarry (andesite dome); 83A, Batu Payong andesite lava flow and 84A, Kobal quarry andesite dome (Swauger et al., 1995). The results are shown in Figure 116 and the light REE-enriched patterns are typical of calc-alkaline island-arc intermediate rocks.

The plot of incompatible elements (Figure 116B) of the Miocene volcanic rocks is identical in all respects to that of the volcanic rocks of the Dent Peninsula, as shown in Figure 111. This indicates that the Miocene volcanism of the Tawau–Wullersdorf district is one and the same as that of the Tunku Formation of the Dent Peninsula. They are both integral parts of the same volcanic arc and the magmas shared the same source. By dramatic contrast, the younger (Pliocene–Pleistocene) lavas of the Tawau–Wullersdorf region have a different source (Figure 116C) and a major tectonic reorganization separates the Miocene from the later basaltic-andesites, that are similar to the Pliocene lavas of the Mostyn district.

XXIV.1.2.2. Petrography

The more basic andesites contain phenocrysts of plagioclase (forming 35–55% of total), clinopyroxene, orthopyroxene and magnetite. More evolved andesites contain hornblende. Biotite is present in the dacites with hornblende and plagioclase, while clinopyroxene occurs in small amounts in the groundmass. In some samples, there are xenocrysts of altered clinopyroxene and hornblende (Chiang, 2002). Lim (1988) notes that there are three common andesite types: hornblende andesite, hypersthene

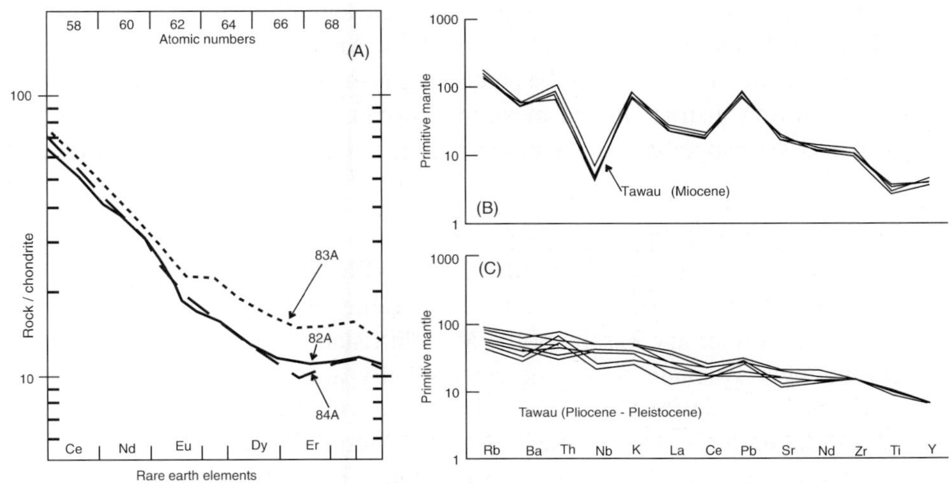

Figure 116. Tawau District, (A) rare earths (from Swauger et al. (1995). Right: Incompatible elements plot of Tawau lavas, from Chiang (2002); (B) Miocene lavas; (C) Pliocene–Pleistocene lavas.

andesite and quartz andesite. There are also agglomerates, dacites and silicified volcanics. Columnar jointing is well displayed along the Sungai Balung (Lim, 1981).

Agglomerate at Batu Payong is of sub-rounded blocks of hornblende andesite in a pebbly tuffaceous matrix. The Mount Wullersdorf volcanic rocks have been hydrothermally altered to quartz–sericite and chlorite–epidote assemblages (Lim, 1981).

Hypersthene diorite forms a line of hills near Tawau: Mount Gemok, Middle Hill and Kukusan Hill. It is grey, medium-grained and only slightly porphyritic. The phenocrysts are andesine, hypersthene and clinopyroxene. Hornblende has been pseudomorphed (Kirk, 1968). However, Kukusan Hill pluton (82A) is not homogeneous, and contains andesite tuff breccia, composed of andesite and diorite clasts, euhedral feldspar, altered hornblende and pyroxene in a glass matrix devitrified to a mixture of smectite and zeolites (Swauger et al., 1995).

Microdiorite porphyry forms a stock at Bukit Kawa (Lim, 1981). It contains phenocrysts of augite, plagioclase and hypersthene in a microcrystalline matrix. Microdiorite dykes also cut the Kalumpang Formation and dacitic volcanic rocks in the Wullersdorf region. Microgranodiorite porphyry forms a stock on Bukit Bald. Phenocrysts of hornblende, biotite and plagioclase occur in a microcrystalline matrix.

XXIV.1.3. Pliocene to Pleistocene volcanism

The younger volcanic flows of the district occur in two main areas: From Gunung Tiger Tree southwards to the Burut Estate including Table Estate, and Bukit Quoin southwards to the Morris Waterfall, including Sungai Balung and Bukit Bald (Figure 114). They are not adequately dated. By analogy with the Mostyn district where two dated tholeiitic basalt samples give an age of 2.95 ± 0.23 Ma (Pliocene), one might assume that the tholeiitic basalts of the Tawau area, which are chemically identical to those of Mostyn, are also Pliocene. The two young radiometric ages (1.62 and 6.40 Ma) are not of tholeiitic basalt, but are respectively from Bukit Andrassy and Gunung Maria. A future priority research study should be precise argon–argon dating of the different lava types of the Tawau district. In the meantime, because of the exact similarity with Mostyn, the rocks mapped as olivine basalt in Figure 114 are taken as Pliocene.

XXIV.1.3.1. Geochemistry

A selection of whole rock analyses of the Pliocene–Pleistocene volcanic rocks of the Tawau and Mount Wullersdorf area is given in Table 28. For comparison, selected analyses for the similar age basalts of the Mostyn Estate, near Kunak is also given.

These younger volcanic rocks characteristically are restricted to the fields of basalt and basaltic-andesite and are considerably less evolved than the Miocene volcanic rocks that underlie them in the Tawau and Mount Wullersdorf area.

All available whole-rock analyses (Kirk, 1968; Lim, 1981; Chiang, 2002) have been plotted on a K_2O versus SiO_2 diagram (Figure 117). Both the Mostyn Estate and the younger Tawau district volcanics are of restricted SiO_2 contents. The

Table 28. Selected analyses of Pliocene low-potassium volcanic rocks of Kunak, Tawau and Wullersdorf areas

Specimen	a	b	c	d	e	f	g	h	i	j
Source	3	3	1	2	2	1	1	3	3	3
	Mostyn Estate, near Kunak					Tawau and Mount Wullersdorf area				
SiO_2	50.99	52.48	52.64	52.70	53.00	52.57	52.87	54.02	55.10	55.63
TiO_2	2.09	2.07	2.02	0.76	2.12	1.70	1.80	1.86	1.86	2.09
Al_2O_3	15.49	14.63	16.17	15.5	14.00	15.57	13.69	14.96	14.96	15.25
Fe_2O_3	12.08	11.86	1.44	4.28	2.71	4.52	4.82	10.63	10.58	11.22
FeO	a	a	8.71	3.70	8.90	4.97	6.62	a	a	a
MnO	0.17	0.15	0.14	0.18	0.16	0.16	0.14	0.15	0.14	0.15
MgO	6.79	6.65	5.78	3.75	5.53	5.91	5.76	5.76	5.93	4.43
CaO	7.90	7.78	7.60	8.09	7.35	8.14	7.78	7.57	7.26	7.11
Na_2O	3.29	3.29	3.25	3.56	2.84	3.50	3.37	3.18	3.09	3.57
K_2O	0.59	0.81	0.75	0.30	0.35	1.85	0.75	1.16	0.91	0.55
H_2O+			1.27	2.50	0.80	0.88	1.24			
H_2O-			0.03	0.90	0.40	0.11	1.19			
CO_2				2.58	<0.01					
P_2O_5	0.25	0.24	0.31	0.10	0.18	0.33	0.26	0.26	0.22	0.23
Total	99.66	99.97	100.11	98.80	98.60	100.21	100.29	99.56	100.04	

Data Source: 1, Kirk (1962); 2, Swauger et al. (1995) and 3, Chiang (2002).
a = basalt, SBK61, Tingkayu Camp, 1 km past guardhouse [Loss on Ignition = 0.13]. b = basalt, SBK31, Sungai Madai waterfall on main road [Loss on Ignition = -0.70]. c = Olivine basalt, NB6138, Mostyn Estate near Kunak. d = basaltic tuff, 94SB-80A, Mostyn Estate. e = vesicular basalt, 94SB-73A, Mostyn Estate f = olivine basalt, NB6286, Quoin Hill, NE of Tawau. g = olivine basalt, NB 6042, Burut Estate, north of Tawau. h = olivine basalt, SBK6, Tiger Hill quarry [Loss on ignition = -0.39]. i = basaltic-andesite, SBK71, Table Estate, Payong Quarry, near Tawau. (loss on Ignition = -0.10). j = basaltic andesite, SBK92, Morris Waterfall, Kawa Estate. [Loss on Ignition =-0.53].
[a] total iron as Fe_2O_3.

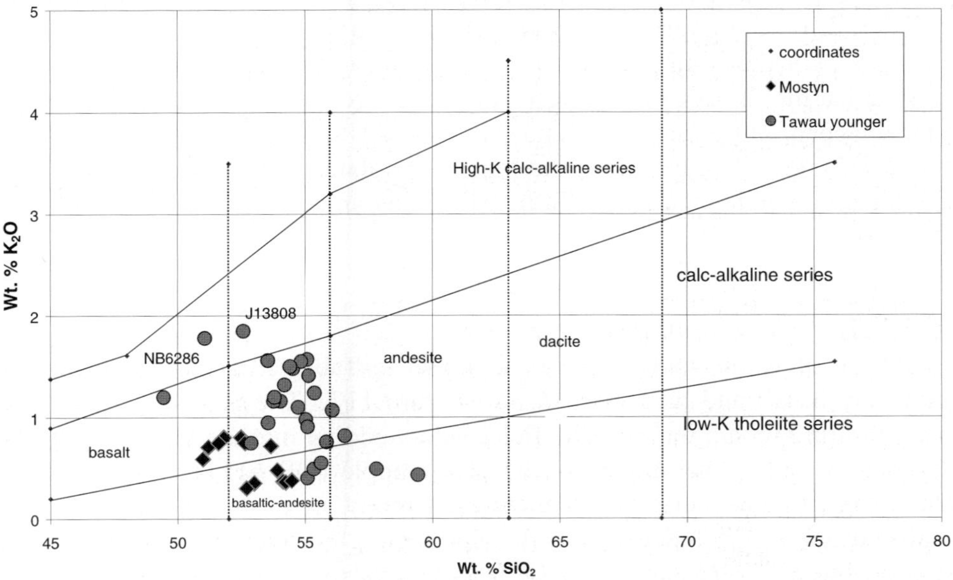

Figure 117. K_2O versus SiO_2 for Mostyn and Tawau younger volcanics.

Mostyn Estate samples are all basaltic-andesite. Whereas the Mostyn Estate volcanics extend from the tholeiite field slightly into the calc-alkaline series, the younger Tawau–Mount Wullersdorf volcanics extend into higher potassium values. Indeed Lim (1981) refers to the two samples NB6286 and J1308 (Figure 117) as belonging to the category high-K basalt. They are from the neighbourhood of Bukit Quoin (Figure 114).

The great contrast between the Miocene (older) and the Pliocene–Pleistocene (younger) volcanism of the Tawau region is well demonstrated by the contents of incompatible elements (Figure 116). The Miocene lavas have a pattern identical to the lavas of the Dent Peninsula. By contrast, the younger Tawau volcanics have a pattern very similar to the Mostyn volcanics. The younger volcanics are unrelated to the older and represent a complete change in the tectonic process.

XXIV.1.3.2. Petrography

The lava flows are both vesicular and non-vesicular. Phenocrystys are commonly plagioclase, olivine and magnetite. Augite and hypersthene are common in the groundmass. Cinder cones may occur and they overlie the flows. The basaltic-andesites show good columnar jointing. Bukit Payong is a relict volcanic cone.

XXIV.2. MOSTYN ESTATE, KUNAK

The basalts underlie most of the Estate. Soil is thin but good and float samples are everywhere. The flat-lying nature of the basalt sequence is well displayed at the waterfall on the main Tawau road. Deeper within the estate is a relict volcanic cone. On the south side there is a hot spring and constructed pool. It is assumed that the Pliocene–Pleistocene basalts are related to 130° trending faults. Radar imagery shows that the Orchid Plateau of Ulu Segama is flat topped and Fitch (1955) described basalts at the foot of the plateau. Only two samples have been dated by whole rock K:Ar (Table 24) giving an Upper Pliocene age of 2.95 ± 0.23 Ma.

XXIV.2.1. Geochemistry

A selection of whole-rock analyses is shown in Table 28. The lava flows are exclusively basalt and basaltic-andesite (Figure 117). They are low-potassium, mostly plotting within the tholeiite field, but extending into the lower part of the calc-alkaline field. The single non-basaltic sample (NB 6111) is quoted from Kirk (1962). It is a dacite tuff occurring distinctly south of the Mandai Estate Basalts. It may be included within the Kalumpang Formation volcanics. I have left it in Figure 117 because of doubt of its relationship. It is nevertheless low in potassium.

The rare-earth plots are shown in Figure 118. The horizontal-concave patterns with relatively low REE contents (< 10–20* chondrite) are indicative of MORB. The ^{87}Sr / ^{86}Sr ratios of the Mostyn lavas range widely from 0.7045 to 0.7063 (Chiang, 2002). The Incompatible trace elements (Figure 118) suggest the Mostyn

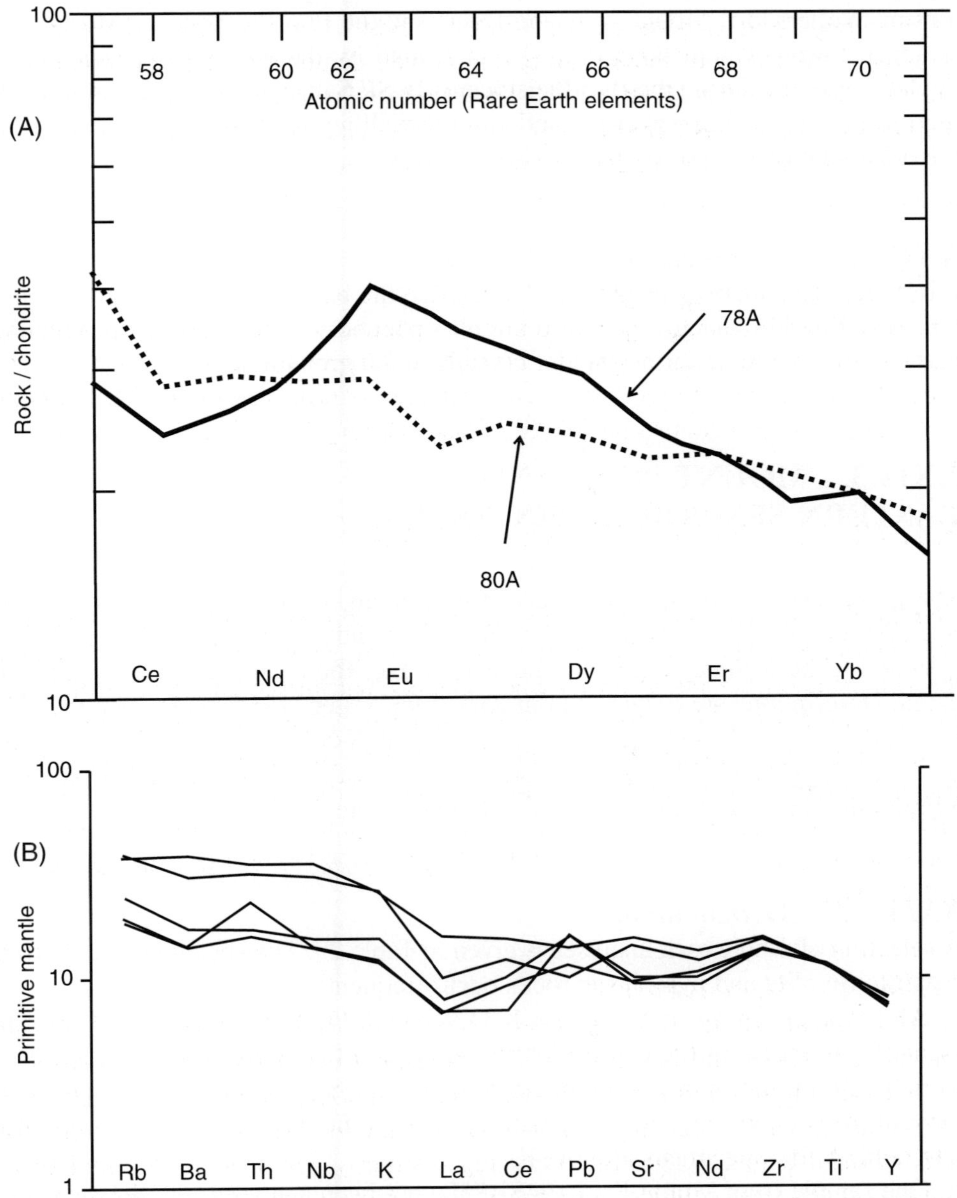

Figure 118. Mostyn Estate, near Kunak: (A) rare earths (from Swauger et al., 1995); (B) Incompatible elements from Chiang (2002).

lavas were derived from an amphibole-bearing spinel lherzolite mantle source (Chiang, 2002). They are not subduction- or plume-related. Lithospheric delamination is a feasible mechanism for their origin, making use of reactivated basement faults.

Chiang (2002) summarizes her studies of SE Sabah as follows: "A dramatic change in character of the mantle source between the Late Miocene [older Tawau–Wullersdorf, Mount Pock and Dent] and the Plio-Pleistocene [Mostyn and younger Tawau–Mount Wullersdorf] is indicated by the within-plate character of the lavas erupted during the Plio-Pleistocene. In SE Sabah, the magmatism is attributed to melting of lithospheric mantle and upwelling asthenosphere as a result of the delamination of an unstable lithospheric root".

XXIV.2.2. Petrography

Large fresh phenocrysts of olivine characterize the basalts. Orthopyroxene characterises the basaltic-andesites. There are also phenocrysts of plagioclase and magnetite. Clinopyroxene forms smaller crystals in the groundmass.

XXIV.3. MOUNT POCK AND EASTERN SEMPORNA PENINSULA

The most easterly volcanic rocks occur in Gunong Pock and the eastern part of Semporna Peninsula, including some offshore islands such as Pulau Gaya. All are assumed to be Middle to Upper Miocene in age, though none has been directly dated. Descriptions are to be found in Kirk (1962), Lee (1988) and Chiang (2002). Pyroclastic hypersthene andesite breccia forms high cliffs and large cliff exposures on Pulau Gaya. Hypersthene basalt forms pillow lava on Manampili Island south of Semporna. The pillows range from 1 to 6 m diameter.

XXIV.3.1. Geochemistry

A selection of whole rock analyses is given in Table 29. The specimens range from basalt to rhyolite and pyroclastic rocks occur frequently.

All the analyses are plotted on a K_2O versus SiO_2 diagram (Figure 119). The main characteristic of the eastern Semporna area is that there is a complete series from basalt through andesite and dacite to rhyolite. The potassium contents indicate most of the samples occupy the upper calc-alkaline field into the high-K calc-alkaline field. A few specimens are low-K.

One sample from Mount Pock (94SB85B) has been analysed for RRE. The plot is shown in Figure 120. It displays a typical shape of subduction-related volcanic arcs of calc-alkaline character. It is however, noteworthy that this specimen is very

Table 29. Selected whole rock w% analyses of Mount Pock and eastern Semporna Peninsula

Sample	a	b	c	d	e	f	g	h	i	j
Source	3	2	2	3	3	3	2	3	1	3
SiO_2	49.43	49.75	53.79	57.25	61.01	66.68	67.47	69.28	70.7	71.10
TiO_2	1.00	1.05	2.30	0.99	0.84	0.44	1.41	0.27	0.40	0.47
Al_2O_3	19.55	17.82	19.57	16.96	16.76	16.11	10.91	15.95	13.30	14.34
Fe_2O_3	11.68	6.86	2.76	9.75	6.91	4.49	1.89	2.82	1.88	3.69
FeO	*	4.21	3.17	*	*	*	0.79	*	1.30	*
MnO	0.21	0.10	0.21	0.15	0.19	0.12	0.03	0.14	0.09	0.10
MgO	3.94	2.21	4.05	3.73	2.49	1.64	1.52	1.18	0.61	0.67
CaO	11.24	8.09	8.86	6.71	5.76	3.69	3.80	3.85	1.43	2.08
Na_2O	1.94	3.22	2.70	4.30	3.76	3.04	3.16	3.67	3.73	2.08
K_2O	0.42	1.24	1.36	0.13	1.42	3.39	3.24	2.32	3.82	4.28
H_2O+		2.49	0.85				3.00		0.80	
H_2O-		1.71	0.20				1.00		0.30	
CO_2		tr							0.21	
P_2O_5	0.08	1.09	0.14	0.24	0.32	0.20	2.02	0.15	0.08	0.10
total	99.49	99.84	99.98	100.21	99.47	99.81	100.24	99.64	98.60	99.74

Data Source: 1, Swauger et al. (1995); 2, Kirk (1968) and 3, Chiang (2002).

a = basalt from pyroclastics, SBK 96, south of Gunung Sigalona (loss on Ignition = 3.83). b = basalt breccia, NB6220, Pulau Manampili near Semporna. c = basaltic-andesite, SP 122, Bukit Sigalong near Semporna. d = andesite, SBK 112, Mount Pock lower Timbangan river (loss on Ignition = 2.00). e = andesite, SBK 17, Mount Pock north of Sungai Gading [Loss on Ignition = 1.27]. f = dacite, SBK 106, Semporna abandoned quarry (loss on ignition = 4.13). g = dacite, SP 124, near Mount Conner, Semporna. h = dacite, SBK 27, Semporna, Kalumpang area, (loss on Ignition = 3.07). i = rhyodacite lava, 94SB85B, Mile 52 quarry, Mount Pock, j = rhyolite, SBK 24, Semporna road cut [Loss on Ignition = 1.00].

* total iron expressed as Fe_2O_3.

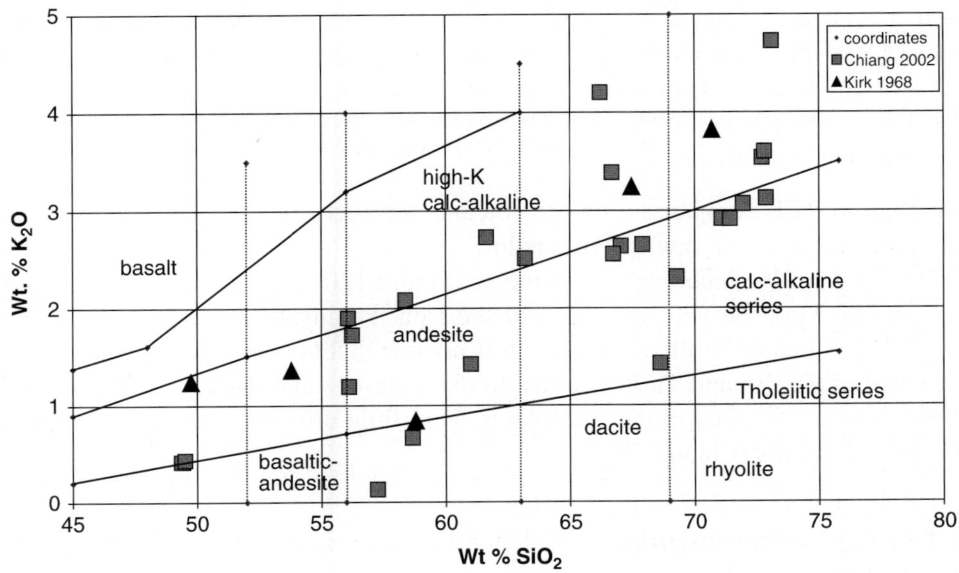

Figure 119. Plot of the Mount Pock and Semporna volcanics.

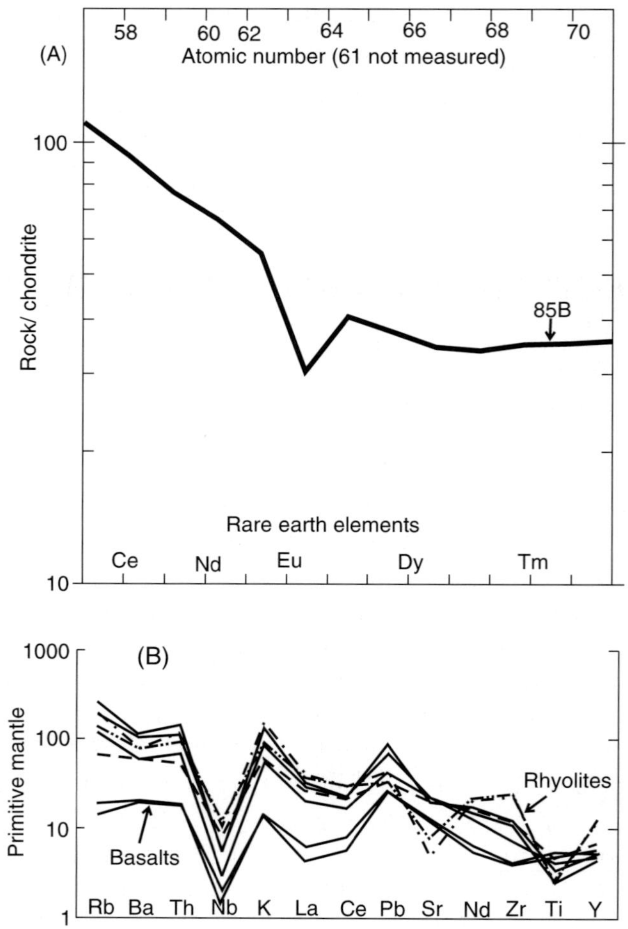

Figure 120. Mount Pock and eastern Semporna Peninsula: (A) rare earth elements (Swauger et al., 1995); (B) incompatible elements (Chiang, 2002).

enriched in REE compared with the volcanic arc rocks of the Dent Peninsula and the Miocene rocks of Tawau–Wullersdorf.

The plots of the incompatible elements (Figure 120) show trends very similar to the Middle Miocene volcanic rocks of the Dent Peninsula and the older volcanic rocks of the Tawau–Wullersdorf area. It is notable that the more basic rocks are less enriched in Rb, Ba and Th in contrast to the andesites and dacites which are more than 10 times enriched in these elements. Very little difference is shown in the Zr, Ti and Y contents (Chiang, 2002).

XXIV.3.2. Petrography

The basalts have phenocrysts of plagioclase, clinopyroxene and magnetite. The andesites also have them and in addition have orthopyroxene. The percentage of orthopyroxene and clinopyroxene decreases in the dacites. Rhyolites have phenocrysts only of plagioclase and magnetite.

Generally the rocks are all somewhat altered and the pyroxenes replaced by calcite. The plagioclase is also clouded by alteration products. It appears that silicificatioin has contributed to the most acid varieties because they contain phenocrysts that are more typical of more basic rocks (Chiang, 2002).

XXIV.4. TECTONIC SETTING OF SE SABAH VOLCANISM

XXIV.4.1. Miocene subduction-related volcanism

The principal characteristics and possible tectonic setting were summarized by Chiang (2002):

1. The volcanic rocks range compositionally from basalt to dacite and are predominantly andesite. Rhyolite compositions appear to have resulted from silicification.
2. The rocks present an Island-Arc character in the field and are closely associated with shallow-water tuffaceous strata.
3. Plagioclase, clinopyroxene, orthopyroxene, magnetite, hornblende and biotite are the main phenocrysts. Crystallization was at shallow crustal levels due to fractional crystallization.
4. The magmas resulted from partial melting of a mantle wedge that was very similar to only slightly depleted compared to the MORB source.
5. Fluids derived from the subducted oceanic lithosphere triggered melting during the Middle to Upper Miocene (Figure 121).
6. Trace element modelling suggests a small degree of auto-assimilation, in an open system during fractional crystallization, affected magmas from the Miocene Tawau–Wullersdorf and Dent volcanic centres, whereas Mount Pock lavas interacted with a wider range of crustal lithologies, leading to greater variability in the ^{87}Sr / ^{86}Sr and trace element concentrations.
7. The influence of fluids derived from the subducted lithosphere diminished from Mount Pock, through Tawau, to the Dent volcanic centres. This is consistent with the Dent volcanic centre lying farthest away from the volcanic front and suggests that the Miocene volcanism of SE Sabah resulted from NW-directed subduction of the Celebes Sea from a trench lying parallel to the Sulu Archipelago. A similar model was also proposed by Hutchison (1992a).
8. There is an increase in the magnitude of the recycled sediment component from Mount Pock to Tawau, through to Dent, consistent with models in which sediment melting at depth is greater than the dehydration of altered oceanic crust.

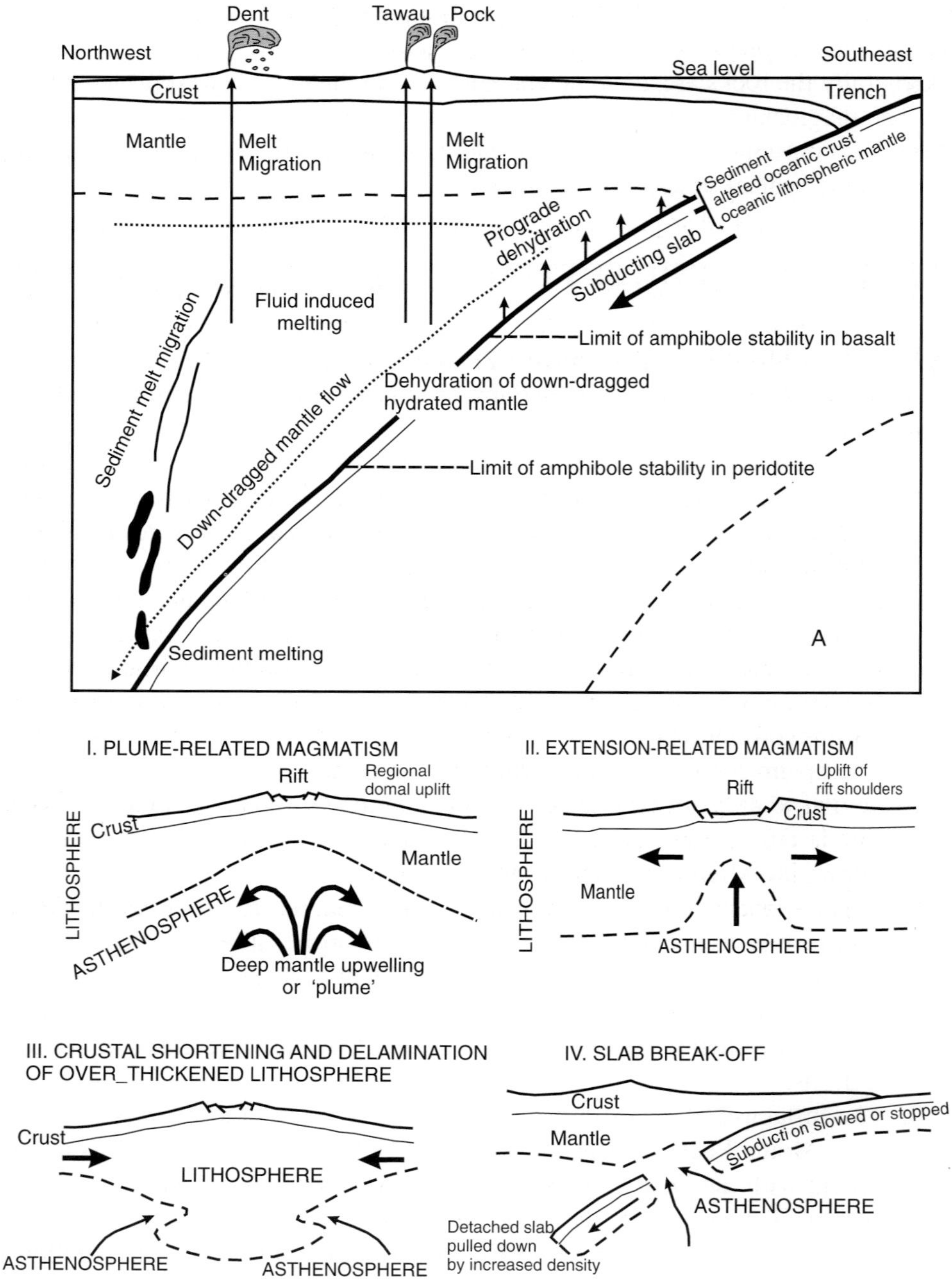

Figure 121. (A) Model of the subduction zone for the Miocene volcanic rocks of SE Sabah. I to IV: Alternative models for melting in intra-plate settings applicable to the Pliocene–Pleistocene volcanic rocks of SE Sabah. Redrawn from Chiang (2002).

XXIV.4.2. Pliocene–Pleistocene rift-related volcanism

The principal features and possible tectonic setting are summarized as follows by Chiang (2002):

1. The Pliocene–Pleistocene basalts and basaltic andesites differ greatly from the foregoing Miocene volcanic rocks.
2. They originated as partial melts of an enriched mantle similar to within-plate basalts.
3. The normalized patterns of the incompatible elements and some of the trace element ratios suggest the lavas were derived from shallow melting of a trace-element-enriched spinel lherzolite source. Some Mostyn basalts were derived from larger degrees of melting of an amphibole-bearing spinel lherzolite source.
4. The elevated radiogenic Sr isotope ratios of some samples suggest that the mantle source was different from that of the Miocene lavas.
5. The elevated ^{87}Sr / ^{86}Sr signature could have resulted from contamination by heterogeneous continental crust during differentiation, contamination of mantle source by a younger heterogeneous enriched component, or contamination of the mantle source by an older heterogeneous component isolated from the convecting upper mantle for between 100 and 750 Ma.
6. The trace element characteristics resemble ocean island basalts from the Philippine Sea Plate and extension-related basalts found in the interior of the East Asian continent and its continental margin.
7. The Sr isotope signature resembles that of extension-related basalts occurring in the continental margin of East Asia.
8. These lavas may be related to heating of the lithospheric mantle and asthenospheric upwelling resulting from delamination of over-thickened lithosphere (Figure 121).

Chapter XXV
Mount Kinabalu Granitoids

Mount Kinabalu is unique in SE Asia, not only because of its impressive elevation of 4101 m but also because of its youth and unusual geological setting. The most comprehensive description is by Jacobson (1970). Its petrology has been detailed by Vogt and Flower (1989) and geochemistry by Chiang (2002). The batholith is exposed over an area of 155 km^2. The bulk of the intrusion appears to be of hornblende quartz monzonite and related lithologies. Detailed mapping and sampling (Jacobson, 1970) have been confined to the southern part of the massif because of virtually impossible accessibility elsewhere.

XXV.1. EMPLACEMENT AGE AND COOLING HISTORY

The following specimen locality sites (Figure 122) lie along the mountain path from the Mount Kinabalu Park Headquarters to the summit: MK1 Mount Kinabalu at 13,455 ft elevation; MK2 Mount Kinabalu at 12,000 ft elevation; MK3 Mount Kinabalu at 10,000 ft elevation; MK4 Mount Kinabalu at 9700 ft elevation. In addition, specimens were collected from a small satellite pluton: 6A Tamparuli Quarry, 6° 7.5′N; 116° 19.9′E.

XXV.1.1. K:Ar dating

Three samples (MK1, MK4 and 6A) were dated by Tom Bills at the Geochron laboratory, Cambridge, MA, and listed by Swauger et al. (2000). The data are given in Table 30, together with two whole-rock analyses reported by Rangin et al. (1990). The previous K:Ar age determinations have been listed by Jacobson (1970). The majority are considered unreliably low, except for their maximum value of 10 Ma for biotite. They are therefore not incorporated here.

The blocking temperatures for the K:Ar system are: hornblende 500 ± 50°C, muscovite 350 ± 50°C, biotite 300 ± 25°C, plagioclase 200 ± 50°C, and K-feldspar 175 ± 50°C (Figure 123). Therefore discordant ages reflect protracted cooling whereas concordant ages indicate rapid cooling usually of volcanic rocks. The exact meaning of a whole-rock K:Ar age cannot be strictly defined. The rocks that constitute the Mount Kinabalu intrusions represent a composite of plagioclase, K-feldspar, biotite and hornblende, all of which have different blocking temperatures. However the rock specimens are mainly constituted of plagioclase and alkali feldspar, so that whole-rock ages would date a time when the granitoids cooled below 200–175°C.

The dates of Table 30 are plotted on a temperature–age diagram (Figure 123). For each date, its position and standard error are fixed on the *x*-axis, but its position on the

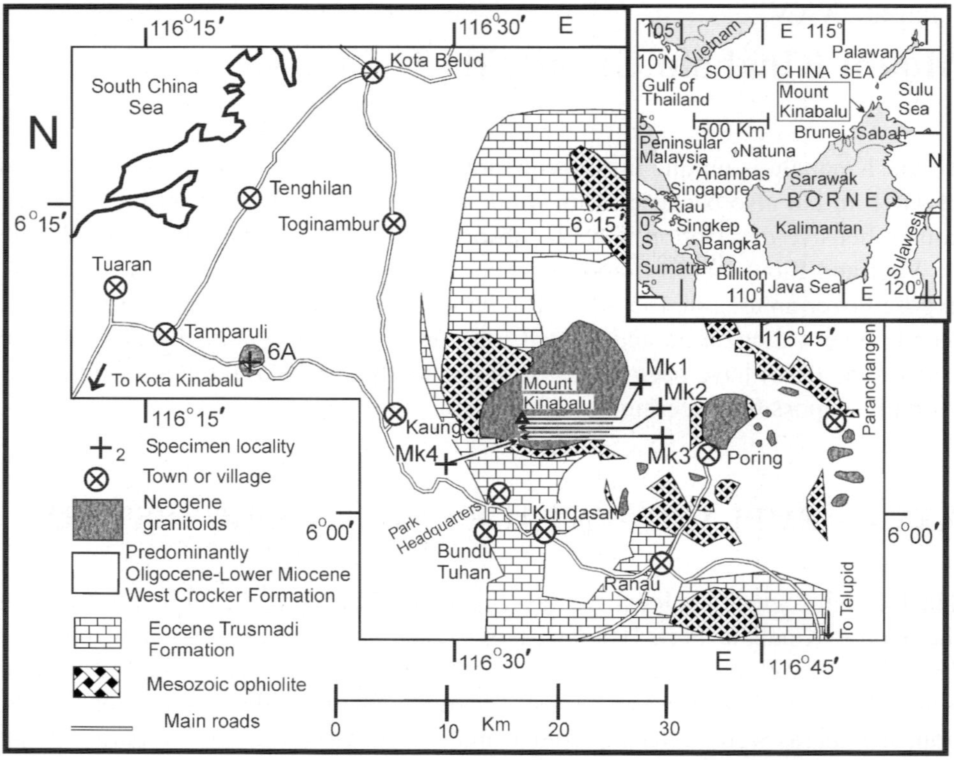

Figure 122. Simplified geology of the Mount Kinabalu district (based on Jacobson, 1970). Sample localities for age dating are from Swauger et al. (2000).

Table 30. The modern K:Ar age determinations of Mount Kinabalu (Swauger et al., 1995, 2000)

Specimen number	Mineral	wt.% K	Age (Ma)
MK1	Hornblende	0.80	13.7 ± 0.7
MK4	Hornblende	0.74	10.8 ± 0.5
6A	Biotite	K_2O wt.%	10.3 ± 0.3
*6A	Whole rock	4.03	6.84 ± 0.34
*6A	Whole rock	3.21	6.43 ± 0.32

* From Rangin *et al.* (1990).

y-axis is not precisely known, but must lie within the temperature range of the blocking temperature (for hornblende and biotite). For whole-rock K:Ar, the blocking temperature, being a complicated average of K-feldspar, plagioclase, biotite and hornblende, should lie at a temperature <200°C since most of the potassium is contained in alkali feldspar and biotite. Whole-rock K:Ar dating is not particularly useful for dating slowly cooled plutons and is most appropriate for fast-cooled volcanic rocks. The temperature significance of a biotite age, 300±25°C, could in some instances be identical to the zircon fission-track age, the partial annealing of which extends from 350°C to 200°C.

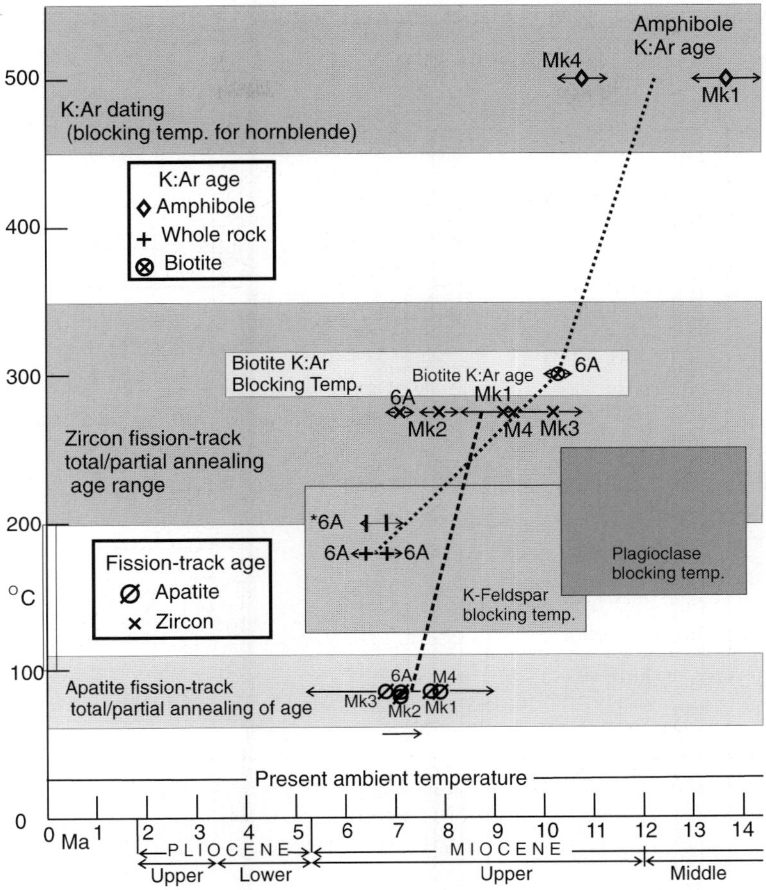

Figure 123. Temperature–age diagram for the cooling history of the Mount Kinabalu pluton, based on the data of Table 30 and Figure 124.

In constructing the diagram, the data points have been placed vertically in the mid-position of the blocking temperature range and a theoretical cooling curve is drawn through the average values. For the age range 500–300°C, a cooling rate of 102°C Ma^{-1} is suggested. From 300°C to 180°C a slower cooling rate of 32°CMa^{-1} is suggested. It may be concluded that the pluton was emplaced in the late Middle Miocene and slowly cooled over a period >6 my through most of the Upper Miocene. The pluton is accordingly wholly coarse-grained and cooling resulted from uplift and exhumation to give the present outcrop.

XXV.1.2. Fission-track dating

Four samples from the upper part of Mount Kinabalu and one from Tamparuli quarry have been analysed for zircon and apatite fission-track dating (Swauger et al., 1995). The histograms of the results are given in Figure 124. The averages of the results from the five rocks are: for Zircon 8.8±1.2 Ma and for Apatite 7.6±0.6 Ma, all

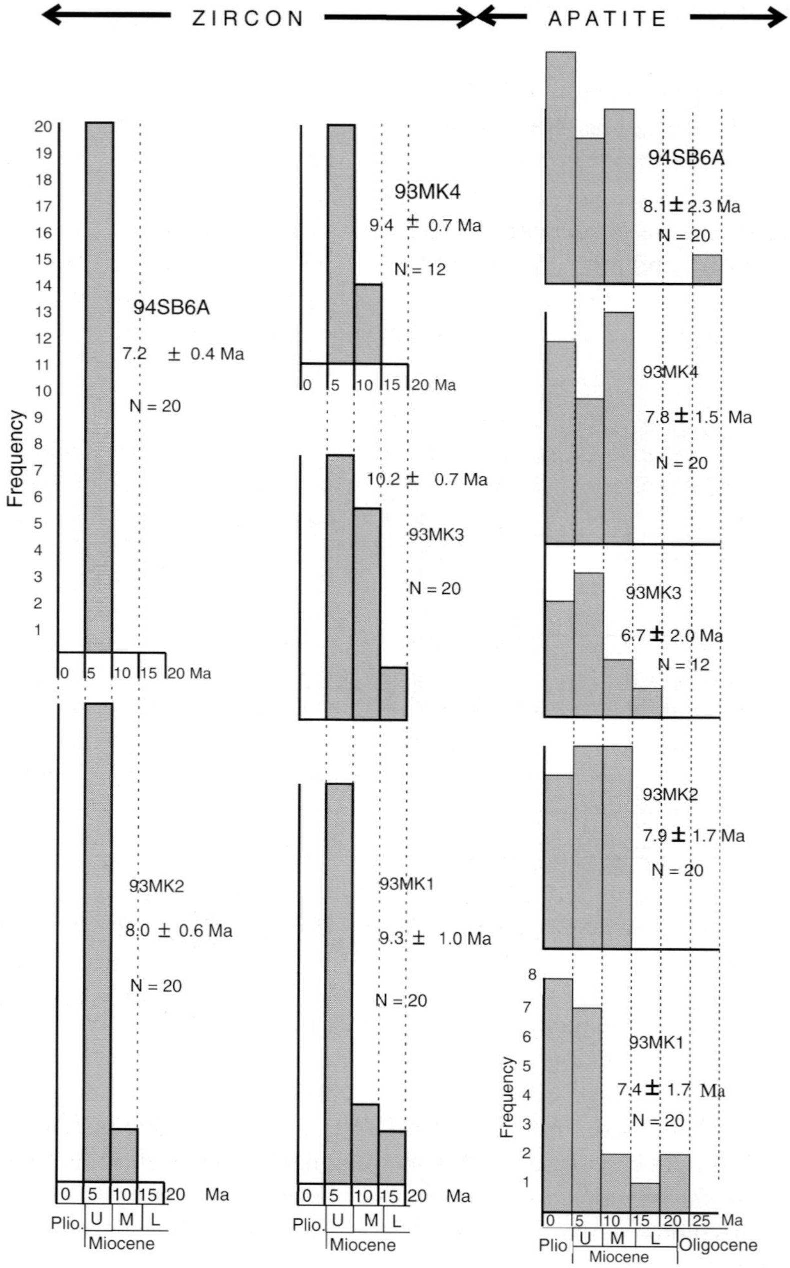

Figure 124. Histograms (5 Ma interval) of apatite and zircon ages from Mount Kinabalu intrusives.

Upper Miocene and consistent with the K:Ar radiometric dates as shown on the cooling curve of Figure 123. They represent the time when the already solidified pluton cooled from 300°C to 100°C.

XXV.2. PETROGRAPHY

XXV.2.1. Hornblende-quartz monzonite

This light-coloured rock is the most abundant of the granitoids lithologies. It is porphyritic with phenocrysts of hornblende, plagioclase and alkali feldspar. The ferro-edenitic hornblende is euhedral to subhedral making up 25–35% modal percent (Figure 125). Biotite is subhedral to anhedral. The plagioclase phenocrysts range from An_{70} to An_{35} and form 25–40% of the rock (Vogt and Flower, 1989). The alkali feldspar is Or_{93} to Or_{82}. Quartz is anhedral and occurs in the groundmass. Accessory minerals are magnetite, apatite, zircon, ilmenite, epidote and pyrite.

XXV.2.2. Biotite quartz monzodiorite

These are greyish medium-grained rocks showing weak flow lineation of the biotite and hornblende phenocrysts. Plagioclase phenocrysts make up 30–45% of the rock (Figure 125). Alkali feldspar and quartz are confined to the groundmass (Vogt and Flower, 1989).

XXV.2.3. Hornblende-biotite quartz monzodiorite

Euhedral to subhedral edenitic hornblende makes up 20–30% of these grey rocks. Biotite forms subhedral crystals. Some samples contain $En_{33}Fs_{19}Wo_{48}$ clinopyroxene. Plagioclase phenocrysts form 25–30% of the rock, ranging from An_{59} to An_{29} (Vogt and Flower, 1989). Alkali feldspar and quartz are anhedral (Figure 125).

XXV.2.4. Pyroxene-quartz monzodiorite

These light-coloured porphyritic rocks contain phenocrysts and groundmass plagioclase showing weak flow lineation. Clinopyroxene occurs as euhedral to subhedral phenocrysts forming 15–35% of the rock, with an average composition $En_{33}Fs_{27}Wo_{47}$. Biotite forms phenocrysts. The groundmass is of subhedral plagioclase, alkali feldspar and quartz.

XXV.2.5. Hornblende-biotite quartz monzonite porphyry

This grey rock shows flow lineation of the phenocrysts. Biotite and hornblende form euhedral to subhedral phenocrysts. Subhedral plagioclase makes 15% of the rock. The groundmass grain size is 0.05–0.2 mm and is composed of quartz, alkali feldspar and plagioclase.

XXV.2.6. Aplite

Aplite is light coloured and porphyritic. Plagioclase phenocrysts are euhedral and quartz is anhedral. Biotite is subhedral. The groundmass is of quartz, alkali feldspar and plagioclase.

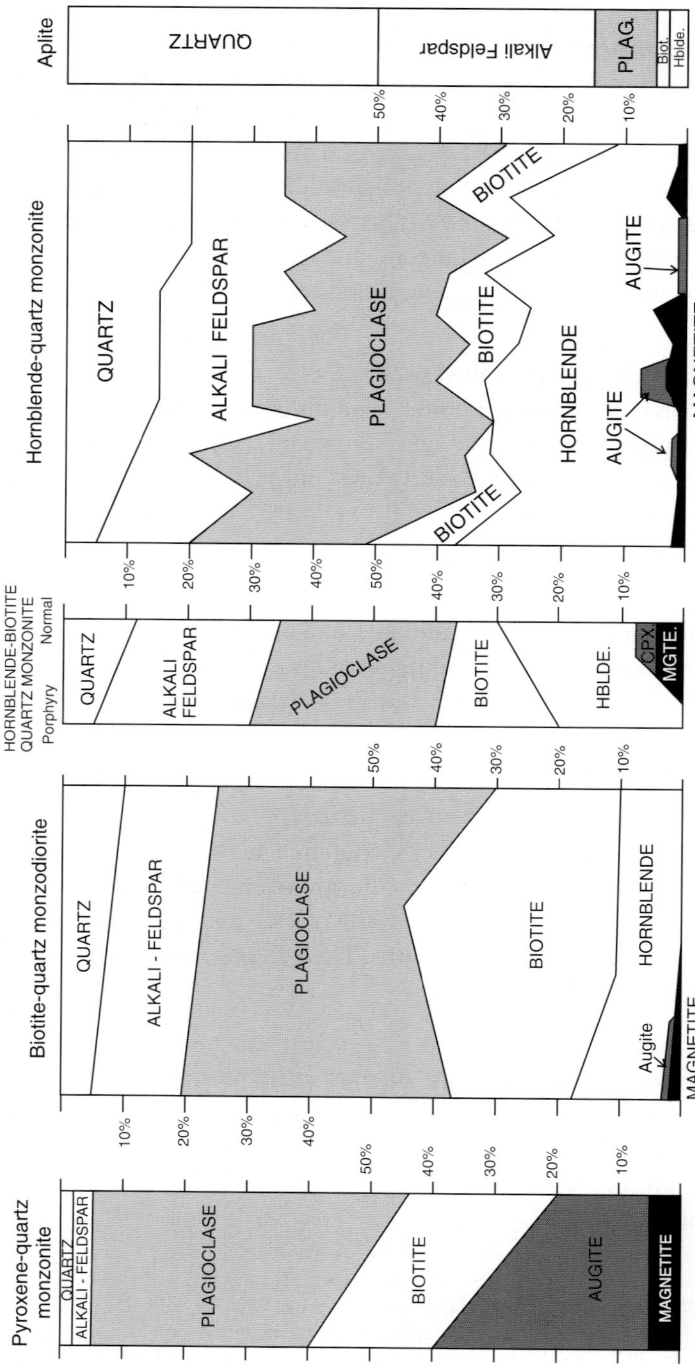

Figure 125. Modal proportions of the various granitoids in the Mount Kinabalu massif and satellite bodies (after Vogt and Flower, 1989).

XXV.2.7. Enclaves

The inclusions are both of mafic igneous and metasedimentary rocks. Igneous inclusions predominate in the biotite quartz monzodiorite and in the hornblende quartz monzonite. They are mafic and contain amphibole, plagioclase, alkali feldspar and accessory minerals. Metasedimentary xenoliths occur mainly in the hornblende quartz monzonite and are quartzite or biotite schist.

XXV.3. WHOLE-ROCK MAJOR ELEMENT GEOCHEMISTRY

A selection of whole-rock analyses is given in Table 31. With the exception of the aplite dyke, the compositions are fairly restricted.

A plot of available analyses on a K_2O versus SiO_2 diagram (Figure 126) shows the restricted nature of the bulk chemistry. Rocks containing higher K_2O contain hornblende and rocks containing less K_2O contain biotite. Intermediate values contain both mafic minerals. Pyroxene-bearing dykes have normal bulk chemistry, but the aplite dykes are highly differentiated.

Table 31. Major whole-rock analyses of selected Mount Kinabalu granitoids

Sample	a	b	c	d	e	f	g	h	i	j
Source	2	3	2	3	2	1	1	1	2	2
SiO_2	57.48	58.88	61.65	62.94	63.40	64.5	64.8	65.0	66.53	76.78
TiO_2	0.834	0.67	0.697	0.58	0.528	0.56	0.73	0.59	0.445	0.095
Al_2O_3	15.48	15.21	14.94	17.13	14.83	15.7	14.7	14.5	14.84	12.51
Fe_2O_3	8.44*	0.88	6.68*	0.75	6.05*				4.98*	0.80*
FeO		5.24		4.16		5.63*	5.88*	5.75*		
MnO	0.155	0.12	0.116	0.03	0.129	0.09	0.11	0.11	0.091	0.013
MgO	5.16	5.33	3.70	2.31	2.39	2.74	3.22	3.00	1.96	0.15
CaO	7.74	6.20	4.56	5.88	4.98	4.01	4.01	4.19	3.45	0.83
Na_2O	2.52	2.35	2.01	3.07	2.71	2.15	2.15	2.09	2.08	2.68
K_2O	1.791	3.79	5.088	1.87	4.678	4.15	4.15	4.58	4.972	5.597
H_2O+		0.43		0.62						
H_2O-		0.21		0.15						
LOI	0.98		1.54		0.34	2.20	0.72	0.95	1.38	0.11
P_2O_5	0.294	0.30	0.301	0.21	0.281	0.24	0.20	0.23	0.221	0.019
Total	100.874	99.61	101.282	99.70	100.316	102.24	100.67	100.99	100.949	99.584

LOI, Loss on ignition.

Data Source: 1, Vogt and Flower (1989); 2, Chiang (2002) and 3 Jacobson (1970).

Notes: a = SBK122, biotite quartz monzodiorite, between Low's and St. John's Peaks, summit of Mount Kinabalu. b= J6044 hornblende microgranodiorite. Mount Kinabalu South Peak. c = SB118, hornblende-quartz monzonite, Poring, Kipungit waterfall. d = J6088, adamellite, Gunong Nungkok. e = SBK120 Hornblende-quartz monzonite. South from Poring. f = J6054 pyroxene-quartz monzonite, g = J6317, hornblende-biotite quartz monzonite porphyry, south of Low's Peak. h = J6316, hornblende-quartz monzonite. i = SBK130 hornblende-quartz monzonite, Mount Kinabalu down from the Villosa Shelter. j = SBK124 aplite dyke near two big boulders on summit plateau.

* Total iron expressed as either FeO or Fe_2O_3.

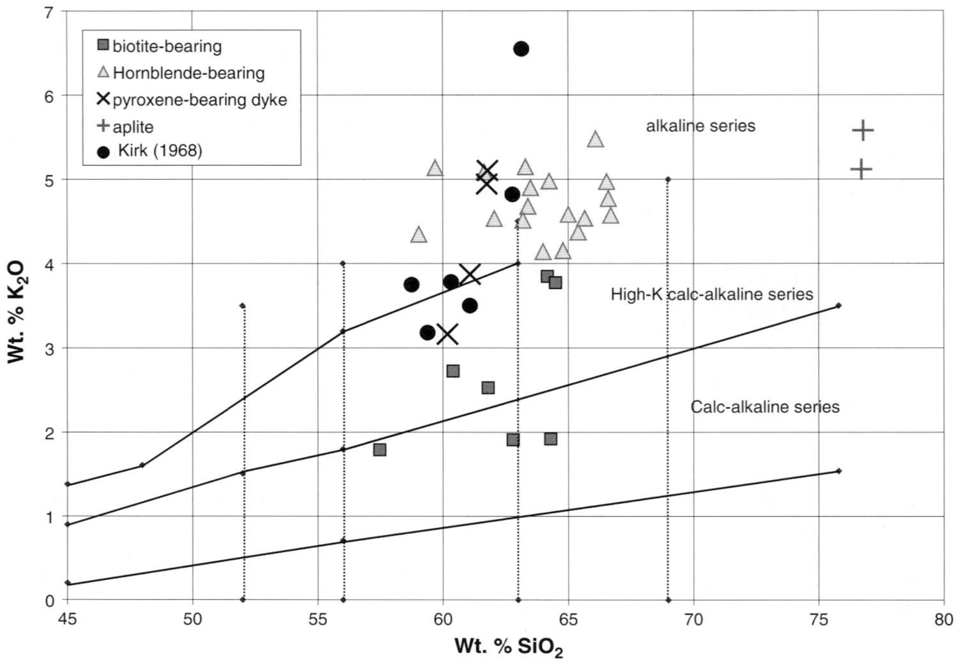

Figure 126. Whole rock wt% K_2O versus SiO_2 for analysed rocks of the Mount Kinabalu plutonic suite.

The metaluminous nature of the Kinabalu rocks, the large proportion of biotite and hornblende phenocrysts, and accessory magnetite, suggest they can be classified as I-type granitoids as defined by Chappell and White (1974). Magnetic susceptibility shows that they are magnetite-bearing and hence a product of anatexis of an igneous protolith.

XXV.4. WHOLE-ROCK TRACE ELEMENT GEOCHEMISTRY

The summary of trace element analyses by Chiang (2002) and Vogt and Flower (1989) is given in Figure 127. High and low K_2O lithologies are distinguished by magmaphile trace elements, high K_2O hornblende quartz monzonite being relatively enriched in Rb, Sr and Th and depleted in Nb.

Chondrite-normalized rare earth elements (REE) distributions (Figure 127) are between ~7 and ~90 times chondrite values. The rocks are preferentially enriched in light rare earth elements (LREE), and there is a moderate Eu anomaly. The trace element variation patterns are typical of island arcs with enrichment in large ion lithophile elements (LILE) and LREE (Chiang, 2002). However, Chiang (2002) concluded from trace element modelling that the Kinabalu intrusives were not formed by subduction-induced partial melting of the mantle. They have a subduction signature inherited from pre-collisional subduction events.

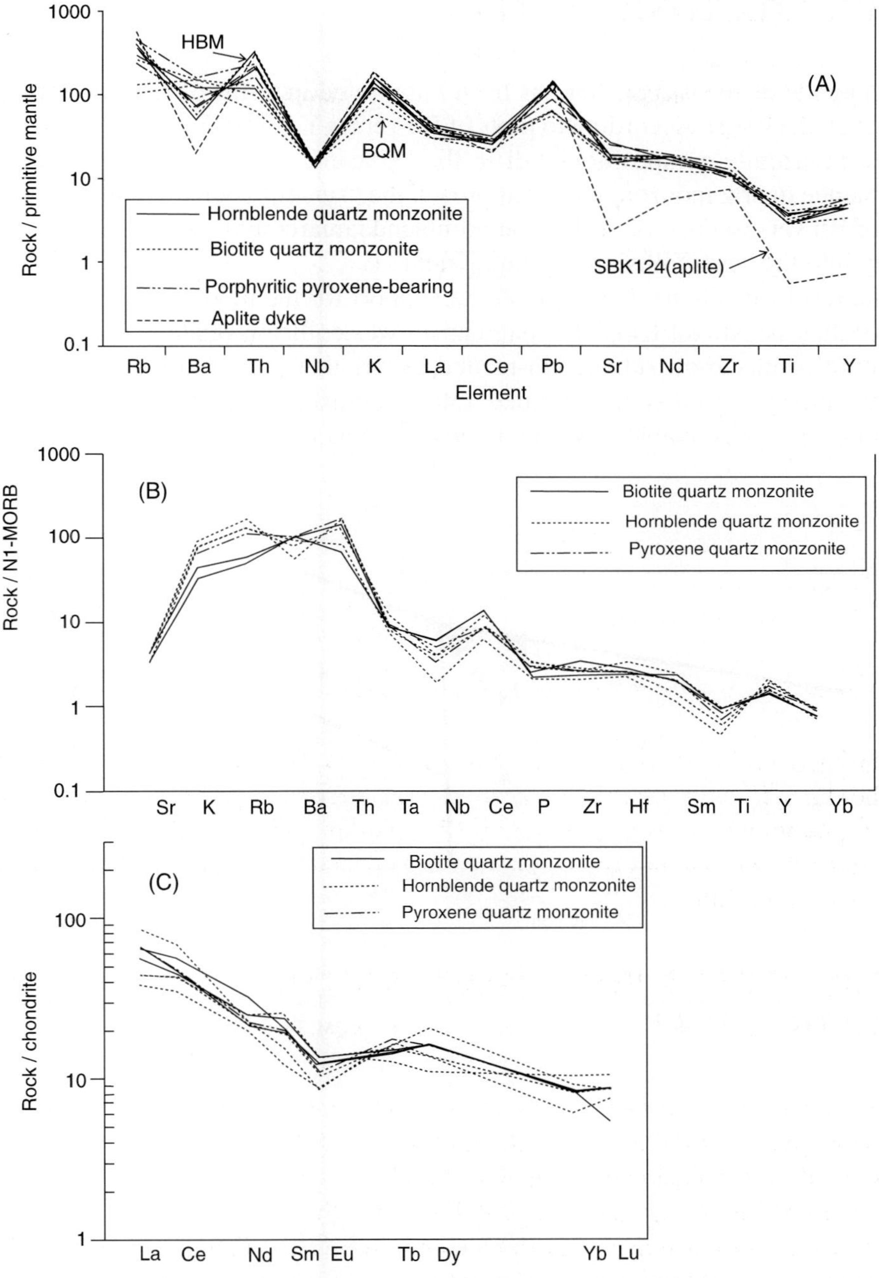

Figure 127. Multi-element variation diagram for the Kinabalu biotite quartz monzonite, hornblende quartz monzonite and both pyroxene-bearing and aplite dykes, normalized to primitive mantle composition. (A) redrawn from Chiang (2002); (B) redrawn from Vogt and Flower (1989); (C) rare earth element distribution for Kinabalu granitoids, redrawn from Vogt and Flower (1989).

XXV.5. TECTONIC SETTING

Magmaphile element character has been interpreted in terms of tectonic setting by Pearce et al. (1984). According to plots of Rb versus Y + Yb (Figure 128), the biotite quartz monzonite compositions fall in the "volcanic arc granite" field, while the hornblende quartz monzonite plots approach the "syn-collisional granite" field. The plot of Rb versus Yb + Ta pushes the hornblende quartz monzonite and aplite more firmly into the "syn-collisional granite" field.

The results of Figure 128 yield strong support for the model of Hutchison et al. (2000) that the Mount Kinabalu magmatism was collision-related, and the magma signature results from subduction-related events that preceded the collision. The subduction system was transformed into a collision setting as the Dangerous Grounds plate was unable to subduct because of its continental composition.

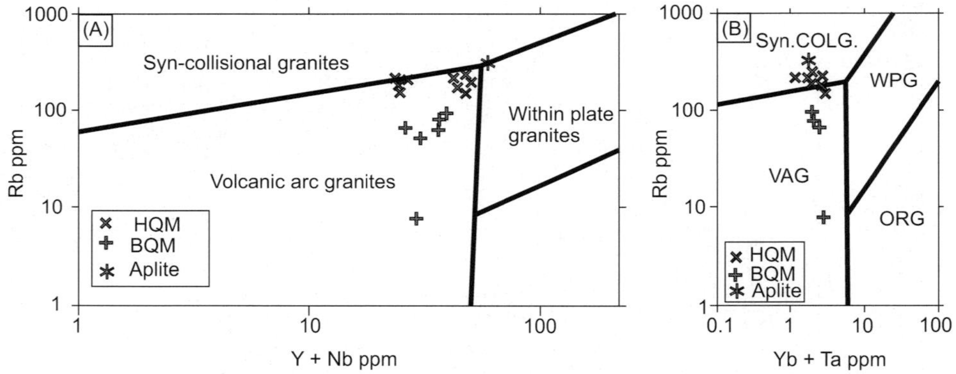

Figure 128. Pearce et al. (1984) Discriminant diagrams for tectonic setting . (A) Combined data. The (Y + Nb) values of Chiang (2002) are systematically lower than those of Vogt and Flower (1989). (B) Data only from Vogt and Flower (1989). HQM, Hornblende quartz monzonite. BQM, Biotite quartz monzodiorite.

Chapter XXVI
The East Baram Delta

The Baram Delta, centred on Negara Brunei Darussalam, extends north-eastwards off shore Sabah. It extends outwards from the coast as far as the NW Borneo Trough (Figure 129). Onland outcrops in Sabah are well exposed on Labuan Island and the Klias Peninsula.

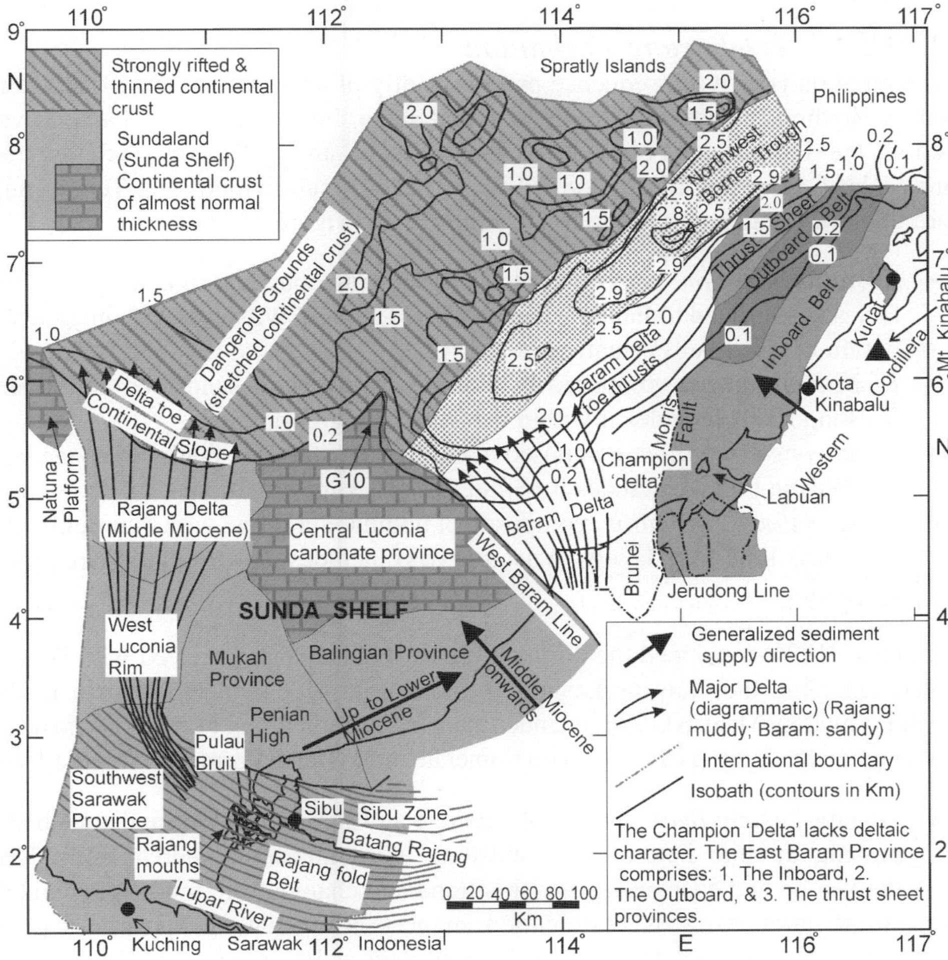

Figure 129. The Baram Delta, centred on Brunei Darussalam, extends eastwards offshore as far as the NW Borneo Trough. Onland outcrops occur on Labuan Island (after Hutchison, 2004).

XXVI.1. LABUAN ISLAND

The island is of a relatively simple anticline whose axis plunges to the NE. The core of the island is composed of Upper Oligocene–Lower Miocene Temburong Formation, unconformably overlain by Middle–Upper Miocene Belait Formation. The basal unconformity is represented by a conglomerate strike-ridge extending from Kubong Bluff to Tanjong Layang-Layang (Mazlan, 1994). The western limb of the anticline is occupied by coastal outcrops of Belait Formation, dipping north-westwards, with outcrops extending from Tanjong Layang-Layang to Bethune Head (Figure 130). The eastern limb is of similar Belait Formation, dipping north-eastwards, and exposed on the coastal tract NE of the airport runway (Figure 130).

XXVI.1.1. Temburong Formation

The formation is a flysch sequence predominantly of argillaceous rocks characterized by rhythmic repetitions of siltstone and shale. From the nearby Padas River section, the Temburong Formation is known to be an argillaceous facies of the sandy West Crocker Formation. Wilson (1964) subdivided the Temburong Formation of Labuan into the following sub-facies (Figure 130):

- *Nosong Formation:* composed of sandstone with abundant lignitic films. This facies is restricted to Papan Island (Figure 130) and Wilson (1964) included it within the Crocker Formation.
- *Kiam Sam Series:* composed of alternating layers of sandstone and shale. Wilson (1964) includes it within the Temburong Formation.
- *Limbayong Formation:* composed of claystone with some thick sandstone layers. Wilson (1964) includes it within the Crocker Formation.
- *Upper Sabong Formation:* composed of nodular claystone with thin sandstone and some limestone layers. Wilson (1964) includes it within the Temburong Formation.

Mazlan (1994) remapped the northern part of Labuan and disagreed with Wilson (1964) regarding the "Layang-Layang" beds. Wilson (1964) included them in the Belait Formation. Mazlan (1994) includes them in the Temburong Formation and takes the conglomerate ridge as the basal conglomerate of the Belait Formation (Figure 130).

- *Layang-Layang Beds:* consist of siltstone and shale with thin sandstone beds, passing up into heterolithic sandstone with mudstone intercalations. The thickening and coarsening-upwards trend and change from low-angle parallel stratification to cross-stratification suggests a shallowing upwards depositional environment. The Layang-Layang Beds at Tanjong Layang-Layang and along Jalan OKK Daud show variable strike directions and a higher degree of deformation than the overlying Belait Formation. Overturned beds were also observed near Taman Layang-Layang. This suggests a deformation phase before deposition of the overlying Belait Formation. The Layang-Layang

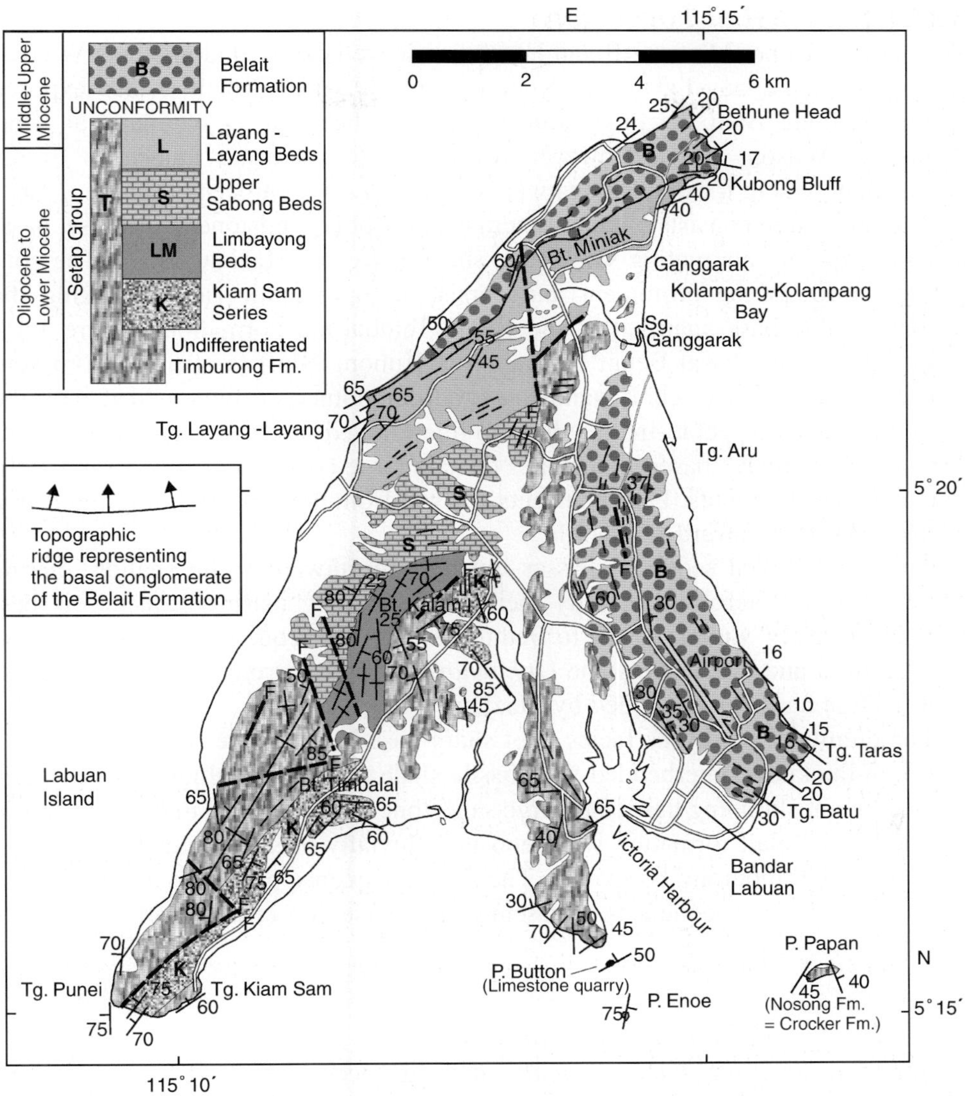

Figure 130. Geological map of Labuan Island (based on Wilson (1964) with modifications by Mazlan (1994)). With permission from the Department of Minerals and Geosciences Malaysia.

Beds do not represent the Setap Shale, which is absent from Labuan Island (Brondijk, 1963).

XXVI.1.1.1. Age

Rocks mapped on Labuan Island are remarkably devoid of fossil content. An age of Tcd to Te_5 (Lower Oligocene to Lower Miocene) is inferred from correlation with similar facies on Klias Peninsula and the Padas River of the adjacent mainland (Wilson, 1964).

XXVI.1.2. Belait Formation

The Middle–Upper Miocene Belait Formation is well exposed along the NW coast from Tanjong Layang-Layang to Kubong Bluff (Figure 130). It is also exposed around Tanjong Batu, on Pulau Button, and on the adjacent mainland at Klias Peninsula (Wilson, 1964). However, the best outcrops occur in nearby Brunei Darussalam, where it extends downwards into the Lower Miocene (Sandal, 1996).

The Formation consists of conglomerate and pebbly sandstone at the base, passing upwards into alternating sandstone, shale and coal (Mazlan, 1997). The conglomerate forms the prominent and persistent topographic ridge, marking the unconformable base against the underlying Temburong Formation (Figure 130). About 80% of the basal Belait Formation at Kubong Bluff is of medium-to very coarse-grained fluvial pebbly sandstone and conglomerate. Interbedded with the conglomerates are pebble-free medium to fine-grained sandstones and minor mudstones. Some erosive-based sandstones contain large coalified driftwood and well-rounded coal fragments. Palaeocurrents derived from the cross bedding indicate a depositional flow towards the north.

The basal fluvial sequence passes upwards (northwards) into shallow marine deposits via a heterolithic muddy facies. The shallow marine sequence is best exposed from Kg. Lubok Piasau to Bethune Head. The sequence has yielded some Foraminifera and they have some *Ophiomorpha.* Wave-formed hummocky cross-stratification has been described by Mazlan (1997).

The dominant sandstone facies is of sharp-based sand bodies ranging from 0.3 to 1.5 m thickness, interbedded with fissile shale up to 11 m thickness in places. Soft-sediment slumping has led to large sandstone balls enclosed in the mudstones.

The basal Belait formation was deposited fluvially over an eroded landsurface composed of Temburong Formation. The fluvial sequence passes up into transgressive shallow marine sequences represented by coarsening-upwards offshore shales and shoreface sandstone.

XXVI.1.2.1. Palaeontology and age

Foraminifera are sparse. *Globigerinoides* of Tf_1 age occur at Kubong Bluff. One sample from the Klias Peninsula contained *Ammobaculites*, *Elphidium*, *Nonion*, *Rotalia* and *Textularia* of Tf age. The limestone of Pulau Burong contained the following fossils, indicating a Tf_1 (Middle Miocene) age: *Perculina*, *Miogypsina*, *Flosculinella globulosa* (Rutten), *Lepidocyclina* spp. and *Gypsina* spp. From Brunei Darussalam, the fossils indicate an age range from Lower Miocene (Te_5) to Upper Miocene (Tf_3).

XXVI.2. SIPITANG–PANTAI DISTRICT

This district of western Sabah preserves deeply eroded onland synclines of what are considered to represent the Proto-Champion Delta of offshore eastern Brunei. The main sandy formation is the Meligan Formation, that pre-dates the Belait Formation (Figure 131).

XXVI.2.1. Meligan Formation

The Lower Miocene Meligan Formation is composed mainly of massive cross-bedded sandstone. In the type section along the Meligan River, some 40 km due south of

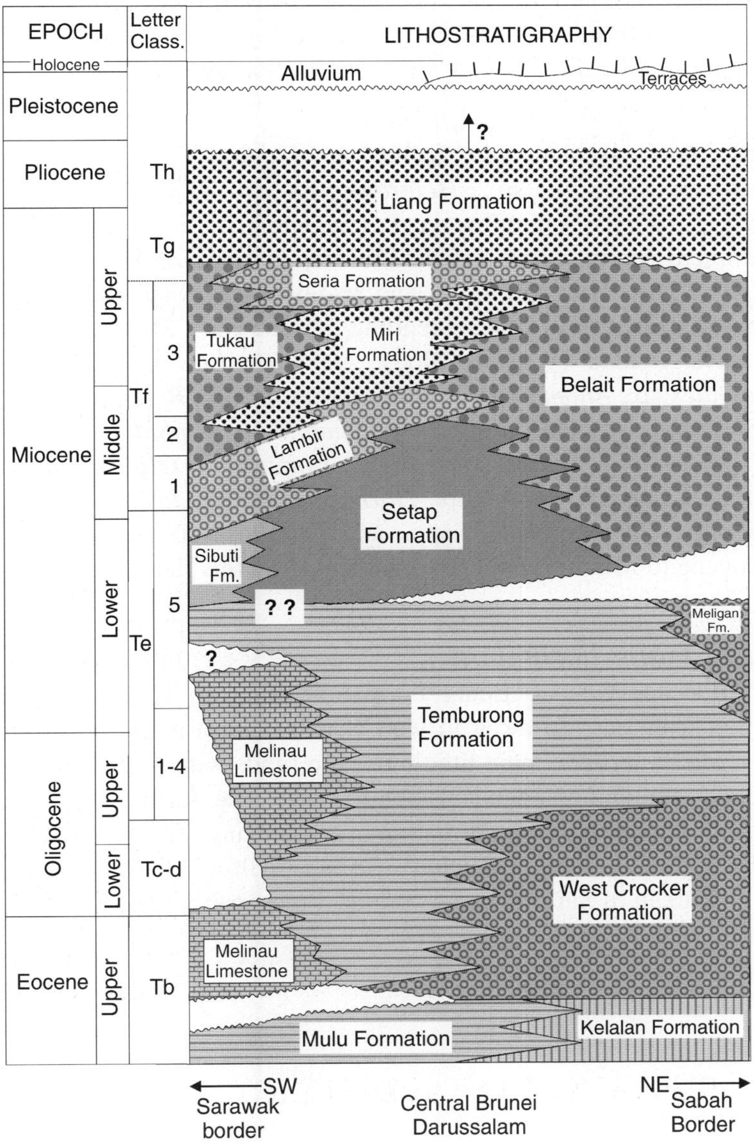

Figure 131. Lithostratigraphic chart across Negara Brunei Darussalam from Miri (in Sarawak) to Sipitang (in Sabah) (after Sandal, 1996). With permission from Brunei Shell Petroleum Company.

Sipitang (Figure 90), a thick shale occurs between a lower buff sandstone and an upper white sandstone (Wilson, 1964).

The outcrops in the Sipitang–Pantai area (Figure 90) are of deeply eroded synclinal outliers offering good outcrops of hard cross-bedded and laminated white sandstone, especially at Tanjong Marintaman. The Meligan Formation is unconformably overlain in places by the Liang Formation.

The Pulun Limestone forms a small hill on the Pulun River. It is a grey biohermal limestone grading downwards into marly shale. It has yielded valuable age-diagnostic Foraminifera.

XXVI.2.1.1. Palaeontology and age

Limestones have yielded the following fauna: *Austrotrillina howchini* (Schlumberger), *Gypsina* sp., *Heterostegina* sp., *Lepidocyclina* spp., *L.* (*Eulepidina*) cf. *bridgei* Schlumberger, *L.* (*Eulepidina*) cf. *formose* Schlumberger, *L.* cf. *richthofeni* Smith, *L.* cf. *inflate* Provale, *Miogypsina* (*Miogypsinoides*) *de haarti* van der Vlerk, *Miogypsina* sp., and *Miogypsinoides* sp. The age range indicated is Te_3–Tf_1 (Lower Miocene, extending into the early Middle Miocene).

Shales have yielded (Wilson, 1964): *Ammobaculites* sp., *Ammodiscus* sp., *Ammobaculoides* sp., *Anomalina* spp., *Bathysiphon* spp., *Bigenerina* sp., *Bolivina* sp., *Cibicides* spp., *Cristellaria* spp., *Cyclammina* sp., *Elphidium* sp., *Glandulina* sp., *Globigerina* sp., *G. subcretacea* Lomnicki, *Globigerinoides* spp., *Globigerinoides*. spp., *Gyroidina* sp., *Hormosina* sp., *Nodosaria* sp., *Nonion* sp., *Operculina* spp., *Quinqueloculina* spp., and *Textularia* sp. An age of Te_5 to Tf_1 (Lower Miocene to Middle Miocene) is indicated by this fauna.

XXVI.2.2. Liang Formation

The Late Miocene–Pliocene Liang Formation outcrops sporadically along the coastal region from east Sarawak to west Sabah, with a major occurrence in the Tutong area of Brunei. The main lithologies are sands with occasional conglomerates and clays containing abundant lignite beds. The formation was deposited unconformably upon the Belait Formation near Limbang. The type section is in the Lumut Hills east of the Seria Oilfield of Brunei (Wilford, 1961). The coastal outcrops east of Mukah in Sarawak have yielded a reasonably good Pliocene fauna.

Chapter XXVII

Brunei Darussalam: Geological Link Between Sabah and Sarawak

Negara Brunei Darussalam has a convoluted boundary with Sarawak and its geology has much in common with western Sabah and eastern Sarawak (Figure 132).

Brunei is dominated by two major synclines that plunge and widen to the north and die out towards the south. The largest is the Belait Syncline and the better known Berakas Syncline. South of the latter across the border in Sarawak is the Limbang Syncline (Figure 132). The Berakas and Belait Synclines are separated by the N–S trending Jerudong Anticline that continues southwards towards the Mulu area and north-eastwards along the Morris Fault offshore Sabah.

The geological map (Figure 132) illustrates the difficulty of field mapping in a region where several formations are difficult, if not impossible to differentiate. Thus east of Lamudin, the eastern limb of the Belait Syncline is mapped composed of Belait Formation to the south and Miri Formation along strike northwards. The distinction between these two formations is obscure: the Belait Formation contains more lignite than the Miri Formation. The stratigraphic chart of Brunei and adjacent Sarawak and Sabah shows that the two formations are predominantly of the same age and represent only facies changes, generally subtle. Therefore the distinction is unnecessary.

The characteristically shaly Setap Shale Formation interdigitates with the Belait Formation (Figure 132). However, in the Temburong District the mapped Setap Shale is indurated and has undergone low grade metamorphism, in sharp contrast to the unconformably overlying Belait Formation. Accordingly this so-called "Old Setap Shale" is more appropriately referred to as the Temburong Formation (Brondijk, 1963).

The Seria Formation continues westwards from Lumut, but the region west from Lumut and Belait lacks outcrops. The flat-lying region around Seria contains the still-producing major onland oilfield. Outcrops of the Upper Miocene Seria Formation and the overlying Upper Miocene to Pliocene Liang Formation are confined to the coastal region, but they thicken seawards offshore.

Only the Seria Formation is described here, since the others have been described above from western Sabah and eastern Sarawak.

XXVII.1. SERIA FORMATION

The Upper Miocene Seria Formation outcrops in the lower Belait to Tutong areas and is laterally continuous into the subsurface, notably in the Seria oilfield. Total thickness approaches 2 km. The lower part has a higher content of poorly consolidated

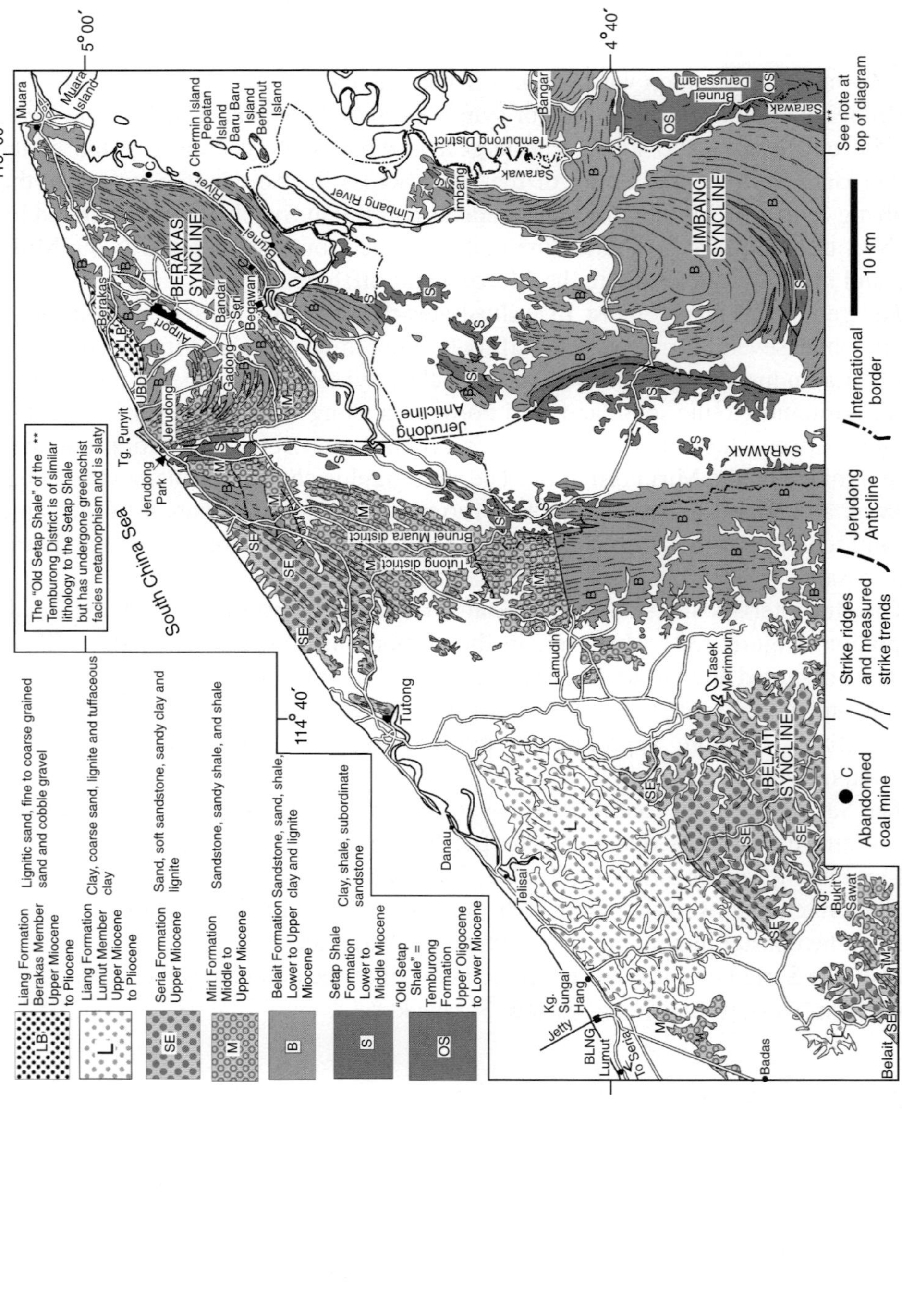

Figure 132. Geologicl map of the major part of Negara Brunei Darussalam (redrawn from Sandal, 1996). Unshaded areas represent surficial Pleistocene to Recent alluvial deposits. With permission from Brunei Shell Petroleum Company.

sand and sandstone. The bedding is generally less regular than in the Miri Formation. Ironstone nodules are common (Sandal, 1996). The formation conformably overlies the Miri Formation and shales-out northwards into the Setap Shale Formation. It is overlain semi-conformably by the Liang Formation. In east Sarawak, the Seria and Miri Formations pass into the coarser Tukau Formation (Figure 131).

XXVII.1.1. Age

The marine microfauna is not age-diagnostic (Liechti et al., 1960). It consists of the following three ecozones: *Textularia* 3, *Triloculina* 16 and *Triloculina* 18. It also belongs to the *Stenochlaena Laurifolia* palynological zone. Its accepted age is upper Tf –lower Tg, Upper Miocene.

Chapter XXVIII
Offshore Brunei and Sabah

The principal features of the region are illustrated in Figure 133. The Baram Delta continues north-eastwards from the West Baram Line and extends outwards from the coast as far as the prominent linear deep known as the NW Borneo Trough. Across the Morris Fault, the offshore area is subdivided into the Inboard Belt, the Outboard Belt, and the "Thrust Sheet" province.

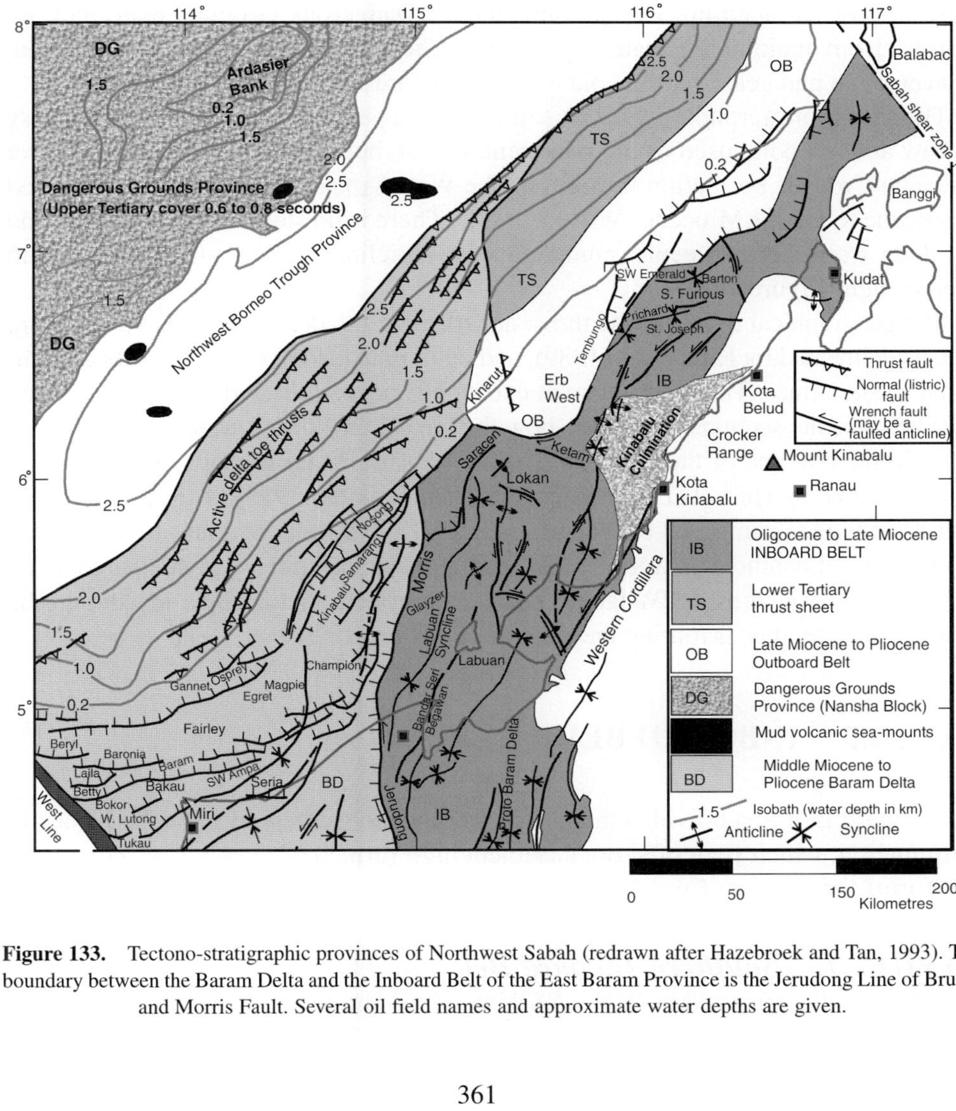

Figure 133. Tectono-stratigraphic provinces of Northwest Sabah (redrawn after Hazebroek and Tan, 1993). The boundary between the Baram Delta and the Inboard Belt of the East Baram Province is the Jerudong Line of Brunei and Morris Fault. Several oil field names and approximate water depths are given.

XXVIII.1. BARAM DELTA TOE FOLD-THRUST ZONE

For a considerable time, this feature of the front of the delta was wrongly interpreted as an accretionary prism, thereby making the NW Borneo Trough the related trench of a NW-facing system. The cross section of Bol and Van Hoorn (1980), drawn through Mount Kinabalu, presented the popular model that the Trough was the trench. The delta toe (not described as such) was the accretionary prism, and Mount Kinabalu the volcanic arc. Later Hazebroek and Tan (1993) showed that the folded and thrust sediments of the delta toe (Figure 136) also characterize the toe of the Niger Delta. Such features are not a product of compression tectonics; rather they result from sedimentary-structural processes. Nevertheless, these features are not universally present in the front of all major deltas. The toe sediments (Figure 136) are not 'scraped-up' by the 'subducting' Dangerous Grounds plate, but are derived from mainland Sabah, representing the outer limit of delta deposition. Convergence had ceased at this plate margin before the Baram Delta existed.

The sea-floor morphology has no similarity to a passive margin. The shelf is very narrow and not composed of old continental crust, but of uplifted Eocene to Lower Miocene Crocker Formation turbidites. The West Crocker Formation is best dated Oligocene to Lower Miocene (Wilson, 1964). There is no slope/rise distinction, and the delta front slopes steeply, complete with anticlinal ridges, towards the 2 km deep trough (Figure 136).

The geographical extent of the thrust anticlines of the delta toe is shown in Figure 133, as delineated by Hinz et al. (1989), and determined from a large number of additional seismic lines. The delta does not extend NE beyond the Jerudong Line (Figure 133) and seismic sections no longer show the typical fold-thrust system of a delta toe zone. In the absence of the Baram Delta front a seismically irresolvable feature has been identified by Hinz et al. (1989) and Hinz and Schlüter (1985) as an allochthonous overthrust wedge composed of mélange (Figure 137). It still continues to be debated whether this allochthonous wedge could represent the accretionary prism paired to the NW Borneo Trough as a NW-facing subduction system that became inactive in the Lower Miocene, thus favouring the view of Hamilton (1979).

XXVIII.2. INBOARD BELT

The Inboard Belt is subdivided into a southern and northern section by the Kinabalu Culmination, which is an offshore basement high formed of the Crocker Formation that forms the Western Cordillera.

XXVIII.2.1. Southern Inboard Belt

This belt is characterized by NNE–SSW trending anticlines with steep flanks and strongly faulted crests (Sabah Ridges), spaced 5–20 km apart (Mazlan et al.,

1999b). The Labuan–Paisley Syncline contains up to 4 km of stage IVc deep marine sands and shelfal deltaic sediments. The anticlinal cores are commonly of stage III shales (Figure 134).

Transpression is suggested by flower structures and reverse faults steepening with depth (Bol and van Hoorn, 1980). Large-scale left-lateral strike-slip displacement of up to 100 km has been proposed by Hazebroek and Tan (1993) across the Morris Fault and Jerudong Line. However, Hutchison (1994) has shown that this interpretation, based upon contrasting facies on either of the Line, is not necessarily valid because of contrasting depositional water depths across the Line. Intense deformation occurred in the Late Miocene resulting in the Shallow Regional Unconformity (SRU), which is a peneplain surface with only a shallow cover of younger sediments (Figure 135).

XXVIII.2.2. Northern Inboard Belt

This belt lies north of the Kinabalu Culmination and is characterized by an intersecting network of N-, E- and SE-trending compressional ridges and synclines (Mazlan et al., 1999b). They terminate against the NE-trending Emerald Fault Zone to the west and the Crocker fold-belt to the east. The SRU is less prominent in this belt.

In late Miocene time (stages $IV_{D\&E}$ of Sabah = late cycle V to early cycle VI of the Baram Delta), Rice-Oxley (1991) showed that the Morris Fault and the boundary between the Inboard and Outboard belts represented a very steep continental slope, with land on the SE side rapidly giving way north-westwards (and westwards in Brunei) to bathyal conditions (Figure 135). This time (stage $IV_{D\ \&\ E}$) coincides with the SRU of Bol and Van Hoorn (1980) and Tan and Lamy (1990). It was a time of maximum uplift and seaward extension of the Sabah Orogen. Levell and Kasumajaya (1985) have documented that in the Morris Fault area, the continental slope (normal fault) had a westwards downthrow of as much as 1.1 km from the land. Voluminous erosion products off the Western Cordillera were fluvially brought to the coastline and slid westwards down the unstable slope as slump scars into the outer neritic to bathyal Baram Delta. A growth index of about 2.0 characterizes the Lower Pliocene section westwards across the slope. A coalescing series of elongate spoon-shaped unconformities has been mapped along approximately 150 km of the 250 km long Upper Miocene shelf (Figure 135).

The scars are interpreted as submarine slumps. Most retain a smooth morphology, but some may have been modified by turbidity currents, as an example the source of the Tembungo field oil-bearing reservoirs. One of the scars has been described as olistostrome along the Jerudong Line onshore (Hutchison, 1994). Typically 1–5 km^3 of sediments were redeposited down-slope in any one slump scar episode (Levell and Kasumajaya, 1985). Drills have shown that they are composed on deep–water claystone.

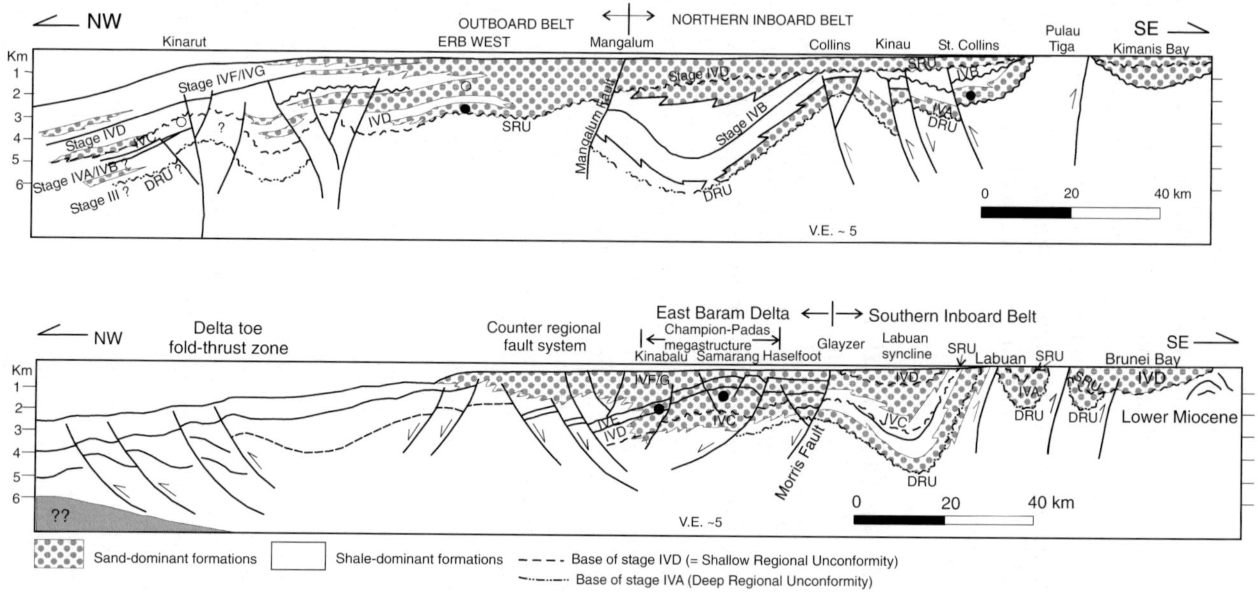

Figure 134. Regional cross-sections across the South China Sea Inboard Belt towards the NW Borneo Trough. Modified after Tan and Lamy (1990) and Hazebroek and Tan (1993).

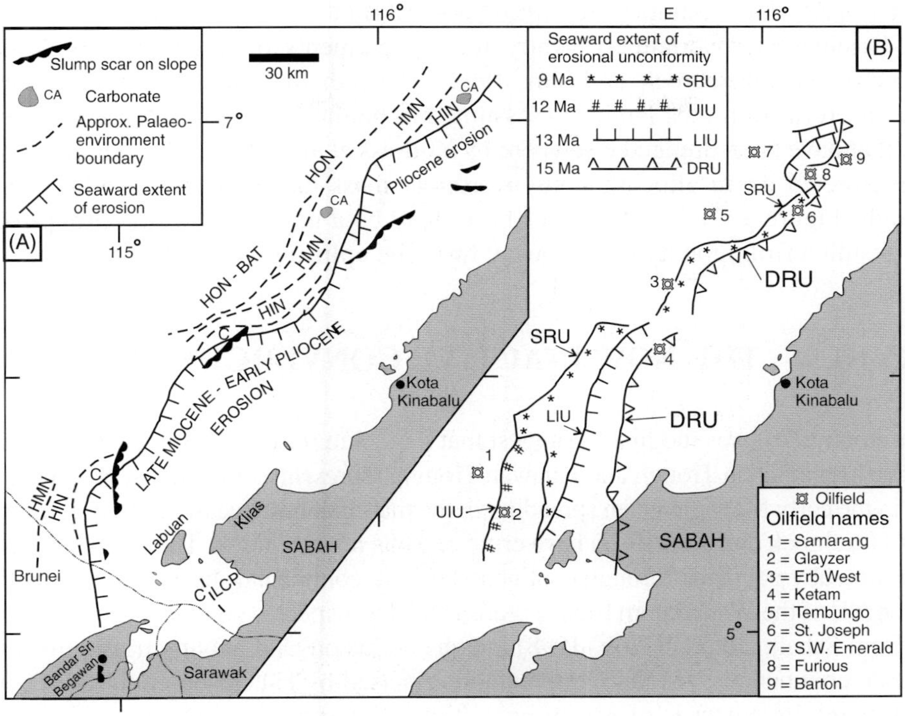

Figure 135. The East Baram Delta and NW Sabah. (A) Upper Miocene palaeo-environments for stage IV E/F (after Rice-Oxley, 1991), showing also positions of identified slump scars along the Upper Miocene shelf edge, from Levell and Kasumajaya (1985) and Hutchison (1994). (B) Seaward limit of the important regional unconformities that successively migrated away from the uplifting Western Cordillera, based on Levell (1987). Palaeo-environments: LCP = lower coastal plain, C = coastal, HIN = holomarine inner neritic, HMN = holomarine middle neritic, HON = holomarine outer neritic, BAT = bathyal. Unconformities: DRU = deep regional unconformity, LIU = Lower intermediate, UIU = Upper intermediate, SRU = shallow regional unconformity. Redrawn after Hutchison (1996b).

XXVIII.3. OUTBOARD BELT

This belt occurs west of the northern Inboard Belt. It is structurally complex and contains both extensional and compressional structural features. The former deformation is marked by large NE–SW trending down-to-basin normal faults, and the latter by wrench-induced features in the Tembungo and Kinarut areas and by clay diapirism along its western margin (Hazebroek and Tan, 1993). The province is a depocentre of stages IVD, E, F and G sediments prograding north-westwards from shallow to deep marine.

XXVIII.4. THE "THRUST SHEET" BLOCK

Hinz et al. (1989) proposed the term "Thrust Sheet" for a block of chaotic seismic facies that is bounded to the NW and SW by steep thrust faults and to the SE by normal faults that separate it from the Outboard Belt (Figure 133). The chaotic seismic basement is overlain by gently folded sediments of possible early Middle Miocene age (Hazebroek and Tan, 1993). Beneath the chaotic seismic facies a coherent reflection dips landward, assumed to be the Oligocene–Miocene carbonates that have been dredged elsewhere by Kudrass et al. (1986). The "Thrust Sheet" is interpreted as an allochthonous overthrust mass that borders the NW Borneo Trough (Figure 137). The "Thrust Sheet" may be a nappe of Rajang Group rocks that resulted from gravity sliding away from the uplifting Western Cordillera.

XXVIII.5. THE NOW INACTIVE CONVERGENT MARGIN

Hamilton (1979) was the first to suggest that the 2 km deep NW Borneo Trough, also known as the Sabah Trough and Palawan Trough, represents a trench that became inactive in the early Miocene when spreading in the marginal basin ceased. Many geologists (e.g. Hazebroek and Tan, 1993) have criticized this interpretation. The normal progression from abyssal plain, through continental rise, to continental shelf is absent from the region NE of the West Baram Line (Figure 136). The expected continental shelf is missing from Sabah, although Holt (1998) interpreted his on land gravity measurements to indicate continental crust, an interpretation criticised by Hutchison et al. (2001). The outcropping basement is of Mesozoic ophiolite, interpreted as remnants of a 'Proto South China Sea' (Hutchison et al., 2000). Its activity is shown by its drape of Miocene and younger strata and by modern GPS data indicating that Sundaland, extending from Indo-china to Brunei and central Indonesia, behaves as a stable rigid block moving east with respect to Eurasia at 12 ± 3 mm a^{-1} (Michel et al., 2001).

XXVIII.6. RELATIONSHIP WITH THE DANGEROUS GROUNDS

The Dangerous Grounds rift sequence is capped by a carbonate platform that presents a good seismic reflecting horizon that dips beneath the toe zone of the Baram Delta and beneath the tectonic front of the allochthonous wedge before seismic resolution is lost (Figure 137). The Baram Delta front has grown progressively into the NW Borneo Trough, created by the southeastwards-dipping Dangerous Grounds. Whether or not this deep was the original pre-Middle Miocene trench has continued to be debated, for example by Hazebroek and Tan (1993). Change from a convergent plate margin to a collision zone is constrained by fission-track uplift data and a succession of unconformities, including the DRU, indicating uplift of the Western Cordillera of Sabah (Hutchison et al., 2000). Alternatively the NW Borneo Trough may be regarded as a foredeep, resulting from continental lithosphere descending beneath the collision zone. The pro-delta muds of the delta front represent a suitable décollement surface.

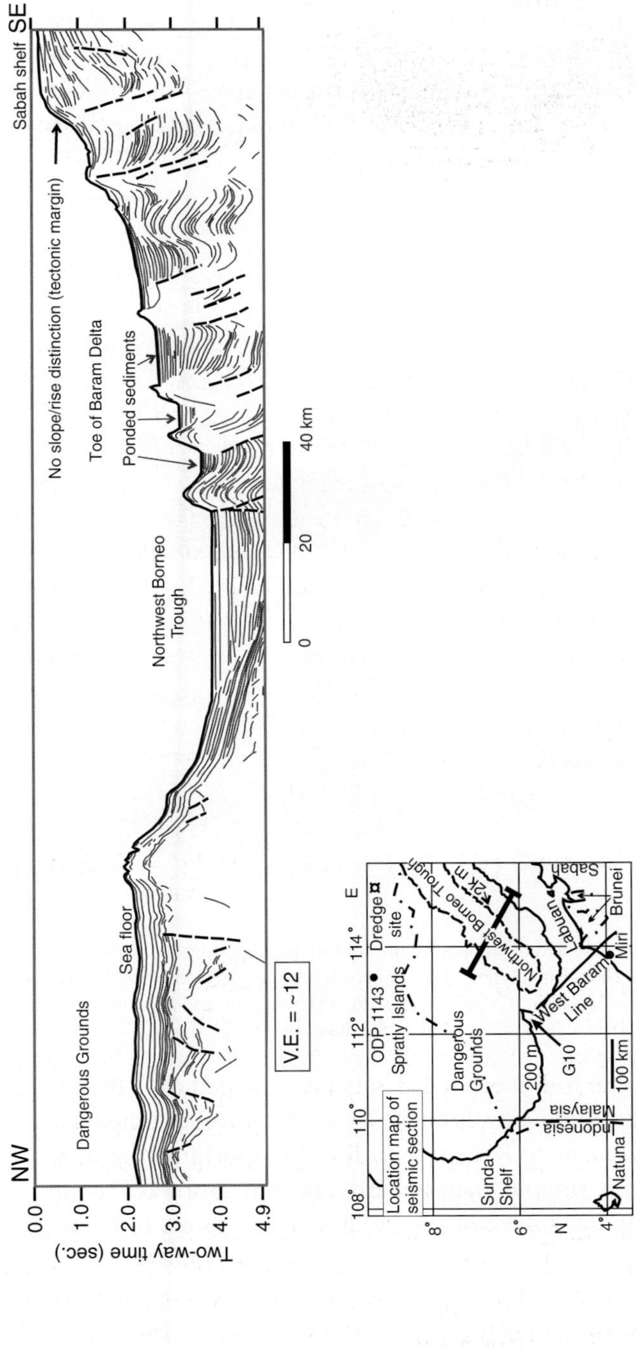

Figure 136. Interpreted seismic section across the NW Borneo Trough form the Dangerous Grounds onto the active toe of the Baram Delta (from Hutchison, 2004).

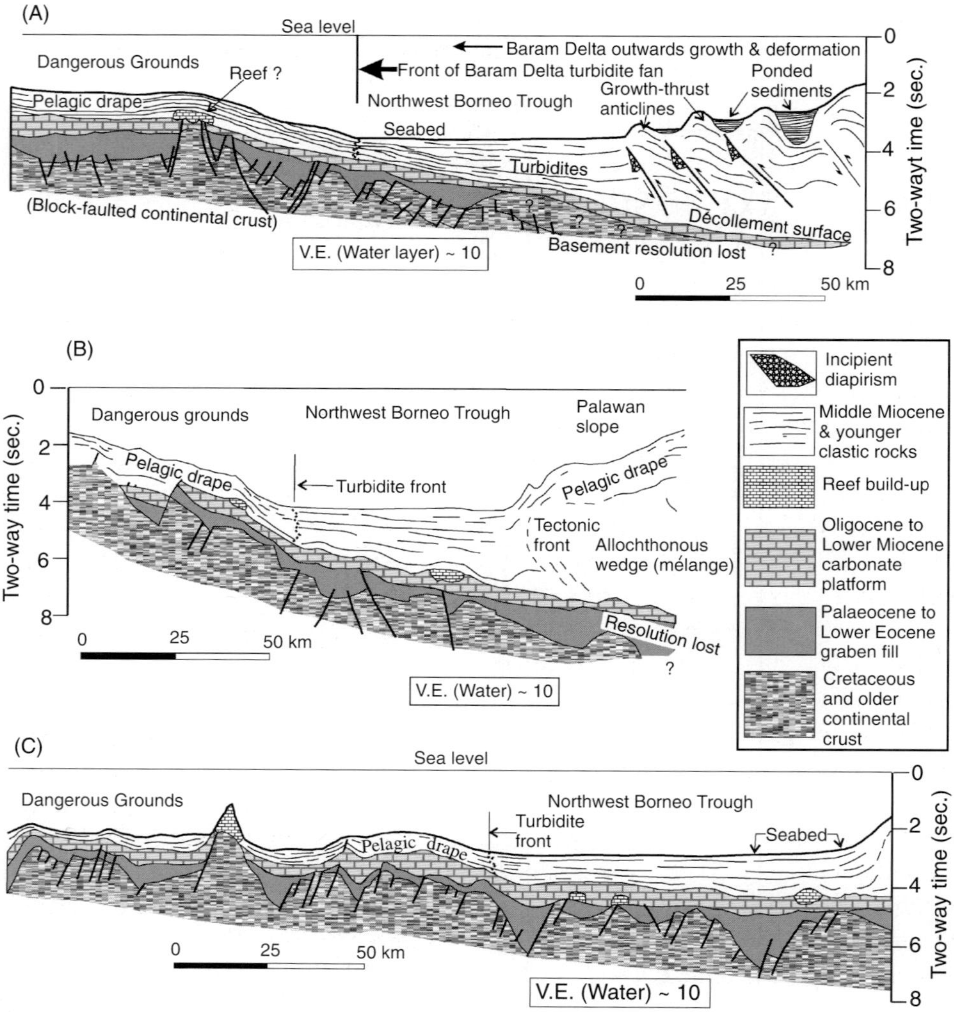

Figure 137. Geological interpretations of three seismic profiles from the Dangerous Grounds across the NW Borneo Trough, after Hutchison (1996a). (A) Composite section based on Hinz et al. (1989). (B) Section adjacent to the Kudat Peninsula of Sabah. (C) Section close to the Reed Bank and adjacent Palawan. (B, C) After Hinz and Schlüter (1985).

Upper Miocene to Recent sediments have draped over the allochthonous wedge and filled up the foredeep (?trench) to form a flat-bottomed linear trough under 2 km of water. Pliocene to Recent turbidite flows, originating in nearby Sabah, experienced a difficult sinuous path towards the NW Borneo Trough. The anticlines of the Baram Delta toe zone caused the turbidite currents to be confined between the sea-floor ridges, before filling up the intervening synclines and breaking through to a deeper level (Grant, 2003). Thus, as in Figure 136, the synclinal areas between the sea-floor anticlines are progressively filled by young flat-lying ponded sediments. The NW Borneo Trough is the ultimate pond.

Numerous seismic sections across the NW Borneo Trough, e.g. Figure 136, show that the high edifices of the Dangerous Grounds would have presented difficulty for continuing subduction.

XXVIII.7. TECTONIC MODEL

Figure 138 summarizes the scenario proposed by Hutchison et al. (2000). The very thick sandy turbidites of Oligocene to Lower Miocene age, provenanced from western Borneo and named the West Crocker Formation (Wilson, 1964), are thought to have filled and overwhelmed the original trench, following the model of Westbrook et al. (1988) for the presently active Barbados Ridge Complex of the Caribbean Sea. The rocks of the original trench have been deeply buried and metamorphosed to glaucophane-epidote facies and subsequently exhumed to form outcrops in central Sabah near Telupid (Hutchison et al., 2000). The high sedimentary influx caused the trench to migrate far to the NW. The result was that the West Crocker Formation was allochthonously thrust out over the buried continental shelf and attenuated continental crust of the Dangerous Grounds (Figure 138). The present NW Borneo Trough may be referred to as a foredeep, but some geologists prefer to retain the term fossil trench.

The dramatic uplift of the Western Cordillera of Sabah is attributed to isostatic rebound following cessation of any subduction or underthrusting. Apatite fission track studies show that maximum exhumation was in the Middle to Upper Miocene (8–15 Ma) and the Mount Kinabalu monzonite pluton is dated 10–13.7 Ma (Hutchison et al., 2000). Trace element tectonic discriminant plots of Mount Kinabalu monzonites and monzodiorites lie in the border zone between volcanic-arc and syn-collisional granitoids (Vogt and Flower, 1989; Chiang, 2002), thereby supporting the subduction-collisional model of Figure 138. The young mountain chain of the Western Cordillera lies close to the coastal plain, and the rivers cannibalized the mountain outcrops of West Crocker turbidites to form the oil-prolific Baram Delta.

The model shown in Figure 138 is based on extensive mapping of Sabah resulting in the interpretation that this part of Borneo is not continental but is built upon a basement of Mesozoic ophiolite. Holt (1998), however, interpreted the gravity data to imply that Sabah was constructed of continental crust, a view strongly opposed by Hutchison et al. (2001). A non-continental crust and the turbiditic nature of the main mountain belt (Western Cordillera), combined with very young uplift, and the island-arc character of the south-eastern Miocene–Pliocene volcanic arcs (Chiang, 2002), all support the interpretation of the Sabah margin being formerly of active convergence.

The south-western half of Palawan Island, also dominated by turbidites upon an ophiolite basement, has the same foredeep as in Sabah (C, Figure 137). However,

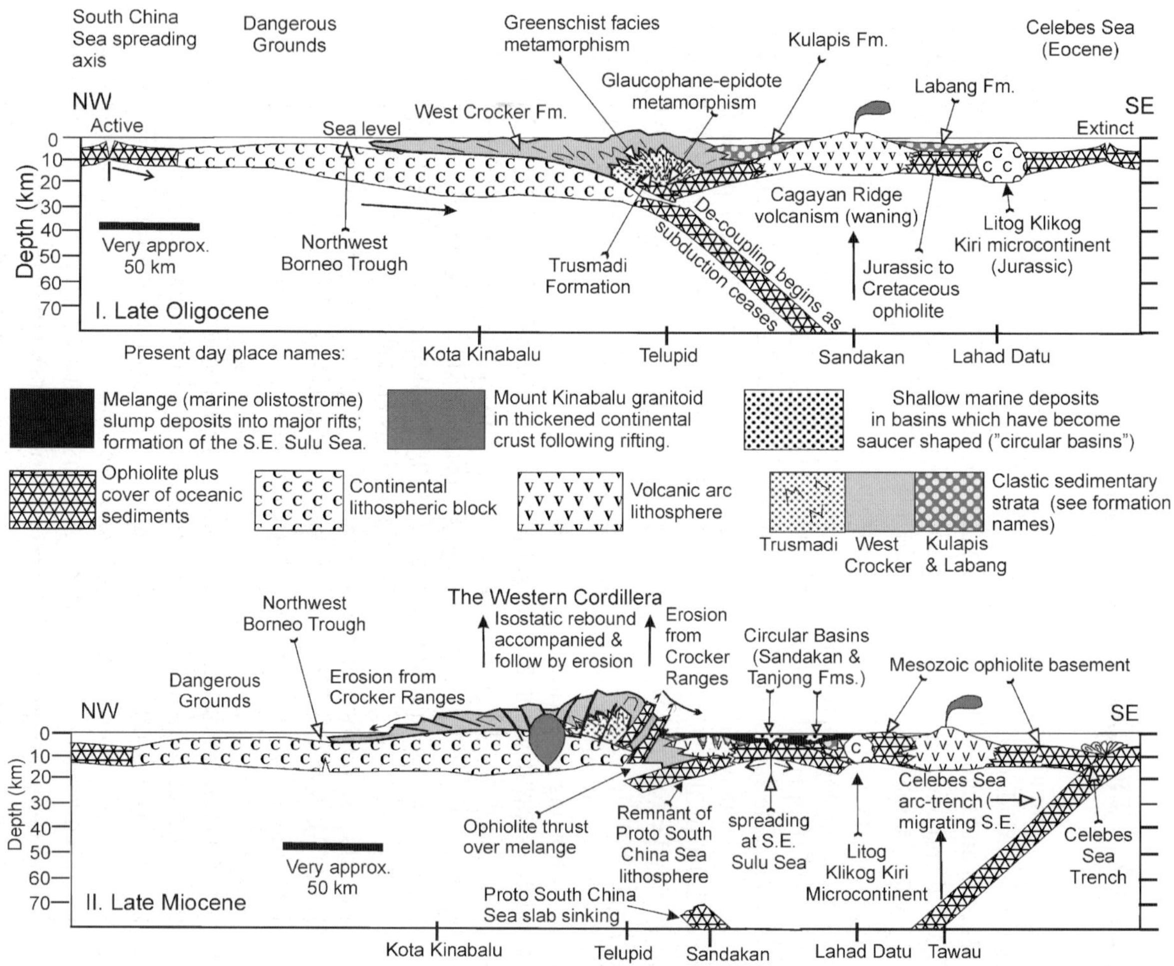

Figure 138. Diagrammatic cross-sections to suggest a tectonic scenario for the Tertiary evolution of Sabah (after Hutchison et al., 2000). (by permission of the Geological Society of London)

farther to the NE across the Ulugan Fault, the Mesozoic and older continental terrain of the Dangerous Grounds has not been underthrust and forms the outcropping continental terrain of NE Palawan and the Calamian Islands (Mitchell and Leach, 1991, Hutchison, 1996b).

XXVIII.8. GRAVITY MEASUREMENTS

Holt (1998) carried out an on-land gravity survey of eastern Sabah (Figure 139). A similar survey could not be carried out over the Western Cordillera because of insufficient accuracy of the contour elevations. The model of Figure 138 implies that Sabah is underlain by a Mesozoic ophiolitic basement. However Holt (1998) and Milsom and Holt (2001) maintain that their gravity data indicate a continental basement for Sabah.

The high-gravity values are centred around Darvel Bay and the Telupid area (Figure 139) where there are extensive outcrops of ophiolite. The low values are centred around the Meliau Basin that represents a great thickness of Middle to Upper Miocene low-density sediments. The gravity map (Figure 139) therefore correlates well with the outcropping geology.

Milsom and Holt (2001) make a valid argument that the large ophiolite massifs of Sabah need to be supported at the surface by underlying low-density crust, and

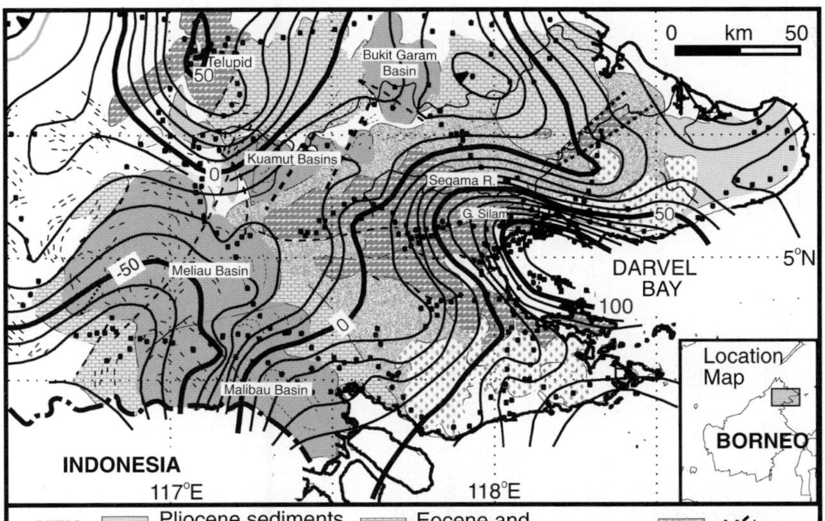

Figure 139. Bouguer gravity in eastern Sabah. Contour interval 10 mGal. Heavy dots show gravity station locations and, in the Upper Segama, indicate forestry roads used in the survey (from Milsom and Holt, 2001). (by permission of the Geological Society of London)

there are sporadic outcrops of Mesozoic calc-alkaline granites within the large Segama River ophiolite that may represent a continental basement supporting the ophiolite. These represent the only evidence of a possible continental basement. The outcropping geology of Sabah indicates a Mesozoic ophiolite basement covered by deep-water sediments (Figure 71). Only in the Miocene were shallow water quartz-rich sediments brought into the uplifted area (Hutchison et al., 2000, 2001).

Chapter XXIX

Mineral Deposits

Sabah is not a heavily mineralized country. It had one active porphyry copper mine at Mamut, now closed. There had been a long unfruitful search for economic gold deposits.

XXIX.1. CHROMITE CONCENTRATIONS IN OPHIOLITE

Ophiolite sequences occupy a considerable part of eastern Sabah. The chromitite concentrations in the peridotite and serpentinized peridotite are too small for commercial mining (Bailey, 1963; Hutchison and Dhonau, 1971). The reason for the absence of large deposits from Sabah, as compared with the Zambales of nearby Philippines, is that only the upper layers of the Sabah ophiolite are exposed, and the chromitite layers are thin and podiform. They represent the lower ultrabasic cumulate part of the gabbro layer. By contrast the rich Zambales deposits occur well into the mantle sequence.

Hutchison (1972) made a systematic study of the Sabah chromite. It shows a range of Al/Cr substitution and the unit cell (a, Å) edge varies systematically thus:

$$\text{Al cations per unit cell} = 407.3 - 48.6a$$

Reflectivity (%) at 590 nm wavelength varies linearly with composition, thus:

$$\text{Al cations per unit cell} = 27.93 - 1.82 \times \text{reflectivity}$$

Hardness ranges from 1200 to 1500 VH.

XXIX.2. NICKEL LATERITE

Nickel laterite has been mapped in the Tavai Plateau and the Bidu Bidu Hills.

XXIX.3. CYPRUS-TYPE DEPOSITS

The search has been made for Cyprus-type sulphide deposits in Sabah (Lee and Weber, 1986; Weber and Lee, 1990). Geochemical exploration was followed by drilling in the eastern flanks of the Bidu-Bidu Hills. Massive sulphides were encountered. The ore occurs as pockets associated with shale and spilite of the upper parts of the Lower Cretaceous ophiolite complex (Chert–Spilite Association). At Sualog, the massive sulphide bodies are closely associated with intensely chloritized and zeolitized pillow lavas of the ophiolite sequence. They are associated with quartz veining and silicification leading to an interpretation as 'stockwork-type', frequently found underlying massive sulphide deposits of submarine exhalative origin (Weber and Lee, 1990). They are considered to be syn-volcanic and genetically related to the basalt of the ophiolite suite. The volcanic rocks are tholeiitic and Ti, Cr and Ni plots indicate an ocean floor origin of MORB characteristic (Muff, 1990). The massive sulphide, which readily oxidizes to sulphate upon exposure of only a few weeks, occurs at the top of the basalt near or at its contact with overlying shale. The ore minerals are predominantly pyrite, accompanied by marcasite, pyrrhotite, chalcopyrite, sphalerite, bornite, covellite, chalcocite, chromite and magnetite (Muff, 1990).

The small Leadstar deposit has been evaluated on the eastern flank of the Bidu-Bidu Hills, near the serpentinite conglomerate and sandstone locality described by Hutchison and Tungah Surat (1991), but production was delayed indefinitely in the absence of environmental approval. A similar massive sulphide body occurs at the contact of overlying shale and basalt at Kiabau (Mylius, 1990).

XXIX.4. MAMUT PORPHYRY COPPER DEPOSIT

The Mamut porphyry copper mine (Figure 140), now closed, was unusual for two reasons: the source rock is adamellite porphyry and not diorite as in the Philippines, and the plutons were emplaced hypabysally into Palaeogene turbiditic rocks (Trusmadi Formation) and Lower Cretaceous ophiolite, and not into coeval volcanic rocks that are totally unknown from the region (Hutchison, 1996b).

The average age of the porphyry stocks is 9 Ma (Upper Miocene). The total reserves of the deposit were 179 Mt of Cu grade 0.476% (Akiyama, 1984). Other metal contents were Pb, 102 ppm, Zn, 103 ppm, Ag, 2.5 gt^{-1}; Au, 0.66 gt^{-1}; Mo, 7.15 ppm and S, 2.28%. The ore body was only about one-seventh the size of the Atlas Mine of Cebu. The beneficiated ore was trucked to Kota Belud and shipped to Japan for refining. Viability of the mine depended upon the content of precious metals.

The distribution of mineralization is unique. About 50% of the mineralization was hosted by the source adamellite porphyry, 30% was hosted by the serpentinized peridotite country rock, and the remaining 20% by the fine-grained sandstones of the Trusmadi Formation (Figure 140) (Kosaka and Wakita, 1978). There is a post-mineralization barren granodiorite porphyry which contains xenoliths of the earlier mineralized pluton. The hydrothermal alteration is interesting, for the serpentinite has been altered to an assemblage of talc, antigorite, tremolite, chlorite and biotite, with minor phlogopite. Silicification and deposition of biotite and sericite are characteristic of the other mineralized rocks.

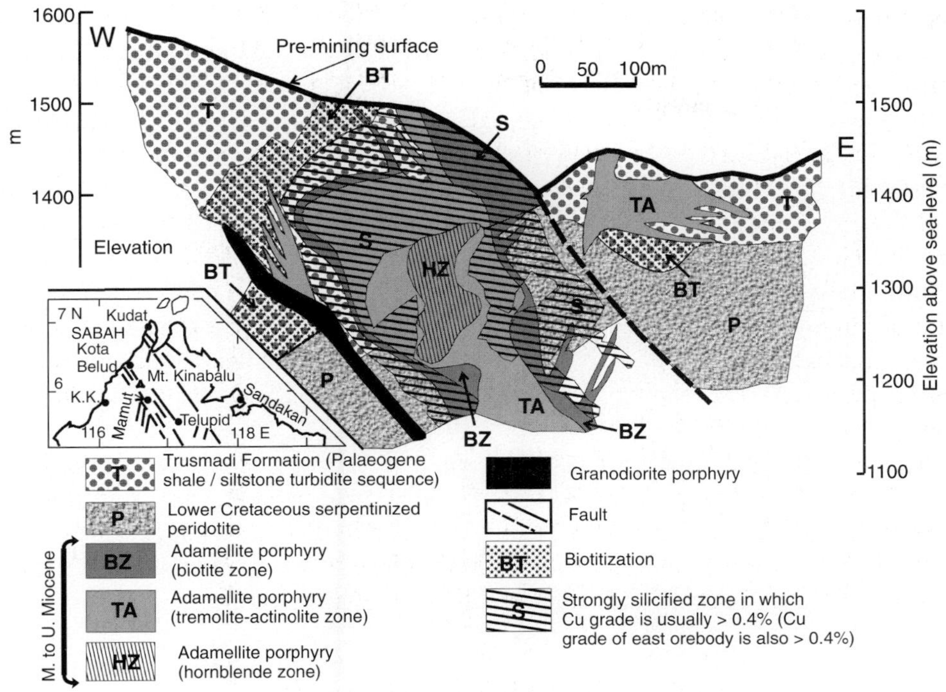

Figure 140. Mamut porphyry copper mine of the Mount Kinabalu region of Sabah (after Kosaka and Wakita, 1978). The mineralized rocks are adamellite porphyry, siltstone and serpentinite. From Hutchison, C. (1996) South-East Asian Oil, Gas Coal and Mineral Deposits. By permission from Oxford University Press.

XXIX.5. OTHER OCCURRENCES

The strongly silicified Mount Wullersdorf area of the Semporna Miocene–Pliocene volcanic arc contains chalcopyrite in association with galena and sphalerite in quartz veins. These occurrences are also associated with minor silver and gold; the latter has been panned from rivers draining the volcanic arc.

Chapter XXX

Coal Deposits

Only three major coal deposits have been mined. Both were historically important. Labuan Island was originally an integral part of Sabah, but has been designated a federal territory. There was also an important mine at Muara in Brunei Darussalam.

XXX.1. SILIMPOPON COAL MINE

The mine lay 24 km NW inland from the head of Cowie Harbour, the waterway of Tawau. Four coal seams occur in the eastern part of the field, three in the western part. Only one, the Queen Seam of the eastern part, was mined. At its thickest it was 170 cm thick, decreasing over a distance of 6.5 km to a thickness of 48 cm. At its thinner extensions, the coal became shaley and uneconomic (Collenette, 1954). Mining initially was by opencast, then by inclines. Shafts were recommended but never implemented. The underground extension was determined by drilling by the Cowie Harbour Coal Company that owned and operated the mine.

Mining began in 1904 and the mine closed in 1932. During this period, total production was in excess of 1, 370, 535 t. A railway carried the coal to a wharf 7.3 km distant. Lighters of 500 t capacity were then towed to a coaling station on Sebatik Island or to Sandakan.

The coal was black glossy high rank sub-bituminous. It was in good condition, did not disintegrate upon exposure to weather. It had good coking qualities, but the yield was low, and ash and sulphur contents were high.

Mining ceased because of water infiltration from the inclines. The weak roofs to the coal seam failed, causing falls that spontaneously heated leading to production of acid water from the alum shales of the roof falls. The acid water destroyed pipes and pumps.

XXX.2. LABUAN

Coal was mined in Labuan Island for 60 years, but only about half a million tonnes had been produced by the time mining ceased in 1912. The mined areas extended inland west-south-west (WSW) along the anticlinal axis from Kubang Bluff. In 1887, 40 t of coal were produced for *S.S. Phlegethon*. The peak of production was in 1896 with 47,000 t. The coal seams ranged from 2 to 3 m, but thinned westwards (Wilson, 1964). Mining was discontinued because of a series of accidents and underground fires.

Most of the production was of bright non-coking coal, with a small proportion of dull coal. They were consistently lignite to sub-bituminous. Coal was investigated at Weston, but never produced.

XXX.3. BROOKETON COLLIERY, MUARA, BRUNEI

Almost all of the approximately 600×10^3 t of coal produced in Brunei Darussalam between 1888 and 1924 came from the Brooketon Colliery (Wilford, 1961). Brooketon Colliery was situated 2.4 km WNW of Muara, where there is a safe deep-water anchorage, to which the mine was connected by rail. The coal was mined under European supervision in 1883. In 1888 it was leased to Rajah Charles Brooke and operated by the Sarawak Government from 1889 to 1924, when annual outputs varied between 10,000 and 25,000 t.

The mine was opencast until all overburden was removed. Adits were driven along the almost vertical seams: vertical dips are the rule along the eastern limb of the Berakas Syncline. The mine was closed in 1924 because of heavy financial losses, because only about 10% of the coal could be extracted as large coal barriers had to be left in place to prevent flooding and fires. Four and possibly six seams were present, two of which were 8 m thick. The seams were lenticular and interbedded with carbonaceous shale and thick massive sandstone that in some outcrops shows deeply scoured bases, rootlet beds and ferruginous shale bands (Sandal, 1996).

Chapter XXXI
Petroleum

By the end of 1997 a total of 188 wells had been drilled (Figure 141). About 90% of them were drilled in the South China Sea and the most of the remainder in the Sandakan sub-basin. Several abortive wells had been drilled before the establishment of Petronas, south of the Dent and Semporna peninsulas.

The production and reserves of oil are tabulated in Table 32. There is no production from the Sandakan sub-basin but reserves are tabulated. The figures for gas are given in Table 33. The prospects for hydrocarbons around the Spratly Islands has been discussed by Blanche and Blanche (1997), but the problems appear to be insurmountable: dispute of ownership and deep water in excess of 2 km.

Negara Brunei Darussalam had a daily oil production of 160,000 barrels for many years and this was increased to 200,000 barrels in 1998. Table 34 gives details of the various main fields extracted from Sandal (1996).

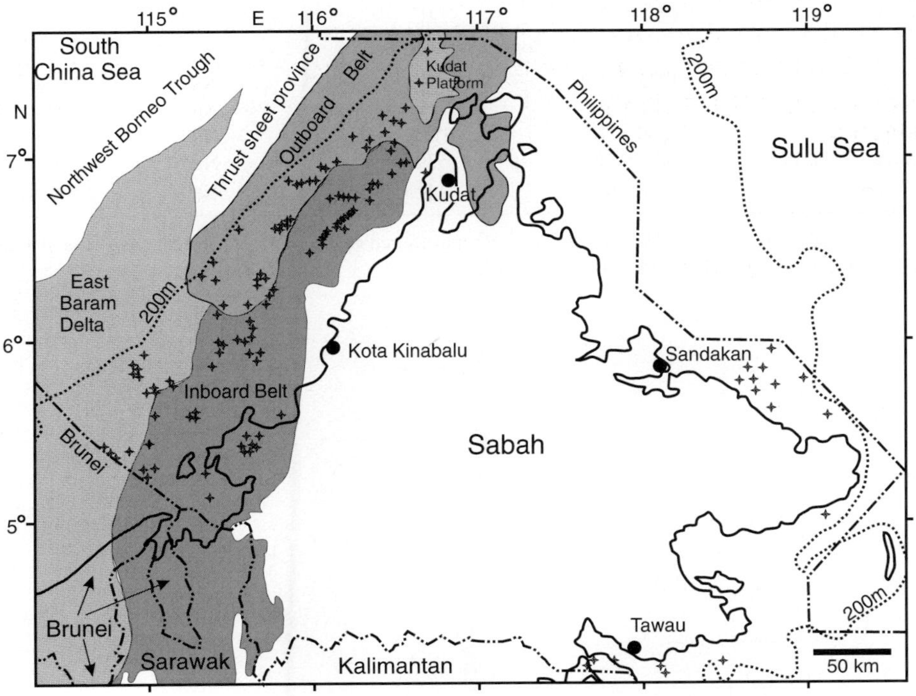

Figure 141. Wells drilled offshore Sabah before 1998 (from Mohd. Idrus, 1999). With permission from Petronas.

Table 32. Discovered oil resources in Sabah in barrels at 1.1.1998 (Mohd. Idrus, 1999)

Basin province	Oil initially in place	Estimated ultimate recovery	Production to 1.1.1998	Reserves at 1.1.1998
East Baram Delta	1280×10^6	608×10^6	367×10^6	235×10^6
Inboard Belt	1280×10^6	384×10^6	191×10^6	202×10^6
Outboard Belt	608×10^6	236×10^6	122×10^6	106×10^6
Sandakan sub-basin	32×10^6	12×10^6	Nil	17×10^6
Total for Sabah	3.2×10^9	1.24×10^9	680×10^6	560×10^6

1 barrel = 158.987 l, ~0.14 t.

Table 33. Discovered natural gas resources in standard cubic feet, as on 1.1.1998

Basin province	Gas initially in-place	Estimated ultimate recovery	Production to 1.1.1998	Reserves as at 1.1.1998
		Gas associated with oil		
East Baram Delta	3210×10^9	1960×10^9	493×10^9	1538×10^9
Inboard Belt	749×10^9	280×10^9	59×10^9	246×10^9
Outboard Belt	5671×10^9	4130×10^9	298×10^9	3751×10^9
Sandakan sub-basin	1070×10^9	630×10^9	Nil	615×10^9
Sabah total	10.7×10^{12}	7.0×10^{12}	850×10^9	6.15×10^{12}
		Non-associated gas		
East Baram Delta	1775×10^9	1078×10^9	Nil	$1\,078 \times 10^9$
Inboard Belt	213×10^9	98×10^9	Nil	98×10^9
Outboard Belt	4260×10^9	3185×10^9	Nil	$3\,185 \times 10^9$
Sandakan sub-basin	852×10^9	539×10^9	Nil	539×10^9
Sabah Total	7.1×10^{12}	4.9×10^{12}	Nil	4.9×10^{12}

1 ft^3 = 0.02831685 m^3 = 28.31685 litres. Standard ft^3 is measured at 70°F (21.11°C) under a pressure of 1 atm. 1 atm = 101.325 kilopascals. (1 Pascal = 1 Ne m^{-2}).

Table 34. Petroleum data, up to 1/1/1996, on the major oilfields of Brunei, from Sandal (1996)

Field	Wells drilled	1995 Average daily production	Cumulative production	Ultimate recoverable Resources (estimate)	At % recovery
Seria	774	2400 m^3/day	>162 × 10^6 m^3	175 × 10^6 m^3 oil	38
				46 × 10^9 m^3 gas	90
Rasau	27	1000 m^3/day oil	3.3 × 10^6 m^3 oil	6.5 × 10^6 m^3 oil	30
		0.5×106 m^3/day gas		3.6 × 10^9 m^3 gas	63
Enggang	3	100 m^3/day oil		1.4 × 10^6 m^3 oil	39
		10 000 m^3/day gas		0.16 × 10^9 m^3 gas	42
S.W.Ampa	279	6950 m^3/day oil		128 × 10^6 m^3 oil	36
		1460 m^3/day condensate		35 × 10^6 m^3 condensate	42
		21×10^6 m^3/day gas		345 × 10^9 m^3 gas	86
Fairley	63	1110 m^3/day oil		31 × 10^6 m^3 oil	36
		210 m^3/day condensate		3.7 × 10^6 m^3 condensate	46
		3.1×10^6 m^3/day gas		56 × 10^9 m^3 gas	80
Egret	9			5.4 × 10^6 m^3 oil	34
		Not yet in production		2.18 × 10^6 m^3 condensate	33
				20.7 × 10^9 m^3 gas	69
Gannet	9			0.71 × 10^6 m^3 oil	22
		130 m^3/day condensate		2 × 10^6 m^3 condensate	51
		1.69 × 10^6 m^3/day gas		16.5 × 10^9 m^3 gas	79
Champion	282	10 000 m^3/day liquids		128 × 10^6 m^3 oil	27
				2.18 × 10^6 m^3 condensate	34
		1.2 × 10^6 m^3/day gas		27.7 × 10^9 m^3 gas	55
Peragam	2			1.17 × 10^6 m^3 condensate	52
				7.7 × 10^9 m^3 gas	72
Iron Duke	14	900 m^3/day liquids		9.3 × 10^6 m^3 oil	42
				0.98x10^6 m^3 condensate	44
		0.6 × 10^6 m^3/day gas		15.5 × 10^9 m^3 gas	77
Bugan	1	Gas field			
Selangkir	1	Gas field			
Magpie	33	1500 m^3/day liquids		21.1 × 10^6 m^3 oil	44
				3.1 × 10^9 m^3 gas	58
Osprey	4	Not yet developed		10.2 × 10^9 m^3 gas	55
Jerudong	9	Shut down in 1962	0.1 × 10^6 m^3 oil		

References

Abdul Manaf, M. & Wong, R.H.F. (1995) Seismic sequence stratigraphy of the tertiary sediments, offshore Sarawak deepwater area, Malaysia. *Geol. Soc. Malay. Bull.*, **37**, 345–361.

Adams, C.G. (1959) Eocene and Miocene Foraminifera from limestone and shale in the middle Baram Valley, Sarawak. *British Borneo Geological Survey Annual Report*, no. **1959**, pp. 64–77.

Adams, C.G. (1965) The Foraminifera and stratigraphy of the Melinau Limestone, Sarawak, and its importance in Tertiary correlation. *Quart. J. Geol. Soc. London*, **121**, 283–338.

Adams, C.G. & Haak, R., (1962) The stratigraphical succession in the Batu Gading area, middle Baram, north Sarawak. In: Haile, N.S. (1962). The geology and mineral resources of the Suai-Baram area, north Sarawak. *Geological Survey Department British Territories in Borneo, Memoir* **13**, pp.141–150.

Adams, C.G. & Kirk, H. J.C. (1962) The Madai-Baturong Limestone member of the Chert–Spilite Formation. *Geol. Mag.*, **99** (4), 289–303.

Adams, C.G. & Wilford, G.E. (1972) On the age and origin of the Keramit and Selidong limestones, Sarawak, East Malaysia. *Geological Papers* **1**, Geological Survey of Malaysia, Kuala Lumpur pp.28–42.

Agostinelli, E., Mohamad Raisuddin b. Ahmad Tajuddin; Antonielli, E. & Mohamad b. Mohd. Aris (1990) Miocene–Pliocene paleogeographic evolution of a tract of Sarawak offshore between Bintulu and Miri. *Geol. Soc. Malay. Bull.*, **27**, 117–135.

Aitchison, J.C. (1994) Early Cretaceous (pre-Albian) radiolarians from blocks in Ayer Complex mélange, eastern Sabah, Malaysia, with comments on their regional tectonic significance and the origins of enveloping mélanges. *J. Southe. Asian Earth* Sci., **9**, 255–262.

Akiyama, Yoshio (1984) A case history—exploration, evaluation and development of the Mamut porphyry copper deposit. *Geol. Soc. Malay. Bull.*, **17**, 237–255.

Allen, A.W. (1952) The geology of the Lundu district, west Sarawak. Annual Report for 1952 of the Geological Survey Department, British Terrirories in Borneo, pp.41–85.

Amiruddin, & Trail, D.S. (1989) *Nangapinoh 1:250,000 Quadrangle. West Kalimantan, Geological Data Record*, Geological Research and Development Centre, Bandung.

Aubouin, J. (1965) *Geosynclines* (*Developments in Geotectonics*, vol. **1**), Elsevier, Amsterdam.

Aumento, F., Loncarevic, B.D. & Ross, D.I. (1971) Hudson Geotraverse: geology of the Mid-Atlantic Ridge at 45° N. *Philos. Trans. Roy. Soc. London* A, **268**, 623–650.

Bailey, P.S. (1963) *The Ultramafics of the Silam–Beeston Range, Darvel Bay, North Borneo*. M.Sc. thesis, University of Durham.

Balaguru, A. (2001) *Tectonic Evolution and Sedimentation of the Southern Sabah Basin, Malaysia.* Ph.D. thesis, University of London, CD-Rom.

Balaguru, A. & Nichols, G. (2004) Tertiary stratigraphy and basin evolution, southern Sabah (Malaysian Borneo). *J. Asian Earth Sci.*, **23**, 537–554.

Balaguru, Allagu, Nichols, Gary & Hall, Robert. (2003) The origin of the circular basins of Sabah, Malaysia. *Geol. Soc. Malay. Bull.*, **46**, 333–351.

Basir, Jasin (1992) Significance of radiolarian cherts from the Chert–Spilite Formation, Telupid, Sabah. *Geol. Soc. Malay. Bull.*, **31**, 67–83.

Basir, Jasin (1996) Late Jurassic to Early Cretaceous Radiolaria from chert blocks in the Lubok Antu mélange, Sarawak. *J. Southeast Asian Earth Sci.*, **13** (1), 1–11.

Basir, Jasin (2000) *Significance of Mesozoic Radiolarian Chert in Sabah and Sarawak.* Proceedings Annual Geological Conference *2000* (In: Teh, G.H. Pereira J.J. & Ng T.F. Eds.), Geological Society of Malaysia, pp. 123–130.

Basir, Jasin & Aziman, M (1996) Some Lower Cretaceous radiolaria from the Serabang complex, Sarawak. *Warta Geol., Geol. Soc. Malay.*, **22** (2), 61–65.

Basir, Jasin & Haile, N.S. (1993) Some radiolaria from the chert block of Lubok Antu Mélange, Sarawak. *Warta Geol., Geol. Soc. Malay.*, **19** (5), 205–209.

Basir, Jasin, Sanudin Tahir & Abdul, Rahim Hj. Samsuddin (1985) Lower Cretaceous radiolaria from the Chert–Spilite Formation, Kudat, Sabah. *Newslett. Geol. Soc. Malay.* (*Warta Geol.*), **11**, 161–162.

Basir, Jasin & Taj Madira Taj Ahmad (1995) Some Paleogene planktonic Foraminifera from the Lubok Antu Complex, Sarawak. *Newslett. Geol. Soc. Malay.* (*Warta Geol.*), **21** (3), 147–151.

Basir, Jasin, Uyop, Said & Ang Duen Woei (1996) Discovery of Early Jurassic Radiolaria from the tuff sequence, near Piching, west Sarawak. *Newslett. Geol. Soc. Malay.* (*Warta Geol.*), **22** (5), 343–347.

Bates, J.A. (1996) Overpressuring in the Kutai Basin: Distribution, origins and implications for the petroleum system. *Proc. Indones. Petrol. Assoc.*, **25**, 93–116.

Bayliss, D.D. (1966) Foraminifera from the Bau Limestone Formation, Sarawak, Malaysia. *Geological Survey, Borneo Region Malaysia, Annual Report for 1965*, pp.173–195. Kuching.

Beattie, D. (1986) *Gravity Modelling of a Mafic-Ultramafic Association, Darvel Bay, Sabah.* B.Sc. thesis, Dalhousie University, Canada.

Beauvais, L. & Fontaine, H. (1990) Corals from the Bau Limestone Formation, Jurassic of Sarawak, Malaysia. In: Fontaine, H. (Ed.) *Ten Years of CCOP Research on the Pre-Tertiary of East Asia, CCOP/TP* **20**, pp. 209–239, CCOP, Bangkok.

Beddoes, L.R. (1976) *The Balabac sub-basin, southwestern Sulu Sea, Philippines*, Offshore Southeast Asia Conference, Southeast Asian Petroleum Exploration Society, paper 15, Singapore.

Bellon, H. & Rangin, C. (1991) Geochemistry and isotope dating of the Cenozoic volcanic arc sequences around the Celebes and Sulu Seas. In: Silver, E.A., Rangin, C. & von Breymann, M.T. (Eds.), *Proceedings of the Ocean Drilling Program Scientific Results*,Vol. 124, pp. 321–338.

Bergman, S.C., Hutchison, C.S., Swauger, D.A. & Graves, J.E. (2000) K:Ar ages and geochemistry of the Sabah Cenozoic volcanic rocks. *Geol. Soc. Malay. Bull.*, **44**, 165–171.

Bignell, J.D. (1972) *The Geochronology of the Malayan Granites,* D.Phil. thesis, University of Oxford.

Bladon, G.M., Pieters, P.E. & Supriatna, S. (1989) *Catalogue of Isotopic Ages Commissioned by the Indonesia–Australia Geological Mapping Project for Igneous and Metamorphic Rocks in Kalimantan*. Geological Research and Development Centre, Bandung.

Blanche, J.B. & Blanche, J.D. (1997) An overview of the hydrocarbon potential of the Spratly Islands archipelago and its implications for regional development. In: Fraser, A.J., Matthews, S.J. & Murphy, R.W. (Eds.), *Petroleum Geology of Southeast Asia.* Geological Society of London, Special Publication, 126, pp. 293–310.

Bol, A.J. & van Hoorn, B. (1980) Structural styles in western Sabah offshore. *Geol. Soc. Malay. Bull.*, **12**, 1–16.

Briais, A., Patriat, P. & Tapponnier, P. (1993) Updated interpretation of magnetic anomalies and seafloor spreading stages in the South China Sea: Implications for the Tertiary tectonics of Southeast Asia. *J. Geophys.* Res., **98**, 6299–6328.

Brondijk, J.F. (1963) Reclassification of part of the Setap Shale Formation as Temburong Formation. *British Borneo Geological Survey Annual Report for 1962*, pp. 56–60.

Brondijk, J.F. (1964) The Danau Formation in Sarawak and Sabah. *Annual Report for 1963, Geological Survey of Malaysia, Borneo Region*, pp. 167–178.

Bruggen, Gerrit Ter (1935) De Eocene Fyllietformatie in Centraal-Borneo. Dissertation, University of Delft. [English translation by Haile, N.S. (1955). The Eocene Phyllite Formation in Central Borneo, *Geol. Surv. Dept. British Territories in Borneo, Bulletin*, Vol. 2, Kuching, pp. 39–124.]

Bureau of Mines (1976) A review of oil exploration and stratigraphy of sedimentary basins of the Philippines. *United Nations ESCAP-CCOP Technical Bulletin*, **10**, 55–99. Bangkok.

Campbell, C.J. (1956) Geology of the Usun Apau area. *British Borneo Geological Survey Annual Report* 1956, pp. 86–120.

Chambers, J.L.C. & Daley, T.E. (1997) A tectonic model for the onshore Kutai Basin, East Kalimantan. In: Fraser, A.J., Matthews, S.J. & Murphy, R.W. (Eds.), *Petroleum Geology of Southeast Asia*, Geological Society of London, Special Publication, **126**, 375–393.

Chappell, B.W. & White, A.J.R. (1974) Two contrasting granite types. *Pac. Geol.*, **8**, 173–174.

Chiang, Kai Kim (2002) *Geochemistry of the Cenozoic igneous rocks of Borneo and Tectonic Implications*. Ph.D. thesis, University of London, 364pp.

Chua Beng Yap (1980) *Sedimentology and Structure of the Crocker Formation, Kampung Balabak Area, Sabah, East Malaysia*. B.Sc. (hons.) thesis, Dept. of Geology, University of Malaya.

Clennell, M.B. (1991) The origin and tectonic significance of mélanges in eastern Sabah, Malaysia. *J. Southeast Asian Earth Sci.*, **6**, 407–429.

Clennell, M.B. (1992) *The Mélanges of Sabah, Malaysia*. Ph.D. thesis, University of London.

Clennell, M.B. (1996) Far-field and gravity tectonics in Miocene basins of Sabah, Malaysia. Hall, R. and Blundell, D. (eds.), *Tectonic evolution of Southeast Asia*, Geological Society of London Special Publication 106, pp. 307–320.

Clift, P.D., Lin J. & Barckhausen, U. (2002) Evidence of low flexural rigidity and low viscosity lower continental crust during continental break-up in the South China Sea. *Mar. Petrol. Geol.*, **19** (8), 951–970.

Cobbing, E.J., Pitfield, P.E.J., Darbyshire, D.P.F. & Mallick, D.I.J. (1992) The granites of the South-East Asian tin belt. *Overseas Memoir* 10, British Geological Survey, Keyworth, Nottinghamshire.

Coleman, J.M., Gagliano, S.M. & Smith, W.G. (1970) Sedimentation in a Malaysian High Tide tropical delta. In: Morgan, J.P. & Shaver, R.H. (Eds.) *Deltaic Sedimentation Modern and Ancient*, Society of Economic Palaeontologists and Mineralogists, Special Publication **15,** 185–197.

Coleman, R.G. (1977) *Ophiolites: Ancient Oceanic Lithosphere?* Springer, New York.

Collenette, P. (1954) The coal deposits and a summary of the Geology of the Silimpopon area, Tawau District, Colony of North Borneo. *Geological Survey Department, British Territories in Borneo, Memoir* 2, 74pp.

Collenette, P. (1958) The geology and mineral resources of the Jesselton–Kinabalu area, North Borneo. *Geological Survey Department, British Territories in Borneo, Memoir* **6**.

Collenette, P. (1963) A physiographic classification of North Borneo. *J. Trop. Geogr.*, **17**, 28–33.

Collenette, P. (1965) The geology and mineral resources of the Pensiangan and Upper Kinabatangan area, Sabah, Malaysia. *Geological Survey of Malaysia, Borneo Region, Memoir*, Vol. 12, 150p.

Collenette, P. (1966) The Garinono Formation, Sabah, Malaysia. *Borneo Region Malaysian Geological Survey Annual Report for 1965*, 161–167.

Cummings, R.H. (1962). Limestones of the Terbat Formation, West Sarawak. Annual Report for 1961, Geological Survey Department, British Territories in Borneo, Kuching, 36–48.

Defant, M.J. & Drummond, M.S. (1990) Derivation of some modern arc magmas by melting of young subducted lithosphere. *Nature*, **347**, 662–665.

de Silva, S. (1986a) Petrology and geochemical analysis of coal from the south Mukah-Balingian region, Sarawak. *Newsl. Geol. Soc. Malay.*, 12 (4), 169.

de Silva, S. (1986b) Stratigraphy of south Mukah–Balingian region, Sarawak. *Newsl. Geol. Soc. Malay.*, **12** (5), 215–219.

de Silva, Sriyanee. (1987) The sedimentology of the "Begrih–Liang" Formation [Sarawak]. *Newsl. Geol. Soc. Malay.*, **13** (4), 165–167.

de Silva, Sriyanee. (1988) The telogenetic formation of a redbed [Mukah Road, Sarawak]. *Newsl. Geol. Soc. Malay.*, **14** (1), 1–10.

Doutch, H.F. (1992) Aspects of the structural histories of the Tertiary sedimentary basins of East, Central and West Kalimantan and their margins. *BMR J. Aust. Geol. Geophys.*, **13**, Canberra, 237–250.

Drummond, M.S., Defant, M.J. & Kepezhinskas, P.K. (1996) The petrogenesis of slab derived trondhjemite-tonalite-dacite/adakite magmas. *Trans. Roy. Soc. Edin. Earth Sci.*, **87**, 205–215.

Epting, M. (1980) Sedimentology of Miocene carbonate buildups, Central Luconia, offshore Sarawak. *Geol. Soc. Malay. Bull.*, **12**, 17–30.

Ewart, A. (1982) The mineralogy and petrology of Tertiary–Recent orogenic volcanic rocks: with special reference to the andesite-basaltic compositional range. Thorpe, P.S. (ed.) *Andesites: Orogenic Andesites and Related Rocks*, Wiley, Chichester, pp.25–95.

Fitch, F.H. (1955) The geology and mineral resources of part of the Segama Valley and Darvel Bay area, Colony of North Borneo. *British Borneo Geological Survey Memoir* 4.

Fitch, F.H. (1958) The geology and mineral resources of the Sandakan area and parts of the Kinabatangan and Labuk Valleys, North Borneo. *British Borneo Geological Survey, Memoir*, 9.

Fontaine, H. (1990) The Terbat Formation of Sarawak (Malaysia): a very peculiar limestone. In: Fontaine, H. (Ed.), *Ten Years of CCOP Research on the Pre-Tertiary of East Asia, CCOP/TP* **20**, CCOP Technical Secretariat, Bangkok, pp. 173–181.

Fontaine, H. & Ho W. K. (1989) Note on the Madai–Baturong Limestone, Sabah, East Malaysia; discovery of *Caprinidae* (rudists). *CCOP Newsletter*, **14** (3 and 4), 27–32, Bangkok.

Fontaine, H., Daud R. & Updesh S. (1990) Discovery of an Upper Triassic limestone basement in the Malay Basin, offshore Peninsular Malaysia: regional implications. *J. Southeast Asian Earth Sci.*, **4** (3), 219–232.

Fontaine, H. & Workman, D.R. (1978) *Review of the geology and mineral resources of Kampuchea, Laos and Vietnam*. In: Nutalaya, P. (Ed.), Proceedings of the 3rd Regional Conference on Geology and Mineral Resources of Southeast Asia, Bangkok Nov. 1978, Asian Institute of Technology, Bangkok, pp. 541–603.

Fuller, M., Ali, J.R., Moss, S.J., Frost, G.M., Richter, B. & Mahfi, A. (1999) Paleomagnetism of Borneo. *J. Asian Earth Sci.*, **17**, 3–24.

Fuller, M., Haston, R. Lin Jin-Lu, Richter, B., Schmidtke, E. & Almasco, J. (1991) Tertiary paleomagnetism of regions around the South China Sea. *J. Southeast Asian Earth Sci.*, **6**, 161–184.

Gallagher, K., Brown, R. & Johnson, C. (1998) Fission track analysis and its applications to geological problems. *Annu. Rev. Earth Planet. Sci.*, **26**, 519–572.

Gass, I.G. & Smewing, J.D. (1973) Intrusion, extrusion and metamorphism at constructive margins: evidence from the Troodos Massif, Cyprus. *Nature*, **242**, 26–29.

Gatinsky, Y.G., Hutchison, C.S., Nguyen, Minh, N. N. & Tri, T.V. (1984) *Tectonic Evolution of Southeast Asia*. 27th International Geology Congress Reports no. 5, Colloquium 5, Tectonics of Asia, Moscow, pp. 225–240.

Gatinsky, Y.G. & Hutchison, C.S. (1986) Cathaysia, Gondwanaland and the Paleotethys in the evolution of continental Southeast Asia. Geol. Soc. Malay. Bull., **20,** 179–199.

Gonguet, C. (2001) Strike-slip structural style in the Sarawak offshore: a case study. *Petroleum Geology Conference & Exhibition 2001, Programme & Abstracts of Papers*, Geological Society of Malaysia, Kuala Lumpur, p.78.

Grant, C.J. (2003) The Pink Fan: a classic deep-marine canyon-fill complex, Block G, NW Sabah. *Geol. Soc. Malay. Bull.*, **47,** 85–94.

Graves, J.E. Hutchison, C.S., Bergman, S.C. & Swauger, D.A. (2000) Age and MORB geochemistry of the Sabah ophiolite basement. *Geol. Soc. Malay. Bull.*, **44,** 151–158.

Hageman, H. (1987) Palaeobathymetrical changes in NW Sarawak during Oligocene to Pliocene. *Geol. Soc. Malay. Bull.*, **21,** 91–102.

Haile, N.S. (1954) The geology and mineral resources of the Strap and Sadong valleys, West Sarawak, including the Klingkang Range coal. *Geological Survey Department, British Territories in Borneo, Memoir* 1.

Haile, N.S. (1957) The geology and mineral resources of the Lupar and Saribas valleys, West Sarawak. *Geological Survey Department British Territories in Borneo, Memoir* 7.

Haile, N.S. (1962) The geology and mineral resources of the Suai-Baram area, north Sarawak. *Geological Survey Department British Territories in Borneo, Memoir* 13.

Haile, N.S. (1969) Geosynclinal theory and the organizational pattern of the North-west Borneo Geosyncline. *Quart. J. Geol. Soc. London*, **124,** 171–194.

Haile, N.S. (1974) Borneo. In: Spencer, A.M. (Ed.), *Mesozoic–Cenozoic Orogenic belts: Data for Orogenic studies*, Geological Society of London, special publication, **4,** 333–347.

Haile, N.S. (1996) Note on the Engkilili Formation and the age of the Lubok Antu Mélange, West Sarawak, Malaysia. *Newslett. Geol. Soc. Malay.* (*Warta Geol.*), **22** (2), 67–70.

Haile, N.S. & Ho Chee Kwong (1991) *Geological field guide, Sibu–Miri Traverse, Sarawak (24 September–1 October 1991)*. PETRONAS Petroleum Research Institute, Kuala Lumpur 37pp.

Haile, N.S. & Kirk, H.J.C. (1956) Cretaceous of the Baleh headwaters and volcanic rocks of the Nieuwenhuis Mountains. *British Borneo Geological Survey Annual Report for 1956*, 78–86.

Haile, N.S., McElhinny, M.W. & McDougall, I. (1977) Palaeomagnetic data and radiometric ages from the Cretaceous of West Kalimantan (Borneo), and their significance in interpreting regional structure. *J. Geol. Soc. London*, **133,** 133–144.

Haile, N.S. & Wong, N.P.Y. (1965) The geology and mineral resources of Dent Peninsula, Sabah. *Geological Survey, Borneo Region, Malaysia, Memoir* 16.

Hall, R. (1996) Reconstructing Cenozoic SE Asia. In: Hall, R. & Blundell, D. (Eds.), *Tectonic evolution of Southeast Asia*. Geological Society of London Special Publication no. 106, 153–184.

Hall, R. (2002) Cenozoic geological and plate tectonic evolution of SE Asia and the SW Pacific: computer-based reconstructions, model and animations. *J. Asian Earth Sci.*, **20,** 353–431.

Hamilton, W. (1979) Tectonics of the Indonesian region. *U.S. Geological Survey Professional Paper*, no. 1078.

Hancock, W.G. (1964) *The Mount Tavai Peridotite, North Borneo*. Ph.D. thesis, University of Durham.

Haq, B.V., Hardenbol, J. & Vail, P.R. (1987) Chronology of fluctuating sea-levels since the Triassic. *Science*, **235,** 1156–1167.

Hashimoto, W. & Matsumaru, K. (1977) Orbitolinas from west Sarawak, East Malaysia. In: Kobayashi, T., Toriyama, R. & Hashimoto, W. (Eds.), *Geology and Palaeontology of Southest Asia*, Vol. 18, University of Tokyo Press, Tokyo, pp. 49–57.

Hashimoto, W. & Matsumaru, K. (1981) Larger Foraminifera from Sabah, Malaysia. Part 1. Larger Foraminifera from the Kudat Peninsula, the Gomantong area, and the Semporna Peninsula. *Geol. Palaeontol. Southeast Asia*, 22, 49–54.

Hatton, F. (1885) *North Borneo: Explorations and Adventures on the Equator*. Sampson Low, Marston, Searle and Rivington, London.

Hazebroek, H.P. & Tan, D.N.K. (1993) Tertiary tectonic evolution of the NW Sabah continental margin. Tectonic Framework and Energy Resources of the Western Margin of the Pacific Basin, *Geol. Soc. Malay. Bull.*, 33, 195–210.

Hinz, K. (1983) Reflection seismic record across the Northwest Sulu Sea Basin. Bally, A.W. (ed.) *Seismic Expression of Structural Styles — A Picture and Work Atlas.*, Vol. 3, section 4.2.183, American Association of Petroleum Geologists, Tulsa, OK.

Hinz, K. & Block, M. (1990) Summary of geophysical data from the Sulu and Celebes seas. Rangin, C., Silver, E., Von Breymann, M.Y. et al. *Proceedings of the Ocean Drilling Program, Initial Reports*, Chapter 5, p. 124.

Hinz, K., Block, M., Kudrass, H.R. & Meyer, H. (1991) Structural elements of the Sulu Sea, Philippines. *Geologisches Jaarbuch*, **A127**, 483–506.

Hinz, K., Fritsch, J., Kempter, E.H.K., Mohammad, A.M., Meyer, J., Mohamed, D., Vosberg, H., Weber, J. & Benavidez, J. (1989) Thrust tectonics along the north-western continental margin of Sabah/Borneo. *Geologisches Rundsch.* **78,** 705–730.

Hinz, K. & Schlüter, H.U. (1985) Geology of the Dangerous Grounds South China Sea, and the continental margin of southwest Palawan: results of Sonne cruises SO-23 and SO-27. *Energy*, **10,** 297–315.

Ho, K. F. (1978) Stratigraphic framework for oil exploration in Sarawak. *Geol. Soc. Malay. Bull.*, **10,** 1–13.

Honnorez, J., Mével, C. & Montigny, R. (1984) Occurrence and significance of gneissic amphibolites in the Verma fracture zone, equatorial Mid-Atlantic Ridge. In: Gass, I.G., Lippard, S.J. & Shelton,

A.W. (Eds.), *Ophiolites and Oceanic Lithosphere*, Geological Society of London, Special Publication 13, 121–130.

Holt, R.A. (1998) *The Gravity Field of Sundaland—Acquisition, Assessment and Interpretation*. Ph. D. thesis, Research school of geological and geophysical sciences, Birbeck College and University College, University of London.

Hon, V. (1976) Some analyses of the Serian Volcanics from the Kuap area, Sarawak. *Geological Survey of Malaysia Annual Report for 1975*, 212–220.

Hon, V. (in manuscript) Kuap area, west Sarawak, Malaysia. *Geological Survey of Malaysia, Report*, **7**.

Honza, E., John, J. & Banda, R.M. (2000) An imbrication model for the Rajang accretionary complex in Sarawak, Borneo. *J. Asian Earth Sci.*, **18,** 751–759.

Huchon, P., Nguyen. T.N.H. & Chamot-Rooke, N. (2001) Propagation of continent break-up in the south-western South China Sea. In: Wilson, R.C.L., Beslier, M.-O., Whitmarsh, R.B., Froitzheim, N. & Taylor, B. (Eds.), *Non-Volcanic Rifting of Continental Margins: A Comparison of Evidence from Land and Sea*. Geological Society of London, Special Publication **187,** 31–50.

Hutchison, C.S. (1972) Alpine-type chromite in North Borneo, with special reference to Darvel Bay. *Am. Mineral.*, **57**, 835–856.

Hutchison, C.S. (1975) Ophiolite in Southeast Asia. *Geol. Soc. Am. Bull.*, **86,** 797–806.

Hutchison, C.S. (1978) Ophiolite metamorphism in northeast Borneo. *Lithos*, **11,** 195–208.

Hutchison, C.S. (1982) Indonesia. Thorpe, R.S. (ed.). *Andesites: Orogenic Andesites and Related Rocks*, Wiley, Chichester, pp. 207–224.

Hutchison, C.S. (1983) *Economic Deposits and their Tectonic Setting*, The MacMillan Press, London and Basingstoke.

Hutchison, C.S. (1989) *Geological Evolution of South-East Asia*, Oxford University Press, England, pp. 368.

Hutchison, C.S. Ed. (1991) *Studies in East Asian Tectonics and Resources (SEATAR), Crustal Transect. VII. Jawa–Kalimantan–Sarawak–South China Sea*. Committee for Co-ordination of Joint Prospecting for mineral resources in Asian offshore areas (CCOP) (Bangkok) and Intergovernmental oceanographic commission (IOC) (Paris), CCOP/TP 26.

Hutchison, C.S. (1992a) The Southeast Sulu Sea, a Neogene marginal basin with outcropping extensions in Sabah. *Geol. Soc. Malay. Bull.*, **32,** 89–108.

Hutchison, C.S. (1992b) The Eocene unconformity in Southeast and East Sundaland. *Geol. Soc. Malay. Bull.*, **32,** 69–88.

Hutchison, C.S. (1993) Gondwanaland and Cathaysian blocks, Palaeotethys sutures and Cenozoic tectonics in South-east Asia. *Geologisches Rundsch.*, **82,** 388–405.

Hutchison, C.S. (1994) Mélange on the Jerudong Line, Brunei Darussalam, and its regional significance. *Geol. Soc. Malay. Bull.*, **36,** 157–161.

Hutchison, C.S. (1996a) The 'Rajang accretionary prism' and 'Lupar Line' problem of Borneo. In: Hall, R. & Blundell, D. (Eds.), *Tectonic Evolution of Southeast Asia. Geol. Soc. of London Special Publication* **106,** 247–261.

Hutchison, C.S. (1996b) *South-East Asian Oil, Gas, Coal and Mineral Deposits.* Oxford Monographs on geology and geophysics, 36, Clarendon Press, Oxford.

Hutchison, C.S. (2001) Sundaland half-grabens of Sarawak — implications. *Petroleum Geology Conference & Exhibition 2001. Programme & Abstracts of papers*, Geological Society of Malaysia, Kuala Lumpur, pp. 53–56.

Hutchison, C.S. (2002) *Geology and Topography of the Outer Limit of the South China Sea Continental Shelf of Malaysia.* Report prepared for the Minerals and Geoscience Department, Malaysia, and the Continental Shelf Technical Working Group, Kuala Lumpur.

Hutchison, C.S. (2004) Marginal basin evolution: The southern South China Sea. *Mar. Petrol. Geol.*, **21,** 1129–1148.

Hutchison, C.S. & Dhonau, T.J. (1969) Deformation of an Alpine ultramafic association in Darvel Bay, East Sabah, Malaysia. *Geol. Mijnbouw*, **48,** 481–493.

Hutchison, C.S., Banda, R.M. & Unya, A. (in preparation) *Geology of Sarawak*. Memoir of the Department of Minerals & Geoscience Malaysia, Sarawak.

Hutchison, C.S., Bergman, S.C., Swauger, D.A. & Graves, J.E. (2000) A Miocene collisional belt in north Borneo: Uplift mechanism and isostatic adjustment quantified by thermochronology. *J. Geol. Soc. London*, **157,** 783–793.

Hutchison, C.S., Bergman, S.C., Swauger, D.A. & Graves, J.E. (2001) Discussion of a Miocene collisional belt in north Borneo: Uplift mechanism and isostatic adjustment quantified by thermochronology. *J. Geol. Soc. London*, **158,** 398–400.

Hutchison, C.S. & Dhonau, T.J. (1971) An Alpine association of metabasites and ultrabasic rocks in Darvel Bay, East Sabah, Malaysia. *Overseas Geol. Miner. Res.*, **10,** 287–308.

Hutchison, C.S. & Tungah, Surat (1991) Sabah serpentinite sandstone and conglomerate. *Geol. Soc. Malay. Newslett.* (*Warta Geol.*), **17,** 59–64.

Ida, S.A. (1982) *Geology of the Ranau Area, Sabah, East Malaysia.* B.Sc. (Hons.) thesis, Department of Geology, University of Malaya.

Imai, A. & Ozawa, K. (1991) Tectonic implications of the hydrated garnet peridotites near Mt Kinabalu, Sabah, East Malaysia. *J. Southeast Asian Earth Sci.*, **6** (3/4), 431–445.

Irvine, T.N. & Baragar, W.R.A. (1971) A guide to the chemical classification of common igneous rocks. *Can. J. Earth Sci.*, **8,** 523–548.

Ishibashi, T. (1982) Upper Jurassic and Lower Cretaceous ammonites from Sarawak, Borneo, East Malaysia. In: Kobayashi, T., Toriyama, R. & Hashimoto, W. (Eds.), *Geology and Palaeontology of Southeast Asia*, Vol. 23, University of Tokyo Press, Tokyo, pp. 65–75.

Ismail, Che Mat Zin (1994) Dent Group and its equivalent in the offshore Kinabatangan area, East Sabah. *Geol. Soc. Malay. Bull.*, **36,** 127–143.

Ismail, Che Mat Zin (2000) *Stratigraphic Position of the Rangsi Conglomerate in Sarawak*, Proceedings Annual Geological Conference 2000, Geological Society of Malaysia, pp. 131–136.

Ismail Che Mat Zin & Tucker, M.E. (1999) An alternative stratigraphic scheme for the Sarawak Basin. *J. Asian Earth Sci.*, **17,** 215–232.

Jacobson, G. (1970) Gunung Kinabalu area, Sabah, Malaysia. *Geological Survey of Malaysia, Report,* **8.**

Johansson, M. (1999) Facies analysis of the Plateau Sandstone (Eocene to early Miocene?), Bako National Park, Sarawak, Malaysia. *J. Asian Earth Sci.*, **17,** 233–246.

Johari, D. & Mohamad, F.A. (1994) Geology of the Merit-Pila coal basin, Sarawak, Malaysia. *Am. Assoc. Petrol. Geol. Bull.*, **78,** 1149.

Johnston, J.C. & Walls, P.J. (1974) Geology of the Telupid area, Sabah. *Annual Report of the Geological Survey of Malaysia for 1973*, 213–220.

Jongmans, W.J. (1951) Fossil plants of the island of Bintan (with a contribution by J.W. Adam). *Koninkl. Nederl. Akad. Wetenschappen, Proc. Ser.* B, **54** (2), 183–190, plates 1–3.

Kamaluddin Hassan (1980) *Sedimentology, Structural and Stratigraphy of the Tamparuli area, Crocker Formation, Sabah, East Malaysia.* B.Sc. (Hons.) thesis, Department of Geology, University of Malaya.

Kanno, S. (1978) Brackish molluscan fauna (Upper Eocene) from the Silantek Formation in West Sarawak, Malaysia. *Geol. Palaeontol. Southeast Asia*, **19,** 103–112.

Karig, D.E. (1972) Remnant arcs. *Geol. Soc. Am. Bull.*, **83,** 1057–1068.

Keij, A.J. (1963) *Distichoplax* in Sarawak and North Borneo. *Brit. Borneo Geol. Surv. Bull.*, **4,** 153–160.

Khalid, Ali Alshebani, Wan, Hasiah Abdullah & Abdul Hadi Abd. Rahman. (2003) Biomarker characterisation and thermal maturity evaluation of Ganduman Formation, Sahabat area, Dent Peninsula, Sabah, Malaysia. *Geol. Soc. Malay. Bull.*, **46,** 461–466.

Kho, C.H. (1966) The Bintulu coalfield: Prospecting results. *Annual report for 1965 of the Geological Survey, Borneo Region, Malaysia*, pp. 40–68. Kuching.

Kirk, H.J.C. (1957) The geology and mineral resources of the Upper Rajang and adjacent areas. *Geological Survey Dept. British Territories in Borneo, Memoir* 8.

Kirk, H.J.C. (1962) The geology and mineral resources of the Semporna Peninsula, North Borneo. *British Borneo Geological Survey Memoir* 14.

Kirk, H.J.C. (1964) Igneous rocks of Sabah and Sarawak. *Borneo Region Malaysia Geological Survey Annual Report for 1963*, pp. 82–94.

Kirk, H.J.C. (1968) *The Igneous Rocks of Sarawak and Sabah*. Geological Survey, Borneo Region, Malaysia, Bulletin 5.

Kirst, P. (1976) *Petrology of the Metamorphic Rocks from the Equatorial Mid-Atlantic Ridge and Fracture zone*. Ph.D. thesis, University of Miami.

Kon'no, E. (1972) Some Late Triassic plants from the southwestern border of Sarawak, east Malaysia. In: Kobayashi, T. & Toriyama, R. (Eds.), *Geology and Palaeontology of Southeast Asia*, Vol. 10, University of Tokyo Press, pp. 125–178, Plates XXIV–XXXII.

Koopman, A. (1996) Regional geological setting. In: Sandal, S.T. (Ed.), *The Geology and Hydrocarbon Resources of Negara Brunei Darussalam*, Syabas, Bandar Seri Begawan, pp. 49–63.

Koopmans, B.N. (1967) Deformation of the metamorphic rocks and the Chert–Spilite Formation in the southern part of the Darvel Bay area, Sabah. *Geological Papers 1966*, Geological Survey, Borneo Region, Malaysia, Bulletin **8**, 14–24.

Koopmans, B.N. & Stauffer, P.H. (1967) Glacial phenomena on Mount Kinabalu, Sabah. In: Collenette, P. (Ed.), Geological Papers 1966. *Geological Survey, Borneo Region, Malaysia, Bulletin* **8**, 25–35.

Kosaka, H. & Wakita, K. (1978) Some geologic features of the Mamut porphyry copper deposit, Sabah, Malaysia. *Econ. Geol.*, **73**, 618–627.

Kudrass, H.R., Wiedicke, M., Cepek, P., Kreuzer, H. & Muller, P. (1986) Mesozoic and Cainozoic rocks dredged from the South China Sea (Reed Bank area) and Sulu Sea and their significance for the plate-tectonic reconstructions. *Mar. Petrol. Geol.*, **3**, 19–30.

Le Pichon, X. & Sibuet, J.C. (1981) Passive margins: a model of formation. *J. Geophys. Res.*, **86** (B5), 3708–3720.

Lee, D.T.C. (1970) *Sandakan Peninsula, Eastern Sabah, Eastern Malaysia*, Geological Survey of East Malaysia, Report no. 6.

Lee, D.T.C. (1988) *Gunong Pock Area, Semporna Peninsula, Sabah, Malaysia*, Geological Survey of Malaysia Borneo Region, Report 9.

Lee Chai Peng (2003) The Madai-Baturong Limestone in eastern Sabah and its new interpretation as a seamount. *Geol. Soc. Malay. Bull.*, **46**, 161–165.

Lee, C.S. & McCabe, R. (1986) The Banda–Celebes–Sulu Basin: A trapped piece of Cretaceous-Eocene oceanic crust? *Nature*, **322**, 51–54.

Lee, D.T.C. & Weber, H.S. (1986) Base metal exploration in Sabah. *Geol. Soc. Malay. Bull.*, **19**, 405–419.

Leong Khee Meng. (1974) The geology and mineral resources of the Upper Segama Valley and Darvel Bay area, Sabah, Malaysia. *Geol. Surv. Malay., Memoir* **4**, (revised).

Leong Khee Meng. (1977) New ages from radiolarian cherts of the Chert–Spilite Formation. *Geol. Soc. Malay. Bull.*, **8**, 109–111.

Leong Khee Meng. (1978) The "Sabah Blueschist Belt"—A preliminary note. *Newslett. Geol. Soc. Malay. (Warta Geol.)*, **4**, 45–50.

Leong Khee Meng. (1998) Sabah crystalline basement: "Spurious" radiometric ages? continental? *Warta Geol.* (*Newslett Geol. Soc. Malay.*), **24**, 5–8.

Leong Khee Meng. (1999) Geological setting of Sabah. *The Petroleum Geology and Resources of Malaysia.* Petroliam Nasional Berhad (PETRONAS), Kuala Lumpur, chapter 21, pp. 475–497.

Leong Khee Meng & Azlina Anuar (1999) Northeast Sabah Basin. *The Petroleum Geology and Resources of Malaysia*, Petroliam Nasional Berhad (PETRONAS), Kuala Lumpur, chapter 23, pp. 545–569.

Levell, B.K. (1987). The nature and significance of regional unconformities in the hydrocarbon-bearing Neogene sequences offshore West Sabah. *Geol. Soc. Malay. Bull.*, **21**, 55–90.

Levell, B.K. & Kasumajaya, A. (1985). Slumping at the late Miocene shelf-edge offshore West Sabah: A view of a turbidite basin margin. *Bull. Geol. Soc. Malay.*, **18**, 1–29.

Liechti, P., Roe, F.W. & Haile, N.S. (1960). The geology of Sarawak, Brunei and the western part of North Borneo. *Geological Survey Department, British Territories in Borneo, Bulletin*, 3, Kuching.

Lim Peng Siong (1981). *Wullersdorf Area, Sabah, Malaysia*. Geological Survey of Malaysia, Report 15, 106pp.

Lim Peng Siong (1988). Geology and geothermal potential of the Tawau area, Sabah. *Geological Survey of Malaysia, annual report for 1987*, pp. 402–413.

Mahendran, B. Ganesan (1997). *Geology and Hydrocarbon Potential of the Offshore Western Sarawak Shelfal Area*. Proceedings of the ASCOPE '97 Conference, Kuala Lumpur, 24–27 November 1997, Vol. 2, pp. 131–169.

Mahendran S. Ganesan (1980). *Sedimentology, Stratigraphy and Structure of the Crocker Formation in the Kampung Bongol Area, Sabah, East Malaysia*. B.Sc. (hons.) thesis, Dept. of Geology, University of Malaya.

Mazlan b. Haji Madon (1994). The stratigraphy of northern Labuan, NW Sabah Basin, East Malaysia. *Geol. Soc. Malay. Bull.*, **36**, 19–30.

Mazlan b. Haji Madon (1997). Sedimentological aspects of the Temburong and Belait Formations, Labuan (offshore west Sabah, Malaysia). *Geol. Soc. Malay. Bull.*, **41**, 61–84.

Mazlan b. Hj. Madon. (1999a). Basin types, tectono-stratigraphic provinces, and structural styles. *The Petroleum Geology and Resources of Malaysia*, Petroliam Nasional Berhad (Petronas), Kuala Lumpur, pp. 79–111.

Mazlan b. Hj. Madon. (1999b). North Luconia Province. *The Petroleum Geology and Resources of Malaysia*, chapter 19, Petroliam Nasional Berhad (PETRONAS), Kuala Lumpur, pp. 443–454.

Mazlan b. Hj. Madon & Abolins, P. (1999). Balingian province. *The Petroleum Geology and Resources of Malaysia*, Petroliam Nasional Berhad (Petronas), Kuala Lumpur, pp. 343–367.

Mazlan b. Hj. Madon & Redzuan b. Abu Hassan. (1999). Tatau Province. *The Petroleum Geology and Resources of Malaysia*. Petroliam Nasional Berhad (PETRONAS), Kuala Lumpur, pp. 413–426.

Mazlan b. Hj. M., Abolins, P., Mohammad J. b. H. & Mansor b. A. (1999a). Malay Basin. *The Petroleum Geology and Resources of Malaysia*, Petroliam Nasional Berhad (PETRONAS), Kuala Lumpur, pp. 173–217.

Mazlan, b. Hj. Madon, Leong Khee Meng & Azlina Anuar (1999b). Sabah Basin. *The Petroleum Geology and Resources of Malaysia*, Petroliam Nasional Berhad (PETRONAS), Kuala Lumpur, pp. 501–542.

McKenzie, D.P. (1978). Some remarks on the development of sedimentary basins. *Earth Planet. Sci. Lett.*, **40**, 25–32.

Metcalfe, I. (1985). Lower Permian conodonts from the Terbat Formation. *Warta Geol.: Newslett. Geol. Soc. Malay.*, **11** (1), 1–4.

Michel, G.W., Yue, Q.Y., Sheng, Y.Z., Reigber, C., Becker, M., Reinhart, E., Simons, W., Ambrosius, B., Vigny, C., Chamot-Rooke, N., Le Pichon, X., Morgan, P. & Matheussen, S. (2001) Crustal motion and block behaviour in SE-Asia from GPS measurements. *Earth and Planetary Science Letters*, **187**, 239–244.

Milsom, J. and Holt, R. (2001) Discussion of a Miocene collisional belt in north Borneo: Uplift mechanism and isostatic adjustment quantified by thermochronology. *J. Geol. Soc. London*, **158**, 396–400.

Minerals Yearbook (2000) *Malaysian Minerals Yearbook 1999,* Minerals and Geoscience Department Malaysia, Ministry of Primary Industries, Kuala Lumpur.

Mitchell, A.H.G. & Leach, T.M. (1991) *Epithermal Gold in the Philippines: Island Arc Metallogenesis, Geothermal Systems and Geology.* Academic Press, London, 457pp.

Miyashiro, A., Shido, F. and Ewing, M. (1971) Metamorphism in the Mid-Atlantic Ridge near 24° and 30° N. *Philos. Trans. Roy. Soc. London* A, **268**, 589–603.

Mohammad Yamin b. Ali & Abolins, P. (1999) Central Luconia Province. *The Petroleum Geology and Resources of Malaysia*, Petronas, Kuala Lumpur, pp. 371–392.

Mohd. Idrus b. Ismail (1999) Petroleum resources, Sabah. *The Petroleum Geology and Resources of Malaysia*, Petronas, Kuala Lumpur, pp. 593–602.

Mohd Idrus b. Ismail, Abdul Rahman Eusoff, Abdul Manaf Mohamad, Sahalan Abd Aziz & Barney Mahendran (1995) *The Geology of Sarawak Deepwater and Surrounding Areas*. Proceedings of the AAPG-GSM International Conference 1994, Bulletin of the Geological Society of Malaysia, 37, pp. 165–178.

Mohd. Idrus b. Ismail & Redzuan b. Abu Hassan (1999) Tinjar Province. *The Petroleum Geology and resources of Malaysia.* Petronas, Kuala Lumpur, pp. 395–410.

Molengraaf, G.A.F. (1902) *Borneo Expedition: Geological Explorations in Central Borneo (1893–94)* (English revised edition). Society for the Promotion of the Scientific Exploration of the Dutch Colonies. E.J. Brill & Kegan Paul, Trench; Trubner & Co., Leyden and Amsterdam.

Morley, R.J. (1991) Tertiary stratigraphic palynology in Southeast Asia: Current status and new directions. *Geol. Soc. Malay. Bull.*, **28**, 1–36.

Moss, S.J. (1998) Embaluh Group turbidites in Kalimantan: Evolution of a remnant oceanic basin in Borneo during the Late Cretaceous and Palaeogene. *J. Geol. Soc. London*, **155**, 509–524.

Muff, R. (1990) Geological, geochemical and ore microscopic investigations of the massive cupriferous sulphide occurrences of West Sualog, in the Bidu Bidu ophiolite complex, Sabah, East Malaysia. *Geologisches Jaarbuch*, **B74**, 65–95.

Muller, J. (1968) Palynology of the Pedawan and Plateau Sandstone Formations (Cretaceous–Eocene) in Sarawak, Malaysia. *Micropaleontology*, **14** (1), 1–37.

Mylius, H-G. (1990) The geological setting of a cupriferous sulphide mineralization in the Kiabau area of the Bidu Bidu Hills, Sabah, East Malaysia. *Geologisches Jaarbuch*, **B74**, 97–114.

Nakai, I., Adachi, H., Matsubara, S., Kato, A., Masutomi, K., Fujiwara, T. & Nagashima, K. (1978) Sarabauite, a new oxide sulphide mineral from the Sarabau Mine, Sarawak. *Am. Mineral.*, **63**, 715–719.

Newton-Smith, J. (1967) Bidu-Bidu Hills area, Sabah: Explanation of sheet 5-117-2 and part of 5-117-1. *Borneo Region Malaysia Geological Survey Report* **4**.

Nissen, S.S., Hayes, D.E., Buhl, P., Diebold, J., Bochu, Y., Weijun, Z. & Yongqin, C. (1995) Deep penetration seismic soundings across the northern margin of the South China Sea. *J. Geophys. Res.*, **100** (B11), 22407–22433.

Noad, J.J. (1998) *The Sedimentary Evolution of the Tertiary of Eastern Sabah, Northern Borneo*. Ph.D. thesis, Research school of Geological and Geophysical Sciences, Birbeck College and University College, University of London, 457pp.

Noad, J.J. (2001) The Gomantong Limestone of eastern Borneo: A sedimentological comparison with the near-contemporaneous Luconia Province. *Palaeogeogr. Palaeoclimatol. Palaeoecol.* **175** (1-4), 273–302.

Nuraiteng Tee Abdullah & Kushairi Hj. Abang (1987) Pedawan Formation of the Penrissen Area, Sarawak: A revision of its upper age limit. *Geol. Soc. Malay. Newslett.*, **13** (2), 43–50.

Nuttall, C.P. (1965) Report on the Haile collection of fossil mollusca from the Plio-Pleistocene Togopi Formation, Dent Peninsula, Sabah, Malaysia. In: Haile, N.S. & Wong, N.P.Y. (Eds.), *The Geology and Mineral Resources of Dent Peninsula, Sabah, Geological Survey, Borneo Region, Malaysia, Memoir* 16, pp. 155–192.

Omang, Shariff, .A.K. (1993) *Petrology, Geochemistry and Structural Geology of the Darvel Bay Ophiolite, Sabah, Malaysia*. Ph.D. thesis, Royal Holloway, University of London.

Omang, Shariff, A.K. (1995) Petrology and geochemistry of the mantle-sequence peridotite of the Darvel Bay Ophiolite, Sabah, Malaysia. *Geol. Soc. Malay. Bull.*, **38**, 31–48.

Omang, Shariff, A.K. & Barber, A.J. (1996) Origin and tectonic significance of the metamorphic rocks associated with the Darvel Bay ophiolite, Sabah, Malaysia. Hall, R. & Blundell, D. (Eds.), Tectonic evolution of Southeast Asia. *Geological Society of London Special Publication* **106**, 263–279.

Omang, Shariff, A.K. and Sanudin Haji Tahir (1995) Cretaceous and Neogene lavas of Sabah — origin and tectonic significance. *Geol. Soc. Malay. Bull.*, **38**, 21–30.

Othman Ali Mahmud & Salahuddin b. Saleh (1999) Petroleum Resources, Sarawak. *The Petroleum Geology and resources of Malaysia*, Petroliam Nasional Berhad (PETRONAS), Kuala Lumpur, pp. 457–472.

Othman Ali Mahmud, Tjia, H.D. & Mohd. Idrus Ismail (2001) *Interpretation of Newly Acquired Aerogravity Data Enhances the Prospectivity of the Tinjar Province, onshore Sarawak.* Geological Society of Malaysia Annual Geological Conference 2001, June 2001 In: G.H. Teh, Mohd. Shafeea Leman & T.F. Ng (Eds.), pp. 19–26.

Pearce, J.A. & Cann, J.R. (1973) Tectonic setting of basic volcanic rocks determined using trace element analysis. *Earth Planet. Sci. Lett.*, **19**, 290–299.

Pearce, J.A., Harris, N.B.W. & Tindle, A.G. (1984) Trace element discrimination diagrams for the interpretation of granitic rocks. *J. Petrol.*, **25**, 956–983.

Peccerillo, A. & Taylor, S.R. (1976) Geochemistry of Eocene calc-alkaline volcanic rocks from the Kastamonu area, northern Turkey. *Contrib. Mineral. Petrol.*, **58**, 63–81.

PETRONAS (1999) *The Petroleum Geology and Resources of Malaysia*, Petroliam Nasional Berhad (Petronas), Kuala Lumpur, 665pp.

Pieters, P.E. & Supriatna, S. (1990) *Geological Map of the West, Central and East Kalimantan area (Colour Version), 1:1,000,000*, Geological Research and Development Centre, Bandung; Bureau of Mineral Reasources, Canberra.

Pimm, A.C. (1965) *Serian Area, West Sarawak, Malaysia.* Geological Survey Borneo Region, Malaysia, Report, **3**, Kuching.

Pimm, A.C. (1967a) Triassic volcanic rocks in East and West Malaysia. Geological Papers 1966. In: P. Collenette, (Ed.), *Geological Survey, Borneo Region Malaysia, Bulletin*, 8, pp. 36–40, Kuching.

Pimm, A.C. (1967b) Bau mining district. West Sarawak, East Malaysia. Part II. Krokong. *Geological Survey Borneo Region Malaysia, Bulletin* 7.

Poldervaart, A. & Gilkey, A.K. (1954) On clouded plagioclase. *Am. Mineral.*, **39**, 75–91.

Posewitz, T. (1892) *Borneo: its Geology and Mineral Resource.* (translated from the German by F.H. Hatch), E. Stanford, London.

Priem, H.N.A., Boelrijk, N.A.I.M., Bon, E.H., Hebeda, E.H., Verdurman, E.A.T. & Verschure, R.H. (1975) Isotope geochronology in the Indonesian tinbelt. *Geol. Mijnbouw*, **54**, 61–70.

Prouteau, G., Maury, R.C., Pubellier, M., Cotton, J. & Bellon, H. (2001) Le magmatisme post-collisionnel du Nord-Ouest de Bornéo, produit de la fusion d'un fragment de croûte océanique ancré dans le manteau supérieur. *Bull. Soc. Géol. France*, **172** (3), 319–332.

Rangin, C. (1989) The Sulu Sea, a back-arc basin setting within a Neogene collision zone. *Tectonophysics*, **161**, 119–141.

Rangin, C., Bellon, H., Benard, F., Letouzey, J., Muller, C. & Sanudin, T. (1990) Neogene arc-continent collision in Sabah, Northern Borneo (Malaysia). *Tectonophysics*, **183**, 305–319.

Reinhard, M. & Wenk, E. (1951) Geology of the Colony of North Borneo, *Geological Survey Department of the British Territories in Borneo, Bulletin*, 1, His Majesty's Stationery Office, London.

Rice-Oxley, E.D. (1991) Palaeoenvironments of the Lower Miocene to Pliocene sediments in offshore NW Sabah area. *Geol. Soc. Malay. Bull.*, **28**, 165–194.

Rusmana, E. & Pieters, P.E. (1989) *Preliminary Geological Map of the Sambas/Siluas sheet, Kalimantan*, Geological Research and Development Centre, Bandung.

Rutter, O. (1922) *British North Borneo: An Account of its History, Resources and Native Tribes.* Constable & Co., London.

Ryall, P.J.C. & Beattie, D. (1996) A gravity high in Darvel Bay. *Geol. Soc. Malay. Bull.*, **39**, 113–122.

Sandal, S.T. Ed. (1996) *The Geology and Hydrocarbon Resources of Negara Brunei Darussalam*, Brunei Shell Petroleum Co. and Brunei Museum, Seria, Brunei 243pp.

Sanderson, G.A. (1966) Presence of Carboniferous in West Sarawak. *Am. Assoc. Petrol. Geol. Bull.*, **50**, 578–580.

Sandwell, D.T. & Smith, H.F. (2000) *Marine gravity anomaly from satellite* altimetry, version 7.2, may be downloaded from the Internet and modified.

Sarkar, S.S. (1973) The extension of Tethyan Lower Cretaceous to Sarawak, East Malaysia. *Geol. Soc. Malay. Newslett.*, **45**, 4–5.

Schmidtke, E.A., Fuller, M.D. & Haston, R.B. (1990) Paleomagnetic data from Sarawak, Malaysia (Borneo) and the Late Mesozoic and Cenozoic tectonics of Sundaland. *Tectonics*, **9**, 123–140.

Shipboard Scientific Party (2000) Site 1143. In: Wang, P., Prell, W.L., Blum, P. et al. (Eds.), *Proc. ODP, Init. Repts.*, 184, pp. 1–103. [CD-R] available from: Ocean Drilling Program, Texas A&M University, College Station, Texas 77845-9547, U.S.A.

Sillitoe, R.H & Bonham, H.F. (1990) Sediment-hosted gold deposits: Distal products of magmatic-hydrothermal systems. *Econ.* Geol., **18**, 157–161.

Silver, E.A. et al. (1989a) Origins of marginal basins. *Nature*, **338**, 380–381.

Silver, E.A. et al. (1989b) Leg 124 researchers drill marginal basins. *Geotimes*, **May 1989**, 15–17.

Spooner, E.T.C. & Fyfe, W.SW. (1973) Sub sea-floor metamorphism, heat and mass transfer. *Contrib. Mineral. Petrol.*, **42**, 287–304.

Stauffer, P.H. (1967) Studies in the Crocker Formation, Sabah. *Borneo Reg. Malay. Geol. Surv. Bull.*, **8**, 1–13.

Stauffer, P.H. & Lee, D.T.C. (1972) Sedimentology of the Sandakan Formation, east Sabah. *Geol. Surv. Malay., Geol. Pap.*, **1**, 10–17.

Steinmann, G. (1906) Geologische Beobachtungen in den Alpen (II). Die Schardtsche Überfaltungstheorie und die geologische Bedeutung der Tiefseeabsätze und den ophiolithischen Massengesteine. *Ber. Naturforsch. Gesell. Freiburg Breisgau*, **16**, 1–49.

Stephens, E.A. (1956) The geology and mineral resources of the Kota Belud and Kudat area, north Borneo. *Geological Survey Department British Territories in Borneo Memoir* **5**, Kuching.

Stewart, S.A. (1999) Seismic interpretation of circular geological structures. *Petrol. Geosci.*, **5**, 273–285.

Sukamto, R., Supriatna, S., Trail, D. & Peters, P. (1988) Collaborative regional geochemical survey in Kalimantan: Results of gold analyses. *Research Newsletter*, Vol. 8, Bureau of Mineral Resources, Canberra, pp. 4–6.

Supriatna, S., Margono, U., Sutrisno, De Keyser, F. & Langford, R.P. (1989) *Sanggau, 1:250,000 Quadrangle, West Kalimantan*, Geological Data Record, Geological Research and Development Centre, Bandung.

Swauger, D., Bergman, S.C., Graves, J. Hutchison, C.S., Surat, T., Morillo, A.P., Benavidez, J.J. & Pagado, E.S. (1995) *Tertiary Stratigraphic, Tectonic, and Thermal History of Sabah, Malaysia: Results of a 1994, Ten Day Reconnaissance Field Study and Laboratory Analyses*. ARCO International Oil and Gas Company, ARCO Exploration & Production Technology, TSR 95-0036 (unpublished).

Swauger, D.A., Hutchison, C.S., Bergman, S.C. & Graves, J.E. (2000) Age and emplacement of the Mount Kinabalu pluton. *Geol. Soc. Malay. Bull.*, **44**, 159–163.

Tamura, M. & Hon, V. (1977) Upper Jurassic Bivalves from the Kedadom Formation of Sarawak, Malaysia. In: Kobayashi, T., Toriyama, R. & Hashimoto, W. (Eds.), *Geology and Palaeontology of Southeast Asia*, Vol. 18, University of Tokyo Press, Tokyo, pp. 33–47.

Tan, D. N. K. (1978) Lower Cretaceous age for the chert in Lupar Valley, West Sarawak. *Warta Geol., Geol. Soc. Malay.*, **4** (6), 173–176.

Tan, D.N.K. (1979) Lupar Valley, West Sarawak Malaysia. *Geological Survey of Malaysia Report*, **13.**

Tan, D.N.K. (1982) The Lubok Antu Mélange, Lupar Valley, west Sarawak: A lower Tertiary subduction complex. *Geol. Soc. Malay. Bull.*, **15**, 31–46.

Tan, D.N.K. (1984) Palaeocurrents in the Tertiary sedimentary deposits in western Sarawak. *Geol. Soc. Malay. Bull.*, **17**, 257–264.

Tan, D.N.K. (in manuscript) Kuching area, west Sarawak. *Geological Survey of Malaysia, Report*, **16**.

Tan, D.N.K. & Hon, V. (1984) *Geological field-guidebook, West Sarawak, Malaysia*, Geological Survey of Malaysia, Kuching, 44 pp.+ map.

Tan, D.N.K., Abdul Hadi b. Abd. Rahman, Azlina Anuar, Boniface Bait & Chow Kok Tho (1999) West Baram Delta. *The Petroleum Geology and Resources of Malaysia*, Petroliam Nasional Berhad (Petronas), Kuala Lumpur, pp. 293–341.

Tan, D.N.K. & Lamy, J.M. (1990) Tectonic evolution of the NW Sabah continental margin since the late Eocene. *Geol. Soc. Malay. Bull.*, **27**, 241–260.

Tate, R.B. (1991) Cross-border correlation of geological formations in Sarawak and Kalimantan. *Geol. Soc. Malay. bull.*, **28**, 63–95.

Tate, R.B. (2001) *Geological map of Borneo Island*, CD-Rom published by the Geological Society of Malaysia, Kuala Lumpur.

Tate, R.B. & Hon, V. (1991) The oldest rocks in Borneo: A note on the Tuang Formation, West Sarawak and its importance in relation to the presence of "Basement" in West Borneo. *Warta Geol. (Newslett. Geol. Soc. Malay.)*, **17** (5), 221–224.

Taylor, B. & Hayes, D.E. (1983) Origin and history of the South China Sea Basin. In: Dennis E. Hayes (Ed.) *The Tectonic and Geologic Evolution of Southeast Asian Seas and Islands: Part 2*, American Geophysical Union, Geophysical Monograph, 27, 23–56.

Tham Kum Choong (1984) *The General Geology and Basinal Studies of the Tertiary Deposits in the Bukit Garam District, Sabah, East Malaysia*. Unpubl. B.Sc. (hons.) thesis, Dept. of Geology, University of Malaya.

Ting Ching Soon (1992) Jurassic–Cretaceous palaeogeography of the Jagoi-Serikin area as indicated by the Bau Limestone Formation. *Geol. Soc. Malay. Bull.*, **31**, 21–38.

Tjia, H.D. (1988) Accretion tectonics in Sabah: Kinabalu Suture and East Sabah accreted terrane. *Geol. Soc. Malaysia Bull.*, **22**, 237–251.

Tjia, H.D. (1999) Pre-Tertiary hydrocarbon potential. *The Petroleum Geology and Resources of Malaysia*, Petroliam Nasional Berhad (PETRONAS), Kuala Lumpur, chapter 26, pp. 605–635.

Tjia, H.D., Borhan Sidi & Teoh Chuen Lye. (1987) Superimposed deformations and vergence of Lower Tertiary sediments near Tatau, Sarawak. *Geol. Soc. Malay.* Bull., **21**, 251–271.

Tjia, H.D., Komoo, I., Lim P.S. & Tungah Surat (1990) The Maliau Basin, Sabah: Geology and tectonic setting. *Geol. Soc. Malay. Bull.*, **27**, 261–292.

Tongkul, F. (1994) The geology of northern Sabah, Malaysia: Its relationship to the opening of the South China Sea Basin. *Tectonophysics*, **235**, 131–147.

Tongkul, F. (1997) Polyphase development in the Telupid area, Sabah, Malaysia. Proceedings of GEOSEA 1995. *J. Asian Earth Sci.*, **15**, 175–184.

Tongkul, F. (in press) Structural style of Kudat Peninsula, Sabah. *Geol. Soc. Malay. Bull.*

Tumanda, F.P., Banda, R.M. & Lau, J.J. (1993) Cretaceous radiolaria from Lundu-Sematan and Lupar Valley, Sarawak, Malaysia. IGCP. 321 *Gondwana Dispersion and Asian Accretion*, 3rd Edition, Vol. 43. International Symposium, Kuala Lumpur, Malaysia, Abstract, p.70.

Ujiié, H. (1970) Miocene Foraminiferal faunas from the Sandakan Formation, North Borneo. In: Kobayashi, T. & Toriyama, R. (Eds.,) *Geology and Palaeontology of Southeast Asia*, Vol. 8, University of Tokyo Press, Tokyo, pp. 165–185.

Ujiié, H. (1977) New species and subspecies of Benthonic Foraminifera from the Miocene Sandakan Formation, North Borneo. Kobayashi, T., Toriyama, R. and Hashimoto, W. (eds.) *Geology and Palaeontology of Southeast Asia*, Vol. 18, University of Tokyo Press, pp. 87–102.

Vachard, D. (1990) A new biozonation of the limestones from Terbat area, Sarawak, Malaysia. In: Fontaine, H. (Ed.) *Ten Years of CCOP Research on the Pre-Tertiary of East Asia*. CCOP/TP **20**, CCOP Technical Secretariat, Bangkok, pp. 183–208.

Van Bemmelen, R.W. (1939) De geologie van het Westelijke en Zuidelijk deel van he Westerafdeeling van Borneo. *Jaarbuch van het Mijnwezen in Nederlandsch-Indie*, **1939**, 189–319.

Van Bemmelen, R.W. (1949) *The geology of Indonesia*, **1A**: *General geology of Indonesia and adjacent archipelagos*, **2**: *Economic geology*, **1B**: *Portfolio and index*, first edition, Government Printing Office, The Hague. (1970) second edition, Martinus Nijhoff, The Hague.

Vogt, E.T & Flower, M.F.J. (1989) Genesis of the Kinabalu (Sabah) granitoids at a subduction-collisional junction. *Contrib. Mineral. Petrol.*, **103**, 493–509.

Vozenin-Serra, C. (1977). *Contribution à l'étude de la Paléoflore du Sud-Est Asiatique (Cambodge, Laos, Vietnam)*. Thesis, University of Paris, 310pp.

Waite, S.T. (1940) *Report on the geology of Sarawak and Brunei (Compilation)* [unpublished], GR 305, August 1940.

Wan Hasiah Abdullah (1997) Common liptinitic constituents of Tertiary coals from the Bintulu and Merit-Pila coalfield, Sarawak and their relation to oil generation to coal. *Geol. Soc. Malay. Bull.*, **41**, 85–94.

Weber, H.S. & Lee, D.T.C. (1990) Mineral resources investigation in Sabah, East Malaysia, 1980–1984, selected results and conclusions. *Geologisches Jaarbuch*, **B74**, 3–29.

Westbrook, G.K., Ladd, J.W., Buhl, P., Bangs, N. & Tiley, G. (1988) Cross section of an accretionary wedge: Barbados Ridge complex. *Geology*, **76**, 631–635.

Whitney, P.R. (1972) Spinel inclusions in plagioclase of metagabbros from the Adirondack Highlands. *Am. Mineral.*, **57**, 1429–1436.

Wilford, G.E. (1955) The geology and mineral resources of the Kuching–Lundu area, West Sarawak, including the Bau mining district. *Geological Survey Department, British territories in Borneo, Memoir* 3.

Wilford, G.E. (1961) The geology and mineral resources of Brunei and adjacent parts of of Sarawak, with descriptions of the Seria and Miri Oilfields. *British Borneo Geological Survey, Memoir* 10.

Wilford, G.E. (1964) The geology of Sarawak and Sabah caves *Geological Survey, Borneo region, Malaysia, Bulletin*, 6.

Wilford, G.E. (1968) Notes on the geomorphology of Sabah. P. Collenette & J. Goh (eds.), Geological Papers 1967. *Geological Survey Borneo Region, Malaysia, Bulletin* **9**, 1–22.

Wilford, G.E., Collenette, P. & Kirk, H.J.C. (1967) Radio-carbon determinations on specimens from Sabah. *Geological Survey, Borneo Region Malaysia, Annual Report for 1966*, pp. 83–85.

Wilford, G.E. & Kho, C.H. (1965) Penrissen area, West Sarawak, Malaysia. *Geological Survey Borneo Region, Malaysia, Report*, 2, Kuching, 195pp. + map.

William, A.G. Lambiase, J.J.; Back, S. & Mohd. Khalid Jamiran (2003) Sedimentology of the Jalan Salaiman and Bukit Melinsung outcrops, western Sabah: Is the Crocker Formation an analogue for Neogene turbidites offshore? *Geol. Soc. Malay. Bull.*, **47**, 63–75.

Williams, P.R. & Harahap, B.H. (1987) Preliminary geochemical and age data for postsubduction intrusive rocks, northwest Borneo. *Aust. J. Earth Sci.*, **34**, 405–415.

Williams, P.R., Johnston, C.R., Almond, R.A. & Simamora, W.H. (1988) Late Cretaceous to Early Tertiary structural elements of West Kalimantan. *Tectonophysics*, **148**, 279–297.

Wilson, R.A.M. (1961) The geology and mineral resources of the Banggi island and Sugut River area, North Borneo. *British Borneo Geological Survey Memoir* 15.

Wilson, R.A.M. (1964) The geology and mineral resources of the Labuan and Padas Valley area, Sabah, Malaysia. *Geological Survey of the Borneo Region of Malaysia, Memoir* 17.

Wirasantosa, S., Watkins, J.S. and White, G. (1992) *Cenozoic Structure and Stratigraphy of the Eastern Continental Shelf and Upper Slope of Vietnam*. Symposium on Tectonic Framework and Energy Resources of the Western Margin of the Pacific Basin, Kuala Lumpur, 29 Nov–2 Dec 1992: Programme and Abstracts of Papers, 37, Circum Pacific Council and Geological Society of Malaysia.

Wolfenden, E.B. (1960) The geology and mineral resources of the Lower Rajang valley and adjoining areas, Sarawak. *Geological Survey Dept. British territories in Borneo, Memoir* 11.

Wolfenden, E.B. (1965) Bau mining district, West Sarawak, Malaysia. Part I. Bau. *Geological Survey Borneo Region, Malaysia, Bulletin* 7.

Wolfenden, E.B. & Haile, N.S. (1963) Sematan and Lundu area, West Sarawak. *Geological Survey Dept. British Territories in Borneo, Report* 1.

Wong, Robert H.F. (1993) Sequence stratigraphy of the Middle Miocene–Pliocene southern offshore Sandakan Basin, East Sabah. *Geol. Soc. Malay. Bull.*, **33**, 129–142.

Worth, W.J. (1940) Notes on mineral research in North Borneo. *Asia. Rev. London*, **36**, 1–136.

Xia Kan-yuan & Zhou Di (1993) The geophysical characteristics and evolution of northern and southern margins of the South China Sea. *Geol. Soc. Malay. Bull.*, **33**, 223–240.

Yan, A.T.W. (1979) *Petrology and geochemistry of the ophiolite suite, Lower Bole area, Lahad Datu, eastern Sabah.* B.Sc. (Hons.) thesis, Dept. of geology, University of Malaya.

Yanagida, J. & Lau, J. (1978) Jurassic and Middle Cretaceous Terebratulidae from the Bau Limestone Formation in West Sarawak, Malaysia. Kobayashi, T., Toriyama, R. & Hashimoto, W. (eds.), *Geology and Palaeontology of Southeast Asia*, Vol. 19, University of Tokyo Press, pp. 35–47.

Zeijlmans van Emmichoven, C.P.A. (1938) Korte schets van de Geologie van Centraal-Borneo. *De Ingenieur in Nederlandsch-Indie*, **IV**, 135–148.

Zeijlmans van Emmichoven, C.P.A. (1939) De geologie van het centrale en oostelijke deel van de westerafdeeling van Borneo. *Jaarbuch van het Mijnwezen in Nederlandsch-Indie 'Verlandelingen'*, **Batavia, 1939**, 8–186 [English translation by Haile, N.S. (1955). Geology of the central and eastern parts of the western division of Borneo. *Geological Survey Department, British Territories in Borneo, Bulletin*, **2**, 159–272.

Zeijlmans van Emmichoven, C.P.A. & Ubaghs, J.G.H. (1936) Beschouwingen over den veronderstelden eocenen ouderdom van de geheele "Oude-Leiformatie" in Centraal Borneo. De Ingenieur Ned. Indië, 3, (Section 4), 37–45 [English translation by Haile, N.S. (1955). Consideration of the supposed Eocene age of the whole "Old Slate Formation" in Central Borneo, *Geological Survey Department, British Territories in Borneo, Bulletin* 2, Kuching, pp. 125–138.

Zeiller, R. (1902) Flore fossile des gîtes de charbon du Tonkin. *Étude des Gîtes Min. France*, 1–328, plates 1–56 and A–F, Paris.

Index

Capitalized entries are author names. Bold entries are major sections.
Fig. = figure, Tb. = table